SO-AYG-706

Canadian Mathematical Society
Société mathématique du Canada

Editors-in-Chief
Rédacteurs-en-chef
K. Dilcher
K. Taylor

Advisory Board
Comité consultatif
P. Borwein
R. Kane
S. Shen

CMS Books in Mathematics
Ouvrages de mathématiques de la SMC

Petr Hájek, Vicente Montesinos Santalucía,
Jon Vanderwerff and Václav Zizler

Biorthogonal Systems
in Banach Spaces

 Springer

Petr Hájek
Mathematical Institute of the
  Czech Academy of Sciences
Žitná 25
Praha 1, 11567
The Czech Republic
hajek@math.cas.cz

Jon Vanderwerff
Mathematics and Computer Science
  Department
La Sierra University
4500 Riverwalk Parkway
Riverside, CA 92515
USA
jvanderw@lasierra.edu

Vicente Montesinos Santalucía
Departamento de Matemática Aplicada
E.T.S.I. Telecomunicación
Instituto de Matemática Pura y Aplicada
Universidad Politécnica de Valencia
C/Vera, s/n. 46022 - Valencia
Spain
vmontesinos@mat.upv.es

Václav Zizler
Mathematical Institute of the
  Czech Academy of Sciences
Žitná 25
Praha 1, 11567
The Czech Republic
zizler@math.cas.cz

Editors-in-Chief
Rédacteurs-en-chef
K. Dilcher
K. Taylor
Department of Mathematics and Statistics
Dalhousie University
Halifax, Nova Scotia B3H 3J5
Canada
cbs-editors@cms.math.ca

ISBN: 978-0-387-68914-2   e-ISBN: 978-0-387-68915-9

Library of Congress Control Number: 2007936091

Mathematics Subject Classification (2000): 46Bxx

©2008 Springer Science+Business Media, LLC
All rights reserved. This work may not be translated or copied in whole or in part without the written permission of the publisher (Springer Science+Business Media, LLC, 233 Spring Street, New York, NY 10013, USA), except for brief excerpts in connection with reviews or scholarly analysis. Use in connection with any form of information storage and retrieval, electronic adaptation, computer software, or by similar or dissimilar methodology now known or hereafter developed is forbidden.
The use in this publication of trade names, trademarks, service marks, and similar terms, even if they are not identified as such, is not to be taken as an expression of opinion as to whether or not they are subject to proprietary rights.

Printed on acid-free paper.

9 8 7 6 5 4 3 2 1

springer.com

To Paola, Danuta, Judith, and Jarmila

Preface

The main theme of this book is the relation between the global structure of Banach spaces and the various types of generalized "coordinate systems"—or "bases"—they possess. This subject is not new; in fact, it has been investigated since the inception of the study of Banach spaces. The existence of a nice basis in a Banach space is very desirable. Bases are not only very useful in many analytic calculations and various constructions but can also be used to classify Banach spaces. The long-standing hope of having such a system in every Banach space was shattered first by Enflo's construction of a separable Banach space without a Schauder basis and, more recently, by the work of Argyros, Gowers, Maurey, Schlumprecht, Tsirelson, and others that has produced hereditarily indecomposable Banach spaces and, in particular, Banach spaces containing no unconditional Schauder basic sequence. In light of these results, the classical rich structural theory of various special classes of Banach spaces, such as $\mathcal{L}_p$ spaces, separable $C(K)$ spaces, or Banach lattices, to name a few separable classes, as well as nonseparable weakly compactly generated or weakly countably determined (Vašák) spaces—and the coordinate systems they possess—increases in value, importance, and complexity. Of course, in order to obtain more general results, one has to weaken the analytic properties of the desired systems.

In this book, we systematically investigate the concepts of Markushevich bases, fundamental systems, total systems, and their variants. The material naturally splits into the case of separable Banach spaces, as is treated in the first two chapters, and the nonseparable case, which is covered in the remainder of the book.

Our starting point is that every separable Banach space has a fundamental total biorthogonal system. This was proved by Markushevich, and hence today such systems are called *Markushevich bases*. However, there are now several significantly stronger versions of this result. Indeed, using Dvoretzky's theorem combined with orthogonal transformation techniques, Pełczyński and, independently, Plichko, obtained $(1 + \varepsilon)$-bounded Markushevich bases in every separable Banach space. More recently, Terenzi has constructed several versions of strong Markushevich bases in all separable spaces that, in particular, allow one to recover every vector from its coordinates using permutations and blockings. These results, together with some background material, are treated in Chapter 1.

In Chapter 2, we present some classical as well as some recent results on the universality of spaces. This includes basic material on well-founded trees, applications of the Kunen-Martin theorem, theorems of Bourgain and Szlenk, and a thorough introduction to the geometric theory of the Szlenk index.

Chapter 3's material is preparatory in nature. In particular, it presents some results and techniques dealing with weak compactness, decompositions, and renormings that are useful in the nonseparable setting.

Chapter 4 focuses on the existence of total, fundamental, or, more generally, biorthogonal systems in Banach spaces. Among other things, we give Plichko's characterization of spaces admitting a fundamental biorthogonal system, the Godefroy-Talagrand results on representable spaces, a version under the clubsuit axiom ($\clubsuit$) of Kunen's example of a nonseparable $C(K)$ space without any uncountable biorthogonal system, and finally the recent result of Todorčević under Martin's Maximum (MM) axiom on the existence of a fundamental biorthogonal system for every Banach space of density ω_1. These latter results are typically obtained by using powerful infinite combinatorial methods—in the form of additional axioms in ZFC.

Many Banach spaces with nice structural and renormability properties can be classified according to the types of Markushevich bases they possess. Chapters 5 and 6 present, in detail, characterizations of several important classes of Banach spaces using this approach. This concerns spaces that are weakly compactly generated, weakly Lindelöf determined, weakly countably determined (Vašák), and Hilbert generated, as well as some others.

Chapter 7 deals with the class of spaces possessing long unconditional Schauder bases and their renormings. In particular, elements of the Pełczyński and Rosenthal structural theory of spaces containing $c_0(\Gamma)$, ℓ_∞, and operators that fix these spaces are discussed. The Pełczyński, Argyros, and Talagrand circle of results on the containment of $\ell_1(\Gamma)$ in dual spaces is also included.

The concluding chapter, Chapter 8, is devoted to some applications of biorthogonal and other weaker systems. Among other things, it presents some results on the existence of support sets, the theory of norm-attaining operators, and the Mazur intersection property.

It is our hope that the contents of this book reflect that nonseparable Banach space theory is a flourishing field. Indeed, this is a field that has recently attracted the attention of researchers not only in Banach space theory but also in many other areas, such as topology, set theory, logic, combinatorics, and, of course, analysis. This has influenced the choice of topics selected for this book. We tried to illustrate that the use of set-theoretical methods is, in some cases, unavoidable by showing that some important problems in the structural theory of nonseparable Banach spaces are undecidable in ZFC.

Given the breadth of this field and the diverse areas that impinge on the subject of this book, we have endeavored to compile a large portion of the relevant results into a streamlined exposition—often with the help of simplifications of the original proofs. In the process, we have presented a large variety of techniques that should provide the reader with a good foundation for future research. A substantial portion of the material is new to book form, and much of it has been developed in the last two decades. Several new results are included.

Unfortunately, for reasons of space and time, it has not been possible to include all relevant results in the area, and we apologize to all authors whose important results have been left out. Nevertheless, we believe that the present text, together with [ArTod05], [DGZ93a], [Fab97], [JoLi01h, Chap. 23 and 41], [MeNe92], [Negr84], and the introductory [Fa~01], will help the reader to gain a clear picture of the current state of research in nonseparable Banach space theory.

We especially hope that this book will inspire some young mathematicians to choose Banach space theory as their field of interest, and we wish readers a pleasant time using this book.

Acknowledgments

We are indebted to our institutions, which enabled us to devote a significant amount of time to our project. We therefore thank the Mathematical Institute of the Czech Academy of Sciences in Prague (Czech Republic), the Department of Mathematics of the University of Alberta (Canada), the Universidad Politécnica de Valencia (Spain) and La Sierra University, California (USA). For their support we thank the research grant agencies of Canada, the Czech Republic, and Spain (Ministerio de Universidades e Investigación and Generalitat Valenciana). In particular, this work was supported by the following grants: NSERC 7926 (Canada), Institutional Research Plan of the Academy of Sciences of the Czech Republic AV0Z10190503, IAA100190502, GA ČR 201/04/0090 and IAA100190610 (Czech Republic), and Projects BFM2002-01423, MTM2005-08210, and the Research Program of the Universidad Politécnica de Valencia (Spain).

We are grateful to our colleagues and students for discussions and suggestions concerning this text. We are especially indebted to Marián Fabian, Gilles Godefroy, Gilles Lancien, and Stevo Todorčević, who provided advice, support and joint material for some sections.

We thank the Editorial Board of Springer-Verlag, in particular editors Karl Dilcher and Mark Spencer, for their interest in this project. Our gratitude extends to their staff for their help and efficient work in publishing this text. In particular, we thank the copyeditor for his/her excellent and precise work that improved the final version of the book.

Above all, we thank our wives, Paola, Danuta, Judith, and Jarmila, for their understanding, patience, moral support, and encouragement.

Prague, Valencia, and Riverside.

September, 2006

Petr Hájek

Vicente Montesinos

Jon Vanderwerff

Václav Zizler

Contents

Standard Definitions, Notation, and Conventions

We will work with real Banach spaces in this book. We denote by $B_{(X,\|\cdot\|)}$ or simply B_X, the closed unit ball of a Banach space X under the norm $\|\cdot\|$; that is, $B_{(X,\|\cdot\|)} := \{x \in X;\ \|x\| \leq 1\}$. Similarly, the unit sphere $S_{(X,\|\cdot\|)}$, or simply S_X, is $\{x \in X;\ \|x\| = 1\}$. Unless stated otherwise, the topological dual space X^* (i.e., the space of all continuous linear functionals on X) is considered endowed with the canonical dual supremum norm; i.e., for $f \in X^*$, $\|f\| := \sup\{f(x);\ x \in B_{(X,\|\cdot\|)}\}$. We will use interchangeably $f(x)$, $\langle x, f \rangle$, or $\langle f, x \rangle$ for the action of an element $f \in X^*$ on an element $x \in X$. Whenever convenient, the space X will be assumed to be canonically a subspace of X^{**}. The w^*- (or weak*-) topology on X^* denotes the topology of pointwise convergence on the elements in X. The w-topology denotes the weak topology of X. We write c_0 and ℓ_p for $c_0(\mathbb{N})$ and $\ell_p(\mathbb{N})$, respectively, where $\mathbb{N}$ denotes the set of all positive integers. By a *copy* of a space we will usually mean an isomorphic copy of this space. By an *operator* we always mean a bounded linear operator. Unless stated otherwise, the word *subspace* will be understood to mean a closed linear subspace, and the fact that Y is a subspace of X will be denoted by $Y \hookrightarrow X$. The closure of a set A in a topological space (E, τ) is denoted by $\overline{A}^\tau$, or just by $\overline{A}$ if the topology τ is understood. Convergence in the topology τ will be denoted sometimes as $\xrightarrow{\tau}$. The *density* of a topological space T is the smallest cardinal $\aleph$ such that T has a dense subset of cardinality $\aleph$. The *span* of a subset A of a linear space X, denoted $\mathrm{span}(A)$, is the linear hull of A. The closed linear span of a set A in a topological vector space (E, τ) is denoted by $\overline{\mathrm{span}}^\tau(A)$ or, if the topology τ is understood, just by $\overline{\mathrm{span}}(A)$. For a completely regular topological space T, the space βT denotes the *Čech-Stone compactification* of the space T.

Cardinal numbers are usually denoted by $\aleph$, while ordinal numbers are denoted by α, β, etc. With the symbol $\aleph_0$ we denote the cardinal number of $\mathbb{N}$, and $\aleph_1$ is the first uncountable cardinal. Similarly, ω is the ordinal number of $\mathbb{N}$ under its natural order, and ω_1 is the first uncountable ordinal. The symbol c is used for the cardinal of the continuum.

In this book we use, unless stated otherwise, Zermelo-Fraenkel set theory with the axiom of choice (ZFC).

A *Schauder basis* is a sequence (x_n) in X such that every element $x \in X$ can be uniquely written as $x = \sum_{n=1}^{\infty} a_n x_n$ for some real numbers a_n, $n \in \mathbb{N}$. It then follows that the canonical projections $P_n(x) := \sum_{i=1}^{n} f_i(x) x_i$, where $f_i(x) = a_i$, $i = 1, 2 \ldots, n$, $n \in \mathbb{N}$, are uniformly bounded linear projections. If $\sum_{n=1}^{\infty} a_{\pi(n)} x_{\pi(n)}$ converges to x for every permutation π of $\mathbb{N}$, we speak of an *unconditional basis*. The basis is called *shrinking* if $\overline{\mathrm{span}}\{f_n; n \in \mathbb{N}\} = X^*$.

For information on Schauder bases, we refer the reader to [LiTz77], [Sing70b], [Sing81], [Woj91], [Fa˜01], and [JoLi01h]. In [JoLi01h, Chaps. 1, 7, 14, 41], several more general notions—the approximation property, finite-dimensional decompositions, etc.—are discussed.

1

Separable Banach Spaces

In this chapter, we introduce the basic definitions concerning biorthogonal systems in Banach spaces and discuss several results, mostly in the separable setting, related to this structure. When searching for a system of coordinates to represent any vector of a (separable) Banach space, a natural approach is to consider the concept of a Schauder basis. Unfortunately, not every separable Banach space has such a basis, as was proved by Enflo in [Enfl73]. However, all such spaces have a Markushevich basis (from now on called an M-basis), a result due to Markushevich himself that elaborates on the basic Gram-Schmidt orthogonal process. It will be proved in Chapter 5 that many nonseparable Banach spaces also possess M-bases, even with some extra features, allowing actual computations and opening a way to classification of Banach spaces.

We begin this chapter by introducing in Section 1.1, those special properties of biorthogonal systems that will be used later. Every n-dimensional normed space has a biorthogonal system $\{x_j; x_j^*\}_{j=1}^n$ that, in some sense, has an optimal location: every vector x_j and every functional x_j have norm 1. This system in called an Auerbach basis. The infinite-dimensional counterpart of the former result is an open problem. In Section 1.2, a preliminary approach to this question is presented. Section 1.3 deals with the basic Markushevich construction, a building block for subsequent developments. The control on the size of vectors and functionals in a biorthogonal system leads to the concept of a bounded M-basis. The question of its existence in every separable Banach space was open for some time and solved in the positive by Ovsepian and Pełczyński [OvPe75], a result later sharpened by Pełczyński and, independently, by Plichko. This, together with some hints on a possibly negative solution of the Auerbach system problem in separable Banach spaces, is presented in Section 1.4. Strong M-bases, a natural concept in view of the kind of convergence of the partial sums of developments generated by Schauder bases and Fourier series, are defined, and their existence in every separable Banach space is discussed in Section 1.5, where Terenzi's results, in a utility-grade version, are presented. We follow also Terenzi's approach in extending bounded M-bases from subspaces to bounded M-bases in overspaces in Section

1.6, preceded by the general extension theorem of Gurarii and Kadets and followed by V.D. Milman's result on extensions in directions of quasicomplements. The chapter ends with a discussion, which will be enlarged in Chapter 8, on ω-independent families.

1.1 Basics

This section introduces some basic definitions that will be used throughout this book. Although this chapter deals primarily with separable Banach spaces, general (possibly uncountable) biorthogonal systems are presented in order to cover the nonseparable case as well. It is shown that the minimality of a family of vectors is equivalent to the biorthogonal behavior. We also collect some simple facts about decompositions of a space with a biorthogonal system. Fundamentality, totality and the shrinking or norming character of a biorthogonal system are also defined.

Definition 1.1. *Let X be a Banach space. Let Γ be a nonempty set. A family $\{(x_\gamma, x_\gamma^*)\}_{\gamma \in \Gamma}$ of pairs in $X \times X^*$ is called a* biorthogonal system *in $X \times X^*$ if $\langle x_\alpha, x_\beta^* \rangle = \delta_{\alpha,\beta}$, where $\delta_{\alpha,\beta}$ is the Kronecker δ, for all $\alpha, \beta \in \Gamma$. For simplicity, a biorthogonal system in $X \times X^*$ will be denoted by $\{x_\gamma; x_\gamma^*\}_{\gamma \in \Gamma}$. A family $\{x_\gamma\}_{\gamma \in \Gamma} \subset X$ is called a* minimal system *if there exists a family $\{x_\gamma^*\}_{\gamma \in \Gamma} \subset X^*$ such that $\{x_\gamma; x_\gamma^*\}_{\gamma \in \Gamma}$ is a biorthogonal system.*

In the case where $\Gamma = \mathbb{N}$, we shall use the notation $\{x_n; x_n^*\}_{n=1}^\infty$ (resp. $\{x_n\}_{n=1}^\infty$) instead of $\{x_n; x_n^*\}_{n \in \mathbb{N}}$ (resp. $\{x_n\}_{n \in \mathbb{N}}$). By the Hahn-Banach theorem, a family $\{x_\gamma\}_{\gamma \in \Gamma} \subset X$ is a minimal system if and only if, for every $\gamma \in \Gamma$, $x_\gamma \notin \overline{\text{span}}\{x_\alpha; \alpha \in \Gamma, \alpha \neq \gamma\}$.

The following simple facts will be used frequently throughout this book.

Fact 1.2. *If $\{x_\gamma; x_\gamma^*\}_{\gamma \in \Gamma}$ is a biorthogonal system in $X \times X^*$, if $x \in X$ and $\gamma \in \Gamma$ are such that $\langle x, x_\gamma^* \rangle \neq 0$, and if for some finite set $A \subset \Gamma$*

$$\left\| \sum_{\alpha \in A} \langle x, x_\alpha^* \rangle x_\alpha - x \right\| < \frac{|\langle x, x_\gamma^* \rangle|}{\|x_\gamma^*\|},$$

then $\gamma \in A$.

Proof. If $\gamma \notin A$, then

$$|\langle x, x_\gamma^* \rangle| = \left| \left\langle \left(\sum_{\alpha \in A} \langle x, x_\alpha^* \rangle x_\alpha - x \right), x_\gamma^* \right\rangle \right| \leq \|x_\gamma^*\| \cdot \left\| \sum_{\alpha \in A} \langle x, x_\alpha^* \rangle x_\alpha - x \right\|. \quad \square$$

Fact 1.3. *Let X be a Banach space. Let $\{x_\gamma; x_\gamma^*\}_{\gamma \in \Gamma}$ be a biorthogonal system in $X \times X^*$. Let A be a nonempty finite subset of Γ and B its complement. Then, denoting by $\oplus$ the topological direct sum:*

(i) $X = \text{span}\{x_a; a \in A\} \oplus \{x_a^*; a \in A\}_\perp$.
(ii) $\{x_b^*; b \in B\}_\perp = \text{span}\{x_a; a \in A\} \oplus \{x_\gamma^*; \gamma \in \Gamma\}_\perp$.

Proof. (i) Given $x \in X$, we have $x - \sum_{a \in A}\langle x, x_a^*\rangle x_a \in \{x_a^*; a \in A\}_\perp$, so $X = \text{span}\{x_a; a \in A\} + \{x_a^*; a \in A\}_\perp$. Obviously $\text{span}\{x_a; a \in A\} \cap \{x_a^*; a \in A\}_\perp = \{0\}$, so the sum above is a topological direct sum.

(ii) Given $x \in \{x_b^*; b \in B\}_\perp$, it is obvious that $x - \sum_{a \in A}\langle x, x_a^*\rangle x_a \in \{x_\gamma^*; \gamma \in \Gamma\}_\perp$. Then

$$\{x_b^*; b \in B\}_\perp \subset \text{span}\{x_a; a \in A\} + \{x_\gamma^*; \gamma \in \Gamma\}_\perp \ (\subset \{x_b^*; b \in B\}_\perp).$$

It follows that $\{x_b^*; b \in B\}_\perp = \text{span}\{x_a; a \in A\} + \{x_\gamma^*; \gamma \in \Gamma\}_\perp$; moreover, $\text{span}\{x_a; a \in A\} \cap \{x_\gamma^*; \gamma \in \Gamma\}_\perp = \{0\}$, so the sum above is a topological direct sum. $\qquad\square$

Definition 1.4. *A family $\{x_\gamma\}_{\gamma \in \Gamma}$ of vectors in the Banach space X is called fundamental if $\overline{\text{span}}\{x_\gamma\}_{\gamma \in \Gamma} = X$.*

In the case of a fundamental minimal system $\{x_\gamma\}_{\gamma \in \Gamma}$ in X, there exists a unique system $\{x_\gamma^*\}_{\gamma \in \Gamma}$ (and it is called its system of *functional coefficients*) in X^* such that $\{x_\gamma; x_\gamma^*\}_{\gamma \in \Gamma}$ is a biorthogonal system. The corresponding biorthogonal system $\{x_\gamma; x_\gamma^*\}_{\gamma \in \Gamma}$ is also called *fundamental*. Whenever convenient, we will use the abbreviated notation $\{x_\gamma\}_{\gamma \in \Gamma}$ in the case of a fundamental biorthogonal system $\{x_\gamma; x_\gamma^*\}_{\gamma \in \Gamma}$.

Fact 1.5. *Let X be a Banach space. Let $\{x_\gamma; x_\gamma^*\}_{\gamma \in \Gamma}$ be a biorthogonal system in $X \times X^*$. Then the following are equivalent:*

(i) *$\{x_\gamma; x_\gamma^*\}_{\gamma \in \Gamma}$ is fundamental.*
(ii) *For some nonempty finite set $A \subset \Gamma$ and its complement B, we have $\text{span}\{x_a^*; a \in A\} = \{x_b; b \in B\}^\perp$.*
(iii) *For some nonempty finite set $A \subset \Gamma$ and its complement B, we have $X = \text{span}\{x_a; a \in A\} \oplus \overline{\text{span}}\{x_b; b \in B\}$.*

Moreover, if (i) *holds, then* (ii) *and* (iii) *hold for every nonempty finite set $A \subset \Gamma$.*

Proof. (i)$\Rightarrow$(ii) Take any nonempty finite set $A \subset \Gamma$. From (i) in Fact 1.3, it follows that $X^* = \{x_a; a \in A\}^\perp \oplus \text{span}\{x_a^*; a \in A\}$. From the fundamentality of the system, $\{x_a; a \in A\}^\perp \cap \{x_b; b \in B\}^\perp = \{0\}$. Moreover, $\text{span}\{x_a^*; a \in A\} \subset \{x_b; b \in B\}^\perp$, so we get (ii).

(ii)$\Rightarrow$(iii) From (ii) we get $\{x_a^*; a \in A\}_\perp = \overline{\text{span}}\{x_b; b \in B\}$. Now use (i) in Fact 1.3.

(iii)$\Rightarrow$(i) is trivial.

The last statement is a consequence of the two following observations: (a) (ii) for some nonempty finite set A implies (iii) for the same A, and (b) (i)$\Rightarrow$(ii) for every nonempty finite set A. $\qquad\square$

Definition 1.6. *A biorthogonal system* $\{x_\gamma; x_\gamma^*\}_{\gamma \in \Gamma}$ *in* $X \times X^*$ *is called* total *if* $\overline{\mathrm{span}}^{w^*}\{x_\gamma^*; \gamma \in \Gamma\} = X^*$.

Definition 1.7. *A fundamental and total biorthogonal system* $\{x_\gamma; x_\gamma^*\}_{\gamma \in \Gamma}$ *in* $X \times X^*$ *is called a* Markushevich basis *for* X, *henceforth called an* M-basis *in this book. Whenever convenient, we will use the abbreviated notation* $\{x_\gamma\}_{\gamma \in \Gamma}$ *for an M-basis in* X.

Fact 1.8. *Let* X *be a Banach space and let* $\{x_\gamma; x_\gamma^*\}_{\gamma \in \Gamma}$ *be a fundamental biorthogonal system in* $X \times X^*$. *Let* $(B_i)_{i \in I}$ *be a family of subsets of* Γ.

(i) *If* $\bigcap_{i \in I} B_i = \emptyset$ *and* $\{x_\gamma; x_\gamma^*\}_{\gamma \in \Gamma}$ *is an M-basis in* $X \times X^*$, *then* $\bigcap_{i \in I} \overline{\mathrm{span}}\{x_\gamma; \gamma \in B_i\} = \{0\}$.

(ii) *If* $A_i := \Gamma \setminus B_i$ *is finite for every* $i \in I$ *and* $\bigcap_{i \in I} \overline{\mathrm{span}}\{x_\gamma; \gamma \in B_i\} = \{0\}$, *then* $\{x_\gamma; x_\gamma^*\}_{\gamma \in \Gamma}$ *is an M-basis in* $X \times X^*$.

Proof. (i) Let $x \in \bigcap_{i \in I} \overline{\mathrm{span}}\{x_\gamma; \gamma \in B_i\}$. Fix $\gamma_0 \in \Gamma$. There exists $i \in I$ such that $\gamma_0 \notin B_i$. As $x \in \overline{\mathrm{span}}\{x_\gamma; \gamma \in B_i\}$, we get $\langle x, x_{\gamma_0}^* \rangle = 0$. This happens for every $\gamma_0 \in \Gamma$, so $x = 0$ since $\{x_\gamma; x_\gamma^*\}_{\gamma \in \Gamma}$ is an M-basis.

(ii) Let x be a nonzero element of X. There exists $i \in I$ such that $x \notin \overline{\mathrm{span}}\{x_\gamma; \gamma \in B_i\}$. From (iii) in Fact 1.5, we can write $x = y + z$, where $0 \neq y \in \mathrm{span}\{x_\gamma; \gamma \in A_i\}$ and $z \in \overline{\mathrm{span}}\{x_\gamma; \gamma \in B_i\}$. We can find $\gamma \in A_i$ such that $\langle y, x_\gamma^* \rangle (= \langle x, x_\gamma^* \rangle) \neq 0$. Thus $\{x_\gamma; x_\gamma^*\}_{\gamma \in \Gamma}$ is an M-basis. $\square$

Definition 1.9. *A subset* N *of the dual* X^* *of a Banach space* $(X, \|\cdot\|)$ *is called* λ-norming, *for some* $0 < \lambda \le 1$, *if* $\||\cdot\||$ *defined on* X *by* $\||x\|| := \sup\{\langle x, x^* \rangle; x^* \in N \cap B_{(X^*,\|\cdot\|)}\}$, $x \in X$, *is a norm satisfying* $\lambda\|x\| \le \||x\|| \ (\le \|x\|)$. *If* N *is* λ-norming *for some* $0 < \lambda \le 1$, N *is just called* norming.

Definition 1.10. *Let* X *be a Banach space. A biorthogonal system* $\{x_\gamma; x_\gamma^*\}_{\gamma \in \Gamma}$ *in* $X \times X^*$ *is called* λ-norming *(for some* $0 < \lambda \le 1$*) if* $\overline{\mathrm{span}}^{\|\cdot\|}\{x_\gamma^*\}_{\gamma \in \Gamma}$ *is a* λ-norming *subspace. If* $\{x_\gamma; x_\gamma^*\}_{\gamma \in \Gamma}$ *is* λ-norming *for some* $0 < \lambda \le 1$, *the system is just called* norming.

Remark 1.11. 1. *Every infinite-dimensional separable Banach space contains a fundamental system that is not an M-basis.* Indeed, let $\{y_i\}_{i=1}^\infty$ be a linearly independent system in X such that $\overline{\mathrm{span}}\{y_i\}_{i=1}^\infty = X$. Pick $x \in X \setminus \mathrm{span}\{y_i\}_{i=1}^\infty$. Using the Hahn-Banach theorem, we find $g_i \in X^*$ such that $g_i(x) = 0$, $g_i(y_i) = 1$, and $g_i(y_j) = 0$ for $j = 1, \ldots, i-1$. As in Lemma 1.21, we find a biorthogonal system $\{x_i; f_i\}_{i=1}^\infty$ such that $\mathrm{span}\{x_i\}_{i=1}^\infty = \mathrm{span}\{y_i\}_{i=1}^\infty$ and $\mathrm{span}\{f_i\}_{i=1}^\infty = \mathrm{span}\{g_i\}_{i=1}^\infty$. Then $\overline{\mathrm{span}}\{x_i\}_{i=1}^\infty = X$ and $\{f_i\}_{i=1}^\infty$ does not separate points of the space X.

2. *Every separable Banach space* X *has a 1-norming M-basis* (see Theorem 1.22). The problem of the existence of a norming M-basis in every *weakly compactly generated* Banach space (WCG) (i.e., a space with a weakly compact and linearly dense subset), is still open. Partial negative results are given in [Gode95], [Vald94] and [VWZ94] (see Theorems 5.21, 5.22, and 5.23).

3. *A norming subspace of X^* is w^*-dense in X^*. However, not every w^*-dense subspace of X^* is norming.* In fact, if X is a separable Banach space such that X^{**}/X is infinite-dimensional, there exists a subspace $N \subset X^*$ that is w^*-dense and not norming (see, e.g., [Fa~01, Exer. 3.40]). From this and from Theorem 1.22, it follows that every separable space contains an M-basis that is not norming.

Definition 1.12. *Let X be a Banach space. A biorthogonal system $\{x_\gamma; x_\gamma^*\}_{\gamma \in \Gamma}$ in $X \times X^*$ is called* shrinking *whenever $X^* = \overline{\mathrm{span}}^{\|\cdot\|}\{x_\gamma^*; \gamma \in \Gamma\}$.*

Definition 1.13. *Let X be a Banach space. A biorthogonal system $\{x_\gamma; x_\gamma^*\}_{\gamma \in \Gamma}$ in $X \times X^*$ is called* boundedly complete *if given a bounded sequence (y_n) in $\overline{\mathrm{span}}\{x_\gamma; \gamma \in \Gamma\}$ such that $a_\gamma := \lim_n \langle y_n, x_\gamma^* \rangle$ exists for all $\gamma \in \Gamma$, then there exists an element $y \in \overline{\mathrm{span}}\{x_\gamma; \gamma \in \Gamma\}$ such that $\langle y, x_\gamma^* \rangle = a_\gamma$ for all $\gamma \in \Gamma$.*

Remark 1.14. It is shown in [PeSz65] that there exists a separable Banach space X with a normalized unconditional basis $\{e_n\}_{n \in \mathbb{N}}$ such that $\{e_n; n \in \mathbb{N}\} \cup \{0\}$ is weakly compact and the basis is not shrinking.

1.2 Auerbach Bases

It is well known that a finite-dimensional Banach space X has a Hamel basis. It can be taken as normalized (i.e., all their vectors have norm 1). However, without additional effort, there is no control over the size of the functional coefficients. The following theorem shows that it is possible to choose the Hamel basis such that both their vectors and the functionals have norm 1. The construction has a clear geometric meaning: a parallelepiped of maximal volume is inscribed in the closed unit ball of X. The endeavor to reproduce this behavior in infinite-dimensional Banach spaces has been a recurrent theme in Banach space theory, and this theme will be analyzed later in this book. In this direction, this section also presents a basic construction due to Krein, Krasnosel'skiĭ, and Milman—which is in a sense opposite to Mazur's approach for building a basic sequence in every Banach space—and gives Day's procedure for producing infinite Auerbach systems in every Banach space. However, the goal of having an Auerbach basis in every separable space remains unrealized.

Definition 1.15. *Let X be a Banach space. A biorthogonal system $\{e_\gamma; e_\gamma^*\}_{\gamma \in \Gamma}$ in $X \times X^*$ such that $\|e_\gamma\| = \|e_\gamma^*\| = 1$ for all $\gamma \in \Gamma$ is called an* Auerbach system *in X. If it is, moreover, an M-basis, it is called an* Auerbach basis *for X.*

Theorem 1.16 (Auerbach). *Every finite-dimensional Banach space contains an Auerbach basis.*

Proof. Let $\{x_i\}_{i=1}^n$ be an algebraic basis of X. Given a finite sequence $(u_i)_{i=1}^n$ of vectors in X, let $\det(u_1, u_2, \ldots, u_n)$ be the determinant of the matrix whose

j-th column consists of the coordinates of u_j in the basis $\{x_i\}_{i=1}^n$. The function $|\det|$ is continuous on the compact set $B_X \times \ldots \times B_X$, so it attains its supremum at some $(e_1, e_2, \ldots, e_n) \in B_X \times \ldots \times B_X$.

Determinants are multilinear mappings of their columns. Then $\|e_i\| = 1$, $i = 1, 2, \ldots, n$. Moreover, the vectors $\{u_i\}_{i=1}^n$ are linearly independent if and only if $\det(u_1, u_2, \ldots, u_n) \neq 0$. Then the vectors $\{e_i\}_{i=1}^n$ are linearly independent.

For $i = 1, 2, \ldots, n$, let us define $e_i^* \in X^*$ by

$$\langle x, e_i^* \rangle := \frac{\det(e_1, \ldots, e_{i-1}, x, e_{i+1}, \ldots, e_n)}{\det(e_1, \ldots, e_n)} \quad \text{for all } x \in X.$$

Then $e_i^* \in B_{X^*}$ and $\langle e_k, e_i^* \rangle = \delta_{k,i}$ for all $1 \leq k, i \leq n$, so $\{e_i; e_i^*\}_{i=1}^n$ is an Auerbach basis. $\qquad \square$

Corollary 1.17. *Let X be a Banach space, and let Y be an n-dimensional subspace. Then there exists a linear projection P from X onto Y such that $\|P\| \leq n$.*

Proof. Let $\{e_i; e_i^*\}_{i=1}^n$ be an Auerbach basis for Y. Extend e_i^* to an element of S_{X^*} (still denoted by e_i^*) for all i and define

$$P(x) := \sum_{i=1}^n \langle x, e_i^* \rangle e_i \quad \text{for all } x \in X.$$

Then $P : X \to Y$ is a linear projection and $\|P(x)\| \leq n\|x\|$, $\forall x \in X$. $\qquad \square$

Theorem 1.16 means in particular that in two-dimensional spaces, there always exist *monotone* Schauder bases (i.e., bases where the canonical projections have norm 1). This is no longer true for three-dimensional spaces; Bohnenblust proved in [Bohn41] that *there is a three-dimensional Banach space that does not admit any monotone Schauder basis.* It is still unknown if every infinite-dimensional Banach space contains an infinite-dimensional Banach space with a monotone Schauder basis.

Definition 1.18. *Let X be a Banach space. We will say that $x \in X$ is orthogonal to $y \in X$ (and write $x \perp y$) if $\|y\| \leq \|y + \lambda x\|$ for all $\lambda \in \mathbb{R}$. If Y is a subspace of X and $x \perp y$ for all $y \in Y$, we will say that x is orthogonal to Y and write $x \perp Y$. Analogously, we will say that Y is orthogonal to x, and write $Y \perp x$, if $y \perp x$ for all $y \in Y$.*

It is obvious that $Y \perp x$ if and only if $\operatorname{dist}(x, Y) = \|x\|$.

The following basic result will be applied several times in this book.

Lemma 1.19 (M.G. Krein, M.A. Krasnosel'skiĭ, and D.P. Milman [KKM48]; see, e.g., [Sing70a] p. 269). *Let E be a normed space and G_1 and G_2 two subspaces such that $\dim G_1 < \infty$ and $\dim G_1 < \dim G_2$. Then there exists $y \in S_{G_2}$ such that $G_1 \perp y$.*

Proof. Without loss of generality, we may and do assume that $\dim G_1 = n$, $\dim G_2 = n + 1$, and $E = \operatorname{span}\{G_1 \cup G_2\}$. Suppose first that the norm of E is strictly convex ($\equiv$ rotund). Then π_{G_1}, the metric projection from E onto G_1 that associates to $x \in E$ elements in G_1 of best approximation, is well defined, single-valued, and continuous; moreover, $\pi_{G_1}(-x) = -\pi_{G_1}(x)$ for every $x \in E$. Assume that the result is false. Then $\pi_{G_1}(g_2) \neq 0$ for every $g_2 \in G_2 \setminus \{0\}$. The mapping $\psi : S_{G_2} \to S_{G_1}$ given by $\psi(g_2) := \pi_{G_1}(g_2)/\|\pi_{G_1}(g_2)\|$ is continuous and $\psi(-g_2) = -\psi(g_2)$ for every $g_2 \in S_{G_2}$. Choose a basis $\{y_{k,i}\}_{i=1}^{n+k-1}$ in G_k and define a homeomorphism $\varphi_k : S_{G_k} \to \Sigma_k$, where Σ_k denotes the unit sphere in ℓ_2^{n+k-1}, by $\varphi(\sum \alpha_i y_{k,i}) := (\alpha_i/(\sum \alpha_i^2)^{1/2})_{i=1}^{n+k-1}$, $k = 1, 2$. Then, $\chi := \Sigma_2 \to \Sigma_1$ given by $\chi := \varphi_1 \circ \psi \circ \varphi_2^{-1}$ is a continuous mapping such that $\chi(-z) = -\chi(z)$ for every $z \in \Sigma_2$, which is impossible by the Borsuk Antipodal Theorem. To deal with the general case, define a norm $\|\cdot\|_2$ in E by $\|x\|_2 = \|\sum_{i=1}^m \alpha_i z_i\|_2 := (\sum_{i=1}^m \alpha_i^2)^{1/2}$, where $\{z_i\}_{i=1}^m$ is a basis of E. Then, for every $\delta > 0$, $\|\cdot\|_\delta := \|\cdot\| + \delta\|\cdot\|_2$ defines a strictly convex equivalent norm on E and $\|\cdot\| \leq \|\cdot\|_\delta \leq (1 + \delta\gamma)\|\cdot\|$, where $\gamma := \max\{\|x\|_2; x \in E, \|x\| = 1\}$. Let y_n be the solution to the problem for the $\|\cdot\|_{1/n}$ norm, $n \in \mathbb{N}$. If y is the limit of a convergent subsequence of (y_n), the vector $y/\|y\|$ solves the problem. $\square$

Theorem 1.20 (Day [Day62]). *Let X be an infinite-dimensional Banach space and let (c_n) be a sequence of positive numbers. Then there exists a countable infinite Auerbach system $\{b_n; b_n^*\}_{n=1}^\infty$ in $X \times X^*$ such that $\{b_n\}_{n=1}^\infty$ is a Schauder basic sequence and $\|P_n\| \leq 1 + c_n$, where $P_n(x) := \sum_{k=1}^n \langle x, b_k^* \rangle b_k$, for $x \in \overline{\operatorname{span}}\{b_n; n \in \mathbb{N}\}$ are the canonical projections associated to the basic sequence.*

Proof. As in Mazur's classical construction of a basic sequence (see, e.g., [Fa~01, Thm. 6.14]), choose $\{\varepsilon_n\}_{n=1}^\infty \subset (0,1)$ such that $\prod_{k=n}^\infty (1 + \varepsilon_k) < (1 + c_n)$ for all $n \in \mathbb{N}$. The construction will be done inductively. Let us start by choosing any $b_1 \in S_X$ and some $b_1^* \in S_{X^*}$ such that $\langle b_1, b_1^* \rangle = 1$. Assume that, for some $n \in \mathbb{N}$, elements $b_i \in S_X$ and $b_i^* \in S_{X^*}$, $i = 1, 2, \ldots, n$, have already been defined such that $\langle b_i, b_j^* \rangle = \delta_{i,j}$, $i, j = 1, 2, \ldots, n$, and $\{b_i\}_{i=1}^n$ is a basic sequence in X with $\|Q_n\| \leq 1 + \varepsilon_n$, where $Q_n : \operatorname{span}\{b_i\}_{i=1}^n \to \operatorname{span}\{b_i\}_{i=1}^{n-1}$ denotes the canonical projection. Pick a finite $\varepsilon_{n+1}/2$-net $\{x_1, \ldots, x_k\}$ for $S_{\operatorname{span}\{b_i\}_{i=1}^n}$ and select $x_1^*, \ldots, x_k^*$ in S_{X^*} such that $\langle x_i, x_i^* \rangle = 1$, $i = 1, 2, \ldots, k$. Now apply Lemma 1.19 to $G_1 := \operatorname{span}\{b_i\}_{i=1}^n$ and $G_2 := \left(\bigcap_{i=1}^k \operatorname{Ker} x_i^*\right) \cap \left(\bigcap_{i=1}^n \operatorname{Ker} b_i^*\right)$. We can then find an element $b_{n+1} \in S_{G_2}$ such that $G_1 \perp b_{n+1}$. The element $b_{n+1}^* \in (\operatorname{span}(G_1, b_{n+1}))^*$ given by $\langle b_{n+1}, b_{n+1}^* \rangle = 1$ and $b_{n+1}^* \restriction G_1 \equiv 0$ has norm 1. Extend it to an element in S_{X^*} still denoted by b_{n+1}^*. At the same time, given $x \in S_{\operatorname{span}\{b_i\}_{i=1}^n}$, there exists $i \in \{1, 2, \ldots, k\}$ such that $\|x - x_i\| \leq \varepsilon_{n+1}/2$. Then

$$\|x + \lambda b_{n+1}\| \geq \|x_i + \lambda b_{n+1}\| - \|x - x_i\|$$

$$\geq \langle x_i + \lambda b_{n+1}, x_i^* \rangle - \|x - x_i\| \geq 1 - \frac{\varepsilon_{n+1}}{2} \geq \frac{1}{1 + \varepsilon_{n+1}}.$$

This completes the inductive step. It is clear now that $\{b_n; b_n^*\}_{n=1}^{\infty}$ is an Auerbach system and at the same time a Schauder basic sequence. Moreover, $\|P_n\| \leq \prod_{k=n}^{\infty}(1 + \varepsilon_k) < 1 + c_n$ for all $n \in \mathbb{N}$. □

For a remark on Theorem 1.20, see Exercise 1.4. The existence of an Auerbach basis for every separable Banach space is an open problem.

1.3 Existence of M-bases in Separable Spaces

We present here the classical Markushevich construction of a (countable) M-basis in every separable Banach space. A careful choice of the proof's ingredients yields norming (and if the dual space is separable, shrinking) M-bases.

Lemma 1.21 (Markushevich [Mark43]). *Let X be an infinite-dimensional Banach space. If $\{z_n\}_{n=1}^{\infty} \subset X$ and $\{z_n^*\}_{n=1}^{\infty} \subset X^*$ are such that* $\mathrm{span}\{z_n\}_{n=1}^{\infty}$ *and* $\mathrm{span}\{z_n^*\}_{n=1}^{\infty}$ *are both infinite-dimensional and*

(M1) $\{z_n\}_{n=1}^{\infty}$ *separates points of* $\mathrm{span}\{z_n^*\}_{n=1}^{\infty}$,

(M2) $\{z_n^*\}_{n=1}^{\infty}$ *separates points of* $\mathrm{span}\{z_n\}_{n=1}^{\infty}$,

then there exists a biorthogonal system $\{x_n; x_n^*\}_{n=1}^{\infty}$ in $X \times X^*$ such that

$$\mathrm{span}\{x_n\}_{n=1}^{\infty} = \mathrm{span}\{z_n\}_{n=1}^{\infty} \text{ and } \mathrm{span}\{x_n^*\}_{n=1}^{\infty} = \mathrm{span}\{z_n^*\}_{n=1}^{\infty}.$$

Proof. Define $x_1 = z_{n_1}$, where n_1 is the first index n such that $z_n \neq 0$. By (M2) there exists $z_{m_1}^*$ such that $\langle x_1, z_{m_1}^* \rangle \neq 0$. Put

$$x_1^* := \frac{z_{m_1}^*}{\langle x_1, z_{m_1}^* \rangle}.$$

Then $\langle x_1, x_1^* \rangle = 1$. Find the first index m (call it m_2) such that $z_m^* \notin \mathrm{span}\{x_1^*\}$. Let $x_2^* := z_{m_2}^* - \langle x_1, z_{m_2}^* \rangle x_1^*$. Then $\langle x_1, x_2^* \rangle = 0$. Obviously $x_2^* \neq 0$; hence, by (M1), we can find z_{n_2} such that $\langle z_{n_2}, x_2^* \rangle \neq 0$. Put

$$x_2 := \frac{z_{n_2} - \langle z_{n_2}, x_1^* \rangle x_1}{\langle z_{n_2}, x_2^* \rangle}.$$

Then $\langle x_2, x_1^* \rangle = 0$ and $\langle x_2, x_2^* \rangle = 1$. Find the first index n (call it n_3) such that $z_n \notin \mathrm{span}\{x_1, x_2\}$, and put $x_3 := z_{n_3} - \langle z_{n_3}, x_1^* \rangle x_1 - \langle z_{n_3}, x_2^* \rangle x_2$. Then $\langle x_3, x_1^* \rangle = 0$, $\langle x_3, x_2^* \rangle = 0$, and $x_3 \neq 0$. Using (M2), find $z_{m_3}^*$ such that $\langle x_3, z_{m_3}^* \rangle \neq 0$. Continue in this way to get $\{x_n; x_n^*\}_{n=1}^{\infty}$ with the required properties. □

Theorem 1.22 (Markushevich [Mark43]). *Every separable Banach space has an M-basis (which can be taken to be 1-norming). If, moreover, X has a separable dual, the M-basis can be taken to be shrinking. Every Banach space X such that (X^*, w^*) is separable has a total biorthogonal system.*

Proof. For the first assertion, take two sets $\{z_n\}_{n=1}^{\infty}$ and $\{z_n^*\}_{n=1}^{\infty}$, the first dense in $(X, \|\cdot\|)$ and the second dense in (B_{X^*}, w^*) (a metrizable compact space), and apply Lemma 1.21. Observe that in this case it is always possible to choose a 1-norming system $\{z_n^*\}_{n=1}^{\infty}$ in X^*, and this gives the assertion in parentheses. If X^* is separable, just take $\{z_n^*\}_{n=1}^{\infty}$ to be $\|\cdot\|$-dense in X^*. For the last statement, take $\{z_n^*\}_{n=1}^{\infty}$ dense in (X^*, w^*) and put $Y := \overline{\mathrm{span}}^{\|\cdot\|}\{z_n^*\}_{n=1}^{\infty}$. Let $\{d_n^*\}_{n=1}^{\infty}$ be a dense set in $(B_Y, \|\cdot\|)$. For each $n \in \mathbb{N}$, select a countable set $M_n \subset B_X$ such that $\|d_n^*\| = \sup\{|\langle x, d_n^*\rangle|; x \in M_n\}$. Let $Z := \overline{\mathrm{span}}\bigcup_{n=1}^{\infty} M_n$ and choose a dense set $\{z_n\}_{n=1}^{\infty}$ in Z. Then $\{z_n\}_{n=1}^{\infty}$ separates points of $\mathrm{span}\{z_n^*\}_{n=1}^{\infty}$ and $\{z_n^*\}_{n=1}^{\infty}$ separates points of $\mathrm{span}\{z_n\}_{n=1}^{\infty}$. Now apply Lemma 1.21. □

1.4 Bounded Minimal Systems

The procedure for constructing M-bases, described in the previous section, does not produce automatically bounded M-bases (i.e., bases $\{x_n; x_n^*\}_{n=1}^{\infty}$ where $\sup\{\|x_n\|.\|x_n^*\|; n \in \mathbb{N}\} < \infty$). The existence of such an M-basis in every separable Banach space was an open problem for many years and was solved in the positive by Ovsepian and Pełczyński. It was later adjusted by Pełczyński and independently Plichko, to produce an "almost" Auerbach (even norming) M-basis in every separable Banach space. Prior to presenting this result, we follow the lead of Davis and Johnson to produce a special Auerbach system in every separable Banach space; together with ideas of Singer, this process gives "almost" Auerbach fundamental (resp. total) systems in every separable Banach space. In order to illustrate the difficulties in obtaining a true Auerbach system in every separable Banach space (answering in the negative a question of Singer [Sing81, Problem 8.2.b]), we also present a result due to Plichko that says that no such system exists in certain separable $C(K)$ spaces if we request that the space generated by the functional coefficients contain the Dirac deltas.

Definition 1.23. *A biorthogonal system* $\{x_n; x_n^*\}_{n=1}^{\infty}$ *for a separable Banach space* X *is called* λ*-bounded for some* $\lambda \geq 1$ *if* $\sup\{\|x_n\|.\|x_n^*\|; n \in \mathbb{N}\} \leq \lambda$. *The biorthogonal system is called* bounded *if it is* λ*-bounded for some* $\lambda \geq 1$.

Remark 1.24. Clearly, by the Hahn-Banach theorem, a biorthogonal system $\{x_n; x_n^*\}_{n=1}^{\infty}$ is bounded if and only if $\{x_n\}_{n=1}^{\infty}$ is a *uniformly minimal system* (i.e., a minimal system such that $\inf_{m \in \mathbb{N}} \mathrm{dist}\left(\frac{x_m}{\|x_m\|}, \overline{\mathrm{span}}\{x_n\}_{n \in \mathbb{N}, n \neq m}\right) \geq K > 0$). In this case, the system $\{x_n\}_{n=1}^{\infty}$ is called more precisely K-*uniformly minimal*, and it is clear that

$$\inf_{m \in \mathbb{N}} \mathrm{dist}\left(\frac{x_m}{\|x_m\|}, \overline{\mathrm{span}}\{x_n\}_{n \in \mathbb{N}, n \neq m}\right) = \left(\sup\{\|x_n\|.\|x_n^*\|; n \in \mathbb{N}\}\right)^{-1}.$$

Using this remark, it is simple to see that *every separable Banach space X contains an unbounded M-basis*. Indeed, let $\{x_n; x_n^*\}_{n=1}^\infty$ be any M-basis in $X \times X^*$ (its existence is guaranteed by Theorem 1.22). We may assume that $\|x_n\| = 1$ for every n. Let us define, for $n \in \mathbb{N}$,

$$v_{2n-1} := x_{2n-1}, \qquad\qquad v_{2n} := x_{2n-1} + \frac{1}{2n}x_{2n},$$

$$v_{2n-1}^* := x_{2n-1}^* - 2nx_{2n}^*, \qquad\qquad v_{2n}^* := 2nx_{2n}^*.$$

It is clear that $\{v_n; v_n^*\}_{n=1}^\infty$ is a biorthogonal system in $X \times X^*$. Moreover,

$$\operatorname{span}\{v_i; 1 \le i \le n\} = \operatorname{span}\{x_i; 1 \le i \le n\},$$
$$\operatorname{span}\{v_i^*; 1 \le i \le n\} = \operatorname{span}\{x_i^*; 1 \le i \le n\}, \ \forall n \in \mathbb{N},$$

so $\{v_n; v_n^*\}_{n=1}^\infty$ is an M-basis in $X \times X^*$. Note that, for all $n \in \mathbb{N}$,

$$1 - \frac{1}{2n} \le \|v_{2n}\|.$$

Then

$$\left\| \frac{v_{2n}}{\|v_{2n}\|} - \frac{x_{2n-1}}{\|v_{2n}\|} \right\| = \frac{1}{\|v_{2n}\|}\frac{1}{2n} \le \frac{1}{1 - 1/2n}\frac{1}{2n} = \frac{1}{2n-1},$$

and so

$$\operatorname{dist}\left(\frac{v_{2n}}{\|v_{2n}\|}, \operatorname{span}\{v_i; 1 \le i \le 2n-1\} \right) \le \frac{1}{2n-1}, \ \forall n.$$

It follows that

$$\lim_n \operatorname{dist}\left(\frac{v_{2n}}{\|v_{2n}\|}, \operatorname{span}\{v_i; i \in \mathbb{N}, \ i \ne 2n\} \right) = 0,$$

and so the M-basis $\{v_n; v_n^*\}_{n=1}^\infty$ is not bounded.

Lemma 1.25 (Davis and Johnson [DaJo73a]). *Let X be a separable Banach space, and set $m_k := \frac{k(k+1)}{2}$, $k = 0, 1, 2, \ldots$. Then X admits a biorthogonal system $\{x_n; x_n^*\}_{n=1}^\infty$ satisfying:*

(i) $\|x_n\| = \|x_n^*\| = \langle x_n, x_n^* \rangle = 1$ *for all $n \in \mathbb{N}$.*
(ii) *For $x \in \overline{\operatorname{span}}\{x_n\}_{n=1}^\infty$, $x = \lim_{k \to \infty} \sum_{i=1}^{m_k} \langle x, x_i^* \rangle x_i$.*
(iii) $\{x_i\}_{i=m_k+1}^{m_{k+1}}$ *is $(1 + \frac{1}{k+1})$-equivalent to the canonical basis of ℓ_2^{k+1}, $k = 0, 1, 2, \ldots$.*
(iv) $\overline{\operatorname{span}}\{x_n; n \in \mathbb{N}\} + \{x_n^*; n \in \mathbb{N}\}_\perp$ *is dense in X.*

Proof. Let $(d_n)_{n=0}^\infty$ be a dense sequence in X with $d_0 = 0$. In order to prove the lemma, it is sufficient to define sequences (x_n) in X and (x_n^*) in X^* and finite sets $\emptyset =: S_0 \subset S_1 \subset S_2 \subset \ldots \subset S_{X^*}$ satisfying (i), (iii), and

(v) $x_{m_k+j} \in \left(S_k \cup \{x_i^*\}_{i=1}^{m_k+j-1} \right)_\perp$, $\quad j = 1, 2, \ldots, k+1, \quad k = 0, 1, 2, \ldots,$

(vi) $x^*_{m_k+j} \in \left(\{d_i\}_{i=0}^k \cup \{x_i\}_{i=1}^{m_k+j-1}\right)^\perp$, $\quad j = 1, 2, \ldots, k+1$, $\quad k = 0, 1, 2, \ldots$,

(vii) For each $k = 0, 1, 2, \ldots$ and $x \in \mathrm{span}\{x_i\}_{i=1}^{m_{k+1}}$ there is $x^* \in S_{k+1}$ such that $\|x\| \leq \left(1 + \frac{1}{k+1}\right) |\langle x, x^* \rangle|$.

Then $\{x_n; x^*_n\}_{n=1}^\infty$ is biorthogonal by (i), (v), and (vi). To get (ii), we can use first (vii) and (v) to obtain that, for any finite sequence (a_i) of scalars,

$$
\left\| \sum_{i=1}^{m_k} a_i x_i \right\| \leq \left(1 + \frac{1}{k}\right) \max_{x^* \in S_k} \left| \left\langle \sum_{i=1}^{m_k} a_i x_i, x^* \right\rangle \right|
$$
$$
= \left(1 + \frac{1}{k}\right) \max_{x^* \in S_k} \left| \left\langle \sum_{i=1}^\infty a_i x_i, x^* \right\rangle \right| \leq \left(1 + \frac{1}{k}\right) \left\| \sum_{i=1}^\infty a_i x_i \right\|. \quad (1.1)
$$

For every $k \in \mathbb{N}$, let $P_{m_k} : \mathrm{span}\{x_n\}_{n=1}^\infty \to \mathrm{span}\{x_n\}_{n=1}^\infty$ be the linear projection given by

$$
P_{m_k} \left(\sum_{i=1}^\infty a_i x_i \right) := \sum_{i=1}^{m_k} a_i x_i,
$$

where (a_i) is any eventually zero sequence of scalars. By (1.1), P_{m_k} is a continuous mapping and $\|P_{m_k}\| \leq \left(1 + \frac{1}{k}\right)$. It then has a (unique) continuous linear extension (denoted again by P_{m_k}) from $Y := \overline{\mathrm{span}}\{x_n\}_{n=1}^\infty$ into itself, and $\|P_{m_k}\| \leq \left(1 + \frac{1}{k}\right)$.

Fix $x \in Y$. Observe first that

$$
P_{m_k}(x) = \sum_{i=1}^{m_k} \langle x, x^*_i \rangle x_i \text{ for every } k \in \mathbb{N}.
$$

To check this, fix $k \in \mathbb{N}$ and take $n \geq m_k$. Then $\{x_n; x^*_n \restriction Y\}_{n=1}^\infty$ is a fundamental biorthogonal system in $Y \times Y^*$. By (iii) in Fact 1.5, we can write $x = \sum_{i=1}^n \langle x, x^*_i \rangle x_i + z_n$, where $z_n \in \overline{\mathrm{span}}\{x_{n+1}, x_{n+2}, \ldots\} \subset Y$. Then, by the continuity of P_{m_k}, we get $P_{m_k}(z_n) = 0$ and then $P_{m_k}(x) := \sum_{i=1}^{m_k} \langle x, x^*_i \rangle x_i$, as stated.

Second, given $\varepsilon > 0$, we can find $k_0 \in \mathbb{N}$ and $y_{k_0} \in \mathrm{span}\{x_n\}_{n=1}^{m_{k_0}}$ such that $\|x - y_{k_0}\| < \varepsilon$. Then, for $k \geq k_0$,

$$
\|x - P_{m_k}(x)\| \leq \|x - y_{k_0}\| + \|y_{k_0} - P_{m_k}(y_{k_0})\| + \|P_{m_k}(y_{k_0}) - P_{m_k}(x)\|
$$
$$
= \|x - y_{k_0}\| + \|P_{m_k}(y_{k_0}) - P_{m_k}(x)\| < \varepsilon + \left(1 + \frac{1}{k}\right)\varepsilon < 3\varepsilon.
$$

This proves that $P_{m_k}(x) \to x$ when $k \to \infty$. This is (ii).

Finally, from (vi) we have, for every $k \in \mathbb{N}$, $d_k \in \{x^*_{m_k+1}, x^*_{m_k+2}, \ldots\}_\perp = \mathrm{span}\{x_i\}_{i=1}^{m_k} \oplus \{x^*_n; n \in \mathbb{N}\}_\perp \subset Y + \{x^*_n; n \in \mathbb{N}\}_\perp$ (here we used (ii) in Fact 1.3), so (iv) holds, too.

It then remains to prove the existence of sequences (x_n) in X and (x^*_n) in X^* such that (i), (iii), (v), (vi), and (vii) hold. This will be done by induction.

To begin, pick x_1 and x_1^* to satisfy (i). Using the compactness of the unit ball of the finite-dimensional space span$\{x_1\}$ and the Hahn-Banach theorem, pick a finite set S_1 in S_{X^*} to satisfy (vii) for $k = 0$. Assume that, for some $k \in \mathbb{N}$, steps 1 to k already produced $\{x_i; x_i^*\}_{i=1}^{m_k}$ and $\{S_i\}_{i=1}^{k}$. For the next step, set $m := 3k + m_{k+1} + 1$ and use the Dvoretzky theorem (see, for example, [Day73, Thm. IV.2.3]) to get an isomorphism $T : Z \to \ell_2^m$ from an m-dimensional subspace $Z \subset (\{x_i^*\}_{i=1}^{m_k} \cup S_k)_\perp$ onto ℓ_2^m equipped with the $\|\cdot\|_2$-norm such that $\|T\| \leq (1 + 1/k)$ and $\|T^{-1}\| = 1$. We shall define $\{x_i\}_{i=m_k+1}^{m_{k+1}}$ in Z and $\{x_i^*\}_{i=m_k+1}^{m_{k+1}}$ in X^* to satisfy (i), (v), (vi), and

(viii) $\{Tx_i\}_{i=m_k+1}^{m_{k+1}}$ is orthogonal in ℓ_2^m

by induction. First of all, observe that

$$\dim Z = m > m_k + k \geq \dim \operatorname{span}\{\{x_i\}_{i=1}^{m_k} \cup \{d_i\}_{i=0}^{k}\}.$$

By Lemma 1.19, we can find $x_{m_k+1} \in S_Z$ (so (v) holds for this vector) such that dist $(x_{m_k+1}, \operatorname{span}\{\{x_i\}_{i=1}^{m_k} \cup \{d_i\}_{i=0}^{k}\}) = 1$. By the Hahn-Banach theorem, we can find $x_{m_k+1}^* \in S_{X^*}$ such that $\langle x_{m_k+1}, x_{m_k+1}^* \rangle = 1$ and (vi) holds for this vector. Assume that, for some $j \in \{2, 3, \ldots, k+1\}$, $\{x_i\}_{i=m_k+1}^{m_k+j-1}$ and $\{x_i^*\}_{i=m_k+1}^{m_k+j-1}$ were already defined to satisfy (i), (v), (vi), and (viii). Let W be the orthogonal complement to $\operatorname{span}\{Tx_i\}_{i=m_k+1}^{m_k+j-1}$ in ℓ_2^m, so $\dim W = m - (j-1)$. Set $G := T^{-1}(W) \cap \left(\operatorname{span}\{x_i^*\}_{i=m_k+1}^{m_k+j-1} \right)_\perp$ and $F := \operatorname{span}\{\{d_i\}_{i=1}^{k} \cup \{x_i\}_{i=1}^{m_k+j-1}\}$. Then $\dim G \geq m - (j-1) - (j-1) = m - 2(j-1) \geq m - 2k = k + m_{k+1} + 1$ and $\dim F \leq k + m_k + j - 1 < k + m_{k+1}$, so $\dim F < \dim G$ and again we can apply Lemma 1.19 to get $x_{m_k+j} \in S_G$ such that dist $(x_{m_k+j}, F) = 1$. Apply the Hahn-Banach theorem one more time to get $x_{m_k+j}^* \in S_{X^*}$, satisfying (i) and (vi). This finishes the finite induction process and gives $\{x_i\}_{i=m_k+1}^{m_{k+1}}$ in Z and $\{x_i^*\}_{i=m_k+1}^{m_{k+1}}$ in X^*.

Now using the compactness of the unit ball of the finite-dimensional space span$\{x_i\}_{i=1}^{m_{k+1}}$ and the Hahn-Banach theorem, pick a finite set $S_{k+1} \supset S_k$ in S_{X^*} to satisfy (vii). This completes step $k+1$. Inductively we get $\{x_n; x_n^*\}_{n=1}^{\infty}$ and $\{S_n\}_{n=1}^{\infty}$. Clearly, they satisfy (i) and (v)–(viii), while (iii) follows from (viii) and the fact that T defined above satisfies $\|T\| \leq (1 + 1/k)$.   □

Although the result in the next corollary will be improved in Theorem 1.27, the method of its proof, which can be traced back to Singer [Sing73], will be used often in this book.

Corollary 1.26. *Let X be a separable Banach space. Then, for every $\varepsilon > 0$, there exists*

(i) *a $(1 + \varepsilon)$-bounded fundamental biorthogonal system in $X \times X^*$; and*
(ii) *a $(1 + \varepsilon)$-bounded total biorthogonal system in $X \times X^*$.*

Proof. (i) Let $\{x_n; x_n^*\}_{n=1}^\infty$ be the biorthogonal system constructed in Lemma 1.25. Fix a sequence (y_n) dense in the unit ball of $\{x_n^*; n \in \mathbb{N}\}_\perp$. Fix $\varepsilon > 0$.

Let us denote $B_0 := \{1\}$ and $B_k := \{m_k + 1, \ldots, m_{k+1}\}$, where $m_k := k(k+1)/2$, $k \in \mathbb{N}$. Arrange the sequence (n) in a matrix by putting consecutive blocks B_n along the inverse diagonals (so B_0 goes to position $(1, 1)$, B_1 to $(1, 2)$, B_2 to $(2, 1)$, B_3 to $(1, 3)$, B_4 to $(2, 2)$, B_5 to $(3, 1)$, and so on).

Fix a row $n \in \mathbb{N}$. Then, for every $N \in \mathbb{N}$, it is possible to find a block $B_{k(n,N)}$ in this row and positive real numbers $a_{n,i}$, $i \in B_{k(n,N)}$ such that

$$\left(\sum_{i \in B_{k(n,N)}} a_i^2 \right)^{1/2} \leq 1, \quad \sum_{i \in B_{k(n,N)}} a_i \geq N.$$

For i in this row and not in $\bigcup_{N \in \mathbb{N}} B_{k(n,N)}$, put $a_i = 1$. Put $w_i := x_i - \varepsilon y_n$ for every i in this row n. Do this for every row n. We obtain a system $\{w_i; x_i^*\}_{i=1}^\infty$, clearly a $(1 + \varepsilon)$-bounded biorthogonal system in $X \times X^*$.

We claim that it is fundamental. Indeed, let $x^* \in \{w_i; i \in \mathbb{N}\}^\perp$, $\|x^*\| = 1$. Then $\langle x_i, x^* \rangle = \varepsilon \langle y_n, x^* \rangle$ for all i in row n, for all $n \in \mathbb{N}$. Fix again a row n. Given $N \in \mathbb{N}$, find $B_{k(n,N)}$ as above. Then

$$2 > 2 \left(\sum_{i \in B_{k(n,N)}} a_i^2 \right)^{1/2}$$

$$\geq \left\| \sum_{i \in B_{k(n,N)}} a_i x_i \right\| \geq \left| \left\langle \sum_{i \in B_{k(n,N)}} a_i x_i, x^* \right\rangle \right|$$

$$= \varepsilon \langle y_n, x^* \rangle \sum_{i \in B_{k(n,N)}} a_i \geq \varepsilon \langle y_n, x^* \rangle N.$$

(The second inequality follows from the 2-equivalence of the block with the ℓ_2-basis; see (iii) in Lemma 1.25.) This is true for every $N \in \mathbb{N}$, so $\langle y_n, x^* \rangle = 0$. As this happens for every $n \in \mathbb{N}$, it follows that $x^* \in \{y_n : n \in \mathbb{N}\}^\perp$, so $x^* \in \{x_n; n \in \mathbb{N}\}^\perp$. Moreover, $x^* \in \{x_n^*\}_\perp^\perp$. From property (iv) in Lemma 1.25, we obtain $x^* = 0$.

(ii) Start again from the system $\{x_n; x_n^*\}_{n=1}^\infty$ given in Lemma 1.25. Fix $\varepsilon > 0$. Note first that, due to (iv) in this lemma, $x_n^* \xrightarrow{w^*} 0$. Let $\{z_n : n \in \mathbb{N}\}$ be a w^*-dense subset of the unit ball of $\{x_n; n \in \mathbb{N}\}^\perp$. Use the matrix of indices defined in part (i) of the proof. For $n \in \mathbb{N}$, let us define

$$w_i^* := x_i^* - \varepsilon z_n^* \text{ if } i \text{ is in row } n.$$

Obviously, $\{x_n; w_n^*\}_{n=1}^\infty$ is a $(1+\varepsilon)$-bounded biorthogonal system in $X \times X^*$. We claim that it is total. Indeed, let $x \in \{w_n^*; n \in \mathbb{N}\}_\perp$. Fix $n \in \mathbb{N}$. Then $\langle x, x_i^* \rangle = \varepsilon \langle x, z_n^* \rangle$ if i belongs to row n. Let $i \to \infty$ in this row. We get $\langle x, z_n^* \rangle = 0$. This holds for every $n \in \mathbb{N}$, so $x \in \overline{\text{span}}\{x_n; n \in \mathbb{N}\}$. Moreover, $\langle x, x_n^* \rangle = 0$ for all $n \in \mathbb{N}$. From (ii) in Lemma 1.25, we get $x = 0$. $\square$

Theorem 1.27 (Pełczyński [Pelc76], Plichko [Plic77]). *Let X be a separable Banach space. Then, for every $\varepsilon > 0$, X has a $(1+\varepsilon)$-bounded M-basis.*

Proof ([Plic77]). Without loss of generality, we may assume $0 < \varepsilon < 1/2$. As in Lemma 1.25, set $m_k := k(k+1)/2$, $k = 0, 1, 2, \ldots$. Let $P_i := \{n \in \mathbb{N};\ m_{i-1} < n \leq m_i\}$, $i \in \mathbb{N}$. Note that card $P_i = i$, $i \in \mathbb{N}$. We shall exhibit (and prove) several features of the biorthogonal system $\{x_n; x_n^*\}$ provided by Lemma 1.25.

(a) dist $(x, \overline{\mathrm{span}}\{x_n; n \in P_{i+1} \cup P_{i+2} \cup \ldots\}) \geq \frac{\|x\|}{2}$ for all $x \in \mathrm{span}\{x_n; n \in P_1 \cup \ldots \cup P_i\}$; henceforth dist $(x, \overline{\mathrm{span}}\{x_n; n \in \mathbb{N} \setminus P_i\}) \geq \frac{\|x\|}{8}$ for all $x \in \mathrm{span}\{x_n; n \in P_i\}$ and for all $i \in \mathbb{N}$.

Proof of (a). Take $x \in \mathrm{span}\{x_n; n \in P_1 \cup \ldots \cup P_i\}$, $y \in \overline{\mathrm{span}}\{x_n; n \in P_{i+1} \cup P_{i+2} \cup \ldots\}$. Check (vii) in the proof of Lemma 1.25: we can find $x^* \in S_i$ such that $\|x\| \leq (1 + 1/i)|\langle x, x^* \rangle|$. Moreover, $y \in (S_i)_\perp$. Then

$$\|x - y\| \geq |\langle x - y, x^* \rangle| = |\langle x, x^* \rangle| \geq \left(1 + \frac{1}{i}\right)^{-1} \|x\| \geq \frac{1}{2}\|x\|.$$

This proves the first part of the statement. For the second part, now take $x \in \mathrm{span}\{x_n; n \in P_i\}$ for $i \neq 1$ and $y \in \mathrm{span}\{x_n; n \in \mathbb{N} \setminus P_i\}$. Then $y = y_1 + y_2$, where $y_1 \in \mathrm{span}\{x_n; n \in P_1 \cup \ldots \cup P_{i-1}\}$ and $y_2 \in \mathrm{span}\{x_n; n \in P_{i+1} \cup P_{i+2} \cup \ldots\}$. From the first part,

$$\|x - y\| = \|x - y_1 - y_2\| \geq \frac{\|y_1\|}{2}.$$

If $\|y_1\|/2 \geq \|x\|/8$, then we are done. On the other hand, if $\|y_1\|/2 < \|x\|/8$, then $\|x - y_1\| \geq (3/4)\|x\|$ and we get, again from the first part,

$$\|x - y\| = \|x - y_1 - y_2\| \geq \frac{\|x - y_1\|}{2} \geq \frac{3}{8}\|x\| \geq \frac{1}{8}\|x\|.$$

This proves (a).

(b) $(1/2)\|x\|_2 \leq \|x\| \leq (3/2)\|x\|_2$, where $x = \sum_{n \in P_i} \alpha_n x_n$ and $\|x\|_2 := \left(\sum_{n \in P_i} \alpha_n^2\right)^{1/2}$.

Proof of b). This follows right away from (iii) in Lemma 1.25.

(c) $\overline{\mathrm{span}}\{\{x_n\}_{n=1}^\infty \cup (\{x_n^*\}_{n=1}^\infty)_\perp\} = X$, $\overline{\mathrm{span}}\{x_n\}_{n=1}^\infty \cap (\{x_n^*\}_{n=1}^\infty)_\perp = \{0\}$.

Proof of (c). The first assertion is (iv) in Lemma 1.25. The second comes from (ii) in the same lemma.

A straightforward consequence of (c) is that $\{x_n; x_n^* \restriction \overline{\mathrm{span}}\{x_n\}_{n=1}^\infty\}_{n=1}^\infty$ is an M-basis in $\overline{\mathrm{span}}\{x_n\}_{n=1}^\infty \times (\overline{\mathrm{span}}\{x_n\}_{n=1}^\infty)^*$. We could then use Theorem 1.45 (which is independent of any boundedness of the M-basis) to extend it to an M-basis of X. However, thanks to (c), we can do it better. Let $q : X \to X/\overline{\mathrm{span}}\{x_n\}_{n=1}^\infty$, the canonical quotient mapping. $q((\{x_n^*\}_{n=1}^\infty)_\perp)$ is dense in $X/\overline{\mathrm{span}}\{x_n\}_{n=1}^\infty$. (To check this, just take $x^\perp$ an element in $(\overline{\mathrm{span}}\{x_n\}_{n=1}^\infty)^\perp = (X/\overline{\mathrm{span}}\{x_n\}_{n=1}^\infty)^*$ such that $\langle x_n, x^\perp \rangle = 0$ for all $n \in \mathbb{N}$.

Then, from (c) it follows that $x^{\perp} = 0$.) Now, Theorem 1.22 allows us to choose an M-basis $\{\hat{y}_m; y_m^*\}_{m=1}^{\infty}$ of the separable Banach space $X/\overline{\text{span}}\{x_n\}_{n=1}^{\infty}$ such that $\hat{y}_m \in q((\{x_n^*\}_{n=1}^{\infty})_{\perp})$ and $y_m^* \in X^{\perp}$ for all $m \in \mathbb{N}$. Select $y_m \in \hat{y}_m$, $y_m \in (\{x_n^*\}_{n=1}^{\infty})_{\perp}$, $m \in \mathbb{N}$. Then $\{x_n, y_m; x_n^*, y_m^*\}_{n,m=1}^{\infty}$ is an M-basis of X. In order to see that, observe first that

$$\text{dist}\,(x_n, \overline{\text{span}}\{x_k, y_m\}_{k=1, k \neq n, m=1}^{\infty}) = 1 \ \ (\text{use } x_n^*)$$

and

$$\text{dist}\,(y_m, \overline{\text{span}}\{x_n, y_k\}_{n=1, k=1, k \neq m}^{\infty}) \neq 0 \ \ (\text{use } y_m^*).$$

Moreover, $\overline{\text{span}}^{w^*}\{x_n^*, y_m^*\}_{n,m=1}^{\infty} = X^*$; this follows from (c). We may and do assume that $\|y_n\| = 1$ for all $n \in \mathbb{N}$.

Let $a_j := \text{dist}\,(y_j, \{y_j^*\}_{\perp})$ for $j \in \mathbb{N}$. We choose a double sequence $(n_j^k)_{j,k=1}^{\infty} \subset \mathbb{N}$ in such a way that for all k and j

$$\frac{6}{\sqrt{n_j^k}} \le \varepsilon, \quad \frac{(n_j^k - 1)\varepsilon}{8} - \frac{2^9}{\varepsilon} > \begin{cases} n_j^{k-1} & \text{for } k \neq 1, \\ \sqrt{\dfrac{n_j^1}{a_j}} & \text{for } k = 1. \end{cases} \quad (1.2)$$

We denote by P_j^k that P_i for which $n_j^k = i$ (so card $P_j^k = n_j^k$). Define, for $n \in \mathbb{N}$,

$$e_n := \begin{cases} x_n & \text{for } n \notin \bigcup_{j,k} P_j^k, \quad \text{(i)} \\ x_n + y_j/\sqrt{n_j^1} & \text{for } n \in P_j^1, \quad \text{(ii)} \\ x_n + \left(\sum_{s \in P_j^{k-1}} x_s\right)/n_j^{k-1} & \text{for } n \in P_j^k,\ k \neq 1. \ \text{(iii)} \end{cases} \quad (1.3)$$

We claim that $\{e_n\}_{n=1}^{\infty}$ is a *minimal fundamental system in X and*, denoting by $\{e_n^*\}_{n=1}^{\infty}$ the associated system of functional coefficients, $\{e_n; e_n^*\}_{n=1}^{\infty}$ is in fact the $(1 + \varepsilon)$-*bounded M-basis we are looking for*.

In order to prove the claim it is enough to check the following statements:
(d) $\{e_n\}_{n=1}^{\infty}$ is fundamental.
(e)
$$\text{dist}\,(e_n, D_n) \ge 1 - \varepsilon/2 \ \text{ for every } n \in \mathbb{N}, \quad (1.4)$$

where $D_n := \overline{\text{span}}\{e_m\}_{m=1, m \neq n}^{\infty}$, $n \in \mathbb{N}$. (This implies that $\{e_n\}_{n=1}^{\infty}$ is a minimal system, hence the existence of an associated system $\{e_n^*\}_{n=1}^{\infty}$ of functional coefficients.)
(f) $\overline{\text{span}}^{w^*}\{e_n^*\}_{n=1}^{\infty} = X^*$.

Then the $(1 + \varepsilon)$-boundedness of $\{e_n; e_n^*\}_{n=1}^{\infty}$ follows. Indeed, fix $n \in \mathbb{N}$. First, note that $\|e_n\| \le 1 + \varepsilon/4$. (This comes from (1.3) and (1.2), plus the estimate given in (b) if $n \in P_j^k$ for some $k \neq 1$ and some j.) If $\|e_n^*\| \le 1$, we are done. If not, and recalling that $0 < \varepsilon < 1/2$, take $\delta > 0$ small enough to get

$$\left(1 + \frac{\varepsilon}{4}\right)\left(\frac{1 + \delta}{1 - \varepsilon/2} + 2\delta\right) < 1 + \varepsilon. \quad (1.5)$$

Let $x \in S_X$ such that $\langle x, e_n^* \rangle \geq \|e_n^*\| - \delta$. From (d) we can find an eventually zero sequence (α_n) of scalars such that

$$\left\| x - \sum_{k=1}^{\infty} \alpha_k e_k \right\| < \delta / \|e_n^*\|. \tag{1.6}$$

Then $|\langle x, e_n^* \rangle - \alpha_n| = |\langle x - \sum_{k=1}^{\infty} \alpha_k e_k, e_n^* \rangle - \alpha_n| < \delta$ (in particular, $\alpha_n \neq 0$). On the other side,

$$\delta > \frac{\delta}{\|x_n^*\|} > \left\| x - \sum_{k=1}^{\infty} \alpha_k e_k \right\| = |\alpha_n| \left\| e_n + \sum_{k=1, k \neq n}^{\infty} \alpha_k e_k - \frac{x}{\alpha_n} \right\|$$

$$\geq |\alpha_n| \left\| e_n + \sum_{k=1, k \neq n}^{\infty} \alpha_k e_k \right\| - \left\| \frac{x}{\alpha_n} \right\|$$

$$\geq |\alpha_n| \left[\left(1 - \frac{\varepsilon}{2}\right) - \frac{1}{|\alpha_n|} \right] = |\alpha_n| \left(1 - \frac{\varepsilon}{2}\right) - 1.$$

(The last inequality comes from (1.4).) From this it follows that $|\alpha_n| < (1 + \delta)(1 - \varepsilon/2)$ and hence $|\langle x, e_n^* \rangle| < (1 + \delta)(1 - \varepsilon/2) + \delta$, so finally $\|e_n^*\| < (1+\delta)(1-\varepsilon/2)+2\delta$. Then, from (1.5) we can conclude that $\|e_n\|.\|e_n^*\| < 1+\varepsilon$ for every $n \in \mathbb{N}$.

To finish the proof, it remains to prove (d), (e), and (f).

Proof of (d). Applying (1.3), property (b), and (1.2) in succession, we obtain

$$\left\| \sum_{k=1}^{t} (-1)^k \left(\sum_{n \in P_j^k} e_n \right) / n_j^k + \frac{1}{\sqrt{n_j^1}} y_j \right\| = \left\| \frac{1}{n_j^t} \sum_{n \in P_j^t} x_n \right\| \leq \frac{3}{2\sqrt{n_j^t}} \overset{t \to \infty}{\longrightarrow} 0.$$

Thus $\{y_j\}_{j=1}^{\infty} \subset \overline{\operatorname{span}}\{e_n\}_{n=1}^{\infty}$. Inductively, from (1.3) we get also that $\{x_n\}_{n=1}^{\infty} \subset \overline{\operatorname{span}}\{e_n\}_{n=1}^{\infty}$. Now, from property (c) above, it follows that $\overline{\operatorname{span}}\{e_n\}_{n=1}^{\infty} = X$.

Proof of (e). Fix $n \in \mathbb{N}$. Choose any element $z \in \operatorname{span}\{e_m\}_{m=1, m \neq n}^{\infty}$. We can write

$$z = \sum_{i=1}^{\infty} z_i, \quad \text{where } z_i \in \operatorname{span}\{e_s; s \in P_i\}, \ i \in \mathbb{N}. \tag{1.7}$$

(The summands in the sum above are eventually zero.)

If $n \notin \bigcup_{j,k=1}^{\infty} P_j^k$, then (1.4) holds for $\langle e_n, x_n^* \rangle = \langle x_n, x_n^* \rangle = 1$ and $\langle e_m, x_n^* \rangle = 0$ if $m \in \mathbb{N}, m \neq n$.

If, on the contrary, $n \in P_j^k$, we have either

$$\|e_n - z\| = \left\| x_n + \frac{y_j}{\sqrt{n_j^1}} - z \right\| \geq \|x_n - z_n\| - \frac{\varepsilon}{6}$$

in case (1.3(ii)) (we used (1.2)) or

$$\|e_n - z\| = \left\| x_n + \left(\sum_{s \in P_j^{k-1}} x_s \right) / n_j^{k-1} - z \right\| \geq \|x_n - z_n\| - \frac{\varepsilon}{4}$$

in case (1.3(iii)) (we used instead condition (b) about the ℓ_2-norm and (1.2)).

In any case, if we could show that $\|x_n - z\| \geq 1 - \varepsilon/4$, (1.4) will hold, and this will finish the proof of the theorem.

Assume, on the contrary, that

$$\|x_n - z\| < 1 - \frac{\varepsilon}{4}. \tag{1.8}$$

By the very definition (1.4), the vectors e_n appearing in expression (1.7) cannot be only of types (i) and (ii) in (1.4) (just evaluate $\langle x_n - z, x_n^* \rangle$), so there exists a term $z_j^{k+1} = \sum_{s \in P_j^{k+1}} \alpha_s e_s$ in the sum (1.7), where some e_s is of type (iii). Note, too, that $k+1$ in the upper script is compulsory, again evaluating $\langle x_n - z, x_n^* \rangle$. Precisely, we get

$$|\langle x_n - z, x_n^* \rangle| = |\langle x_n - z_j^{k+1}, x_n^* \rangle| = \left| \left\langle x_n - \sum_{s \in P_j^{k+1}} \alpha_s e_s, x_n^* \right\rangle \right|$$

$$= \left| \left\langle x_n - \sum_{s \in P_j^{k+1}} \alpha_s \left(x_s + \frac{1}{n_j^k} \sum_{t \in P_j^k} x_t \right), x_n^* \right\rangle \right|$$

$$= \left| \left\langle x_n - \frac{1}{n_j^k} \left(\sum_{s \in P_j^{k+1}} \alpha_s \right) x_n, x_n^* \right\rangle \right| = \left| 1 - \frac{1}{n_j^k} \sum_{s \in P_j^{k+1}} \alpha_s \right| < 1 - \frac{\varepsilon}{4},$$

and hence

$$b_j^k := \frac{1}{n_j^k} \sum_{s \in P_j^{k+1}} \alpha_s > \varepsilon/4. \tag{1.9}$$

1. We show that there exists a term $z_j^k = \sum_{s \in P_j^k} \alpha_s e_s$ in the sum (1.7) for which

$$\left| \sum_{s \in P_j^k} \alpha_s \right| > \begin{cases} n_j^{k-1} & \text{for } k \neq 1, \\ \sqrt{\dfrac{n_j^1}{a_j}} & \text{for } k = 1. \end{cases} \tag{1.10}$$

Indeed, applying (1.8), and properties (a) and (b) successively, we get

$$1 - \varepsilon/4 > \|x_n - z\| = \left\| \left(x_n - \sum_{s \in P_j^k} (b_j^k + \alpha_s) x_s \right) \right.$$

$$\left. - \left(z - b_j^k \sum_{s \in P_j^k} x_s - \sum_{s \in P_j^k} \alpha_s x_s \right) \right\| \geq \frac{1}{8} \left\| x_n - \sum_{s \in P_j^k} (b_j^k + \alpha_s) x_s \right\|$$

$$\geq \frac{1}{16} \left\| x_n - \sum_{s \in P_j^k} (b_j^k + \alpha_s) x_s \right\|_2 \geq \frac{1}{16} \sqrt{\sum_{s \in P_j^k, s \neq n} (b_j^k + \alpha_s)^2},$$

the first inequality is due to the fact that $\left(z - b_j^k \sum_{s \in P_j^k} x_s - \sum_{s \in P_j^k} \alpha_s x_s \right) \in$ span$\{x_m; m \in \bigcup_{r \in \mathbb{N}, r \neq k} P_j^r \}$. Hence $(1 - \varepsilon/4)^2 > 2^{-8} \sum_{s \in P_j^k, s \neq n} (b_j^k + \alpha_s)^2$. Making some algebraic transformations, we have

$$-2b_j^k \sum_{s \in P_j^k, s \neq n} \alpha_s > (n_j^k - 1)(b_j^k)^2 - 2^8 (1 - \varepsilon/4)^2.$$

Dividing the last inequality by $2|b_j^k|$ and using (1.9) and (1.2), we obtain

$$\left| \sum_{s \in P_j^k, s \neq n} \alpha_s \right| > \begin{cases} n_j^{k-1} & \text{for } k \neq 1, \\ \sqrt{\dfrac{n_j^1}{a_j}} & \text{for } k = 1. \end{cases} \tag{1.11}$$

In the sum (1.7), the vector e_n is not allowed, so $\alpha_n = 0$. Then (1.10) follows from the former inequality.

For $k \neq 1$ proceed to Part 2 below, and for $k = 1$ to Part 3 below.

2. Thus

$$|b_j^{k-1}| \geq 1, \tag{1.12}$$

where $b_j^{k-1} := \left(\sum_{s \in P_j^k} \alpha_s \right) / n_j^{k-1}$; put $u_j^{k-1} := b_j^{k-1} \sum_{s \in P_j^{k-1}} x_s$. We use once again arguments similar to those of 1: we show that there exists a term $z_i^{k-1} = \sum_{s \in P_j^{k-1}} \alpha_s e_s$ in the sum (1.7) for which

$$\left| \sum_{s \in P_j^{k-1}} \alpha_s \right| > \begin{cases} n_j^{k-2} & \text{for } k \neq 2, \\ \sqrt{\dfrac{n_j^1}{a_j}} & \text{for } k = 2. \end{cases} \tag{1.13}$$

Indeed, applying (1.8) and properties a) and b) successively, we get

$$1 - \varepsilon/4 > \|x_n - z\| = \left\| x_n - u_j^{k-1} - \sum_{s \in P_j^{k-1}} \alpha_s x_s \right.$$

$$\left. - \left(z - u_j^{k-1} - \sum_{s \in P_j^{k-1}} \alpha_s x_s \right) \right\| \geq \frac{1}{8} \left\| u_j^{k-1} + \sum_{s \in P_j^{k-1}} \alpha_s x_s \right\|$$

$$\geq \frac{1}{16} \left\| u_j^{k-1} + \sum_{s \in P_j^{k-1}} \alpha_s x_s \right\|_2 \geq \frac{1}{16} \sqrt{\sum_{s \in P_j^{k-1}} (b_j^{k-1} + \alpha_s)^2}.$$

Hence $(1 - \varepsilon/4)^2 > 2^{-8} \sum_{s \in P_j^{k-1}} (b_j^{k-1} + \alpha_s)^2$. Making some elementary algebraic transformations, we have

$$-2b_j^{k-1} \sum_{s \in P_j^{k-1}} \alpha_s > n_j^{k-1}(b_j^{k-1})^2 - 2^8(1 - \varepsilon/4)^2.$$

Dividing the last inequality by $2|b_j^{k-1}|$ and using (1.12) and (1.2), we obtain (1.13).

For $k \neq 2$, we go back to the beginning of Part 2, replacing k by $k - 1$; for $k = 2$, we go to Part 3 below.

3. After finitely many steps, we come to the conclusion that sum (1.7) contains a term $z_j^1 = \sum_{s \in P_j^1} \alpha_s x_s + y$, where $y := \left(\sum_{s \in P_j^1} \alpha_s \right) y_j / \sqrt{n_j^1}$ and $\left| \sum_{s \in P_j^1} \alpha_s \right| \geq \sqrt{n_j^1/a_j}$. Since $x_n - (z - y) \in \{y_j^*\}_\perp$, the definition of a_j gives

$$\|x_n - (z - y) - y\| \geq a_j \|y\| \geq 1,$$

and this contradicts (1.8), so that (1.4) holds.

Proof of (f). From Fact 1.8, it follows that

$$\overline{\operatorname{span}}^{w^*} \{e_n^*\}_{n=1}^\infty = X^* \text{ if and only if } \bigcap_{m=1}^\infty \overline{\operatorname{span}} \{e_n\}_{n=m}^\infty = \{0\}.$$

We have $\bigcap_{m=1}^\infty \overline{\operatorname{span}} \{e_n\}_{n=m}^\infty \subset \{\{x_n^*\}_{n=1}^\infty \cup \{y_n^*\}_{n=1}^\infty\}_\perp = \{0\}$, as it follows easily from the definition of $\{e_n\}_{n=1}^\infty$ in (1.3). This proves f).

This concludes the proof. □

Theorem 1.28 (Plichko [Plic86a]). *Let K be an infinite compact metric space. Let $\{k_n\}_{n=1}^\infty$ be a dense subset in K such that for some $n \in \mathbb{N}$, k_n is an accumulation point of $\{k_m; m \in \mathbb{N}, m \neq n\}$. Then $(C(K), \|\cdot\|_\infty)$ fails to have an Auerbach system $\{f_n; \mu_n\}_{n=1}^\infty \subset C(K) \times C(K)^*$ such that $\overline{\operatorname{span}}^{\|\cdot\|} \{\mu_n\}_{n=1}^\infty \supset \{\delta_{k_n}\}_{n=1}^\infty$.*

We shall need the following intermediate result.

Proposition 1.29. *Let K be an infinite compact metric space and let (k_n) be a sequence in K that converges to some point $k_0 \in K$, $k_n \neq k_0$, $n \in \mathbb{N}$. Let $\{f_n; \mu_n\}_{n=1}^{\infty}$ be a total biorthogonal system in $C(K) \times C(K)^*$. Then, if $\{\delta_{k_n}\}_{n=0}^{\infty} \subset \overline{\mathrm{span}}^{\|\cdot\|}\{\mu_n\}_{n=1}^{\infty}$, we have*

$$\sup_{n \in \mathbb{N}} \|f_n\| \cdot \|\mu_n\| > 1.$$

Proof. Suppose

$$\sup_{n \in \mathbb{N}} \|f_n\| \cdot \|\mu_n\| = 1. \tag{1.14}$$

We may and do assume that $\|f_n\| = \|\mu_n\| = 1$ for all $n \in \mathbb{N}$. Fix $0 < \varepsilon < 1/2$. As $\delta_{k_0} \in \overline{\mathrm{span}}^{\|\cdot\|}\{\mu_n\}_{n=1}^{\infty}$, for some $\mu_0 := \sum_{n=1}^{n_0} \alpha_n \mu_n$ with $\|\mu_0\| = 1$ we have $\|\delta_{k_0} - \mu_0\| < \varepsilon$.

Then, denoting by $\mathrm{Var}\, \mu(S)$ the total variation of $\mu \in C(K)^*$ in some Borel set $S \subset K$, we have

$$\varepsilon > \mathrm{Var}\, (\delta_{k_0} - \mu_0)(K)$$
$$= |(\delta_{k_0} - \mu_0)(\{k_0\})| + \mathrm{Var}\, \mu_0(K \setminus \{k_0\}) \geq |1 - \mu_0(\{k_0\})|,$$

and we get $|\mu_0(\{k_0\})| > 1 - \varepsilon$ and $\mathrm{Var}\, \mu_0(K \setminus \{k_0\}) < \varepsilon$.

We shall prove the following claim: $\mu_n(\{k_0\}) = 0$ for all $n > n_0$. Arguing by contradiction, assume that for some $n > n_0$, we have $\mu_n(\{k_0\}) = b \neq 0$. We may take $b > 0$ (if not, change μ_n and f_n to $-\mu_n$ and $-f_n$, respectively). Then $\mathrm{Var}\, \mu_n(K \setminus \{k_0\}) = 1 - b$ and hence

$$\left\| \mu_n - \frac{b}{\mu_0(\{k_0\})} \mu_0 \right\|$$
$$= \mathrm{Var}\, \left(\mu_n - \frac{b}{\mu_0(\{k_0\})} \mu_0 \right)(K \setminus \{k_0\}) + \left| \left(\mu_n - \frac{b}{\mu_0(\{k_0\})} \mu_0 \right)\{k_0\} \right|$$
$$\leq \mathrm{Var}\, \mu_n(K \setminus \{k_0\}) + \left| \frac{b}{\mu_0(\{k_0\})} \mathrm{Var}\, \mu_0(K \setminus \{k_0\}) \right| \leq 1 - b + \frac{b\varepsilon}{1 - \varepsilon} < 1.$$

However, condition (1.14) implies that for any $n > n_0$ and $\alpha \in \mathbb{R}$ we have $\|\mu_n - \alpha\mu_0\| \geq 1$. This proves the claim. Put $G := \overline{\mathrm{span}}^{\|\cdot\|}\{\mu_n\}_{n=1}^{\infty}$, a w^*-dense closed subspace of $C(K)^*$. Let's define a linear functional ϕ on G by $\phi(\mu) := \mu(\{k_0\})$ for all $\mu \in G$. This functional is obviously $\|\cdot\|$-continuous. However, it is not $\sigma(G, C(K))$-continuous: the sequence (δ_{k_n}) is in G and $\sigma(G, C(K))$-converges to $\delta_{k_0} \in G$, while $\phi(\delta_{k_n}) = 0$ for all $n \in \mathbb{N}$ and $\phi(\delta_{k_0}) = 1$. Put $F := \overline{\mathrm{span}}^{\|\cdot\|}\{\mu_n\}_{n=n_0+1}^{\infty} + (\mathrm{span}\{\mu_n\}_{n=1}^{n_0} \cap H)$, where $H := \{\mu \in G; \; \phi(\mu) = 0\}$. Then $F \subset H \subset G$. Moreover, $F + \mathrm{span}\{\mu_0\} = G$ and $\mu_0 \notin H$. (The sum of a closed subspace and a finite-dimensional one is closed; see, for example, [Fa~01, Exer. 5.27].) It follows that $F = H$, so F is $\sigma(G, C(K))$-dense in G. In particular,

$$\mu_0 \in \overline{F}^{\sigma(G,C(K))} = \overline{\mathrm{span}}^{\sigma(G,C(K))}\{\mu_n\}_{n=n_0+1}^{\infty} + \left(\mathrm{span}\{\mu_n\}_{n=1}^{n_0} \cap H\right).$$

We can then write

$$\mu_0 = \mu + \sum_{n=1}^{n_0} b_n \mu_n,$$

where $\mu \in \overline{\mathrm{span}}^{\sigma(G,C(K))}\{\mu_n\}_{n=n_0+1}^{\infty}$ and $b_1, \dots, b_n$ are some real numbers, so

$$0 \neq \mu \in \overline{\mathrm{span}}^{\sigma(G,C(K))}\{\mu_n\}_{n=n_0+1}^{\infty} \cap \mathrm{span}\{\mu_n\}_{n=1}^{n_0},$$

a contradiction with the fact that $\{f_n; \mu_n\}_{n=1}^{\infty}$ is a biorthogonal system. This proves that supposition (1.14) is false. □

Proof. (Theorem 1.28). It is enough to observe that $\overline{\mathrm{span}}^{\|\cdot\|}\{\delta_{k_n}\}_{n=1}^{\infty} \subset C(K)^*$ is a 1-norming subspace (so, in particular, it is w^*-dense in $C(K)^*$), and apply Proposition 1.29. □

Corollary 1.30. *Let K be an infinite compact metric space. Then the space $(C(K), \|\cdot\|_\infty)$ has no shrinking Auerbach system.*

Proof. In K there exists a sequence (k_n) and an element k_0 such that $k_n \to k_0$ and $k_n \neq k_0$ for all $n \in \mathbb{N}$. Should $\{f_n; \mu_n\}_{n=1}^{\infty}$ be a shrinking Auerbach system in $C(K) \times C(K)^*$, we will have $w^*\text{-}\lim_n \delta_{k_n} = \delta_{k_0}$ in $\overline{\mathrm{span}}^{\|\cdot\|}\{\mu_n\}_{n=1}^{\infty} = C(K)^*$, contradicting Proposition 1.29. □

Note that if K is a countable metric compact, then $(C(K), \|\cdot\|_\infty)$ has a shrinking M-basis, as $C(K)^*$ is separable (see [Fa~01, Thm. 10.47]), and so we can apply Theorem 1.22.

It is an open problem whether every separable Banach space has an Auerbach basis. It is easy to see that, under renorming, the answer is positive. This is the content of the following simple result.

Proposition 1.31. *In every separable Banach space $(X, \|\cdot\|)$ and for any $\varepsilon > 0$, there exists an equivalent norm $\||\cdot\||$ on X such that $\|x\| \leq \||x\|| \leq (1+2\varepsilon)\|x\|$ and $(X, \||\cdot\||)$ has an Auerbach basis.*

Proof. Given $\varepsilon > 0$, the existence of a $(1+\varepsilon)$-bounded M-basis $\{e_n; e_n^*\}_{n=1}^{\infty}$ in $(X, \|\cdot\|)$, where $\|e_n\| = 1$ for all $n \in \mathbb{N}$, is proved in Theorem 1.27. The new norm $\||\cdot\||$ on X is given by the Minkowski functional of the set $B := \{x \in X; \|x\| \leq 1, |\langle x, e_n^*\rangle| \leq 1, n \in \mathbb{N}\}$. It is simple to prove that B is the closed unit ball of an equivalent norm satisfying the required inequalities and that $\{e_n; e_n^*\}_{n=1}^{\infty}$ is an Auerbach basis for $(X, \||\cdot\||)$. □

1.5 Strong M-bases

Motivated by aspects of Fourier series when using trigonometrical systems (and by Schauder bases), the special class of strong M-bases was introduced by Ruckle [Ruck70]. Davis and Singer in [DaSi73] explicitly raised the question of the existence of strong M-bases in every separable Banach space. This question was solved in the positive by Terenzi. In fact, in a series of papers, he gave more and more precise strongness conditions, arriving finally at the concept of a uniform minimal system with quasifixed brackets and permutations. We present here Vershynin's proof of Terenzi's result on the existence of a special strong M-basis (produced from every norming M-basis by a particular procedure called flattened perturbation). The existence of strong M-bases in many nonseparable Banach spaces will be discussed in Chapter 4.

Definition 1.32. *Let X be a Banach space. An M-basis $\{x_n; x_n^*\}_{n=1}^{\infty} \subset X \times X^*$ is called* strong *if $x \in \overline{\text{span}}\{\langle x, x_n^* \rangle x_n\}_{n=1}^{\infty}$ for all $x \in X$.*

Note that an M-basis $\{x_n, x_n^*\}_{n=1}^{\infty}$ is strong if and only if every $x \in X$ belongs to $\overline{\text{span}}\{x_n; n \in \mathbb{N}, \langle x, x_n^* \rangle \neq 0\}$ since obviously $\text{span}\{x_n; n \in \mathbb{N}, \langle x, x_n^* \rangle \neq 0\} = \text{span}\{\langle x, x_n^* \rangle x_n; n \in \mathbb{N}\}$. Some other equivalent formulations of strongness are given in Proposition 1.35 below.

Obviously, every Schauder basis of X is a strong M-basis.

Remark 1.33. It follows from the next proposition that *every separable Banach space admits an M-basis that is not a Schauder basis under any permutation.*

Proposition 1.34 (Johnson). *Every separable Banach space admits an M-basis that is not strong.*

Proof. Let $\{e_n; h_n\}_{n \in \mathbb{N}}$ be an M-basis for a separable space X (see Theorem 1.22). We can assume, without loss of generality, that $\sup_n \|e_{2n}\| < \infty$ and $\sup_n \|h_{2n-1}\| < \infty$. Put

$$x_{2n-1} = -2^{n-1} e_{2n-1}, \qquad n = 1, 2, \ldots,$$

$$x_2 = -e_1 + \frac{1}{2} e_2, \quad x_{2n} = 2^{n-2} e_{2n-3} - 2^{n-1} e_{2n-1} + \frac{1}{2^n} e_{2n}, \qquad n = 2, 3, \ldots,$$

$$f_{2n-1} = -\frac{1}{2^{n-1}} h_{2n-1} - 2^n h_{2n} + 2^{n+1} h_{2n+2}, \qquad n = 1, 2, \ldots,$$

$$f_{2n} = 2^n h_{2n}, \qquad n = 1, 2, \ldots.$$

Then $\{x_n; f_n\}$ is a biorthogonal system in X. Moreover,

$$\text{span}\{e_1, e_2, \ldots e_n\} = \text{span}\{x_1, x_2, \ldots x_n\}$$

and

$$\text{span}\{f_1, \ldots, f_{n+3}\} \supset \text{span}\{h_1, \ldots, h_n\}.$$

It follows that $\{x_n; f_n\}$ is an M-basis for X.

Put

$$z = \sum_{i=1}^{\infty} \frac{1}{2^i} e_{2i}.$$

We compute that $f_{2n-1}(z) = 0$ for all n. If $\{x_n; f_n\}$ is a strong M-basis, then $z \in \overline{\mathrm{span}}\{f_{2n}(z)x_{2n}\}$ and thus $z \in \overline{\mathrm{span}}\{x_{2n}\}$. However, put

$$g = h_2 + \sum_{i=1}^{\infty} \frac{1}{2^i} h_{2i-1}.$$

Then $g(z) = \frac{1}{2}$ and $g(x_{2n}) = 0$. Thus z cannot be in $\overline{\mathrm{span}}\{x_{2n}\}$. This contradiction shows that $\{x_n; f_n\}$ is not a strong M-basis. $\qquad\square$

Proposition 1.35. *Let $\{x_n\}_{n=1}^{\infty}$ be an M-basis for a Banach space X. Then the following are equivalent:*

(i) $\{x_n\}$ *is a strong M-basis for X.*
(ii) [PlRe83] $\overline{\mathrm{span}}\{x_n; n \in A\} \cap \overline{\mathrm{span}}\{x_n; n \in B\} = \overline{\mathrm{span}}\{x_n; n \in A \cap B\}$ *for every $A \subset \mathbb{N}$, $B \subset \mathbb{N}$.*
(iii) *For each $A \subset \mathbb{N}$, $B \subset \mathbb{N}$, such that $A \cup B = \mathbb{N}$ and $A \cap B = \emptyset$, then $\overline{\mathrm{span}}\{x_a; a \in A\} = \{x_b^*; b \in B\}_{\perp}$.*

Proof. (i)$\Rightarrow$(ii) Obviously,

$$\overline{\mathrm{span}}\{x_n; n \in A \cap B\} \subset \overline{\mathrm{span}}\{x_n; n \in A\} \cap \overline{\mathrm{span}}\{x_n; n \in B\}.$$

It is equally obvious that, given $x \in \overline{\mathrm{span}}\{x_n; n \in A\} \cap \overline{\mathrm{span}}\{x_n; n \in B\}$, we have $\langle x, x_n^* \rangle = 0$ for all $n \notin A \cap B$. Therefore, from the strongness of the basis, we get $x \in \overline{\mathrm{span}}\{x_n; n \in A \cap B\}$. This proves (ii).

(ii)$\Rightarrow$(i) Assume that, for some $x \in X$, $x \notin \overline{\mathrm{span}}\{x_n; n \in A\}$, where $A := \{n \in \mathbb{N}; \langle x, x_n^* \rangle \neq 0\}$. Let $B := \mathbb{N} \setminus A$. Using (iii), from Fact 1.5 it follows that

for every finite set $F \subset B$, we have $x \in \overline{\mathrm{span}}\{x_n; n \in A \cup (B \setminus F)\}$.   (1.15)

Find a finite set $B_1 \subset B$ and an element $v_1 \in \mathrm{span}\{x_n; n \in A \cup B_1\}$ such that $\|x - v_1\| < 1$. From (1.15) we have $x \in \overline{\mathrm{span}}\{x_n; n \in A \cup (B \setminus B_1)\}$; hence we can find a finite set $B_2 \subset (B \setminus B_1)$ and an element $v_2 \in \mathrm{span}\{x_n; n \in A \cup B_2\}$ such that $\|x - v_2\| < 1/2$. Again, it follows from (1.15) that $x \in \overline{\mathrm{span}}\{x_n; n \in A \cup (B \setminus (B_1 \cup B_2))\}$; hence we can find a finite set $B_3 \subset B \setminus (B_1 \cup B_2)$ and an element $v_3 \in \mathrm{span}\{x_n; n \in A \cup B_3\}$ such that $\|x - v_3\| < 1/3$. Continue in this way to get sequences (v_k) and (B_k). The sequence $(v_{2k-1})_{k=1}^{\infty}$ converges to x; hence $x \in \overline{\mathrm{span}}\{x_n; n \in A \cup B_1 \cup B_3 \cup \ldots\}$. By considering instead the sequence $(v_{2k})_{k=1}^{\infty}$, which also converges to X, $x \in \overline{\mathrm{span}}\{x_n; n \in A \cup B_2 \cup B_4 \cup \ldots\}$. Now use that $(B_n)_{n \in \mathbb{N}}$ is a pairwise disjoint sequence together with (ii) to get $x \in \overline{\mathrm{span}}\{x_n; n \in A\}$, a contradiction.

(ii)$\Rightarrow$(iii) Obviously, $\overline{\mathrm{span}}\{x_a; a \in A\} \subset \{x_b^*; b \in B\}_{\perp}$. In order to prove the reverse inclusion, use Lemma 1.40: let $r(1) < r(2) < \ldots$ be a sequence of representing indices for the fundamental biorthogonal system $\{x_n; x_n^*\}_{n=1}^{\infty}$. Let $x \in \{x_b^*; b \in B\}_{\perp}$. We get $x = \lim_{m \to \infty} \left(\sum_{n=1, n \in A}^{r(m)} \langle x, x_n^* \rangle x_n + v_m \right)$, where $v_m \in \mathrm{span}\{x_n\}_{n=r(m)+1}^{r(m+1)}$. Let $B_m := \{n; \ n \in \mathbb{N}, r(m) < n \leq r(m+1)\} \cap B$,

$m = 1, 2, \ldots$. Then $x \in \overline{\operatorname{span}}\{x_n; n \in A \cup B_2 \cup B_4 \cup \ldots\}$ and, simultaneously, $x \in \overline{\operatorname{span}}\{x_n; n \in A \cup B_1 \cup B_3 \cup \ldots\}$. From (ii) we get $x \in \overline{\operatorname{span}}\{x_n; n \in A\}$.

(iii)$\Rightarrow$(ii) Let A and B be two subsets of $\mathbb{N}$. Let $x \in \overline{\operatorname{span}}\{x_n; n \in A\} \cap \overline{\operatorname{span}}\{x_n; n \in B\}$. Then, if $n \notin A \cap B$, $\langle x, x_n^* \rangle = 0$. From (iii) applied to the sets $A \cap B$ and $\mathbb{N} \setminus (A \cap B)$, we get $x \in \overline{\operatorname{span}}\{x_n; n \in A \cap B\}$. $\square$

Theorem 1.36 (Terenzi [Tere94]). *Every separable Banach space admits a strong M-basis.*

Remark 1.37. In fact, Terenzi proved [Tere98] that every separable Banach space admits an M-basis having a property stronger than being a strong M-basis (he called it a *uniformly minimal basis with quasifixed brackets and permutations*). In particular, this M-basis satisfies that for every $x \in X$, then $x \in \overline{M}$, where $M := \left\{ \sum_{j \in F} \langle x, x_j^* \rangle x_j : F \text{ finite set in } \mathbb{N} \right\}$. This is the same as saying that, for every $x \in X$, there exists an increasing sequence (F_n) of finite subsets of $\mathbb{N}$ so that $x = \lim_n \sum_{k \in F_n} \langle x, x_k^* \rangle x_k$. Indeed, if this last condition holds for some $x \in X$, then obviously $x \in \overline{M}$. The proof of the reverse implication follows easily from Fact 1.2.

We shall give Vershynin's proof of a strengthened version of Theorem 1.36 in Theorem 1.42. Prior to it, we need the following two definitions.

Definition 1.38. *A* partition *of $\mathbb{N}$ into finite sets $(A(j))_{j=1}^\infty$ is called a* block partition *if there exists $1 = m_1 < m_2 < \ldots$ in $\mathbb{N}$ such that $\left(\bigcup_{j=m_k}^{m_{k+1}-1} A(j) \right)_{k=1}^\infty$ is a sequence of successive intervals in $\mathbb{N}$.*

Given a (finite or infinite) biorthogonal system $\{x_n; x_n^*\}_{n \in M}$ in $X \times X^*$, where $M \subset \mathbb{N}$, another biorthogonal system $\{z_n; z_n^*\}_{n \in M}$ in $X \times X^*$ is called a block perturbation *if $\operatorname{span}\{x_m; m \in M\} = \operatorname{span}\{z_m; m \in M\}$ and $\operatorname{span}\{x_m^*; m \in M\} = \operatorname{span}\{z_m^*; m \in M\}$.*

Given a block partition $(A(j))_{j=1}^\infty$ and elements $n(j) \in A(j)$, $j = 1, 2 \ldots$, we introduce a kind of perturbation of a biorthogonal system that will be crucial for our purposes.

Definition 1.39. *A biorthogonal system $\{z_n; z_n^*\}_{n=1}^\infty$ in $X \times X^*$ is called a* flattened perturbation with respect to $(n(j), A(j))_{j=1}^\infty$ *of another biorthogonal system $\{x_n; x_n^*\}_{n=1}^\infty$ in $X \times X^*$ if, for every $j \in \mathbb{N}$,*

(i) $\{z_n; z_n^*\}_{n \in A(j)}$ *is a block perturbation of $\{x_n; x_n^*\}_{n \in A(j)}$, and*
(ii) $\|z_n^* - x_{n(j)}^*\| \leq \varepsilon_j / \|x_{n(j)}\|$ *for $n \in A(j)$, $j = 1, 2, \ldots$,*

where $(\varepsilon_j)_{j=1}^\infty$ is a sequence of positive numbers such that $\sum_{j=1}^\infty \varepsilon_j < \infty$.

Flattened perturbations are easy to construct. Just define, for each $j \in \mathbb{N}$, an invertible operator from $\operatorname{span}\{x_n^*; n \in A(j)\}$ onto itself that sends each x_n^* to some vector close to $x_{n(j)}^*$. This will provide the flattened perturbation sought.

Let us proceed with the proof of Theorem 1.36. This will be done through a series of lemmas. We start with a simple observation. Fix $x \in X$. Define an increasing sequence $(r(m))_{m=1}^{\infty}$ of positive integers by induction: start by taking $r(1) := 1$. Assume that $r(k)$ has already been defined for $k = 1, 2, \ldots, m$. By Fact 1.5, we can write $x = y_{r(m)} + z_{r(m)}$, where $y_{r(m)} = \sum_{n=1}^{r(m)} \langle x, x_n^* \rangle x_n$ and $z_{r(m)} \in \overline{\operatorname{span}}\{x_k; k > r(m)\}$. Therefore we can find $r(m+1) \in \mathbb{N}$ with $r(m+1) > r(m)$ and $v_m \in \operatorname{span}\{x_n\}_{r(m)+1}^{r(m+1)}$ such that $\|z_{r(m)} - v_m\| < 1/m$. This proves that

$$x = \lim_{m \to \infty} \left(\sum_{n=1}^{r(m)} \langle x, x_n^* \rangle x_n + v_m \right),$$

where $v_m \in \operatorname{span}\{x_n\}_{r(m)+1}^{r(m+1)}$ for every m. This simple construction has a drawback: the sequence $(r(m))_{m=1}^{\infty}$ depends on x. That it can be made independent of x is the content of the following lemma.

Lemma 1.40. *Let $\{x_n; x_n^*\}_{n=1}^{\infty}$ be a fundamental biorthogonal system for a Banach space X. Then there exists $r(1) < r(2) < \ldots$ in $\mathbb{N}$ (called representing indices of $\{x_n; x_n^*\}_{n=1}^{\infty}$) with the following property: for every $x \in X$ and for every $m \in \mathbb{N}$, there exists $v_m \in \operatorname{span}\{x_n\}_{n=r(m)+1}^{r(m+1)}$ such that*

$$x = \lim_{m \to \infty} \left(\sum_{n=1}^{r(m)} \langle x, x_n^* \rangle x_n + v_m \right).$$

Obviously, a fundamental biorthogonal system $\{x_n; x_n^*\}_{n=1}^{\infty}$ is a Schauder basis if and only if we can choose $r(m) = m$ for every $m \in \mathbb{N}$ and $v_m \to 0$.

Proof of Lemma 1.40. In order to simplify the notation, for some $\varepsilon > 0$ we shall join two ε-close symbols by $\overset{\varepsilon}{\approx}$. Let's define the sequence $(r(m))$ by induction. Set $r(1) = 1$ and assume that, for some $m \in \mathbb{N}$, elements $r(1) < r(2) < \ldots < r(m)$ have already been defined. A simple compactness argument using a finite net proves that there exists $p(m+1) > r(m)$ such that $\forall z \in \operatorname{span}\{x_n\}_{n=1}^{r(m)}$ with $\|z\| \leq 2 + \sum_{n=1}^{r(m)} \|x_n\|.\|x_n^*\|$. Then

$$\operatorname{dist}\left(z, \operatorname{span}\{x_n\}_{n=r(m)+1}^{\infty} \right)^{1/m} \overset{\varepsilon}{\approx} \operatorname{dist}\left(z, \operatorname{span}\{x_n\}_{n=r(m)+1}^{p(m+1)} \right).$$

Then set $r(m+1) := p(m+1)$. (This procedure of first introducing $p(m+1)$ and then setting $r(m+1) = p(m+1)$ will be justified as a notational device in the next lemma, where two different steps will really be needed.) In this way, we define $(r(m))_{m=1}^{\infty}$. We shall prove that it is a sequence of representing indices.

Let $x \in B_X$ and let $0 < \varepsilon < 1$. Find $m \in \mathbb{N}$ big enough that $1/m < \varepsilon$ and for some $\hat{x} \in \operatorname{span}\{x_n\}_{n=1}^{r(m)}$ we have $x \overset{\varepsilon}{\approx} \hat{x}$. Let $z := \hat{x} - \sum_{n=1}^{r(m)} \langle x, x_n^* \rangle x_n \in \operatorname{span}\{x_n\}_{n=1}^{r(m)}$. Then

$$\|z\| \le \|\hat{x}\| + \sum_{n=1}^{r(m)} \|x_n\|.\|x_n^*\| \le 1 + \varepsilon + \sum_{n=1}^{r(m)} \|x_n\|.\|x_n^*\| \le 2 + \sum_{n=1}^{r(m)} \|x_n\|.\|x_n^*\|;$$

therefore

$$\mathrm{dist}\left(z, \mathrm{span}\{x_n\}_{n=r(m)+1}^{\infty}\right) \overset{\varepsilon}{\approx} \mathrm{dist}\left(z, \mathrm{span}\{x_n\}_{n=r(m)+1}^{r(m+1)}\right).$$

Let $x' := x - \sum_{n=1}^{r(m)} \langle x, x_n^* \rangle x_n \overset{\varepsilon}{\approx} z$. By Fact 1.5, $x' \in \mathrm{span}\{x_n\}_{n=r(m)+1}^{\infty}$, and hence

$$0 = \mathrm{dist}\left(x', \mathrm{span}\{x_n\}_{n=r(m)+1}^{\infty}\right) \overset{3\varepsilon}{\approx} \mathrm{dist}\left(x', \mathrm{span}\{x_n\}_{n=r(m)+1}^{r(m+1)}\right).$$

We can then find $v_m \in \mathrm{span}\{x_n\}_{n=r(m)+1}^{r(m+1)}$ such that $x' \overset{3\varepsilon}{\approx} v_m$ and therefore $x \overset{3\varepsilon}{\approx} \sum_{n=1}^{r(m)} \langle x, x_n^* \rangle x_n + v_m$. This estimate holds for indices from m on, so we get the conclusion.   □

Lemma 1.41. *Let X be a Banach space. Let $\{x_n; x_n^*\}_{n=1}^{\infty}$ be a norming M-basis in $X \times X^*$. Then a sequence $(r(m))_{m=1}^{\infty}$ of representing indices for this basis can be chosen in such a way that the following property holds: Given $x \in X$ such that, for some sequence $m_1 < m_2 < \dots$ in $\mathbb{N}$, the series*

$$\sum_{k=1}^{\infty} \sum_{n=r(m_k)+1}^{r(m_k+1)} \langle x, x_n^* \rangle x_n \quad \text{converges}, \tag{1.16}$$

then, setting $r(m_0) = 0$, we have

$$x = \sum_{k=0}^{\infty} \sum_{n=r(m_k)+1}^{r(m_{k+1})} \langle x, x_n^* \rangle x_n.$$

Proof. We shall define $(r(m))_{m=1}^{\infty}$ again by induction. Start by setting $r(1) = 1$ as in Lemma 1.40. Suppose $r(1) < r(2) < \dots r(m)$ were already defined for some $m \in \mathbb{N}$. Then use the construction and the notation in Lemma 1.40 to produce $p(m+1)$. The M-basis is norming, so there exists $c > 0$ such that $\forall v \in X$ there exists $x^* \in S_{\overline{\mathrm{span}}\{x_n^*\}_{n=1}^{\infty}}$ such that $\langle v, x^* \rangle \ge 2c\|v\|$. Again a simple compactness argument proves that there exists $r(m+1) > p(m+1)$ such that

$$\forall v \in \mathrm{span}\{x_n\}_{n=1}^{p(m+1)}, \exists \, x^* \in S_{\mathrm{span}\{x_n^*\}_{n=1}^{r(m+1)}} \text{ such that } \langle v, x^* \rangle \ge c\|v\|. \tag{1.17}$$

Now the argument in Lemma 1.40 proves that $(r(m))_{m=1}^{\infty}$ is a sequence of representing indices for the M-basis.

We shall prove that this sequence satisfies the requirement. Fix $x \in X$. Assume that an increasing sequence $(m_k)_{k=1}^{\infty}$ in $\mathbb{N}$ satisfies (1.16). Subtracting the convergent series $\sum_{k=1}^{\infty} \sum_{n=r(m_k)+1}^{r(m_k+1)} \langle x, x_n^* \rangle x_n$ from x (again call the result x), we may and do assume that $\langle x, x_n^* \rangle = 0$, $r(m_k) + 1 \leq n \leq r(m_k + 1)$, and $k \in \mathbb{N}$. Assume also, without loss of generality, that $\|x\| \leq 1$. Fix $\varepsilon > 0$. Then, from the construction of $(r(m))$ and $(p(m))$, we can find k big enough that $x \overset{\varepsilon}{\approx} \sum_{n=1}^{r(m_k)} \langle x, x_n^* \rangle x_n + v_{m_k}$ for some $v_{m_k} \in \text{span}\{x_n\}_{n=r(m_k)+1}^{p(m_k+1)}$. Observe that

$$v_{m_k} \overset{\varepsilon}{\approx} x - \sum_{n=1}^{r(m_k)} \langle x, x_n^* \rangle x_n = x - \sum_{n=1}^{r(m_k+1)} \langle x, x_n^* \rangle x_n,$$

so $\langle v_{m_k}, x^* \rangle \overset{\varepsilon}{\approx} 0$ for $x^* \in \text{span}\{x_n^*\}_{n=1}^{r(m_k+1)}$. From (1.17), it follows that $c\|v_{m_k}\| \leq \varepsilon$, so $x \overset{\varepsilon(1+1/c)}{\approx} \sum_{n=1}^{r(m_k)} \langle x, x_n^* \rangle x_n$. This estimate holds for indices from k on. This completes the proof. $\square$

Theorem 1.42 (Terenzi [Tere90], [Tere94], Vershynin [Vers00]). *Let X be a separable Banach space and let $\{x_n\}_{n=1}^{\infty}$ be a norming M-basis. Then there is a block partition $(A(j))_{j=1}^{\infty}$ and numbers $n(j) \in A(j)$, $j = 1, 2, \ldots$, such that each flattened perturbation of $\{x_n\}_{n=1}^{\infty}$ with respect to $(n(j), A(j))_{j=1}^{\infty}$ is a strong M-basis. In particular, X has a strong M-basis that is also norming.*

Proof (Vershynin, [Vers00]). By Lemma 1.41, a sequence of representing indices $(r(m))_{m=1}^{\infty}$ exists with the property stated there. We can always assume that $r(1) = 1$. Let us define by induction the block partition $(A(j))_{j=1}^{\infty}$ and numbers $n(j) \in A(j)$, $j \in \mathbb{N}$. At each successive step, we shall add some new $A(j)$'s whose union will form an interval in $\mathbb{N}$ (i.e., a block, ending at some representing index, called a *block bound*), and succeeding the previously constructed blocks. At the same time, numbers $n(j) \in A(j)$ will be defined.

Let us start by taking $r(1) = 1$ as the first block bound, $A(1) = \{1\}$ and $n(1) = 1$. Assume that $(n(j), A(j))_{1 \leq j \leq j_0}$ have already been defined, $\bigcup_{j=1}^{j_0} A(j)$ filling an interval $[1, r(m)]$ in $\mathbb{N}$ for some representing index $r(m)$ (another block bound).

For $r(m) + 1 \leq j \leq r(m + 1)$, put $d_j := m + j - r(m)$. Let

$$E(j) := \{j\} \cup \{r(d_j) + 1, \ldots, r(d_j + 1)\}.$$

$\{E(j)\}_{r(m)+1 \leq j \leq r(m+1)}$ is a disjoint family of sets whose union is the new block, which starts at $r(m) + 1$ and ends at the next block bound, precisely $r(d_{r(m+1)} + 1)$. Define

$$A(j_0 + j - r(m)) := E(j), \quad n(j_0 + j - r(m)) := j$$

for $r(m) + 1 \leq j \leq r(m + 1)$. This completes the construction of the induction step.

Observe that the sequence $n(1) < n(2) < \ldots$ so defined fills precisely the set $I := \{1\} \cup \bigcup_{r(m) \text{ a block bound}} \{r(m) + 1, \ldots, r(m + 1)\}$. Let $j : I \to \mathbb{N}$

be the one-to-one function defined by $j(n(j)) = j$, $j \in \mathbb{N}$ (j just successively enumerates the elements in I).

Let $\{z_n; z_n^*\}_{n=1}^\infty$ be any flattened perturbation of $\{x_n; x_n^*\}_{n=1}^\infty$ with respect to $(n(j), A(j))_{j=1}^\infty$ and $(\varepsilon_j)_{j=1}^\infty$ a sequence of positive numbers such that $\sum_{j=1}^\infty \varepsilon_j < \infty$. We shall prove that $\{z_n\}_{n=1}^\infty$ is a strong M-basis. To that end, pick any $x \in S_X$. We have to show that $x \in \overline{\text{span}}\{\langle x, z_n^*\rangle z_n\}_{n=1}^\infty$.

There are two possibilities:

(A) There exists a sequence of block bounds $r(m_1) < r(m_2) < \ldots$ such that, for every $k \in \mathbb{N}$,

$$\|\langle x, x_n^*\rangle x_n\| \leq \varepsilon_{j(n)}$$

for $r(m_k) + 1 \leq n \leq r(m_k + 1)$.

(B) There exists a block bound $r(m_0)$ such that for all bigger block bounds $r(m)$ we can find n_0 (one for each $r(m)$) with $r(m) + 1 \leq n_0 \leq r(m + 1)$ verifying

$$\|\langle x, x_{n_0}^*\rangle x_{n_0}\| > \varepsilon_{j(n_0)} \text{ and } \|\langle x, x_n^*\rangle x_n\| \leq \varepsilon_{j(n)}, \ n_0 < n \leq r(m+1). \quad (1.18)$$

(If it happens that $n_0 = r(m + 1)$, then the second property is empty.)

In case (A), observe that

$$\sum_{k=1}^\infty \sum_{n=r(m_k)+1}^{r(m_k+1)} \varepsilon_{j(n)} \leq \sum_{j=1}^\infty \varepsilon_j < \infty,$$

hence Lemma 1.41 applies and we get, setting $r(m_0) = 0$,

$$x = \sum_{k=0}^\infty \sum_{n=r(m_k)+1}^{r(m_{k+1})} \langle x, x_n^*\rangle x_n.$$

As $\{z_n; z_n^*\}_{n=r(m_k)+1}^{r(m_{k+1})}$ is a block perturbation of $\{x_n; x_n^*\}_{n=r(m_k)+1}^{r(m_{k+1})}$, we get that $x \in \overline{\text{span}}\{\langle x, z_n^*\rangle z_n\}_{n=1}^\infty$, and this finishes the proof of case (A).

Assume now that case (B) holds. Let $\Omega := \{n \in \mathbb{N}; \langle x, z_n^*\rangle = 0\}$. We shall prove $x \in \overline{\text{span}}\{z_n; n \in \mathbb{N} \setminus \Omega\}$. Fix $\varepsilon > 0$ and let $r(m)$ be a block bound greater than $r(m_0)$. Let $n_0 = n_0(m)$ be the corresponding integer given in case (B).

Claim. $E(n_0) \subset \mathbb{N} \setminus \Omega$.

Suppose for a moment that the claim is false. Then we can find $\langle x, z_n^*\rangle = 0$ for some $n \in E(n_0) = A(j(n_0))$. Then, by the definition of a flattened perturbation, we would have $|\langle x, x_{n_0}^*\rangle| \leq \varepsilon_{j(n_0)}/\|x_{n_0}\|$, a contradiction to (1.18), and the claim is proved.

Use Lemma 1.40 to find an element $v \in \text{span}\{x_n; n \in E(n_0)\}$ such that, setting $\Gamma := \{1, \ldots, r(m)\} \cup \bigcup_{j=r(m)+1}^{n_0-1} E(j)$ (in the case where $n_0 = r(m)+1$, the second union in this formula should be empty), we have

$$x \overset{\varepsilon}{\approx} \sum_{n=1}^{r(d_{n_0})} \langle x, x_n^* \rangle x_n + v = \sum_{n \in \Gamma} \langle x, x_n^* \rangle x_n + (\langle x, x_{n_0}^* \rangle x_{n_0} + v) + \sum_{n=n_0+1}^{r(m+1)} \langle x, x_n^* \rangle x_n$$

(in the case where $n_0 = r(m + 1)$, the last summand in the previous formula should be 0). First of all, $\sum_{n \in \Gamma} \langle x, x_n^* \rangle x_n$ belongs to $\overline{\mathrm{span}}\{z_n; n \in \mathbb{N} \setminus \Omega\}$. Indeed, $\{z_n; z_n^*\}_{n \in \Gamma}$ is a block perturbation of $\{x_n; x_n^*\}_{n \in \Gamma}$; thus $\sum_{n \in \Gamma} \langle x, x_n^* \rangle x_n = \sum_{n \in \Gamma} \langle x, z_n^* \rangle z_n$. Second, $(\langle x, x_{n_0}^* \rangle x_{n_0} + v) \in \mathrm{span}\{x_n; n \in E(n_0)\}$; by the claim, $(\langle x, x_{n_0}^* \rangle x_{n_0} + v) \in \mathrm{span}\{x_n; n \in \mathbb{N} \setminus \Omega\}$. Third, $\left\| \sum_{n=n_0+1}^{r(m+1)} \langle x, x_n^* \rangle x_n \right\| \leq \sum_{n=n_0+1}^{\infty} \varepsilon_{j(n)}$ due to (1.18). This last quantity is less than ε if m (and therefore $n_0 = n_0(m)$) was chosen large enough. This finishes the proof.   □

Corollary 1.43. *Every Banach space X such that X^* is separable has an M-basis that is strong, and its dual coefficients form a strong M-basis for the dual.*

Proof. First of all, X has a shrinking M-basis. This follows from Theorem 1.22. Obviously this basis is norming. From Theorem 1.42, there is a flattened perturbation $\{y_n\}_{n=1}^{\infty}$, which is strong (and, of course, an M-basis). Let $\{y_n^*\}_{n=1}^{\infty}$ be the corresponding system of functional coefficients; it is a 1-norming M-basis in X^*. Applying Theorem 1.42 again we get another flattened perturbation $\{z_n^*\}_{n=1}^{\infty}$, which is a strong M-basis of X^*. The corresponding system of functional coefficients in X^{**} in fact lies in X. Accordingly, call it $\{z_n\}_{n=1}^{\infty}$. This is again a strong M-basis in X whose system of functional coefficients $\{z_n^*\}_{n=1}^{\infty}$ forms a strong M-basis in X^*.   □

1.6 Extensions of M-bases

This section begins by presenting an early result on extensions of M-bases from subspaces to overspaces in the case of separable Banach spaces; this was shown by Gurarii and Kadets. It is desirable to achieve the extension by preserving special features of the M-basis. We present here Terenzi's result for the case of bounded M-bases, providing bounds for the vectors and the functional coefficients. This strengthening of the general extension theorem is particularly interesting because in some cases it can provide a tool for constructing projections onto subspaces, as will be seen later in the book. An M-basis in a Banach space X naturally decomposes the space in couples of quasicomplemented subspaces. In a certain way, there is a reciprocal of this result: it is possible to extend an M-basis in a subspace of a separable Banach space X to the whole space in the direction of a given quasicomplement. We provide here Plichko's proof of this result due to V.D. Milman.

Definition 1.44. *Let $Y \hookrightarrow X$, let $\{y_\alpha; g_\alpha\}_{\alpha \in \Lambda}$ be an M-basis of Y and let $\{x_\gamma; f_\gamma\}_{\gamma \in \Gamma}$ be an M-basis of X such that $\{y_\alpha\}_{\alpha \in \Lambda} \subset \{x_\gamma\}_{\gamma \in \Gamma}$. We say that*

$\{x_\gamma; f_\gamma\}_{\gamma \in \Gamma}$ *is an* extension *of* $\{y_\alpha; g_\alpha\}_{\alpha \in \Lambda}$ *(i.e., the latter can be extended to* X*). Note that due to the linear density of* $\{y_\alpha\}$ *in* Y*, we automatically obtain that if* $y_\alpha = x_\gamma \in X$*, then* f_γ *is an extension of* g_α *from* Y *to* X*.*

Theorem 1.45 (Gurarii and Kadets [GuKa62]). *Let* Z *be a closed subspace of a separable Banach space* X*. Any Markushevich basis (resp., fundamental biorthogonal system)* $\{x_i; f_i\}_{i=1}^\infty$ *in* $Z \times Z^*$ *can be extended to a Markushevich basis (resp. fundamental biorthogonal system) in* $X \times X^*$*.*
   More precisely, there are $z_j \in X$*,* $g_j \in X^*$*,* $j \in \mathbb{N}$*, and extensions of* f_i *to functionals on* X*,* $i \in \mathbb{N}$ *such that* $\{\{x_i\}_{i=1}^\infty \cup \{z_j\}_{j=1}^\infty; \{f_i\}_{i=1}^\infty \cup \{g_j\}_{j=1}^\infty\}$ *is a Markushevich basis (resp. fundamental biorthogonal system) of* X*.*

Proof. Assume that $\{x_i; f_i\}_{i=1}^\infty$ is an M-basis (the case of a fundamental biorthogonal system follows the same lines). Extend all f_i onto X and denote these extensions by $\tilde{f}_i$, $i \in \mathbb{N}$. Let $\{\hat{y}_j; \phi_j\}_{j=1}^\infty$ be a Markushevich basis of X/Z (see Theorem 1.22). For all j, choose $y_j \in \hat{y}_j$ and define $\phi_j(x) = \phi_j(\hat{x})$ for $x \in X$. Note that $\phi_j(x_i) = 0$ for all i. We have $\overline{\text{span}}\{\{x_i\} \cup \{y_j\}\} = X$, and $\{\tilde{f}_i\} \cup \{\phi_j\}$ is a family separating points of X.

Put $z_j = y_j - \sum_{i=1}^j \lambda_{ij} x_i$ and $\psi_i = \tilde{f}_i - \sum_{j=1}^i \lambda_{ij} \phi_j$, where $\lambda_{ij} = \tilde{f}_i(y_j)$ for $i \neq j$ and $\lambda_{ii} = \frac{1}{2} \tilde{f}_i(y_i)$, $i \in \mathbb{N}$, $j \in \mathbb{N}$. Then $\{\{x_i\}_{i=1}^\infty \cup \{z_j\}_{j=1}^\infty; \{\psi_i\}_{i=1}^\infty \cup \{\phi_j\}_{j=1}^\infty\}$ is a Markushevich basis of X that extends $\{x_i; f_i\}_{i=1}^\infty$. Indeed, from the definition of z_j and ψ_i, it is clear that $\overline{\text{span}}\{x_i, z_j; i \in \mathbb{N}, j \in \mathbb{N}\} = X$, $\{\psi_i\}_{i=1}^\infty \cup \{\phi_j\}_{j=1}^\infty$ is separating points of X, and ψ_i extend f_i onto X, $i \in \mathbb{N}$. It is routine to check that the system is biorthogonal. □

In order to extend a bounded (i.e., uniformly minimal) M-basis from a subspace of a separable Banach space to the whole space (Theorem 1.50), we shall use the following results.

Given a closed subspace X of a Banach space E and an element $e \in E$, $e + X$ denotes both a subset of E and an element of E/X.

Theorem 1.46 (Terenzi [Tere83]). *Let* $K > 0$*,* $\varepsilon > 0$*,* $\{x_n\}_{n=1}^\infty$ *be a* K*-uniformly minimal system in a Banach space* E*, and* (y_n) *be a sequence in* E*. Then, setting* $X := \overline{\text{span}}\{x_n\}_{n=1}^\infty$*, there exists a subspace* $Z := \overline{\text{span}}\{z_n\}_{n=1}^\infty$ *of* E *such that* $z_n + X = y_n + X$ *for every* $n \in \mathbb{N}$ *with*

$$\inf_{m \in \mathbb{N}} \text{dist}\left(\frac{x_m}{\|x_m\|}, \overline{\text{span}}\{x_n\}_{n \in \mathbb{N}, n \neq m} + Z\right) > \frac{K}{2} - \varepsilon.$$

In the proof of this result, the following lemma is needed.

Lemma 1.47. *Let* $\{x_n\}_{n=1}^\infty$ *be a* K*-uniformly minimal system in a separable Banach space* E*. Let* V *be a finite-dimensional subspace of* E *with*

$$V \cap \overline{\text{span}}\{x_n\}_{n=1}^\infty = \{0\},$$

and let $\varepsilon > 0$*. Then there exists a natural number* n_ε *such that*

$$\inf_{m > n_\varepsilon} \text{dist} \left(\frac{x_m}{\|x_m\|}, \overline{\text{span}}\{x_n\}_{n \in \mathbb{N}, n \neq m} + V \right) > \frac{K}{2} - \varepsilon.$$

Proof. Suppose the contrary; since we may assume from the beginning that $\|x_n\| = 1$ for all n, there exists an increasing sequence $(n_k)_{k=1}^\infty$ of natural numbers such that

$$\|x_{n_k} + \tilde{x}_k + v_k\| < \frac{K}{2} - \frac{\varepsilon}{2} \text{ for all } k \in \mathbb{N}, \tag{1.19}$$

where $\tilde{x}_k \in \text{span}\{x_i\}_{i=1, i \neq n_k}^{n_{k+1} - 1}$ and $v_k \in V$ for all $k \in \mathbb{N}$. We claim that (v_k) is a bounded sequence. If not,

$$\left\| \frac{x_{n_k} + \tilde{x}_k}{\|v_k\|} + \frac{v_k}{\|v_k\|} \right\| \leq \frac{1}{\|v_k\|} \left(\frac{K}{2} - \frac{\varepsilon}{2} \right) \to 0$$

when $k \to \infty$. As V is finite-dimensional, we can find a subsequence of (v_k) (for simplicity again denoted (v_k)) such that $(v_k/\|v_k\|) \to s \in S_V$. Then $(1/\|v_k\|)(x_{n_k} + \tilde{x}_k) \to s$ and so $s \in X \cap V$, a contradiction. This proves the claim. We can suppose then that (v_k) converges and hence $k_0 \in \mathbb{N}$ exists such that $\|v_{k_0} - v_{k_0+1}\| < \varepsilon$. Note that

$$x_{n_{k_0}} + \tilde{x}_{k_0} \in \text{span}\{x_i\}_{i=1}^{n_{k_0+1} - 1}$$

and, accordingly,

$$\tilde{x}_{k_0+1} - x_{n_{k_0}} - \tilde{x}_{k_0} \in \overline{\text{span}}\{x_i\}_{i=1, i \neq n_{k_0+1}}^\infty.$$

Therefore, from the K-uniform minimality of $\{x_n\}_{n=1}^\infty$, we get

$$K \leq \|x_{n_{k_0+1}} + (\tilde{x}_{k_0+1} - x_{n_{k_0}} - \tilde{x}_{k_0})\|$$
$$= \|(x_{n_{k_0+1}} + \tilde{x}_{k_0+1} + v_{k_0+1}) - (x_{n_{k_0}} + \tilde{x}_{k_0} + v_{k_0}) - (v_{k_0+1} - v_{k_0})\|$$
$$< 2 \left(\frac{K}{2} - \frac{\varepsilon}{2} \right) + \varepsilon = K,$$

a contradiction. $\square$

Proof. (Theorem 1.46). We may and do assume from the beginning that $\|x_n\| = 1$, $n \in \mathbb{N}$, and $\{y_n + X\}_{n=1}^\infty$ is a linearly independent subset of E/X. By hypothesis, there exists a system $\{x_{0,n}^*\}_{n=1}^\infty$ in E^* such that

$$\{x_n; x_{0,n}^*\}_{n=1}^\infty \text{ is biorthogonal and } \|x_{0,n}^*\| \leq 1/K \text{ for every } n \in \mathbb{N}. \tag{1.20}$$

By Lemma 1.47, there exists $n_1 \in \mathbb{N}$ such that

$$\inf_{m > n_1} \text{dist} \left(x_m, \overline{\text{span}}\{x_n\}_{n=1, n \neq m}^\infty + \text{span}\{y_1\} \right) > \frac{K}{2} - \varepsilon,$$

hence there exists $\{x_{1,n}^*\}_{n > n_1}$ in E^* such that

$\{x_n; x^*_{1,n}\}_{n>n_1}$ is biorthogonal and $\{y_1\} \cup \{x_k\}_{k=1}^{n_1} \subset \left(\{x^*_{1,n}\}_{n>n_1}\right)_{\perp}$, (1.21)

$$\|x^*_{1,n}\| < \frac{1}{K/2 - \varepsilon} \text{ for } n > n_1. \tag{1.22}$$

Let us set $x^*_n := x^*_{1,n}$ for $1 \le n \le n_1$, $z_1 := y_1 - \sum_{k=1}^{n_1} \langle y_1, x^*_k \rangle x_k$. Then, by (1.20) and (1.21), we have that $\{x_n; x^*_n\}_{n=1}^{n_1} \cup \{x_n; x^*_{1,n}\}_{n>n_1}$ is biorthogonal, $z_1 + X = y_1 + X$, and $z_1 \in \left(\overline{\text{span}}\{\{x^*_n\}_{n=1}^{n_1} \cup \{x^*_{1,n}\}_{n>n_1}\}\right)_{\perp}$.

Again using Lemma 1.47, we get a natural number $n_2 > n_1$ such that

$$\inf_{m>n_2} \text{dist}\left(x_m, \overline{\text{span}}\{x_n\}_{n=1, n\neq m}^{\infty} + \text{span}\{y_2, z_1\}\right) > \frac{K}{2} - \varepsilon,$$

and hence there exists $\{x^*_{2,n}\}_{n>n_2}$ in E^* such that

$\{x_n; x^*_{2,n}\}_{n>n_2}$ is biorthogonal and $\{y_2, z_1\} \cup \{x_k\}_{k=1}^{n_2} \subset \left(\{x^*_{2,n}\}_{n>n_2}\right)_{\perp}$,

$$\|x^*_{2,n}\| < \frac{1}{K/2 - \varepsilon} \text{ for } n > n_2.$$

Set $x^*_n := x^*_{2,n}$ for $n_1 + 1 \le n \le n_2$ and put $z_2 := y_2 - \sum_{k=1}^{n_2} \langle y_2, x^*_k \rangle x_k$. Then, as before, we have that $\{x_n; x^*_n\}_{n=1}^{n_2} \cup \{x_n; x^*_{2,n}\}_{n>n_2}$ is biorthogonal, $z_2 + X = y_2 + X$ and $z_2 \in \left(\overline{\text{span}}\{\{x^*_n\}_{n=1}^{n_2} \cup \{x^*_{2,n}\}_{n>n_2}\}\right)_{\perp}$.

Again use Lemma 1.47 to obtain a natural number $n_3 > n_2$ such that

$$\inf_{m>n_3} \text{dist}\left(x_m, \overline{\text{span}}\{x_n\}_{n=1, n\neq m}^{\infty} + \text{span}\{y_3, z_1, z_2\}\right) > \frac{K}{2} - \varepsilon.$$

Continue in this way to finally get $\{x^*_n\}_{n=1}^{\infty}$ in E^* and $\{z_n\}_{n=1}^{\infty}$ in E such that $\{x_n; x^*_n\}_{n=1}^{\infty}$ is biorthogonal, $\|x^*_n\| < \frac{1}{K/2-\varepsilon}$, $z_n + X = y_n + X$ for all $n \in \mathbb{N}$, and $\{z_n\}_{n=1}^{\infty} \subset \{x^*_n\}_{\perp}$. □

From this result, we obtain immediately the following corollary.

Corollary 1.48. *Let E be a separable Banach space. Let $\{x_n\}_{n=1}^{\infty}$ be a minimal system in S_E. Let $X := \overline{\text{span}}\{x_n; n \in \mathbb{N}\}$. Let $\{x^*_n\}_{n=1}^{\infty} \subset X^*$ be the set of functional coefficients. Assume that, for some $\lambda \ge 1$, we have $\|x^*_n\| \le \lambda < \infty$ for every $n \in \mathbb{N}$. Then, for every $\varepsilon > 0$ there exists extensions $e^*_n \in E^*$ of x^*_n with $\|e^*_n\| < 2\lambda + \varepsilon$, $n \in \mathbb{N}$, such that $X + (\{e^*_n\}_{n=1}^{\infty})_{\perp}$ is dense in E.*

Proof. Fix $K > 1/(\lambda + \varepsilon/2)$. Then $\{x_n\}_{n=1}^{\infty}$ is a K-uniformly minimal system in X (see Remark 1.24). Choose a sequence $(y_n)_{n=1}^{\infty}$ in E such that $\overline{\text{span}}\{y_n + X\}_{n=1}^{\infty} = E/X$. Choose $\delta > 0$ such that $K - 2\delta > 1/(\lambda + \varepsilon/2)$. By Theorem 1.46, there exists a sequence $(z_n)_{n=1}^{\infty}$ in E such that $z_n + X = y_n + X$ for every $n \in \mathbb{N}$ and

$$\inf_{m \in \mathbb{N}} \text{dist}\left(x_m, \overline{\text{span}}\{x_n\}_{n \in \mathbb{N}, n\neq m} + Z\right) > \frac{K}{2} - \delta,$$

where $Z := \overline{\text{span}}\{z_n\}_{n=1}^{\infty}$. By the Hahn-Banach theorem, there exists a sequence $(\overline{e}_m^*)_{m=1}^{\infty}$ in E^* such that $\langle x, \overline{e}_m^* \rangle = 0$ for all $x \in \overline{\text{span}}\{x_n\}_{n=1, n \neq m}^{\infty} + Z$, $\|\overline{e}_m^*\| = 1$, and $\langle x_m, \overline{e}_m^* \rangle = \text{dist}(x_m, \overline{\text{span}}\{x_n\}_{n \in \mathbb{N}, n \neq m} + Z) =: d_m$ for every $m \in \mathbb{N}$. Set $e_m^* := \overline{e}_m^*/d_m$, $m \in \mathbb{N}$. Then e_m^* extends x_m^* and

$$\|e_m^*\| < \frac{1}{\frac{K}{2} - \delta} < 2\lambda + \varepsilon \text{ for all } m \in \mathbb{N}.$$

Moreover, $X + (\overline{\text{span}}\{e_m^*\}_{m=1}^{\infty})_{\perp}$ is dense in E. Indeed, given $e^* \in E^*$ such that $\langle x_n, e^* \rangle = 0$ for all $n \in \mathbb{N}$ and $\langle x, e^* \rangle = 0$ for all $x \in (\{e_m^*\}_{m=1}^{\infty})_{\perp}$, we have $e^* \in (E/X)^*$ and $\langle z_n, e^* \rangle = \langle y_n, e^* \rangle = 0$ for all $n \in \mathbb{N}$, so $e^* = 0$. □

To prove the extension theorem in the case of bounded M-bases, in addition to Theorem 1.46 and its corollary, Corollary 1.48, we need the following intermediate result.

Lemma 1.49 (Terenzi). *Let E be a separable Banach space. Fix $\varepsilon > 0$. Let $\{x_n; e_n^*\}_{n=1}^{\infty}$ be the biorthogonal system in $E \times E^*$ constructed in Corollary 1.48 from a given biorthogonal system $\{x_n; x_n^*\}$ in $X \times X^*$, where $X := \overline{\text{span}}\{x_n\}_{n=1}^{\infty}$, $\|x_n\| = 1$, and $\|x_n^*\| \leq \lambda$ (so $\|e_n^*\| \leq 2\lambda + \varepsilon$) for all $n \in \mathbb{N}$. Then there exists a sequence $(y_n)_{n=1}^{\infty}$ in S_E, sequences $(z_n^*)_{n=1}^{\infty}$ and $(y_n^*)_{n=1}^{\infty}$ in E^* with $\|z_n^*\| < 3(2\lambda + \varepsilon)$, $\|y_n^*\| < 2$ for all $n \in \mathbb{N}$ such that $\{x_n, y_n; z_n^*, y_n^*\}_{n=1}^{\infty}$ is a biorthogonal system in $E \times E^*$, z_n^* is an extension of x_n^* to E for $n \in \mathbb{N}$, and $(y_n + X)_{n=1}^{\infty}$ is a basic sequence in E/X.*

Proof. The density of $(\{e_n^*\}_{n=1}^{\infty})_{\perp}/X$ in E/X follows from the density of $X + (\{e_n^*\}_{n=1}^{\infty})_{\perp}$ in E. Therefore, by the technique of Mazur (see, e.g., [Fa~01, Thm. 6.14]), we can select

$$\text{a basic sequence } (u_n + X)_{n=1}^{\infty} \text{ in } E/X \text{ such that} \quad (1.23)$$
$$u_n \in (\{e_m^*\}_{m=1}^{\infty})_{\perp} \text{ for every } n \in \mathbb{N}.$$

We shall choose by induction two sequences, $(v_n)_{n=1}^{\infty}$ and $(w_n)_{n=1}^{\infty}$, of vectors in E and two increasing sequences, $(p_n)_{n=1}^{\infty}$ and $(q_n)_{n=1}^{\infty}$, of natural numbers such that, setting $p_0 = q_0 := 0$, $X^{(0)} := X$, and $X^{(n)} := \overline{\text{span}}\{x_k\}_{k=n+1}^{\infty}$ for every $n \in \mathbb{N}$, the following property will be satisfied:

$$v_n = \sum_{k=p_{n-1}+1}^{p_n} a_{n,k} x_k + w_n, \quad v_n \in \text{span}\{u_k\}_{k=q_{n-1}+1}^{q_n}, \quad (1.24)$$

$$\|v_n + X^{(p_{n-1})}\| = \|v_n + X\|, \text{ and}$$

$$\|v_n + X\| \leq \|w_n\| < 2\|v_n + X\| \text{ for all } n \in \mathbb{N}.$$

Start by choosing $v_1 := u_1$ (and then $q_1 = 1$). Select $w_1 \in v_1 + \text{span}\{x_n\}_{n=1}^{\infty}$ such that $\|v_1 + X\| \leq \|w_1\| < 2\|v_1 + X\|$. Then we can write $v_1 = \sum_{k=1}^{p_1} a_{1,k} x_k + w_1$ for some $p_1 \in \mathbb{N}$. Assume that, for some $m \geq 1$, v_n, w_n, p_n, and q_n have been chosen for $n = 1, 2, \ldots, m$ with properties in (1.24).

Set $q_{m+1} := q_m + p_m + 1$. The sequence $(u_n + X)_{n=1}^\infty$ is basic in E/X, so the corresponding functional coefficients are in $X^\perp$. In particular, those with indices from $q_m + 1$ to q_{m+1} are in $(X^{(p_m)})^\perp$. This guarantees that in the space $E/X^{(p_m)}$ the system $\{u_k + X^{(p_m)}\}_{k=q_m+1}^{q_{m+1}}$ is linearly independent. We shall apply Lemma 1.19 to the space $E/X^{(p_m)}$ and the two subspaces $G_1 := \mathrm{span}\{x_k + X^{(p_m)}\}_{k=1}^{p_m}$ and $G_2 := \mathrm{span}\{u_k + X^{(p_m)}\}_{k=q_m+1}^{q_{m+1}}$ for $\dim G_1 = p_m < p_m + 1 = \dim G_2$. This gives a vector $v_{m+1} + X^{(p_m)} \in G_2 \setminus \{0\}$ for which the space G_1 is orthogonal. In particular, $\mathrm{dist}\left(v_{m+1} + X^{(p_m)}, G_1\right) = \|v_{m+1} + X^{(p_m)}\|$. In other words,

$$
\mathrm{dist}\left(v_{m+1} + X^{(p_m)}, G_1\right)
$$
$$
= \inf \left\| \left(v_{m+1} + X^{(p_m)}\right) - \sum_{k=1}^{p_m} \alpha_k \left(x_k + X^{(p_m)}\right) \right\|
$$
$$
= \|v_{m+1} + X\| = \left\| v_{m+1} + X^{(p_m)} \right\|, \tag{1.25}
$$

where the infima are over all scalars α and β, and the infinite sums in fact have only a finite number of nonzero summands. Obviously v_{m+1} can be chosen in $\mathrm{span}\{u_k\}_{k=q_m+1}^{q_{m+1}}$. Now, select $w_{m+1} \in v_{m+1} + \mathrm{span}\{x_n\}_{n=p_m+1}^\infty$ such that $\|v_{m+1} + X^{(p_m)}\| \le \|w_{m+1}\| < 2\|v_{m+1} + X^{(p_m)}\|$. Then we can write $v_{m+1} = \sum_{k=p_m+1}^{p_{m+1}} a_{m+1,k} x_k + w_{m+1}$ for some $p_{m+1} \in \mathbb{N}$. Having in mind (1.25), we get (1.24), and this completes the step $(m+1)$ in the construction.

Let us set

$$
y_n := \frac{w_n}{\|w_n\|} \quad \text{for all } n \in \mathbb{N}. \tag{1.26}
$$

By (1.24), $(w_n + X)_{n=1}^\infty = (v_n + X)_{n=1}^\infty$ is a block sequence of $(u_n + X)_{n=1}^\infty$ and hence, by (1.23) and (1.26), $(y_n + X)_{n=1}^\infty$ is also basic; on the other hand, by (1.24) and (1.26),

$$
\mathrm{dist}\,(y_n, X) = \|y_n + X\| = \frac{\|w_n + X\|}{\|w_n\|} = \frac{\|v_n + X\|}{\|w_n\|} > \frac{1}{2} \quad \text{for every } n \in \mathbb{N}.
$$

By the Hahn-Banach theorem, we can choose $\overline{y}_n^* \in S_{E^*}$ such that $\langle y_n, \overline{y}_n^* \rangle = \mathrm{dist}\,(y_n, X) > 1/2$ and $\langle x, e_n^* \rangle = 0$ for all $x \in X$; set $y_n^* := \overline{y}_n^*/\mathrm{dist}\,(y_n, X)$ for every $n \in \mathbb{N}$. In this way, we get a biorthogonal system $\{y_n; y_n^*\}_{n=1}^\infty$ such that $\|y_n^*\| < 2$ for all $n \in \mathbb{N}$ and $X \subset (\{y_n^*\}_{n=1}^\infty)_\perp$. Put

$$
z_k^* := e_k^* + \frac{a_{n,k} y_n^*}{\|w_n\|} \quad \text{for } p_{n-1} + 1 \le k \le p_n \quad \text{for every } n \in \mathbb{N}. \tag{1.27}
$$

From (1.23), (1.24), (1.26), and (1.27), the system $\{x_n, y_n; z_n^*, y_n^*\}_{n=1}^\infty$ in $E \times E^*$ is biorthogonal and z_n^* extends x_n^* to E for all $n \in \mathbb{N}$. Moreover,

$$
\frac{|a_{n,k}|}{\|w_n\|} = \frac{|\langle w_n, e_k^* \rangle|}{\|w_n\|} = |\langle y_n, e_k^* \rangle| \le \|e_k^*\| \quad \text{for } p_{n-1}+1 \le k \le p_n \text{ for every } n \in \mathbb{N}.
$$

Hence $\|z_n^*\| < 3(2\lambda + \varepsilon)$ for all $n \in \mathbb{N}$. $\qquad\square$

Theorem 1.50 (Terenzi [Tere83]). *Let E be a separable Banach space, X be a subspace of E and $\varepsilon > 0$. Then every λ-bounded M-basis in $X \times X^*$ can be extended to a $(12\lambda + \varepsilon)$-bounded M-basis in $E \times E^*$.*

Proof. We can assume that the λ-bounded M-basis $\{x_n; x_n^*\}_{n=1}^\infty$ in $X \times X^*$ satisfies $\|x_n\| = 1$ for all $n \in \mathbb{N}$. Apply Lemma 1.49 to $\{x_n; x_n^*\}_{n=1}^\infty$ and $\varepsilon/7$ to obtain a biorthogonal system $\{x_n, y_n; z_n^*, y_n^*\}_{n=1}^\infty$ in $E \times E^*$ such that $\|x_n\| = \|y_n\| = 1$, $\|z_n^*\| < 3(2\lambda + \varepsilon/7)$, $\|y_n^*\| < 2$, and z_n^* is an extension of x_n^* to E for all $n \in \mathbb{N}$. Let $Y := \overline{\operatorname{span}}\{x_n, y_n\}_{n=1}^\infty$. Corollary 1.48 applied to the biorthogonal system $\{x_n, y_n; z_n^* \restriction Y, y_n^* \restriction Y\}_{n=1}^\infty$ in $Y \times Y^*$ gives extensions u_n^* (resp. v_n^*) of $z_n^* \restriction Y$ (resp. $y_n^* \restriction Y$) to E, for $n \in \mathbb{N}$, such that $Y + (\{u_n^*\}_{n=1}^\infty \cap \{v_n^*\}_{n=1}^\infty)_\perp$ is dense in E and such that $\|u_n^*\| < 2(3(2\lambda + \varepsilon/7)) + \varepsilon/7$, $\|v_n^*\| < 2(3(2\lambda + \varepsilon/7)) + \varepsilon/7$. Let $Q : E \to E/Y$ be the canonical quotient mapping. The subspace $Q((\{u_n^*\}_{n=1}^\infty \cap \{v_n^*\}_{n=1}^\infty)_\perp)$ is dense in E/Y, so from Theorem 1.27 there exists an M-basis $\{\overline{w}_n; w_n^*\}_{n=1}^\infty$ in $E/Y \times Y^\perp$ such that $\overline{w}_n \in Q((\{u_n^*\}_{n=1}^\infty \cap \{v_n^*\}_{n=1}^\infty)_\perp)$, $\|\overline{w}_n\| < 1$, $\|w_n^*\| < 1 + \varepsilon$, for all $n \in \mathbb{N}$. Take $w_n \in (\{u_n^*\}_{n=1}^\infty \cap \{v_n^*\}_{n=1}^\infty)_\perp$ such that $Q(w_n) = \overline{w}_n$ and $\|w_n\| < 1$ for all $n \in \mathbb{N}$. By scaling, we may assume that $\|w_n\| = 1$ and $\|w_n^*\| < 1 + \varepsilon$ for all $n \in \mathbb{N}$, and $\{Q(w_n), w_n^*\}_{n=1}^\infty$ is an M-basis in $E/Y \times Y^\perp$. It is obvious that

$$\{x_n, y_n, w_n; u_n^*, v_n^*, w_n^*\}_{n=1}^\infty \tag{1.28}$$

is a biorthogonal system in $E \times E^*$. We claim that (1.28) is in fact an M-basis in $E \times E^*$. To prove the claim, first take $e^* \in E^*$ such that $\langle x_n, e^* \rangle = \langle y_n, e^* \rangle = \langle w_n, e^* \rangle = 0$ for all $n \in \mathbb{N}$. Then $e^* \in Y^\perp$ and it vanishes on $\{w_n; n \in \mathbb{N}\}$, a linearly dense subset of E/Y. Then $e^* = 0$. We have proved that the system (1.28) is fundamental. To show that it is also total, let $e \in E$ be such that $\langle e, u_n^* \rangle = \langle e, v_n^* \rangle = \langle e, w_n^* \rangle = 0$ for all $n \in \mathbb{N}$. Then $\langle Q(e), w_n^* \rangle = 0$ for all $n \in \mathbb{N}$. Since $\{\overline{w}_n; w_n^*\}_{n=1}^\infty$ is an M-basis in $E/Y \times Y^\perp$, we get $Q(e) = 0$, so $e \in Y (= X + \overline{\operatorname{span}}\{y_n\}_{n=1}^\infty)$. Then $q(e) \in \overline{\operatorname{span}}\{q(y_n)\}_{n=1}^\infty$, where $q : E \to E/X$ is the canonical quotient mapping. We know that $(q(y_n))$ is a Schauder basis of $\overline{\operatorname{span}}\{q(y_n)\}_{n=1}^\infty$, so

$$q(e) = \sum_{n=1}^\infty \langle q(e), y_n^* \rangle q(y_n) = \sum_{n=1}^\infty \langle e, y_n^* \rangle q(y_n) = \sum_{n=1}^\infty \langle e, v_n^* \rangle q(y_n) = 0$$

and so $e \in X$. Finally,

$$\langle x, x_n^* \rangle = \langle x, e_n^* \rangle = \langle x, z_n^* \rangle = \langle x, u_n^* \rangle = 0$$

for all $n \in \mathbb{N}$ (the second equality thanks to the particular expression (1.27) of z_n^*), and so $e = 0$ because $\{x_n; x_n^*\}_{n=1}^\infty$ is an M-basis in $X \times X^*$. This proves that (1.28) is also total. The system (1.28) is the extension sought. $\qquad\square$

Theorem 1.51 (Singer [Sing74]). *There exists a Banach space with a Schauder basis and a subspace of it also having a Schauder basis and such that no Schauder basis of the subspace is extendable to the entire space.*

Proof. By Enflo's result [Enfl73], there exists a separable Banach space Z without a basis. By a result of J. Lindenstrauss [Lind71a], there exists a separable Banach space X such that X^* has a shrinking basis and Z is isomorphic to X^{**}/X. Then X^{**} has a basis (the functional coefficients of a basis form a basic sequence, this is standard) and X has a shrinking basis ([JRZ71, Thm. 1.4(a)]). However, no basis (x_n) of X can be extended to a basis $((x_n), (x_n^{**}))$ of X^{**} since otherwise the quotient space X^{**}/X would have a basis, namely $(q(x_n^{**}))$, where $q : X^{**} \to X^{**}/X$ is the canonical quotient mapping. □

Quasicomplemented subspaces of a Banach space will be treated in Section 5.7. Here we consider the problem of extending M-bases in the direction of a given quasicomplement.

Definition 1.52. *Let Y and Z be subspaces of a Banach space X. We will say that Y and Z are* quasicomplemented *or that Z is a* quasicomplement *of Y in X if $Y \cap Z = \{0\}$ and $Y + Z$ is dense in X.*

Theorem 1.53 (V.D. Milman). *Let Y and Z be quasicomplemented subspaces of a separable Banach space X. Let $\{y_n\}_{n=1}^\infty$ be an M-basis in Y. Then there exists a sequence (z_n) in Z such that $\{y_n\}_{n=1}^\infty \cup \{z_n\}_{n=1}^\infty$ is an M-basis in X.*

Proof (Plichko). We denote by the same symbol an element $\hat{g} \in Y^*$ and its preimage under the quotient map $X^* \to X^*/Y^\perp = Y^*$. We need the following result.

Lemma 1.54. *Under the conditions of Theorem 1.53, and for the family $\{\hat{g}_n\}_{n=1}^\infty$ in $X^*/Y^\perp$ of coefficient functionals associated to the M-basis $\{y_n\}_{n=1}^\infty$, there are representatives $g_n \in \hat{g}_n$ for which*

$$\overline{\operatorname{span}\{g_n\}_{n=1}^\infty + Z^\perp}^{w^*} \cap Y^\perp = \{0\}. \tag{1.29}$$

Proof. Since X is separable, it is possible to write $Y^\perp \setminus \{0\}$ as the union of a sequence $(K_n)_{n=1}^\infty$ of convex w^*-compact subsets of X^*.

Let us construct elements $x_n \in X$ and representatives $g_n \in \hat{g}_n$ so that, for every n,

(a) x_n separates $G_{n-1} := \operatorname{span}\{g_i\}_{i=1}^{n-1} + Z^\perp$ and K_n,

(b) the restriction $x_n \upharpoonright Y^\perp \notin \operatorname{span}\{x_i \upharpoonright Y^\perp\}_{i=1}^{n-1}$,

(c) $G_n \subset (\operatorname{span}\{x_i\}_{i=1}^n)^\perp$, and

(d) $G_n \cap Y^\perp = \{0\}$.

Start from $n = 1$. Let us separate, by the Hahn-Banach theorem, the w^*-closed subspace $Z^\perp$ and K_1 by a functional $x_1 \in X$. Observe that $Z^\perp \subset x_1^\perp$. Consider two cases.

1. $\hat{g}_1 \cap Z^\perp \neq \emptyset$.

Take as g_1 any element of this intersection. Then $G_1 \subset x_1^\perp$ and $G_1 \cap Y^\perp = \{0\}$.

2. $\hat{g}_1 \cap Z^\perp = \emptyset$.

Then it is easy to show that

$$\operatorname{span}\{\hat{g}_1\} \cap Z^\perp = 0. \tag{1.30}$$

We claim that $x_1^\perp \cap \operatorname{span}\{\hat{g}_1\} \not\subset Y^\perp$. Indeed, assume the opposite. If we have $\langle x_1, g_1 \rangle = 0$ for all $g_1 \in \hat{g}_1$, then $Y^\perp \subset x_1^\perp$ and so x_1 vanishes on the w^*-dense set $Y^\perp + Z^\perp$, a contradiction; on the other hand, if there exists $g_1 \in \hat{g}_1$ such that $\langle x_1, g_1 \rangle \neq 0$, choose, for a given $y^\perp \in Y^\perp$, some λ such that $\langle x_1, \lambda g_1 + y^\perp \rangle = 0$. Then $\lambda g_1 + y^\perp \in Y^\perp$, so $\lambda = 0$ and then $\langle x_1, y^\perp \rangle = 0$. This proves that $Y^\perp \subset x_1^\perp$ and again we reach a contradiction, so the claim is proved. Therefore we can find $g_1 \in x_1^\perp \cap \operatorname{span}\{\hat{g}_1\}$ such that $g_1 \notin Y^\perp$. It is simple to see that in fact g_1 can be taken in $\hat{g}_1$. Then $G_1 \subset x_1^\perp$ and $G_1 \cap Y^\perp = \{0\}$.

Assume that collections $(x_i)_{i=1}^{n-1}$ and $(g_i)_{i=1}^{n-1}$ with conditions (a) to (d) have already been constructed. Using condition (d), separate the (w^*-closed) subspace G_{n-1} and the w^*-compact set K_n by a functional $x \in X$. Precisely, put $\inf\{x(f); f \in K_n\} =: a > 0$ and

$$x \upharpoonright G_{n-1} \equiv 0. \tag{1.31}$$

If $x \upharpoonright Y^\perp \notin \operatorname{span}\{x_i \upharpoonright Y^\perp\}_{i=1}^{n-1}$, put $x_n := x$. In the other case, choose $z \in (G_{n-1})_\perp$ with $\sup\{z(f); f \in K_n\} < a/2$ and $z \upharpoonright Y^\perp \notin \operatorname{span}\{x_i \upharpoonright Y^\perp\}_{i=1}^{n-1}$ (of course, the subspaces Y and Z are assumed to be infinite-dimensional). Put $x_n := x + z$. Obviously, for x_n, conditions (a) and (b) are satisfied.

As for $n = 1$, let us consider two cases.

1. $\hat{g}_n \cap G_{n-1} \neq \emptyset$.

Take as g_n any element of this intersection. The verification of conditions (c) and (d) is trivial.

2. $\hat{g}_n \cap G_{n-1} = \emptyset$.

Then

$$\operatorname{span}\{\hat{g}_i\}_{i=1}^n \cap Z^\perp = \{0\}. \tag{1.32}$$

The intersection $(\operatorname{span}\{x_i\}_{i=1}^n)^\perp \cap \operatorname{span}\{\hat{g}_i\}_{i=1}^n$ cannot contain only elements of $\operatorname{span}\{\hat{g}_i\}_{i=1}^{n-1}$ because, in this case, $(\operatorname{span}\{x_i\}_{i=1}^n)^\perp$, which cut out from $Y^\perp$ a subspace of codimension n (by condition b)), shall cut out from $\operatorname{span}\{\hat{g}_i\}_{i=1}^n$ a subspace of codimension $n+1$ (since $\{y_n; \hat{g}_n\}$ is an M-basis, $\hat{g}_n \notin \operatorname{span}\{\hat{g}_i\}_{i=1}^{n-1}$). This is impossible.

Take an element

$$g_n \in (\operatorname{span}\{x_i\}_{i=1}^n)^\perp \cap \operatorname{span}\{\hat{g}_i\}_{i=1}^n, \tag{1.33}$$

$g_n \notin \operatorname{span}\{\hat{g}_i\}_{i=1}^{n-1}$. Since $(g_i)_{i=1}^{n-1} \subset (\operatorname{span}\{x_i\}_{i=1}^n)^\perp$, we can assume $g_n \in \hat{g}_n$.

Condition (c) follows from (1.31) and (1.33); (d) follows from (1.32). Therefore, elements x_n and representatives $g_n \in \hat{g}_n$, $n \in \mathbb{N}$, satisfying (a) to (d) are constructed. Condition (c) implies that

$$\overline{\operatorname{span}\{g_n\}_{n=1}^\infty + Z^\perp}^{w^*} \subset (\overline{\operatorname{span}}\{x_n\}_{n=1}^\infty)^\perp.$$

This and (a) imply (1.29).   □

We continue with the proof of Theorem 1.53. Let (g_n) be the sequence from Lemma 1.54, and let $Z_0 := (\overline{\text{span}}\{g_n\}_{n=1}^{\infty} + Z^{\perp})_{\perp} \subset Z$. If $x^* \in X^*$ satisfies $x^* \in Z_0^{\perp} \cap Y^{\perp}$, then $x^* \in \overline{\overline{\text{span}}\{g_n\}_{n=1}^{\infty} + Z^{\perp}}^{w^*} \cap Y^{\perp} = \{0\}$ and hence $Y + Z_0$ is dense in X. Then $q(Z_0)$ is dense in $X/Y^{\perp}$, where $q : X \to X/Y^{\perp}$ is the canonical quotient mapping. Take a linearly dense sequence (z_n) in Z_0. Then $(q(z_n))$ is linearly dense in $X/Y^{\perp}$. Use Lemma 1.21 to construct an M-basis $\{\hat{z}_n; , h_n\}$ in $X/Y \times Y^{\perp}$ such that there are representatives $z_n \in \hat{z}_n \cap Z_0$, $n \in \mathbb{N}$, with $\overline{\text{span}}\{z_n\}_{n=1}^{\infty} = Z_0$. The system $\{y_n, z_n; g_n, h_n\}_{n=1}^{\infty}$ is an M-basis in $X \times X^*$. Indeed, $\langle y_n, g_m \rangle = \delta_{n,m}$, $\langle y_n, h_m \rangle = 0$, $\langle z_n, h_m \rangle = \delta_{n,m}$, $\langle z_n, g_m \rangle = \delta_{n,m}$ for all $n, m \in \mathbb{N}$. Moreover, $\{y_n, z_n\}_{n=1}^{\infty}$ is linearly dense in X due to the density of $Y + Z_0$, and $\langle x, g_n \rangle = \langle x, h_n \rangle$ for all $n \in \mathbb{N}$ implies $\langle q(x), h_n \rangle = 0$ for all $n \in \mathbb{N}$, so $q(x) = 0$ and then $x \in Y$. It follows that $x = 0$, so the system is total. $\qquad \square$

1.7 ω-independence

The natural extension of the property of linear independence enjoyed by any finite algebraic basis to the setting of infinite-dimensional Banach spaces is called ω-independence. It is easy to check that every M-basis is an ω-independent family, and Z. Lipecki asked if every ω-independent family in a separable Banach space must be countable. An affirmative answer to this question was provided by Fremlin and Sersouri. An alternative proof of their result (in fact, of a slightly more general one) was given by Kalton, and it is his proof that we present here. Some more results concerning ω-independent systems will be presented in Section 8.2.

Definition 1.55. *A family $\{x_\alpha\}_{\alpha \in \Gamma}$ in a Banach space X is said to be ω-independent if, for every sequence $(\alpha_n)_{n=1}^{\infty}$ in Γ of distinct indices and every sequence of real numbers $(\lambda_n)_{n=1}^{\infty}$, the series $\sum_{n=1}^{\infty} \lambda_n x_{\alpha_n}$ converges to zero in X if and only if all λ_n are zero.*

Because ω-independence clearly implies linear independence, no finite-dimensional space can contain an infinite ω-independent family.

Theorem 1.56 (Fremlin and Sersouri [FrSe88]). *Let X be an infinite-dimensional Banach space. Then X contains a continuous curve C that forms an ℓ_1-independent family; i.e., for every sequence (c_n) in C and every sequence (λ_n) of real numbers with $\sum |\lambda_n| < \infty$, $\sum \lambda_n c_n = 0$ if and only if all λ_n are equal to 0.*

Proof. Let $(x_n)_{n=1}^{\infty}$ be a normalized basic sequence in X. The family $C := \{e_t; t \in (0,1)\}$ given by $e_t := \sum_{p=1}^{\infty} t^p x_p$, $t \in (0,1)$ is well defined and obviously represents a continuous curve C in X. We shall prove that C is ℓ_1-independent. First of all, given a sequence (λ_n) in $\mathbb{R}$ and a sequence (t_n) in $(0,1)$ such that $\sum_{n=1}^{\infty} \lambda_n e_{t_n} = 0$, it is simple to prove that

$$\sum_{n=1}^{\infty} \lambda_n t_n^p = 0 \text{ for all } p \in \mathbb{N}. \tag{1.34}$$

Assume now that $(\lambda_n) \in \ell_1$. Then (λ_n) defines a measure $\mu := \sum_{n=1}^{\infty} \lambda_n t_n \delta_{t_n} \in C[0,1]^*$, where δ_t is the Dirac measure at $t \in [0,1]$. Notice that (1.34) is equivalent to $\int_0^1 t^p d\mu = 0$ for all $p = 0, 1, 2, \ldots$. The set of all polynomials is dense in $C[0,1]$, and hence $\mu = 0$. Since $t_n \neq 0$ for every $n \in \mathbb{N}$, this implies that $\lambda_n = 0$ for every $n \in \mathbb{N}$. $\qquad\square$

Example 1.57. (a) Let $\{x_\alpha; f_\alpha\}_{\alpha \in A}$ be a biorthogonal system. Then $\{x_\alpha\}_{\alpha \in A}$ is ω-independent.

(b) Let $\{x_n; f_n\}_{n \in \mathbb{N}}$ be a fundamental biorthogonal system that is not an M-basis for a separable Banach space X (Remark 1.11). Then there is an $x_0 \in X$ such that $\{x_n\}_{n=0}^{\infty}$ is an ω-independent family, but it is not a minimal system.

Proof. (a) Suppose $\sum_{n=1}^{\infty} a_n x_{\alpha_n} = 0$. Then $a_k = f_{\alpha_k}\left(\sum_{n=1}^{\infty} a_n x_{\alpha_n}\right) = 0$ for all $k \in \mathbb{N}$.

(b) Because $\{x_n\}_{n=1}^{\infty}$ is not an M-basis, there is an $x_0 \in X \setminus \{0\}$ so that $f_n(x_0) = 0$ for all $n \in \mathbb{N}$. Now suppose $\sum_{n=0}^{\infty} a_n x_n = 0$. If $a_0 \neq 0$, then $a_{n_0} \neq 0$ for some n_0, and also

$$x_0 = -\frac{1}{a_0} \sum_{n=1}^{\infty} a_n x_n.$$

Thus $0 = f_{n_0}(x_0) = -a_{n_0}/a_0 \neq 0$, a contradiction. Hence $a_0 = 0$. Then, by part (a), $a_n = 0$ for all $n \geq 1$ as well. Clearly, $\{x_n\}_{n=0}^{\infty}$ cannot be minimal because x_0 is in the closed linear hull of $\{x_n\}_{n=1}^{\infty}$. $\qquad\square$

Theorem 1.58 (Kalton [Kalt89]). *Let X be a Banach space, and let G be a subset of X. Let H be the set of accumulation points of G, and suppose that X is the closed linear span of H. Then, given any $x \in X$ and any sequence of numbers (a_n) with $\sum |a_n| = \infty$ and $\lim a_n = 0$, there is a sequence of signs ϵ_n and distinct elements $g_n \in G$ so that*

$$x = \sum_{n=1}^{\infty} \epsilon_n a_n g_n.$$

Before proving this, we present the following corollary that answers a question of Z. Lipecki.

Corollary 1.59 (Fremlin and Sersouri [FrSe88]). *Suppose X is a separable Banach space. Then every ω-independent family in X is countable.*

Proof. By contradiction, let G be an uncountable ω-independent family in X. Because G is a subset of the separable (metric) space X, it contains an uncountable set H that is dense in itself. Let Y be the closed linear span of H. According to Theorem 1.58, given any sequence (a_n) with $\sum_n |a_n| = \infty$ and $\lim a_n = 0$, we can find signs ϵ_n and distinct $g_n \in G$ so that

$$0 = \sum_{n=1}^{\infty} \epsilon_n a_n g_n,$$

a contradiction. $\square$

Proof of Theorem 1.58. Let (a_n) be a sequence of nonnegative numbers satisfying the condition in the theorem. Let us denote $b_n = \max_{i>n} a_i$. Then the following fact will be used, and its proof is left as an exercise.

Fact 1.60. *Suppose $\alpha \in \mathbb{R}$ and $m \in \mathbb{N}$. Then one can choose signs ϵ_i, $i \geq m+1$ so that if $s_m := \alpha$ and $s_k := \alpha + \sum_{i=m+1}^{k} \epsilon_i a_i$ for $k = m+1, m+2, \ldots$, then $\lim s_k = 0$ and $\sup_{k \in \mathbb{N}} |s_k| \leq \max\{b_m, |\alpha|\}$.*

Next let us define $F(N, \delta)$ for $N \in \mathbb{N}$ and $\delta > 0$ to be the subset of X defined by $x \in F(N, \delta)$ if for any $m \geq N$ we can find $n > m$ and $\epsilon_i = \pm 1$, $h_i \in H$, $i = m+1, \ldots, n$, such that

$$\left\| x + \sum_{i=m+1}^{n} \epsilon_i a_i h_i \right\| < \delta$$

and

$$\left\| x + \sum_{i=m+1}^{k} \epsilon_i a_i h_i \right\| \leq \|x\| + \delta, \quad m+1 \leq k \leq n.$$

Note that the h_i are not required to be distinct.

Now define $F := \bigcap_{\delta > 0} \bigcup_{N \in \mathbb{N}} F(N, \delta)$ and $E := \{x : \alpha x \in F \text{ for all } \alpha \in \mathbb{R}\}$. It is an easy exercise to show that F and hence E are closed.

We now show that $H \subset E$. Indeed, suppose $h \in H$ and $\alpha \in \mathbb{R}$. For arbitrary $\delta > 0$, choose N so large that $b_N \|h\| < |\alpha| \|h\| + \delta$. If $m \geq N$, we choose ϵ_i according to Fact 1.60 applied to α and N, stopping at n where $|s_n|.\|h\| < \delta$. Letting $h_i = h$ for $m+1 \leq i \leq n$, one sees that $H \subset E$.

The next step is to show that E is a linear subspace, and for this we need only to show that if $x, y \in E$, then $x + y \in F$. Indeed, suppose $\delta > 0$ and let $M = \max\{\|x\|, \|y\|\}$. Let s be a positive integer so large that $6M < s\delta$. Then choose N so that $s^{-1}x, s^{-1}y \in F(N, \delta/(4s))$. Now suppose $m \geq N$ and let $p_0 = m$. Then we may inductively define q_k, p_k for $1 \leq k \leq s$, e_i, and h_i for $p_0 + 1 \leq i \leq p_s$ so that $p_{k-1} < q_k < p_k$ for $1 \leq k \leq s$,

$$\left\| s^{-1}x + \sum_{p_{k-1}+1}^{q_k} \epsilon_i a_i h_i \right\| < \frac{\delta}{4s} \quad \text{and} \quad \left\| s^{-1}x + \sum_{p_{k-1}+1}^{j} \epsilon_i a_i h_i \right\| < \frac{4\|x\| + \delta}{4s},$$

for $1 \le k \le s$ and $p_{k-1} + 1 \le j \le q_k$, and

$$\left\| \sum_{q_k+1}^{p_k} \epsilon_i a_i h_i + s^{-1} y \right\| < \frac{\delta}{4s} \qquad \text{and} \qquad \left\| \sum_{q_k+1}^{j} \epsilon_i a_i h_i + s^{-1} y \right\| < \frac{4\|y\| + \delta}{4s},$$

for $1 \le k \le s$ and $q_k + 1 \le j \le p_k$.
Then

$$\left\| x + y + \sum_{m+1}^{p_s} \epsilon_i a_i h_i \right\| < \delta.$$

If $p_{k-1} + 1 \le j \le q_k$, then

$$\left\| x + y + \sum_{m+1}^{j} \epsilon_i a_i h_i \right\| < \frac{s-k+1}{s} \|x + y\| + \frac{(k-1)\delta}{2s} + \frac{4\|x\| + \delta}{4s}.$$

If $q_k + 1 \le j \le p_k$, then

$$\left\| x + y + \sum_{m+1}^{j} \epsilon_i a_i h_i \right\| < \frac{s-k+1}{s} \|x + y\| + \frac{(k-1)\delta}{2s} + \frac{4\|x\| + \delta}{4s} + \frac{4\|y\| + \delta}{4s}.$$

In either case, we conclude that

$$\left\| x + y + \sum_{m+1}^{j} \epsilon_i h_i \right\| < \|x + y\| + \delta.$$

This shows that $x + y \in F(N, \delta)$, and so E is a linear subspace as desired, and consequently $E = X$.

Now fix any $h_0 \in H$ and let $\gamma = \|h_0\|$. Then we claim that for any $x \in X$, $m \in \mathbb{N}$, and $\delta > 0$, we can find $n > m$, $h_i \in H$, $m + 1 \le i \le n$ and $\epsilon_i = \pm 1$ for $m + 1 \le i \le n$ so that

$$\left\| x + \sum_{m+1}^{n} \epsilon_i a_i h_i \right\| < \delta$$

and, for $m + 1 \le j \le n$,

$$\left\| x + \sum_{m+1}^{j} \epsilon_i a_i h_i \right\| < \gamma b_m + \|x\| + \delta.$$

In fact, there exists N so that $x \in F(N, \delta/2)$. To verify the claim, apply Fact 1.60 to $\alpha = 0$ and m to obtain ϵ_i, $m + 1 \le i \le k$, where $k \ge N$ so that

$$\left| \sum_{m+1}^{k} \epsilon_i a_i \right| < \frac{\delta}{2\gamma} \qquad \text{and} \qquad \left| \sum_{m+1}^{j} \epsilon_i a_i \right| < b_m$$

for $m + 1 \leq j \leq k$. Now choose $n > k$ and $h_i \in H$, $\epsilon_i = \pm 1$ for $k + 1 \leq i \leq n$ so that

$$\left\| \sum_{k+1}^{n} \epsilon_i a_i h_i + x \right\| < \frac{\delta}{2} \quad \text{and} \quad \left\| \sum_{k+1}^{j} \epsilon_i a_i h_i + x \right\| < \|x\| + \frac{\delta}{2}$$

for $k + 1 \leq j \leq n$. Now let $h_i = h_0$ for $m + 1 \leq i \leq k$ to substantiate the claim.

Finally, suppose $x \in X$ is fixed, and then let $p_0 = 0$. Because H is the set of accumulation points of G, we may inductively choose p_k, signs ϵ_i, and $g_i \in G$ for $p_{k-1} + 1 \leq i \leq p_k$ so that g_i are distinct for $1 \leq i \leq p_k$,

$$\left\| \sum_{i=1}^{p_k} \epsilon_i a_i g_i - x \right\| < 2^{-k}$$

for $k \geq 1$, and if $p_{k-1} + 1 \leq j \leq p_k$, then

$$\left\| \sum_{i=1}^{j} \epsilon_i a_i g_i - x \right\| < \left\| \sum_{i=1}^{p_{k-1}} \epsilon_i a_i g_i - x \right\| + 2^{-k} + \gamma b_{p_{k-1}} < 4(2^{-k}) + \gamma b_{p_{k-1}}$$

for $k \geq 1$. The series constructed in this fashion converges to x and we are done. $\square$

1.8 Exercises

1.1. A subset $\{x_n\}_{n=1}^{\infty}$ of a Banach space is called *overfilling* if any infinite subset of it is linearly dense. Prove that every separable Banach space contains overfilling sets (see [Klee58]).

Hint (J.I. Lyubich, see [Milm70a, p. 113]). Take an arbitrary linearly dense set $\{x_n\}_{n=0}^{\infty}$ in S_X and form the analytic function $f : \mathbb{R} \to X$ defined by

$$f(\lambda) := \sum_{k=0}^{\infty} \frac{x_k}{k!} \lambda^k, \quad \lambda \in \mathbb{R}.$$

Take any sequence $(\lambda_k)_{k=1}^{\infty}$ in $\mathbb{R}$ such that $\lambda_k \to 0$ and $\lambda_k \neq 0$, $k = 0, 1, 2, \ldots$. Then $\{f(\lambda_k)\}_{k=0}^{\infty}$ is linearly dense in X (it has the same linear span as $\{x_k\}_{k=0}^{\infty}$) and overfilling.

1.2. Assume that $\{x_n\}_{n=1}^{\infty}$ is a fundamental minimal system in X such that it can be partitioned into two infinite M-basic systems $\{x_a : a \in A\}$ and $\{x_b : b \in B\}$ (a system is *M-basic* if it is an M-basis for its closed linear span). Is $\{x_n\}_{n=1}^{\infty}$ an M-basis?

Hint. No, even if $\{x_a : a \in A\}$ and $\{x_b : b \in B\}$ are both Schauder basic sequences. The following example comes from [Sing81, Example III.8.1]: Let $X := \ell_1$ and let $\{e_n; e_n^*\}_{n=1}^\infty$ be its canonical basis. Put $x_0 := e_1 - e_2$, $x_{2n} := e_1 - e_{2n+1} - e_{2n+2}$, $x_{2n-1} := e_1 + e_{2n} + e_{2n+1}$, $x_{2n-1}^* := w^* - \sum_{k=1}^\infty (-1)^{k+1} e_{2n+k}^*$, $n = 1, 2, \ldots$, $x_{2n}^* := w^* - \sum_{k=1}^\infty (-1)^k e_{2n+1+k}^*$, $n = 0, 1, 2, \ldots$. It is easy to see that $\{x_n; x_n^*\}_{n=0}^\infty$ is a biorthogonal system. Moreover, $x_{2n-1} + x_{2n} = 2e_1 + e_{2n} - e_{2n+2}$, $n = 1, 2, \ldots$, whence

$$\frac{1}{2n} \sum_{i=1}^{2n} x_i = \frac{1}{2n} \sum_{j=1}^n (x_{2j-1} + x_{2j}) = \frac{1}{2n} \sum_{j=1}^n (2e_1 + e_{2j} - e_{2j+2})$$

$$= e_1 + \frac{1}{2n} e_2 - \frac{1}{2n} e_{2n+2} \to e_1 \quad \text{as } n \to \infty,$$

and therefore $e_1 \in \overline{\text{span}}\{x_j\}_{j=1}^\infty$. Since $e_2 = e_1 - x_0$ it follows that $e_2 \in \overline{\text{span}}\{x_j\}_{j=0}^\infty$ and hence, inductively, $e_n \in \overline{\text{span}}\{x_j\}_{j=0}^\infty$, $n = 1, 2, \ldots$. This proves that $\{x_n\}_{n=0}^\infty$ is a fundamental minimal system in X. It is not an M-basis since $\langle e_1, x_n^* \rangle = 0$, $n = 0, 1, 2, \ldots$. However, $\{x_{2n-1}\}_{n=1}^\infty$ and $\{x_{2n}\}_{n=0}^\infty$ are basic sequences, both equivalent to the canonical basis of ℓ_1.

1.3. Prove that a fundamental minimal system $\{x_n\}_{n=1}^\infty$ in X is an M-basis if and only if $\{x_a\}_{a \in A}$ and $\{x_b\}_{b \in B}$ are both M-basic systems for every partition of $\mathbb{N}$ into two infinite sets, A and B (compare with Exercise 1.2).

Hint. One direction is easy. For the other, assume that $\{x_n\}_{n=1}^\infty$ is a fundamental minimal system that is not an M-basis. From Fact 1.8, there exists $0 \neq x \in \bigcup_{n=1}^\infty \overline{\text{span}}\{x_k : k > n\}$. Put $m_0 = 0$. Find $m_1 \in \mathbb{N}$ such that $\text{dist}\,(x, \text{span}\{x_n : 1 < n \leq m_1\}) < 1$. Find $m_2 > m_1$ such that $\text{dist}\,(x, \text{span}\{x_n : m_1 < n \leq m_2\}) < 1/2$. Proceed in this way to find an increasing sequence $(m_i)_{i=0}^\infty$ such that $\text{dist}\,(x, \text{span}\{x_n : m_i < n \leq m_{i+1}\}) < 1/(i+1)$, $i = 0, 1, 2, \ldots$. Put $A := \bigcup_{i=1}^\infty I_{2i-1}$, where $I_i := \{m_{i-1}, \ldots, m_i\}$, $i = 1, 2, \ldots$. Neither $\{x_a\}_{a \in A}$ nor $\{x_b\}_{b \in B}$ are M-basic systems.

1.4 ([Day62]). Prove that, in every Banach space X such that (X^*, w^*) is not separable, it is possible to choose $c_n = 0$ for all $n \in \mathbb{N}$ in Corollary 1.20 (so in every such space there exists an Auerbach basic sequence that is at the same time a monotone Schauder basic sequence).

Hint. Change in the proof of Corollary 1.20 the $\varepsilon/2$-net $\{x_i\}_{i=1}^k$ in the unit sphere of $\text{span}\{b_i\}_{i=1}^n$ to a dense countable set $\{x_i\}_{i=1}^\infty$ and choose the corresponding set $\{x_i^*\}_{i=1}^\infty$. The non-w^*-separability of X^* still allows us to ensure that $G_2 := \bigcap_{i=1}^\infty \text{Ker}\, x_i^* \cap \bigcap_{i=1}^n \text{Ker}\, b_i^*$ is an infinite-dimensional subspace of X.

1.5. Show that if $\{x_n\}_{n=1}^\infty$ is a strong M-basis in a Banach space X, then $\{x_n^*\}_{n=1}^\infty$ is a w^*-*strong minimal system* in X^* (i.e., for all $x^* \in X^*$, $x^* \in \overline{\text{span}}^{w^*}\{\langle x_n, x^* \rangle x_n^*\}_{n=1}^\infty$).

Hint. Given $x^* \in X^*$, let us write $\mathrm{supp}(x^*) := \{n \in \mathbb{N}; \langle x_n, x^* \rangle \neq 0\}$. Similarly, if $x \in X$, put $\mathrm{supp}(x) := \{n \in \mathbb{N}; \langle x, x_n^* \rangle \neq 0\}$. Assume that, for some $x^* \in X^*$, $x^* \notin \overline{\mathrm{span}}^{w^*}\{x_n^*; n \in \mathrm{supp}(x^*)\}$. Then, by the separation theorem, we can find $x \in X$ such that $\langle x, x_n^* \rangle = 0$ for all $n \in \mathrm{supp}(x^*)$ and $\langle x, x^* \rangle = 1$. It follows that $\mathrm{supp}(x) \subset \mathbb{N} \setminus \mathrm{supp}(x^*)$. The M-basis (x_n) is strong, and hence there exists a sequence (y_n) in $\mathrm{span}\{x_n; n \in \mathrm{supp}(x)\}$ such that $y_n \to x$. In particular, $\langle y_n, x^* \rangle = 0$ for all $n \in \mathbb{N}$, so $\langle x, x^* \rangle = 0$, a contradiction.

1.6 (Gurarii and Kadets). A biorthogonal system $\{x_n; x_n^*\}_{n=1}^{\infty}$ in $X \times X^*$, where X is a separable Banach space X, is called *convex strong* if, for every $x \in X$, $x \in \overline{\mathrm{conv}}\left\{\sum_{n \in F}\langle x, x_n^* \rangle x_n; F \in \text{ finite subsets of } \mathbb{N}\right\}$. A biorthogonal system $\{x_n; x_n^*\}_{n=1}^{\infty}$ is called a *Steinitz basis* if for each $x \in X$ and $x^* \in X^*$ there exists a permutation π of $\mathbb{N}$ such that $\langle x, x^* \rangle = \sum_{n=1}^{\infty}\langle x, x_{\pi(n)}^* \rangle \langle x_{\pi(n)}, x^* \rangle$. Prove that a bounded fundamental system is convex strong if and only if it is a Steinitz basis.

Hint. One direction follows from the separation theorem. For the other, observe first that for every $x \in X$, $\lim_n \langle x, x_n^* \rangle = 0$. Then proceed as in the proof of Riemann's result on reordering a series of real numbers.

1.7. Verify that the trigonometric system forms a strong M-basis in the subspace of $C[0, 2\pi]$ formed by functions equal at endpoints; note that this system is not a Schauder basis under any permutation.

Hint. Fejér, [LiTz77, p. 43].

1.8. Check the following properties of biorthogonal systems.
   (1) *Schauder bases* $\{x_n\}_{n=1}^{\infty}$: There is a constant $C > 0$, such that for every finite set $F \subset \mathbb{N}$ and every $x \in S_X$ supported on F, then $\mathrm{dist}(x, L) > C$, where $L := \overline{\mathrm{span}}\{x_i; i \in \mathbb{N}, i > \max F \text{ or } i < \min F\}$.
   (2) *Unconditional Schauder basis* $\{x_n\}_{n=1}^{\infty}$: There is a constant $C > 0$ such that for every finite set $F \subset \mathbb{N}$ and every $x \in S_X$ supported by F, then $\mathrm{dist}(x, L) > C$, where $L := \overline{\mathrm{span}}\{x_i; i \in \mathbb{N}, i \notin F\}$.
   (3) *Normalized bounded biorthogonal system* $\{x_n\}_{n=1}^{\infty}$: There is a constant $C > 0$ such that for every $n \in \mathbb{N}$, then $\mathrm{dist}(x_n, L_n) > C$, where $L_n := \overline{\mathrm{span}}\{x_i; i \in \mathbb{N}, i \neq n\}$.
   (4) *Normalized biorthogonal system* $\{x_n\}_{n=1}^{\infty}$: For every $n \in \mathbb{N}$, then we have $\mathrm{dist}(x_n, L_n) > 0$, where $L_n := \overline{\mathrm{span}}\{x_i; i \in \mathbb{N}, i \neq n\}$.

2

Universality and the Szlenk Index

In this chapter, we study two closely related topics, namely the notion of the Szlenk index of a Banach space and the existence of universal Banach spaces with additional properties. Historically, this area of research arose from one of the problems from the Scottish book due to Banach and Mazur. The problem asked whether there exists a separable Banach (resp. reflexive separable) space that contains an isomorphic copy of every separable Banach (resp. reflexive separable) space. The first part was solved by Banach and Mazur themselves in the positive when they showed that $C[0, 1]$ is a separable universal Banach space. The reflexive case was solved negatively by Szlenk using what is now called the Szlenk ordinal index of a Banach space. The value of the index of a separable reflexive space is a countable ordinal, and the index of a subspace is bounded from above by the index of the overspace. The negative solution of Szlenk then consisted of showing that there exists a separable reflexive space with an arbitrarily large countable index.

In the first two sections, we investigate various versions of the universality problem, with emphasis on reflexivity and complementability conditions. One of the main tools used is the theory of well-founded trees on Polish spaces. Using this notion, it is possible to assign to Banach spaces an ordinal index that measures the extent to which the space satisfies certain properties in which we are interested. One of them may be, for example, a containment of a certain subspace, but the technique is very versatile and includes the Szlenk index as a special case. The results include Bourgain's theorem stating that a separable Banach space containing every separable reflexive space is universal for all separable spaces and the Prus, Odell, and Schlumprecht construction of a reflexive separable space universal for all separable superreflexive spaces.

The rest of the chapter is devoted to the development of the geometrical theory of the Szlenk index and its various applications to general universality problems, classification of $C[0, \alpha]$ spaces, and renormings. In particular, we prove the result of Bessaga, Pełczyński, and Samuel that the isomorphism type of $C[0, \alpha]$ for countable α is determined by the space's Szlenk index. The connection to renormings is realized through the w^*-dentability index $\Delta(X)$

and a theorem of Bossard and Lancien that claims the existence of a universal function $\Psi : \omega_1 \to \omega_1$ such that $\Delta(X) \leq \Psi(\mathrm{Sz}(X))$.

2.1 Trees in Polish Spaces

We start by collecting some results on trees on Polish spaces. We refer to Kechris [Kech95] for a full background.

Let A be a nonempty set. Given $n \in \mathbb{N}$, the *power* A^n is the set of all sequences (also called *nodes*) $s := (s(0), \ldots, s(n-1))$ of length n of elements from A. If $m < n$, we let $s|m := (s(0), \ldots, s(m)) \in A^m$. In this situation, we say that $t := s|m$ is an *initial segment* of s and that s *extends* t, writing $t \leq s$. Two nodes are *compatible* if one is an initial segment of the other. For compatible nodes $t \leq s$, we introduce the *interval* $[t, s]$ as the set of all initial segments of s extending t.

Definition 2.1. *Let* $A^{<\omega} = \bigcup_{n=0}^{\infty} A^n$. *A* tree *on* A *is a subset* T *of* $A^{<\omega}$ *closed under initial segments. The relation* $\leq$ *defined above induces a partial ordering on* T.

Whenever convenient, we add a unique minimal element (*root*) to T, defined formally for $n = 0$ and denoted by $\varnothing$, that is the sequence of length zero. A *branch* of T is a linearly ordered subset of T, that is not properly contained in another linearly ordered subset of T. A tree is *well founded* if there is no infinite branch. For a well-founded tree, we inductively define an ordinal sequence of trees (T^α) on A as follows:

$$\begin{cases} T^0 := T, \\ T^{\alpha+1} := \{(x_1, \ldots, x_n); (x_1, \ldots, x_n, x) \in T^\alpha \text{ for some } x \in A\}, \\ T^\alpha := \bigcap_{\beta < \alpha} T^\beta \text{ for a limit ordinal } \alpha. \end{cases}$$

Since T is well founded, (T^α) is a strictly decreasing sequence, and thus $T^\alpha = \emptyset$ for some ordinal α. We define the ordinal index $o(T) := \min\{\alpha; T^\alpha = \emptyset\}$. We also define for $x \in A$

$$T_x := \{(x_1, \ldots, x_n); (x, x_1, \ldots, x_n) \in T\}.$$

For $s = (x_1, \ldots, x_n) \in A^n$, $t = (y_1, \ldots, y_m) \in A^m$, we define the *concatenation* $s^\frown t := (x_1, \ldots, x_n, y_1, \ldots, y_m) \in A^{n+m}$. Given trees $T \subset A^{<\omega}$, $S \subset B^{<\omega}$, we say that $\rho : S \to T$ is a *regular map* if it preserves the lengths of nodes and the partial tree ordering. If there exists an injective regular map from S into T, we say that S is *isomorphic to a subtree* of T. Then we have $o(S) \leq o(T)$. On the other hand, if there exists a surjective regular map from S onto T, then we have $o(S) \geq o(T)$.

Lemma 2.2. (i) $(T_x)^\alpha = (T^\alpha)_x$ *for every ordinal* α.

(ii) $o(T) = \sup_{x \in A}(o(T_x) + 1)$.

Proof. The first statement follows by a standard transfinite induction argument. If $x \in A$ is fixed and $\alpha < o(T_x)$, then $x \in T^{\alpha+1}$. Consequently, $x \in T^{o(T_x)}$, which implies that $o(T) \geq \sup_{x \in A}(o(T_x) + 1)$. The opposite inequality is also clear.   □

Lemma 2.3. *For every $\alpha < \omega_1$, there exists a well-founded tree T_α on $\mathbb{N}^{<\omega}$ satisfying $o(T_\alpha) = \alpha$.*

Proof. By induction on $\alpha < \omega_1$, we are going to construct a sequence (T_α) of trees on $[0, \alpha]^{<\omega}$ with $o(T_\alpha) = \alpha$. This is sufficient (using isomorphisms) since $|\alpha| = \omega$. Put $T_1 = \{(1)\}$. Having defined T_α for all $\alpha < \beta$, we put $T_{\alpha+1} = \{(\alpha + 1), (\alpha + 1)^\frown t : t \in T_\alpha\}$ if $\beta = \alpha + 1$ is nonlimit. Otherwise, choose a sequence $(\alpha_n) \nearrow \beta$ and put $T_\beta = \bigcup_{n=1}^{\infty}(\alpha_n) \cup \bigcup_{n=1}^{\infty}(\alpha_n)^\frown T_{\alpha_n}$. The desired properties follow by a standard argument.   □

It is not hard to prove by induction the following additional minimality property of the family (T_α): every tree S on $\mathbb{N}^{<\omega}$ with $o(S) = \alpha$ contains a subtree isomorphic to T_α [AJO05].

We put $T_n := T \cap A^n$. We say that a tree $T = \bigcup_{n=1}^{\infty} T_n$ on a topological space A is *closed*, if every $T_n \subset A^n$ is closed in the product topology. Similarly, if A is a Polish space, we say that T is *analytic* if every $T_n \subset A^n$ is analytic. We set $\pi_n : A^{n+1} \to A^n$ to be the projection onto the first n coordinates.

Lemma 2.4. *Let T be a closed tree on a Polish space A, and assume that $T_n = \overline{\pi_n(T_{n+1})}$ for all n. Then either $T = \emptyset$ or T is not well founded.*

Proof. Let ρ be the complete metric on A. If $T \neq \emptyset$, then there exists an infinite sequence $(x_1^n, \ldots, x_n^n) \in T_n$ such that $\rho(x_i^n, x_i^{n+1}) < \frac{1}{2^n}$, $1 \leq i \leq n$. Denote $y_k = \lim_{n \to \infty} x_k^n$, $k \in \mathbb{N}$. By our assumptions on T, $(y_1, \ldots, y_n) \in T$ for every n, which means that T is not well founded.   □

Theorem 2.5 (Kunen and Martin; see [Kech95]**).** *If T is a well-founded analytic tree on a Polish space A, then $o(T) < \omega_1$.*

Proof. Let us first assume that T is a closed tree. It is clear that $(T^{\alpha+1})_n = \pi_n((T^\alpha)_{n+1})$. By the separability of A, there exists some $\eta < \omega_1$ such that $\overline{(T^\eta)_n} = \overline{(T^{\eta+1})_n}$ for every $n \in \mathbb{N}$. Thus $\tilde{T} := \bigcup_{n=1}^{\infty} \overline{T_n^\eta}$ is a closed tree on A, satisfying $\tilde{T}_n = \overline{T_n^\eta} = \overline{\pi_n(\tilde{T}_{n+1})}$ for all $n \in \mathbb{N}$. Since T is closed, $\tilde{T} \subset T$, and thus $\tilde{T}$ is well founded. By Lemma 2.4, $\tilde{T} = T^\eta = \emptyset$. The proof for closed trees is complete.

The analytic case. The fact that T is analytic can be stated equivalently as follows. Denote by $\mathcal{N}$ the Polish space $\mathbb{N}^{\mathbb{N}}$, isometric to the set of all irrational numbers. Then there exists for every $n \in \mathbb{N}$ a continuous and onto mapping $\Pi_n : \mathcal{N} \to T_n$.

We now define a tree $\tilde{T}$ on the Polish space $A \times \mathcal{N}^{\mathbb{N}}$ as follows:

$$\tilde{T}_n := \{((a_1, b_1), \ldots, (a_n, b_n)); (a_1, \ldots, a_n) \in T, b_k \in \mathcal{N}^{\mathbb{N}},$$

$$(\forall 1 \le k \le n)(\forall 1 \le j \le k), b_k = (b_k^i)_{i \in \mathbb{N}}, b_k^j \in \Pi_j^{-1}(a_1, \ldots, a_j)\}.$$

It is standard to check that $\tilde{T}$ is indeed a tree, which is, moreover, closed. Also, $\tilde{T}$ is clearly well founded, and $\rho : \tilde{T} \to T$, $\rho((a_1, b_1), \ldots, (a_n, b_n)) = (a_1, \ldots, a_n)$ is a surjective regular map. Consequently, $o(T) \le o(\tilde{T}) < \omega_1$. □

We will also use an equivalent reformulation of the Kunen-Martin theorem, which uses the notion of partial ordering. Let $(A, \prec)$ be a partially ordered set. We can define an ordinal process $(A^{(\alpha)})$ as follows:

1. $A^{(0)} = A$.
2. $A^{(\alpha)} = \{a \in A; (\forall \beta < \alpha)(\exists b \in A^{(\beta)}; a \prec b)\}$.

Based on this, we may introduce $\mathrm{rank}_\prec(a) := \sup\{\alpha + 1; a \in A^{(\alpha)}\}$ and $\mathrm{rank}_\prec(A) := \sup_{a \in A}(\mathrm{rank}_\prec(a) + 1)$. In correspondence with $(A, \prec)$, we introduce the tree $T_A \subset A^{<\omega}$ such that $(a_1, \ldots, a_n) \in T_A$ if and only if $a_1 \prec \cdots \prec a_n$. It is standard to check that $\mathrm{rank}_\prec(A) = o(T_A)$. We say that $\prec$ is *well founded* if and only if there exists no infinite sequence $(a_i)_{i=1}^\infty \subset A$ such that $a_i \prec a_{i+1}$. Clearly, this property is characterized by T_A being well founded.

Theorem 2.6 (Hillard, Dellacherie [Della77]). *Let A be an analytic subset of a Polish space with a well-founded partial ordering $\prec$ such that $\prec \subset A^2$ is an analytic set. Then $\mathrm{rank}_\prec(A) < \omega_1$.*

Proof. Consider the set $\mathcal{N} \times A^2$ with the natural projection $\Pi_2 : \mathcal{N} \times A^2 \to A^2$. Since $\prec \subset A^2$ is analytic, it can be represented by a closed set $S_\prec \subset \mathcal{N} \times A^2$ such that $a \prec b$ iff there exists some $\tilde{a} \in S_\prec$ for which $\Pi_2(\tilde{a}) = (a, b)$. Since A is analytic, there exists a continuous surjective map $\phi(\mathcal{N}) \to A$. We can therefore introduce an analytic partial ordering $(\mathcal{N}, \prec)$ by the property $p, q \in \mathcal{N}$, $p \prec q$ if and only if $\phi(p) \prec \phi(q)$. The analyticity is witnessed by the closed set $(\mathrm{id}^{-1} \times \phi^{-1})(S_\prec) \subset \mathcal{N} \times \mathcal{N}^2$. The new partial ordering preserves the well-foundedness and rank. In particular, in our theorem, we may without loss of generality assume that $A = \mathcal{N}$ is a Polish space. Thus T_A is a well-founded tree on a Polish space. Let us see that T_A is an analytic tree. Consider the analytic set in A^{2n-1} (a set is *analytic* if it is the continuous image of $\mathbb{N}^{\mathbb{N}}$)

$$B_n := \{(a_1, \ldots, a_n, b_1, \ldots, b_{n-1}); a_i \prec b_i, b_i = a_{i+1} \text{ for all } 1 \le i \le n - 1\}.$$

We have

$$(T_A)_n = \{(a_1, \ldots, a_n); a_1 \prec \cdots \prec a_n\} = \pi_n(B_n),$$

which is clearly an analytic set. Thus Theorem 2.5 implies that $o(T_A) < \omega_1$. The rest follows by the remarks above. □

2.2 Universality for Separable Spaces

Definition 2.7. *Let P be a property of a Banach space. We say that a Banach space X is* universal for property P *if every Banach space with property P is isomorphic to a subspace of X.*

We say that X is isometrically universal for property P *if every Banach space with property P is isometric to a subspace of X.*

We say that X is complementably universal for property P *if every Banach space with property P is isomorphic to a complemented subspace of X.*

We say that a Schauder basis $\{x_n\}_{n=1}^\infty$ is universal *for a family of Schauder bases with property P if every member of the family is equivalent to a subsequence of $\{x_n\}_{n=1}^\infty$.*

The fundamental result regarding universality is the Banach-Mazur theorem (see, e.g., [Fa~01, Thm. 5.17] for a proof).

Theorem 2.8 (Banach, Mazur). *The space $C[0,1]$ is isometrically universal for all separable Banach spaces.*

We are going to present more specialized results on the existence of universal spaces regarding in particular the roles of reflexivity and complementability.

Theorem 2.9 (Pełczyński [Pelc69]). *The family of all unconditional Schauder bases has a complementably universal element.*

Proof (Schechtman [Sche75]). Let $(x_n)_{n=1}^\infty$ be a dense sequence in $C[0,1]$. We define a norm $\|\cdot\|_U$ on c_{00} by the formula

$$\|(a_i)\|_U = \sup\left\{\left\|\sum_{i=1}^\infty \varepsilon_i a_i x_i\right\|; |\varepsilon_i| \leq 1\right\}.$$

It is clear that the completion of $(c_{00}, \|\cdot\|_U)$ is a Banach space U with an unconditional Schauder basis, precisely $\{e_i\}_{i=1}^\infty$. Given any unconditional Schauder basis $\{z_n\}_{n=1}^\infty$ of a Banach space Z, the universality of $C[0,1]$ together with the basis perturbation lemma (see, e.g., [Fa~01, Thm. 6.18]) implies that $\{z_n\}_{n=1}^\infty \sim \{x_{i_n}\}_{n=1}^\infty$ for some increasing sequence of integers $(i_n)_{n=1}^\infty$. Clearly, $\{z_n\}_{n=1}^\infty \sim \{x_{i_n}\}_{n=1}^\infty \sim \{e_{i_n}\}_{n=1}^\infty \in U$, which finishes the proof. □

Theorem 2.10 (Pełczyński, Kadets). *There exists a separable Banach space V with a Schauder basis that is complementably universal for the family of all Schauder bases. Moreover, V is complementably universal for all separable Banach spaces with the bounded approximation property (BAP).*

Proof (Schechtman). The basis case is due to Pełczyński [Pelc69]. The BAP case (due to Kadets [Kad71]) follows using a result of Johnson, Rosenthal, and

Zippin [JRZ71] and Pełczyński [Pelc71] (see [LiTz77, Thm 1.e.13]), claiming that every separable Banach space with BAP is complemented in a Banach space with a Schauder basis.

Let $(x_n)_{n=1}^\infty$ be a dense sequence in $C[0,1]$. Let $(T, \leqslant) = \mathbb{N}^{<\omega}$ be the fully branching tree consisting of all finite sequences of integers. We also fix a linear ordering on $(T, \preccurlyeq)$ isomorphic to the usual ordering of $(\mathbb{N}, \leqslant)$, which satisfies the property $s \leqslant t \Rightarrow s \preccurlyeq t$ for all $s, t \in T$. To every node $s = (s(0), \ldots, s(n)) \in T$, we assign an element $x_s = x_{s(n)}$. We proceed to define the space V, which is the completion of $c_{00}(T)$, under the norm

$$\|x\|_V = \sup \left\{ \left\| \sum_{t \in I} a_t x_t \right\| ; I = [\varnothing, s], s \in (T, \leqslant) \right\}.$$

It is clear from the relation $s \leqslant t \Rightarrow s \preccurlyeq t$ that the sequence of unit vectors $(\chi_t)_{t \in (T, \preccurlyeq)}$, where the nodes are linearly ordered using $\preccurlyeq$, is a Schauder basis of V. It is also clear that all subsequences of $(\chi_t)_{t \in (T, \preccurlyeq)}$ whose indexing nodes correspond to a branch of T span a complemented subspace of V.

Given any basic sequence $\{z_n\}_{n=1}^\infty$, the universality of $C[0,1]$ together with the basis perturbation lemma (see, e.g., [Fa~01, Thm. 6.18]) implies that $\{z_n\}_{n=1}^\infty \sim \{x_{t_n}\}_{n=1}^\infty$ for some increasing sequence of integers $(i_n)_{n=1}^\infty$. Clearly, $\{z_n\}_{n=1}^\infty \sim \{x_{i_n}\}_{n=1}^\infty \sim \{\chi_{(i_1, \ldots, i_n)}\} \in V$, where $\{(i_1, \ldots, i_n)\}_{n=1}^\infty$ is a branch of T, which finishes the proof.   □

The next theorem shows that the restriction to spaces with a Schauder basis, or at least some form of approximation property, is necessary in Pełczyński's theorems above.

Theorem 2.11 (Johnson and Szankowski [JoSz76]). *There is no separable Banach space complementably universal for all separable superreflexive spaces.*

Proof (Sketch). The starting point of our proof is the result of Davie, extending the fundamental result of Enflo (see [LiTz77]), claiming that ℓ_p, $2 < p < \infty$, has a subspace E_p (necessarily superreflexive) that fails the *compact approximation property* (CAP). A Banach space X has CAP if the identity operator $\mathrm{Id} \in \mathcal{L}(X)$ (the space of all bounded operators from X into X) lies in $\overline{\mathcal{C}(X)}^\tau$, where $\mathcal{C}(X)$ denotes the space of all compact operators from X into X and τ is the topology of uniform convergence on compact sets in X. By contradiction, we assume that there exists a separable Banach space X complementably universal for all E_p. Since $(2, \infty)$ is an uncountable set, a standard argument using compactness and perturbation provides the existence of an uncountable $S \subset (2, \infty)$, $\{x_1, \ldots, x_N\} \subset X$ and $K, \varepsilon > 0$ such that for all $s \in S$:

1. There exists a K-complemented subspace $E_s \cong Z_s \hookrightarrow X$, $Q_s : X \to Z_s$, $\|Q_s\| < K$.

2. $\{x_1, \ldots, x_N\} \subset Z_s$.

3. For every compact operator, $T \in \mathcal{C}(Z_s)$ such that $T(x_i) = x_i$, $\|T\| > K^2$.

By the proof of Pitt's theorem ([Fa~01, Prop 6.25]), $\mathcal{L}(Z_r, Z_p) = \mathcal{C}(Z_r, Z_p)$ whenever $p < r$, where $\mathcal{L}(X, Y)$ denotes the space of all bounded operators from X into Y. Consider $T = Q_r \circ Q_p \upharpoonright Z_r \in \mathcal{C}(Z_r)$, $p < r$, $p, r \in S$. We get $\|T\| < K^2$, but $T(x_i) = x_i$, a contradiction. $\square$

Let $(X, \|\cdot\|_X)$, and $(Y, \|\cdot\|_Y)$ be Banach spaces. Let $(y_n)_{n=1}^{\infty}$ be a sequence in Y, $Y = \overline{\mathrm{span}}\{y_n; n \in \mathbb{N}\}$, $0 < \varepsilon < 1$. Consider the tree $T := T(X, \{y_n\}_{n=1}^{\infty}, \varepsilon) \subset X^{<\omega}$ consisting of all finite sequences $(x_1, \ldots, x_n) \in X^n$ that satisfy

$$\varepsilon \left\| \sum_{k=1}^{n} a_k y_k \right\|_Y \leq \left\| \sum_{k=1}^{n} a_k x_k \right\|_X \leq \varepsilon^{-1} \left\| \sum_{k=1}^{n} a_k y_k \right\|_Y \quad \text{for all } a_k.$$

Clearly, $o(T(X, \{y_n\}_{n=1}^{\infty}, \varepsilon)) \leq o(T(X, \{y_n\}_{n=1}^{\infty}, \rho))$ whenever $0 < \rho < \varepsilon$. It is clear that T is a closed tree, so by Theorem 2.5 we have the following

Proposition 2.12. *Let $(y_n)_{n=1}^{\infty}$ be a sequence in a Banach space Y such that $Y = \overline{\mathrm{span}}\{y_n; n \in \mathbb{N}\}$, and let X be a Banach space. The following are equivalent.*

(i) $Y \hookrightarrow X$.
(ii) *There exists $\varepsilon > 0$ such that $T = T(X, \{y_n\}_{n=1}^{\infty}, \varepsilon)$ is not well founded.*

Moreover, if X is separable, these conditions are also equivalent to:

(iii) *There exists $\varepsilon > 0$ such that $o(T(X, \{y_n\}_{n=1}^{\infty}, \varepsilon)) = \omega_1$.*

Lemma 2.13. *Let $(X, \|\cdot\|_X)$, $(Y, \|\cdot\|_Y) = \overline{\mathrm{span}}\{y_n; n \in \mathbb{N}\}$ and $(Z_\alpha, \|\cdot\|_\alpha)$, $\alpha < \omega_1$ be separable Banach spaces. Suppose that $o(T(Z_\alpha, \{y_n\}_{n=1}^{\infty}, \varepsilon_\alpha)) \geq \alpha$, for some $\varepsilon_\alpha > 0$, and there exist isomorphic embeddings $i_\alpha : Z_\alpha \hookrightarrow X$ for all $\alpha < \omega_1$. Then $Y \hookrightarrow X$.*

Proof. Without loss of generality, the constants $0 < \varepsilon = \varepsilon_\alpha$ and $K > \|i_\alpha\|, \|i_\alpha^{-1}\|$ are uniform for all α. This implies that $T(Z_\alpha, \{y_n\}_{n=1}^{\infty}, \varepsilon_\alpha)$ is isomorphic to a subtree of $T(i_\alpha(Z_\alpha), \{y_n\}_{n=1}^{\infty}, \frac{\varepsilon}{K})$. Consequently,

$$o(T(X, \{y_n\}_{n=1}^{\infty}, \frac{\varepsilon}{K})) = \omega_1. \qquad \square$$

Theorem 2.14 (Bourgain [Bour80a]). *If a separable Banach space is isomorphically universal for all separable reflexive Banach spaces, then it is isomorphically universal for all separable Banach spaces.*

Proof. Every separable Banach space is contained in the universal space $C[0, 1]$, which has a (normalized, monotone) Schauder basis. In view of Lemma 2.13, it suffices to prove the next lemma.

Lemma 2.15. *Let $\{y_n\}_{n=1}^{\infty}$ be a normalized monotone Schauder basis of a separable Banach space Y. Then, for every $\alpha < \omega_1$, there exists a separable reflexive space Z_α such that $o(T(Z_\alpha, \{y_n\}_{n=1}^{\infty}, \frac{1}{2})) \geq \alpha$.*

Proof. Let S be any well-founded tree on $\mathbb{N}$. A finite system of intervals $\{S_k\}_{k=1}^n$ of S is called *admissible* if $S_k \cap S_l = \emptyset$, and $\min S_k$ and $\min S_l$ are incomparable, whenever $k \neq l$.

We introduce a new Banach space $Z(S,i)$, $i \in \mathbb{N}$, defined as a completion of $c_{00}(S)$ under the norm

$$\|z\|_{S,i} = \sup\left\{\left(\sum_{k=1}^n \left\|\sum_{s \in S_k} a_s y_{|s|+i}\right\|^2\right)^{\frac{1}{2}}\right\}, \quad z = \sum_{s \in S} a_s e_s,$$

where the supremum is taken over all admissible systems of intervals $\{S_k\}_{k=1}^n$ of S. We claim that $Z(S,i)$ is reflexive. We proceed by a countable induction in $o(S)$. If $o(S) = 1$, then S has only branches of length 1, which means that $Z(S,i) \cong \ell_2(S)$ for every $i \in \mathbb{N}$. It is clear that given a tree S and $\tilde{S} = t^\frown S$, for some $t \in \mathbb{N}$, we have that $Z(\tilde{S},i) \cong Z(S,i+1) \oplus \mathbb{R}$. Here we are using that $\{y_n\}$ is a Schauder basis. Next, we write $S = \bigcup_{n \in \mathbb{N}} (n)^\frown S_n$, where S_n are trees satisfying $o(S_n) < o(S)$. It is now clear that $Z(S,i) \cong \sum_{\ell_2} \oplus_{n=1}^\infty \mathbb{R} \oplus Z(S_n, i+1)$, so the inductive step is completed.

We will prove that S is isomorphic to a subtree of $T(Z(S,0), \{y_n\}_{n=1}^\infty, \frac{1}{2})$, which implies that $o(T((Z(S,0), \{y_n\}_{n=1}^\infty, \frac{1}{2}))) \geq o(S)$, and so using the trees T_α, $o(T_\alpha) = \alpha$ leads to the desired conclusion of the proof. The embedding $i: S \to T(Z(S,0), \{y_n\}_{n=1}^\infty, \frac{1}{2})$ is defined as $i(s) = (e_{s_1}, \ldots, e_{s_k})$, where $s = (s_1, \ldots, s_k)$ in the tree S, and $e_{s_i} = \chi_{s_i} \in c_{00}(S)$ is from $Z(S,0)$. We see that $i(\cdot)$ is injective and preserves the lengths of nodes, so it remains to verify that the range of $i(S)$ is indeed contained in $T(Z(S,0), \{y_n\}_{n=1}^\infty, \frac{1}{2})$. Let $s = (s_1, \ldots, s_m) \in S$, $a_i \in \mathbb{R}$, $1 \leq i \leq m$, $z = \sum_{i=1}^m a_i e_{s_i}$. Using the admissibility of S_k, we obtain

$$\|z\|_{S,0} = \sup\left\{\sum_{k=1}^n \left\|\sum_{s \in S_k} a_s y_{|s|}\right\|^2\right\}^{\frac{1}{2}} = \sup_{1 \leq p \leq r \leq m} \left\|\sum_{l=p}^r a_l y_l\right\|.$$

The monotonicity of $\{y_n\}$ implies that $\left\|\sum_{l=p}^r a_l y_l\right\| \leq 2\left\|\sum_{l=1}^m a_l y_l\right\|$. Putting things together, we obtain

$$\left\|\sum_{i=1}^m a_i y_i\right\| \leq \left\|\sum_{i=1}^m a_i e_{s_i}\right\| = \sup_{1 \leq p \leq r \leq m}\left\|\sum_{l=p}^r a_l y_l\right\| \leq 2\left\|\sum_{l=1}^m a_l y_l\right\|.$$

This finishes the proof of Lemma 2.15 $\qquad\qquad\qquad\qquad\qquad\qquad \square$

and therefore that of Theorem 2.14. $\qquad\qquad\qquad\qquad\qquad\qquad\qquad\qquad\quad \square$

The general method of constructing scales of Banach spaces with a growing ordinal index plays a fundamental role in many important results. We refer to [Od04] and [JoLi01h, Chap. 23] for more results and references in this area.

Corollary 2.16 (Szlenk [Szl68]). *There is no separable Asplund (in particular reflexive) Banach space universal for all separable reflexive Banach spaces.*

Recall that a Banach space is called an *Asplund space* if every separable subspace has a separable dual.

The original proof of Szlenk's theorem used the notion of a Szlenk index, which will be investigated in Sections 2.4 to 2.7.

Proposition 2.17 (Bourgain [Bour80a]). *There is no separable super-reflexive Banach space universal for all separable superreflexive Banach spaces.*

Proof (Sketch). In fact, there is no superreflexive Banach space X containing isomorphic copies of all ℓ_p, $1 < p < \infty$. By a standard argument, there exists a constant $K > 0$ and a sequence $(p_n) \searrow 1$ such that $(\ell_p, \|\cdot\|_p)$ is K-finitely representable in X. This easily leads to ℓ_1 being finitely representable in X, a contradiction with X having nontrivial type. $\square$

We are going to prove a result of Prus on the existence of a separable reflexive space with a Schauder basis universal for all superreflexive spaces with a Schauder basis (resp. an FDD). To this end, we need to recall some basic properties of an FDD. A separable Banach space X has a *finite-dimensional decomposition* (an FDD) $\{X_n\}_{n=1}^{\infty}$, where X_n are finite-dimensional subspaces of X, if for every $x \in X$ there exists a unique sequence (x_n) with $x_n \in X_n$, $n \in \mathbb{N}$, such that $x = \sum_{n=1}^{\infty} x_n$. FDD's share many properties with Schauder bases. With a fixed FDD, we will denote $\operatorname{supp}(x) := \{m, \ldots, n\}$ when $x = \sum_{i=m}^{n} x_i$, $x_i \in X_i$, $x_m \neq 0$, $x_n \neq 0$. We will write $x < y$ if x and y are consecutively supported. When working with a fixed FDD, we are going to use the convention that whenever we write $x = \sum_{n=1}^{\infty} x_n$, we automatically assume that $x_1 < x_2 < \ldots$ unless specified otherwise (we do not necessarily assume that $x_n \in X_n$). Let $x = \sum_{i=1}^{n} x_i$ and a sequence $1 = j_1 < j_2 < \cdots < j_l \leq n$ be given and put $y_k = \sum_{i=j_k}^{j_{k+1}-1} x_i$ for $k < l$ and $y_l = \sum_{i=j_l}^{n} x_i$. In this case, we see that $x = \sum_{k=1}^{l} y_k$ and we say that y_k is a *block* of the vector $x = \sum_{i=1}^{n} x_i$ and $x = \sum_{k=1}^{l} y_k$ is a *blocking* of $x = \sum_{i=1}^{n} x_i$. We also say that the blocking $x = \sum_{i=1}^{n} x_i$ is a *refinement* of the blocking $x = \sum_{k=1}^{l} y_k$.

We will also need the notion of skipped blocks. Starting from an FDD $\{X_n\}_{n=1}^{\infty}$, we say that $x_1 < x_2 < \ldots$ are *skipped blocks* if $\max \operatorname{supp}(x_i) + 1 < \min \operatorname{supp}(x_{i+1})$ for all $i \geq 1$.

To a given FDD $\{X_n\}_{n=1}^{\infty}$ of X we associate the *dual FDD* $\{X_n^*\}_{n=1}^{\infty}$, consisting of a sequence of dual spaces to X_n, which is canonically embedded into the dual X^*, in complete analogy with the coefficient functionals of the Schauder basis. For convenience, we will assume that all FDD's are bimonotone (i.e., all partial sum projections P_n together with $I - P_n$ have norm 1).

Definition 2.18. *Let* $1 \leq p, q \leq \infty$. *An FDD* $\{X_n\}_{n=1}^{\infty}$ *of* X *is said to satisfy a* lower p-estimate *if there exists* $A > 0$ *such that*

$$\left\| \sum_{i=1}^{m} x_i \right\| \geq A \left(\sum_{i=1}^{m} \|x_i\|^p \right)^{\frac{1}{p}} \quad \text{whenever } x_1 < \cdots < x_m.$$

Similarly, it is said to satisfy an upper q-estimate *if there exists* $B > 0$ *such that*

$$\left\| \sum_{i=1}^{m} x_i \right\| \leq B \left(\sum_{i=1}^{m} \|x_i\|^q \right)^{\frac{1}{q}} \quad \text{whenever } x_1 < \cdots < x_m.$$

We say that an FDD satisfies a (p, q)-estimate *if both of the above are valid. We also say that an FDD satisfies a* skipped (p, q)-estimate *if the previous statement is valid for all skipped block summations (with uniform constants* A, B*).*

Fact 2.19. *Suppose* $\frac{1}{p} + \frac{1}{p'} = 1$, $1 \leq p, p' \leq \infty$. *If* $\{X_n\}_{n=1}^{\infty}$ *satisfies a lower (resp. upper)* p-estimate, *then* $\{X_n^*\}_{n=1}^{\infty}$ *satisfies an upper (resp. lower)* p'-estimate.

Proof. Suppose first that $\{X_n\}_{n=1}^{\infty}$ has a lower p-estimate. Let $\|\sum_{n=1}^{\infty} x_n^*\| = 1$ ($x_1^* < x_2^* < \dots$) and choose $\|\sum_{n=1}^{\infty} x_n\| < 2$ (supp $x_n = $ supp x_n^*) such that $\langle \sum_{n=1}^{\infty} x_n^*, \sum_{n=1}^{\infty} x_n \rangle = \sum_{n=1}^{\infty} x_n^*(x_n) = 1$. We have, using Holder's inequality,

$$\frac{A}{2} \left(\sum_{n=1}^{\infty} \|x_n\|^p \right)^{\frac{1}{p}} \left\| \sum_{n=1}^{\infty} x_n^* \right\| \leq \frac{1}{2} \left\| \sum_{n=1}^{\infty} x_n^* \right\| \cdot \left\| \sum_{n=1}^{\infty} x_n \right\| \leq 1 =$$

$$= \sum_{n=1}^{\infty} x_n^*(x_n) \leq \left(\sum_{n=1}^{\infty} \|x_n\|^p \right)^{\frac{1}{p}} \left(\sum_{n=1}^{\infty} \|x_n^*\|^{p'} \right)^{\frac{1}{p'}}.$$

This implies $\frac{A}{2} \|\sum_{n=1}^{\infty} x_n^*\| \leq \left(\sum_{n=1}^{\infty} \|x_n^*\|^{p'} \right)^{\frac{1}{p'}}$, which finishes the first part due to homogeneity.

For the second statement, suppose that $\left(\sum_{n=1}^{\infty} \|x_n^*\|^{p'} \right)^{\frac{1}{p'}} = 1$. Using essentially the duality theory for ℓ_p spaces and the bimonotonicity of the FDD, we know that there exists $\sum_{n=1}^{\infty} x_n$ such that

$$\frac{1}{2} \left(\sum_{n=1}^{\infty} \|x_n\|^p \right)^{\frac{1}{p}} \left(\sum_{n=1}^{\infty} \|x_n^*\|^{p'} \right)^{\frac{1}{p'}} \leq \left\langle \sum_{n=1}^{\infty} x_n^*, \sum_{n=1}^{\infty} x_n \right\rangle = \sum_{n=1}^{\infty} x_n^*(x_n) = 1.$$

Since $\|\sum_{n=1}^{\infty} x_n\| \leq B \left(\sum_{n=1}^{\infty} \|x_n\|^p \right)^{\frac{1}{p}}$ and

$$\left\langle \sum_{n=1}^{\infty} x_n^*, \sum_{n=1}^{\infty} x_n \right\rangle \leq \left\| \sum_{n=1}^{\infty} x_n \right\| \cdot \left\| \sum_{n=1}^{\infty} x_n^* \right\|,$$

we again obtain the desired estimate. $\qquad \square$

Theorem 2.20 (James [Jam50]). *Let X be a Banach space with an FDD $\{X_n\}_{n=1}^{\infty}$ satisfying a (p, q)-estimate for some $1 < p, q < \infty$. Then X is reflexive.*

Proof. Using Fact 2.19, we see that both $\{X_n\}_{n=1}^{\infty}$ and its dual FDD $\{X_n^*\}_{n=1}^{\infty}$ satisfy nontrivial upper and lower estimates. Thus $\{X_n\}_{n=1}^{\infty}$ is easily seen to be both shrinking and boundedly complete (the notions are defined analogously to the case of Schauder bases; see also Definitions 1.12 and 1.13), and so James' characterization of reflexivity applies. □

The following theorem is a simple generalization of the results of Gurarii and Gurarii and those of James (formulated for a Schauder basis) (see, e.g., [Fa~01, Thm. 9.25]).

Theorem 2.21 (Gurarii and Gurarii, James *(See [Fa~01, Thm. 9.25])).* *Let X be a superreflexive Banach space with an FDD $\{X_n\}_{n=1}^{\infty}$. Then $\{X_n\}_{n=1}^{\infty}$ satisfies a (p, q)-estimate for some $1 < p, q < \infty$.*

We will need the following general construction. Let X be a space with an FDD $\{X_n\}_{n=1}^{\infty}$, and let $1 < p < \infty$. We let X_p be a completion of the FDD $\{X_n\}_{n=1}^{\infty}$, under a new (nonequivalent, in general) norm

$$\|x\|_p = \sup \left\{ \left(\sum_{i=1}^{n} \|y_i\|^p \right)^{\frac{1}{p}} ; x = \sum_{i=1}^{n} y_i, y_1 < \cdots < y_n \right\}.$$

It is clear that $\{X_n\}_{n=1}^{\infty}$ is an FDD satisfying a lower p-estimate for X_p, but $(X, \|\cdot\|)$ and $(X_p, \|\cdot\|_p)$ are not necessarily isomorphic.

Lemma 2.22 (Prus [Prus83]). *Let $1 \leq q < p < \infty$. Suppose that an FDD $\{X_n\}_{n=1}^{\infty}$ of X satisfies an upper q-estimate. Then an FDD $\{X_n\}_{n=1}^{\infty}$ of X_p satisfies an upper q-estimate as well.*

Proof. Denote by B the upper q-estimate constant of $\{X_n\}_{n=1}^{\infty}$. Let $x = \sum_{i=1}^{n} y_i = \sum_{k=1}^{m} x_k$ be any two blockings. We are going to show that

$$\left(\sum_{k=1}^{m} \|x_k\|^p \right)^{\frac{1}{p}} \leq \left(1 + 2^{\frac{1}{q}} B \right) \left(\sum_{i=1}^{n} \|y_i\|_p^q \right)^{\frac{1}{q}},$$

which implies the statement of the lemma using the definition

$$\|x\|_p = \sup \left\{ \left(\sum_{k=1}^{m} \|x_k\|^p \right)^{\frac{1}{p}} ; x = \sum_{k=1}^{m} x_k \right\}.$$

To this end, consider the coarsest blocking $x = \sum_{j=1}^{l} z_j$ refining both of the given ones. Put $N_1 = \{k; x_k = z_j$ for some $j\}$, $N_2 = \{1, \ldots, m\} \setminus N_1$. We have by the definition of $\|\cdot\|_p$

$$\left(\sum_{k \in N_1} \|x_k\|^p\right)^{\frac{1}{p}} \leq \left(\sum_{j=1}^{l} \|z_j\|^p\right)^{\frac{1}{p}} \leq \left(\sum_{i=1}^{n} \|y_i\|_p^p\right)^{\frac{1}{p}} \leq \left(\sum_{i=1}^{n} \|y_i\|_p^q\right)^{\frac{1}{q}}.$$

Using the upper q-estimate and the coarsest refinement property of $x = \sum_{j=1}^{l} z_j$,

$$\left(\sum_{k \in N_2} \|x_k\|^q\right)^{\frac{1}{q}} \leq B\left(2\sum_{i=1}^{n} \|y_i\|^q\right)^{\frac{1}{q}} \leq B\left(2\sum_{i=1}^{n} \|y_i\|_p^q\right)^{\frac{1}{q}}.$$

Summing up the two inequalities, we get

$$\left(\sum_{k \in N_1 \cup N_2} \|x_k\|^p\right)^{\frac{1}{p}} \leq \left(\sum_{k \in N_1} \|x_k\|^p\right)^{\frac{1}{p}} + \left(\sum_{k \in N_2} \|x_k\|^q\right)^{\frac{1}{q}}$$

$$\leq \left(1 + 2^{\frac{1}{q}} B\right) \left(\sum_{i=1}^{n} \|y_i\|_p^q\right)^{\frac{1}{q}}. \quad \square$$

Lemma 2.23 (Prus [Prus83]). *Let $1 < r < s < \infty$. There exists a reflexive space Y_s^r with an FDD satisfying an (s,r)-estimate, and complementably universal for all separable spaces with an FDD satisfying a (p,q)-estimate for some $r \leq q < p \leq s$.*

Proof. There exists a Banach space U with an FDD $\{U_n\}_{n=1}^{\infty}$, that is complementably universal for all FDD's. This follows by the same argument as in Theorem 2.10 for a Schauder basis. More precisely, for every FDD $\{X_n\}_{n=1}^{\infty}$ there exists a subsequence $\{U_{k_n}\}_{n=1}^{\infty} \sim \{X_n\}_{n=1}^{\infty}$ such that $\overline{\text{span}}\{U_{k_n}; n \in \mathbb{N}\}$ is a complemented subspace of U. Consider the space $U_{r'}$, $(\frac{1}{r} + \frac{1}{r'} = 1)$. It follows by Fact 2.19 that $\{U_n^*\}_{n=1}^{\infty}$ is an FDD for the space $Y^r = \overline{\text{span}}\{U_n^*\} \hookrightarrow U_{r'}^*$, which satisfies an upper r-estimate. Repeating the renorming once more, it follows that $\{U_n^*\}_{n=1}^{\infty}$ is an FDD for Y_s^r, which satisfies an (s,r)-estimate. Let X be a separable space with an FDD $\{X_n\}_{n=1}^{\infty}$ satisfying a (p,q)-estimate for some $r \leq q < p \leq s$. By Theorem 2.20, X is reflexive, so $\{X_n^*\}_{n=1}^{\infty}$ is an FDD for X^*. Since U is complementably universal, $X^* \cong Z = \overline{\text{span}}\{U_{k_n}; n \in \mathbb{N}\}_U$, where Z is complemented in U, and $\{U_{k_n}\}_{n=1}^{\infty} \sim \{X_n^*\}_{n=1}^{\infty}$. It is clear that $X \cong Z^* = \overline{\text{span}}\{U_{k_n}^*; n \in \mathbb{N}\}_{U^*} \hookrightarrow U^*$, and Z^* is complemented. Since $\{U_{k_n}\}_{n=1}^{\infty} \sim \{X_n^*\}_{n=1}^{\infty}$, $\{U_{k_n}\}_{n=1}^{\infty}$ satisfies a lower q'-estimate, $q' \leq r'$. It is easy to see from the definition, that $Z_{r'} \cong Z$. In particular, this implies that Z^* is canonically complemented in Y^r, by means of projection $Q\left(\sum_{n=1}^{\infty} u_n^*\right) = \sum_{n=1}^{\infty} u_{k_n}^*$, which preserves the FDD. Repeating the argument, $\{X_n\}_{n=1}^{\infty}$ satisfies a lower p-estimate, $p < s$, so $X \cong Z^*$ remains a complemented copy inside Y_s^r. $\quad \square$

Recall that a separable Banach space X has the *bounded approximation property* if there is a sequence $(T_n)_{n=1}^{\infty}$ of finite rank operators on X that converges strongly to the identity operator; i.e., $\| \cdot \|$- $\lim_n T_n(x) = x$ for every $x \in X$.

Theorem 2.24 (Prus [Prus83]). *There is a separable reflexive space U_P with a Schauder basis that is complementably universal for the class of all separable superreflexive spaces with the BAP (in particular, with a Schauder basis, resp. FDD).*

Proof. Consider the space $E := \left(\sum \oplus_{n=1}^{\infty} Y_{s_n}^{r_n} \right)_{\ell_2}$ for some sequences $(r_n) \searrow 1$, $(s_n) \nearrow \infty$. By Lemma 2.23, E has an FDD that is complementably universal for all FDD satisfying a (p, q)-estimate for some $1 < p, q < \infty$. In light of the Gurarii-James theorem (Theorem 2.21), E is universal for all superreflexive spaces with an FDD. To obtain the full statement of the theorem, one needs to invoke a classical result (in this setting due to Johnson [John71a] and Prus [Prus83]) claiming that a separable reflexive (resp. superreflexive) space with BAP is isomorphic to a complemented subspace of a reflexive (resp. superreflexive) space with a Schauder basis. We omit the proof of this fact and refer to [LiTz77, Th.1.e.13], for a general version of the result. Now put U_P as a reflexive space with a Schauder basis containing E as a complemented subspace. The result then follows by applying the Johnson-Prus result one more time. $\square$

Recently, Prus' result has been finalized by Odell and Schlumprecht [OdSc06], who removed the BAP assumption: *the reflexive space U_P is universal for all separable superreflexive spaces.* Note that, due to Theorem 2.11, we lose the complementability.

Recently, Godefroy showed in [Gode06] that *if a separable Banach space X contains an isometric copy of every strictly convex separable Banach space, then X contains an isometric copy of ℓ_1 equipped with its natural norm. In particular, the class of strictly convex separable Banach spaces has no universal element.* This provides a negative answer to a question asked by J. Lindenstrauss.

2.3 Universality of M-bases

We prove the nonexistence of a universal M-basis for separable Banach spaces, a result of Plichko based on the fundamental work of Banach (and generalized by Godun and Ostrovskij) on w^*-sequential closures of total subspaces in the dual.

Lemma 2.25. *Let X be a nonreflexive Banach space and $Z \hookrightarrow X^{**}$ be a finite-dimensional subspace such that $Z \cap X = \{0\}$. Then $Z_{\perp} \subset X^*$ is a norming subspace.*

Proof. Since $Z + X$ forms a topological sum in X^{**} (see, e.g., [Fa~01, Exer. 5.27]), we have $\inf_{x \in S_X} \mathrm{dist}(x, Z) = \delta > 0$. Suppose that there exists $x \in S_X$

such that $\sup_{f\in S_{Z_\perp}}|f(x)| < \frac{\delta}{2}$. Let $y \in X^{**}$ be an extension of $x \restriction Z_\perp$ that preserves the norm. Thus $x - y$ restricted to $Z_\perp$ is zero, so $x - y \in Z$. Thus $\mathrm{dist}(x, Z) \le \|y\| < \frac{\delta}{2}$, a contradiction. $\qquad\square$

Definition 2.26. *Let X be a separable Banach space and Y be a linear subspace of X^*. We define $Y_1 = \{x \in X^*, \exists \{y_n\}_{n=1}^\infty \subset Y : y_n \xrightarrow{w^*} x\}$. Inductively, we define $Y_{\beta+1} = (Y_\beta)_1$ for any ordinal β, and $Y_\beta = \bigcup_{\alpha<\beta} Y_\alpha$ if β is a limit ordinal.*

Proposition 2.27. *Let X be a separable Banach space and Y be a total subspace of X^*. Then there is a countable ordinal α such that $Y_\alpha = X^*$.*

Proof. The linear subspace $\tilde{Y} = \bigcup_{\alpha<\omega_1} Y_\alpha \subset X^*$ is sequentially closed in the w^*-topology. Since X is separable, by the Banach-Dieudonné theorem ([Fa˜01], Theorem 4.44), $\tilde{Y}$ is w^*-closed. Since Y is w^*-dense in X^*, we obtain $\tilde{Y} = X^*$. Let $\{f_i\}_{i=1}^\infty$ be w^*-sequentially dense in X^*, where $f_i \in Y_{\alpha_i}$ for $\alpha_i < \omega_1$. Choose β, $\alpha_i < \beta < \omega_1$ for all i. Given $f \in X^*$, it follows that $f \in Y_{\beta+1}$. Thus $X^* = Y_{\beta+1}$. $\qquad\square$

Recall that given a subspace $Y \hookrightarrow X^*$, we denote by

$$r(Y) = \inf_{x\in S_X} \sup_{f\in B_Y} |f(x)| \qquad (2.1)$$

the *(Dixmier) characteristic* of Y. Clearly, $r(Y) > 0$ if and only if Y is norming, and $r(Y) := \sup\{r \ge 0; rB_{X^*} \subset \overline{B_Y}^{w^*}\}$.

Proposition 2.28 (Banach). *Let X be a separable Banach space and $Y \hookrightarrow X^*$ be a closed total subspace. The following are equivalent:*

(i) *Y is norming.*
(ii) *$Y_1 = X^*$.*
(iii) *$Y_2 = Y_1$.*

Proof. (i)$\Rightarrow$(ii) Since X is separable, B_{X^*} is w^*-metrizable and thus $r(Y)B_{X^*}$ lies in Y_1. The use of linearity finishes the proof.

(ii)$\Rightarrow$(iii) is trivial.

(iii)$\Rightarrow$(ii) follows from Proposition 2.27.

(ii)$\Rightarrow$(i) Since X is separable, we have that the w^*-sequential closure of B_Y equals $\overline{B_Y}^{w^*}$. Since $X^* = Y_1 = \bigcup_{n\in\mathbb{N}} n\overline{B_Y}^{w^*}$, we have by Baire's theorem that $\overline{B_Y}^{w^*}$ has a nonempty $\|\cdot\|$-interior. $\qquad\square$

Lemma 2.29 (Godun [Godu77]). *Let X be a separable Banach space, $E \hookrightarrow X$, and $G \hookrightarrow E^*$ be a total subspace such that $G_\alpha \ne G_{\alpha+1}$ for some $\alpha < \omega_1$. Then there exists a total subspace $Y \hookrightarrow X^*$ such that $Y_\alpha \ne Y_{\alpha+1}$.*

Proof. Let $i : E \hookrightarrow X$ be the canonical injection. Set $Y = (i^*)^{-1}(G)$. We have $i^*(E^\perp) = 0$, so $E^\perp \hookrightarrow Y$. Considering separately the cases where $x \in i(Y)$ or

$Q(x) \neq 0$, where $Q : X \to X/E$ is the quotient mapping, we can easily find $f \in Y$, $f(x) \neq 0$, so Y is a total subset. Since i^* is w^*-continuous, $i^*(Y_\alpha) \subseteq G_\alpha$ for all $\alpha < \omega_1$. Assuming that $Y_\alpha = Y_{\alpha+1}$, we obtain from Proposition 2.27 that $Y_\alpha = X^*$. Thus $G_\alpha = Y^*$, a contradiction. $\qquad\square$

Lemma 2.30. *Let X be a separable Banach space and $E \hookrightarrow X$ a closed subspace. Let $Q : X \to Z = X/E$ be the quotient mapping. Then, for every closed subspace $G \hookrightarrow Z^*$ and $k \in \mathbb{N}$,*

$$Q^*(G_k) = (Q^*(G))_k.$$

Proof. The operator Q^* is a w^*-homeomorphism from Z^* onto a w^*-closed subspace $E^\perp \hookrightarrow X^*$. $\qquad\square$

Proposition 2.31 (Godun [Godu78]). *A separable Banach space X is reflexive if and only if $Y = Y_1$ for every closed linear subspace $Y \subset X^*$.*

Proof. The nontrivial implication follows from the fact that every nonreflexive space has a proper norming subspace of the dual, e.g., $Y = \mathrm{Ker}(f)$ for $f \in X^{**} \setminus X$ (see Lemma 2.25). From Proposition 2.28, $Y \neq Y_1 = X^*$. $\qquad\square$

Recall that a Banach space X is called *quasireflexive* if $\dim(X^{**}/X) < \infty$.

Proposition 2.32 (Godun [Godu78]). *A separable Banach space is quasireflexive if and only if $Y_1 = Y_2$ for every closed subspace $Y \hookrightarrow X^*$.*

Proof. ($\Rightarrow$). Assume first that $Y \hookrightarrow X^*$ is a total subspace. Thus $Y^\perp \cap X = \{0\}$, and since X is quasireflexive, $\dim Y^\perp < \infty$. By Lemma 2.25, Y is norming and so by Proposition 2.28, $Y_1 = Y_2$. In the general case, let $Q : X \to X/Y_\perp = Z$ be the quotient mapping. Clearly, $Q^*(Y)$ is a total subspace of Z^*. Recall that *a quotient of a quasireflexive space is quasireflexive.* Indeed, $Z^{**}/Z \cong Q^{**}(X^{**})/Q^{**}(X)$, and the latter space is finite-dimensional. Thus $Y_1 = Y_2 = Z^*$. Using Lemma 2.30, we obtain $Y_1 = Y_2 = (Y_\perp)^\perp$.

($\Leftarrow$) follows from the next theorem. $\qquad\square$

Lemma 2.33 (Godun [Godu77]). *Let X be a separable Banach space, $E \hookrightarrow X^*$ be a total subspace and $\alpha < \omega_1$ be a limit ordinal. If $E_\beta \neq E_{\beta+1}$ for all $\beta < \alpha$, then $E_\alpha \neq E_{\alpha+1}$.*

Proof. Set

$$V_\beta^n := \left\{ f \in X^*; f = w^*\text{-}\lim_{k \to \infty} f_k, \text{ where } f_k \in n B_{E_\beta} \right\}.$$

A standard diagonal argument (using a countable norm-dense sequence from X) shows that V_β^n is norm closed. We have $E_{\beta+1} = \bigcup_{n=1}^\infty V_\beta^n$, so $E_\alpha = \bigcup_{\beta < \alpha} \bigcup_{n=1}^\infty V_\beta^n$. Using Proposition 2.27, if $E_\alpha = E_{\alpha+1}$, then $X^* = E_\alpha$. By the Baire category theorem, there exists $\beta < \alpha$ and n for which V_β^n has a nonempty interior and so $E_{\beta+1} = X^*$, a contradiction. $\qquad\square$

The next theorem was proved by Banach for c_0, then generalized by Godun [Godu77], and finalized by Ostrovskij.

Theorem 2.34 (Ostrovskij [Ost87]). *Let X be a nonquasireflexive separable Banach space. Then, for every countable ordinal α, there is a total subspace $Y \hookrightarrow X^*$ such that $Y_\alpha \neq X^*$, $Y_{\alpha+1} = X^*$.*

Proof. Let X be a separable Banach space, $\dim X^{**}/X = \infty$. By a theorem of Davis and Johnson [DaJo73b], in a nonquasireflexive space, there exists a Schauder basic sequence $\{x_n\}_{n=0}^\infty$ in X with its dual sequence $\{x_n^*\}_{n=0}^\infty \subset X^*$ satisfying

$$\|x_n\| = 1, \ \sup_n \|x_n^*\| \leq M_1 < \infty,$$

$$\sup_{0 \leq j < \infty, j \leq k < \infty} \left\| \sum_{i=j}^k x_{i(i+1)/2+j} \right\| \leq M_1.$$

By Lemma 2.29, we may without loss of generality assume that $\{x_n\}_{n=0}^\infty$ is a Schauder basis of X with the dual sequence $\{x_n^*\}_{n=0}^\infty \subset X^*$. The sequence $\{x_n\}_{n=0}^\infty$ can be reindexed using double indices as follows:

$$\{x_{n,m}\}_{0 \leq n,m}, \text{ where } x_{n,m} = x_{(n+m)(n+m+1)/2+n}.$$

Choose $f_n \in \overline{\left\{ \sum_{m=0}^N x_{n,m} \right\}_{N=0}^\infty}^{w^*} \subset X^{**}$. Clearly, $\|f_n\| \leq M_1$. To continue the proof, we need the following lemma.

Lemma 2.35. *Given any vector $g_0 = af_j + x_{r,s} \in X^{**}$ with $a > 0$ and $r \neq j$, any ordinal $\alpha < \omega_1$, and any set $A \subset \mathbb{N}$ with $j, r \notin A$, there exist*

(i) *a countable set $\Omega(g_0, \alpha, A) \subset X^{**}$,*
(ii) $K(g_0, \alpha, A) := \bigcap_{h \in \Omega(g_0, \alpha, A)} \text{Ker}(h)$, *and*
(iii) $Q(b, g_0, \alpha, A) := \left\{ f; f = bx_{r,s}^* + u, u \in \text{span}\{x_{t,k}^*; t \in A \cup \{j\}\} \right\}$

with the following properties.

(1) $\Omega(g_0, \alpha, A) = \{h; h = a(h)f_{j(h)} + x_{r(h),s(h)}, \ j(h), r(h) \in A \cup \{j, r\}, a(h) > 0, \text{ and } h \neq g_0 \Rightarrow j(h), r(h) \neq r\}$.
(2) $K(g_0, \alpha, A)_\alpha \subset \text{Ker}(g_0)$.
(3) *If $f \in \text{Ker}(g_0) \cap \text{span } Q(b, g_0, \alpha, A)$, then*

$$f \in (Q(b, g_0, \alpha, A) \cap K(g_0, \alpha, A))_\alpha.$$

Proof. For (i), we proceed by induction in α. To step from α to $\alpha + 1$, write $A = \bigcup_{k=0}^\infty A_k$, where A_k are pairwise disjoint and infinite. Suppose $\sum_{i=1}^\infty \varepsilon_i < \infty$, where $\varepsilon_i > 0$. Choose a sequence $(g_n)_{n=1}^\infty \subset X^{**}$, $g_n = \varepsilon_n f_{p(n)} + x_{j,n}$ where $p : \mathbb{N} \to A_0$ is a bijection. Using the inductive hypothesis, we assume that $\Omega(g_n, \alpha, A_n)$ exists for all $n \in \mathbb{N}$. Put $\Omega(g_0, \alpha + 1, A) := \{g_0\} \cup \bigcup_{n=1}^\infty \Omega(g_n, \alpha, A_n)$. It is clear that (1) is true.

Let us show (2). We proceed by induction in β, for all $\beta \leq \alpha + 1$, to prove that $K(g_0, \alpha + 1, A)_\beta \subset \mathrm{Ker}(g_0)$. In the inductive step from β to $\beta + 1$, we need to establish that for every w^*-convergent sequence $(\tilde{y}_i)_{i=1}^\infty \subset M_2 B_{X^*} K(g_0, \alpha + 1, A)_\beta$, we have $\tilde{y} = w^*\text{-}\lim \tilde{y}_i \in \mathrm{Ker}(g_0)$. Using that $\{x_n^*\}_{n=1}^\infty$ is a w^*-basis, let us estimate the coefficients $\{\alpha(i)_n^j\}_{n=1}^\infty$ of $\tilde{y}_i$ using the inductive assumption, in particular $g_n(\tilde{y}_i) = 0$. We have $g_n(\tilde{y}_i) = \alpha(i)_n^j + \varepsilon_n f_{p(n)}(\tilde{y}_i)$, and so

$$|\alpha(i)_n^j| \leq \varepsilon_n M_1 M_2 \text{ for all } n \in \mathbb{N}.$$

Thus $\tilde{y}_i = u_i + v_i$, where the expansion of u_i does not contain $x_{r,s}^*$ or $x_{j,n}^*$, $n \in \mathbb{N}$, and $v_i = a_i x_{r,s}^* + \sum_{n=1}^\infty \alpha(i)_n^j x_{j,n}^*$. By the last estimate, the last summation converges in norm, and so $\tilde{y} = u + v$, where $u = w^*\text{-}\lim u_i$ and $v = w^*\text{-}\lim v_i$.

From $u_i \left(\sum_{k=1}^m \gamma_k x_{j,k} + \delta x_{r,s} \right) = 0$ and analogous relations for u that are true for all m, $\{\gamma_k\}_{k=1}^m$ and δ, we obtain that $g_0(u_i) = g_0(u) = 0$. Since $\tilde{y}_i \in \mathrm{Ker}(g_0)$, we have $v_i \in \mathrm{Ker}(g_0)$. It remains to show that $v \in \mathrm{Ker}(g_0)$. The last claim follows as $v = w^*\text{-}\lim_{i \to \infty} v_i$, $v_i \in \mathrm{Ker}(g_0)$, and $\{v_i\}_{i=1}^\infty$ is a norm-relatively compact set by the estimate above. This finishes the step from β to $\beta + 1$. The limit case for β is clear.

The proof of (3) follows from the inductive hypothesis combined with the following easy observations:

(a) Every vector from $Q(b, g_0, \alpha, A)$ has the form

$$v = b x_{r,s}^* + \sum_{k=1}^m \left(a_k x_{j,k}^* + u_k \right),$$

where $u_k \in \mathrm{span} \left\{ \{x_{t,l}^*\}; t \in A_k \cup \{p(k)\} \right\}$.

(b) Using a sliding hump argument, if $M = Q(b, g_0, \alpha, A) \cap \mathrm{Ker}(g_0)$, then $M_1 = Q(b, g_0, \alpha, A)$.

(c) For every $\{a_k\}_{k=1}^m \subset \mathbb{R}$, $b \in \mathbb{R}$, such that $g_0 \left(b x_{r,s}^* + \sum_{k=1}^m a_k x_{j,k}^* \right) = 0$, we have $Q(b, g_0, \alpha + 1, A) \cap K(g_0, \alpha + 1, A) \supset b x_{r,s}^* + \sum_{k=1}^m Q(a_k, g_k, \alpha, A_k) \cap K(g_k, \alpha, A_k)$.

(d) For every finite system of sets $M_i \subset X^*$, we have $\left(\sum_{i=1}^m M_i \right)_\alpha \supset \sum_{i=1}^m (M_i)_\alpha$.

The inductive step toward a limit ordinal $\alpha < \omega_1$. Choose an increasing sequence $\alpha_i \nearrow \alpha$.

Split $A = \bigcup_{k=0}^\infty A_k$, where A_k are pairwise disjoint and infinite. Suppose $\sum_{i=1}^\infty \varepsilon_i < \infty$, where $\varepsilon_i > 0$. Choose a sequence $(g_n)_{n=1}^\infty \subset X^{**}$, $g_n = \varepsilon_n f_{p(n)} + x_{j,n}$, where $p : \mathbb{N} \to A_0$ is a bijection. Suppose that $\Omega(g_n, \alpha_n, A_n)$ exist due to the inductive hypothesis. To show that put $\Omega(g_0, \alpha, A) = \{g_0\} \cup \bigcup_{n=1}^\infty \Omega(g_n, \alpha, A_n)$ is analogous to the nonlimit case, we omit the details. $\square$

Let us continue with the proof of Theorem 2.34. By (2) in Lemma 2.35, $K(g_0, \alpha, A)_\alpha \neq X^*$. Let us show that $K(g_0, \alpha, A)_{\alpha+1} = X^*$. Since $\{x_n\}_{n=1}^\infty$

is a Schauder basis, we have $f = w^*\text{-}\sum_{k=1}^{\infty} a_k x_k^*$ for every $f \in X^*$. Every functional $\tilde{f}_n = \sum_{k=1}^{n} a_k x_k^* - \frac{1}{a} g_0 \left(\sum_{k=1}^{n} a_k x_k^*\right) x_{j,n}^*$ can be written as $\tilde{f}_n = \tilde{f}_n^1 + \tilde{f}_n^2$, where $\tilde{f}_n^1 \in Q(b, g_0, \alpha, A) \cap \mathrm{Ker}(g_0)$ and $\tilde{f}_n^2 \in \mathrm{span}\{x_{t,k}^*; t \notin A \cup \{j\}, \{t, k\} \neq \{r, s\}\}$, and so $\tilde{f}_n^2 \in K(g_0, \alpha, A)$. Using 3, $\tilde{f}_n^1 \in K(g_0, \alpha, A)_\alpha$. Thus $\tilde{f}_n \in K(g_0, \alpha, A)_\alpha$. Finally, $w^*\text{-}\lim \tilde{f}_n = f$. $\qquad\square$

Theorem 2.36 (Plichko [Plic86a]). *There exists no universal countable M-basis. More precisely, there is no separable Banach space X with an M-basis $\{x_n; f_n\}_{n=1}^{\infty}$ such that for every Banach space Y with an M-basis $\{y_n; g_n\}_{n=1}^{\infty}$ there exists an isomorphism $T : Y \to X$ so that $T(y_n) = x_{k_n}$.*

Proof. Proceeding by contradiction, put $F = \overline{\mathrm{span}}\{f_n; n \in \mathbb{N}\}$. By Theorem 2.34, there is an ordinal $\alpha < \omega_1$ such that $F_\alpha = X^*$. Choose a countable ordinal $\beta > \alpha$ and a total subspace $G \hookrightarrow c_0^*$ so that $c_0^* \neq G_\beta$. By Lemma 1.21, c_0 has an M-basis $\{y_n; g_n\}_{n=1}^{\infty}$ such that $G = \overline{\mathrm{span}}\{g_n; n \in \mathbb{N}\}$. Let $T : c_0 \to X$ be an isomorphism such that $T(y_n) = x_{k_n}$. We have $T^*(F) \subset G$, $T^*(F_\alpha) \subset G_\alpha$, so $T^*(X^*) = c_0^* = G_\alpha$, a contradiction. $\qquad\square$

2.4 Szlenk Index

Historically, the result of Szlenk (Corollary 2.16) was proved in [Szl68] using the notion of the Szlenk index, which is rather geometrical, and defined using the w^*-topology of the dual ball. In this section, we will focus on a variant of the Szlenk index that plays an important role in many applications, most notably the structural theory of $C(K)$ spaces. Our approach will be mostly geometrical, in the spirit of Szlenk's work.

Definition 2.37. *Let X be an Asplund space and $B \subset X^*$ be a w^*-compact subset. Given $\varepsilon > 0$, put $B_\varepsilon^0 := B$. Proceed inductively to define B_ε^α: if α is an ordinal, put*

$$B_\varepsilon^{\alpha+1} := B_\varepsilon^\alpha \setminus \bigcup W \text{ for all } w^*\text{-open subsets } W \text{ of } B_\varepsilon^\alpha \text{ with } \mathrm{diam}\, W < \varepsilon.$$

If α is a limit ordinal, put

$$B_\varepsilon^\alpha := \bigcap_{\beta < \alpha} B_\varepsilon^\beta.$$

Clearly, all B_ε^α are w^*-compact sets. Assume that α is the least ordinal so that $B_\varepsilon^\alpha = \emptyset$. Then we define $\mathrm{Sz}_\varepsilon(B) := \alpha$. We define the Szlenk index $\mathrm{Sz}(B) := \sup_{\varepsilon > 0} \mathrm{Sz}_\varepsilon(B)$. In the case where $B = B_{X^*}$, we abuse the notation slightly by denoting $\mathrm{Sz}_\varepsilon(X) := \mathrm{Sz}_\varepsilon(B)$ and calling $\mathrm{Sz}(X) := \mathrm{Sz}(B)$ the Szlenk index of the space X.

The restriction of the definition of Szlenk index to the class of Asplund spaces is necessary. A well-known characterization of Asplund spaces (see, e.g., [DGZ93a, Thm. 5.2]) claims that a Banach space is Asplund if and only if every w^*-compact subset of the dual has a w^*-open subset of diameter less than ε, for every $\varepsilon > 0$. This is exactly the condition needed for the derivation process to end at some ordinal.

Fact 2.38. *The Szlenk index is an isomorphic invariant. Moreover,* $\mathrm{Sz}(Z) \leq \mathrm{Sz}(X)$ *whenever* Z *is a linear quotient space of* X.

The first statement follows readily from the definition. It is clear that $G \subset B \subset B_{X^*}$ implies that $\mathrm{Sz}(G) \leq \mathrm{Sz}(B)$. In particular, using the natural identifications, we immediately see that $\mathrm{Sz}(Z) \leq \mathrm{Sz}(X)$ whenever Z is a linear quotient space of X. Moreover, we have the next lemma.

Lemma 2.39 (Szlenk [Szl68]). *Let* X *be an Asplund space,* $Y \hookrightarrow X$. *Then* $\mathrm{Sz}(Y) \leq \mathrm{Sz}(X)$.

Proof. Denote $i : Y \to X$ as the embedding operator and fix $\varepsilon > 0$. Suppose that $P \subset B_{X^*}$, $S \subset B_{Y^*}$ are w^*-compact sets with $S \subset i^*(P)$. Then $S^1_\varepsilon \subset i^*(P^1_{\frac{\varepsilon}{2}})$. Indeed, if $s \in S^1_\varepsilon$, then there exists a net $S \ni s_\xi \xrightarrow{w^*} s$ such that $\|s_\xi - s\| \geq \frac{\varepsilon}{2}$. Choose a net $p_\xi \in P$, $i^*(p_\xi) = s_\xi$. Since P is w^*-compact, there exists a w^*-convergent subnet $p_\zeta \xrightarrow{w^*} p \in P$. Clearly, $i^*(p) = s$, and so we have that $\|p_\zeta - p\| \geq \frac{\varepsilon}{2}$. Thus $p \in P^1_{\frac{\varepsilon}{2}}$, and the claim follows. A standard induction argument now yields that $\mathrm{Sz}_\varepsilon(S) \leq \mathrm{Sz}_{\frac{\varepsilon}{2}}(P)$. Applying this result to $P = B_{X^*}$ and $S = B_{Y^*}$, the statement of the lemma follows. $\square$

The following is a useful tool for Szlenk index calculations.

Proposition 2.40 (Lancien [Lanc96]). *Let* X *be a Banach space and* α *an ordinal. Assume that*

$$\forall \varepsilon > 0 \ \ \exists \delta(\varepsilon) > 0 \ \ (B_{X^*})^\alpha_\varepsilon \subset (1 - \delta(\varepsilon)) B_{X^*}.$$

Then

$$\mathrm{Sz}(X) \leq \alpha.\omega.$$

Proof. Let $\varepsilon > 0$. An easy homogeneity argument shows that for any integer n such that $\left(1 - \delta(\varepsilon)\right)^{n-1} > 1/2$,

$$(B_{X^*})^{\alpha.n}_{2\varepsilon}(B_{X^*}) \subset \left(1 - \delta(\varepsilon)\right)^n B_{X^*}.$$

Consequently, there exists an integer N such that $(B_{X^*})^{\alpha.N}_{2\varepsilon} \subset \frac{1}{2} B_{X^*}$. Since $B_{X^*} \setminus \frac{1}{2} B_{X^*}$ contains a translate of $\frac{1}{2} B_{X^*}$, we get that $(\frac{1}{2} B_{X^*})^{\alpha.N}_{2\varepsilon} = \emptyset$ and therefore that $(B_{X^*})^{\alpha.N}_{4\varepsilon} = \emptyset$. This finishes the proof. $\square$

For a cardinal τ, τ^+ denotes the follower cardinal and cof τ stands for the cofinality of τ (i.e., $\min \alpha$; α an ordinal such that there exists a transfinite sequence $(\gamma_i)_{i \in \alpha}$, $\gamma_i \nearrow \tau$). Recall that a cardinal is called *regular* if cof $\tau = \tau$. See, e.g., [Je78].

Lemma 2.41. *Let X be an Asplund space of density τ. Then $\mathrm{Sz}(X) < \tau^+$.*

Proof. To this end, it suffices to show that $\mathrm{Sz}_\varepsilon(X) < \tau^+$ for every $\varepsilon > 0$. Indeed (in ZFC, [Je78, p. 27]), $\tau^+ > \omega$ is a regular cardinal, so $\mathrm{cof}\,\tau^+ > \omega$, and $\mathrm{Sz}_\varepsilon(X) \overset{\varepsilon\to 0}{\longrightarrow} \mathrm{Sz}(X)$. By [DGZ93a, Theorem 5.2], for every w^*-compact subset $K \subset X^*$, K_ε^α has a nonempty w^*-open subset disjoint with $K_\varepsilon^{\alpha+1}$. Thus $K \setminus K_\varepsilon^\alpha$ forms a strictly increasing long sequence of w^*-open subsets of K. Recall that K has a basis of the w^*-topology of cardinality τ, which implies that $K_\varepsilon^\alpha = \emptyset$ for some $\alpha < \tau^+$, as claimed. $\qquad\square$

In particular, $\mathrm{Sz}(X) < \omega_1$ for all separable Asplund spaces.

Lemma 2.42 (Lancien [Lanc96]). *Let X be an Asplund space, α an ordinal, and $\varepsilon > 0$. Then*

$$\frac{1}{2}(B_{X^*})_\varepsilon^\alpha + \frac{1}{2}B_{X^*} \subseteq (B_{X^*})_{\frac{\varepsilon}{2}}^\alpha.$$

Proof. We have that $x^* \in (B_{X^*})_\varepsilon^1$ if and only if there exists a net $B_{X^*} \ni s_\xi \overset{w^*}{\to} x^*$ such that $\limsup_{\xi,\zeta}\|s_\xi - s_\zeta\| \geq \varepsilon$. Given any $z^* = \frac{1}{2}x^* + \frac{1}{2}y^*$, where $x^* \in (B_{X^*})_\varepsilon^1$, and $y^* \in B_{X^*}$, we see that the net $B_{X^*} \ni t_\xi = \frac{1}{2}s_\xi + \frac{1}{2}y^* \overset{w^*}{\to} z^*$, and $\limsup_{\xi,\zeta}\|t_\xi - t_\zeta\| \geq \frac{\varepsilon}{2}$. This completes the proof for $\alpha = 1$. The rest follows by a standard inductive argument. $\qquad\square$

Theorem 2.43 (Lancien [Lanc96]). *Let X be an Asplund space. Then there exists an ordinal α such that $\mathrm{Sz}(X) = \omega^\alpha$. If $\mathrm{Sz}(X) < \omega_1$, then α is countable.*

Proof. We first claim that $\mathrm{Sz}(X) > \omega^\alpha$ implies $\mathrm{Sz}(X) \geq \omega^{\alpha+1}$. Find $\varepsilon > 0$ and $x^* \in B_{X^*}$ such that $x^* \in (B_{X^*})_{2\varepsilon}^{\omega^\alpha}$. Using Lemma 2.42, $0 \in (B_{X^*})_\varepsilon^{\omega^\alpha}$, and again $\frac{1}{2}B_{X^*} \subseteq (B_{X^*})_{\frac{\varepsilon}{2}}^{\omega^\alpha}$. $(\frac{1}{2}B_{X^*})_{\frac{\varepsilon}{2}}^{\omega^\alpha} \subseteq (B_{X^*})_{\frac{\varepsilon}{2}}^{\omega^\alpha \cdot 2}$. Hence $0 \in (B_{X^*})_{\frac{\varepsilon}{2}}^{\omega^\alpha \cdot 2}$. Proceeding inductively, $0 \in (B_{X^*})_{\frac{\varepsilon}{2^n}}^{\omega^\alpha \cdot 2^n}$, so $\mathrm{Sz}(X) \geq \omega^{\alpha+1}$. The claim is proved.

Now let $\alpha = \inf\{\gamma; \mathrm{Sz}(X) \leq \omega^\gamma\}$. If α is a limit ordinal, then $\mathrm{Sz}(X) \geq \sup_{\beta<\alpha}\omega^\beta = \omega^\alpha$. So $\mathrm{Sz}(X) = \omega^\alpha$. If $\alpha = \beta + 1$, the claim implies that again $\mathrm{Sz}(X) = \omega^\alpha$. $\qquad\square$

Theorem 2.44 (Lancien [Lanc96]). *Let X be an Asplund space, $\alpha < \omega_1$. If $\mathrm{Sz}(X) > \alpha$ (resp. $\delta^*(X) > \alpha$), then there exists a separable $Y \hookrightarrow X$ with $\mathrm{Sz}(Y) > \alpha$ (resp. $\delta^*(X) > \alpha$).*

Proof. We define an auxiliary family of trees S_α, $\alpha < \omega_1$, on $\mathbb{N}^{<\omega}$ as follows. $S_0 = \{\emptyset\}$, $S_{\alpha+1} = \{\emptyset\} \cup \bigcup_{n=0}^\infty (n) \cup \bigcup_{n=0}^\infty (n)^\frown S_\alpha$ and $S_\alpha = \{\emptyset\} \cup \bigcup_{n=0}^\infty (n) \cup \bigcup_{n=0}^\infty (n)^\frown S_{\alpha_n}$ for α a limit ordinal, where $\{\alpha_n\} = \{\beta, \beta < \alpha\}$. One can verify inductively that $o(S_\alpha) = \alpha$. Denote $S_\alpha(s) = \{t \in \mathbb{N}^{<\omega}; s^\frown t \in S_\alpha\}$, $h_\alpha(s) = o(S_\alpha(s))$. Fix a bijection $\phi_\alpha : \omega \to S_\alpha$ such that $\phi^{-1} : (S_\alpha, \leq) \to (\mathbb{N}, \leq)$ is order-preserving. We are now going to identify a "skeleton" inside B_{X^*} responsible for the growth of $\mathrm{Sz}(X)$. The following lemma is needed.

Lemma 2.45. *Let $\varepsilon > 0$, $\alpha < \omega_1$, $x^* \in (B_{X^*})^\alpha_\varepsilon$. Then there exists a family $\{x^*_s\}_{s \in S_\alpha} \subseteq B_{X^*}$ and a separable $Y \hookrightarrow X$ such that*

(i) $x^*_\varnothing = x^*$.
(ii) $\forall s \in (S_\alpha)^1$, $\forall n \in \mathbb{N}$, $\|x^*_{s^\frown n} - x^*_s\|_Y > \frac{\varepsilon}{2}$.
(iii) $\forall s \in (S_\alpha)^1$, $x^*_{s^\frown n} - x^*_s \to 0$ in $\sigma(Y^*, Y)$.

Proof. We will construct, by induction in n, $\{x_n\}_{n=1}^\infty \subset B_X$ and $\{x^*_{\phi_\alpha(n)}\}_{n=1}^\infty \subset B_{X^*}$ such that

1. $x^*_{\phi_\alpha(0)} = x^*_\varnothing = x^*$
2. $\forall n \in \mathbb{N}, x^*_{\phi_\alpha(n)} \in (B_{X^*})^{h_\alpha(\phi_\alpha(n))}_\varepsilon$
3. $\forall n \geq 1, (x^*_{\phi_\alpha(n)} - x^*_{s_n})(x_n) > \frac{\varepsilon}{2}$, where $\phi_\alpha(n) = s_n^\frown k_n$ with $k_n \in \mathbb{N}$.
4. $\forall n \geq 2, 1 \leq k \leq n-1, |(x^*_{\phi_\alpha(n)} - x^*_{s_n})(x_k)| \leq \frac{1}{2n}$.

Assume we have constructed $x^*_{\phi_\alpha(k)}$ and x_k, $0 \leq k < n$ satisfying 1–4. There is $i_n < n$ such that $\phi_\alpha(n) = \phi_\alpha(i_n)^\frown k_n$. By the inductive hypothesis, $x^*_{\phi_\alpha(i_n)} \in (B_{X^*})^{h_\alpha(\phi_\alpha(i_n))}_\varepsilon$. Since $h_\alpha(\phi_\alpha(i_n)) \geq h_\alpha(\phi_\alpha(n)) + 1$, we have that $x^*_{\phi_\alpha(i_n)} \in (B_{X^*})^{h_\alpha(\phi_\alpha(n))+1}_\varepsilon$. So, for any w^*-neighborhood V of $x^*_{\phi_\alpha(i_n)}$, we have diam$V \cap (B_{X^*})^{h_\alpha(\phi_\alpha(n))}_\varepsilon > \varepsilon$. In particular, there exists $x^*_{\phi_\alpha(n)} \in (B_{X^*})^{h_\alpha(\phi_\alpha(n))}_\varepsilon$ such that $\|(x^*_{\phi_\alpha(n)} - x^*_{\phi_n(i_n)})\| > \frac{\varepsilon}{2}$, and $\forall 1 \leq k \leq n-1, |(x^*_{\phi_\alpha(n)} - x^*_{\phi_\alpha(i_n)})(x_k)| \leq \frac{1}{2^n}$. We conclude the inductive step by choosing $x_n \in B_X$ such that $(x^*_{\phi_\alpha(n)} - x^*_{\phi_\alpha(i_n)})(x_n) > \frac{\varepsilon}{2}$. To finish the proof of the lemma, put $Y = \overline{\text{span}}\{x_n; n \in \mathbb{N}\}$. $\square$

This finishes the proof of Theorem 2.44. $\square$

Proposition 2.46 (Lancien; see [Lanc06]**).** *Let X be an Asplund space. Then $\text{Sz}(X \oplus X) = \text{Sz}(X)$.*

Proof. It is clearly enough to show that $\text{Sz}(X \oplus X) \leq \text{Sz}(X)$. We may also assume that $X \oplus X$ is equipped with the norm $\|(x, x')\| = \|x\| + \|x'\|$. Then one can easily show that for any A and B weak*-compact subsets of X^* and for any $\varepsilon > 0$,

$$(A \times B)^1_\varepsilon \subset (A \times B^1_\varepsilon) \cup (A^1_\varepsilon \times B). \tag{2.2}$$

On the other hand, a straightforward transfinite induction argument yields that for any C and D weak*-compact subsets of $X^* \times X^*$,

$$\forall \varepsilon > 0, \quad \forall \alpha, \quad (C \cup D)^\alpha_\varepsilon \subset (C^\alpha_\varepsilon \cup D^\alpha_\varepsilon). \tag{2.3}$$

The next step is to show by transfinite induction that for any A and B w^*-compact subsets of X^*,

$$\forall \varepsilon > 0, \quad \forall \alpha \geq 0, \quad (A \times B)^{\omega^\alpha}_\varepsilon \subset (A \times B^{\omega^\alpha}_\varepsilon) \cup (A^{\omega^\alpha}_\varepsilon \times B). \tag{2.4}$$

The case $\alpha = 0$ is given by (2.2). Suppose now that the statement above is true for any $\beta < \alpha$. If α is a limit ordinal, then it is clearly also true for α.

So let us assume that $\alpha = \beta + 1$ and that the statement is true for β. Then it follows from an iterated application of (2.3) that

$$\forall n \in \mathbb{N}, \quad (A \times B)_\varepsilon^{\omega^\beta \cdot n} \subset \bigcup_{k=0}^{n} \left(A_\varepsilon^{\omega^\beta \cdot k} \times B_\varepsilon^{\omega^\beta \cdot (n-k)} \right). \tag{2.5}$$

Therefore, for any $(x^*, y^*) \in (B_{X^*} \times B_{X^*})_\varepsilon^{\omega^{\beta+1}}$, we have

$$\forall n \in \mathbb{N}, \quad \exists k(n) \leq n, \quad x^* \in (B_{X^*})_\varepsilon^{\omega^\beta \cdot k(n)}, \text{ and } y^* \in (B_{X^*})_\varepsilon^{\omega^\beta \cdot (n-k(n))}.$$

If $(k(n))_n$ is unbounded, then $x^* \in (B_{X^*})_\varepsilon^{\omega^{\beta+1}}$. Otherwise, $(n - k(n))_n$ is unbounded and $y^* \in (B_{X^*})_\varepsilon^{\omega^{\beta+1}}$. This finishes the inductive proof of (2.4).

Finally, we conclude the proof of Proposition 2.46 by combining (2.4) and Proposition 2.43. $\qquad\Box$

In order to characterize Asplund spaces X with $\mathrm{Sz}(X) = \omega$, we need the following definition.

Definition 2.47. *The norm $\| \cdot \|$ on a Banach space X is UKK* (uniformly w^*-Kadets-Klee) if for every $\varepsilon > 0$ there is $\delta > 0$ such that for every convergent net $f_\mu \to f$, $f_\mu \in S_{X^*}$ with $\|f_\mu - f\| \geq \varepsilon$, we have $\|f\| \leq 1 - \delta$.*

Theorem 2.48 (Knaust, Odell, and Schlumprecht [KOS99]). *Let X be a separable Banach space. The following are equivalent.*

(i) $\mathrm{Sz}(X) = \omega$.
(ii) X *admits an equivalent UKK* renorming.*
(iii) X *is a quotient of a Banach space Y that has a shrinking FDD $\{Y_n\}_{n=1}^\infty$ such that $\{Y_n^*\}_{n=1}^\infty$ has a $(p, 1)$-estimate for some $p < \infty$.*

In the case where X has a shrinking FDD $\{X_n\}_{n=1}^\infty$, the conditions above are equivalent to the existence of a $(p, 1)$-estimate for $\{X_n^*\}_{n=1}^\infty$, for some $p < \infty$.

For more precise results proved by a "coordinate-free approach", we refer to [GKL01].

Proof (Theorem 2.48). (ii)$\Rightarrow$(i) is clear since we have $(B_{X^*})_\varepsilon^1 \subset (1 - \delta) B_{X^*}$.
(iii)$\Rightarrow$(ii) Renorm Y^* equivalently using the formula

$$\|x^*\|_p = \sup \left\{ \left(\sum_{i=1}^{n} \|y_i^*\|^p \right)^{\frac{1}{p}} ; x = \sum_{i=1}^{n} y_i^*, y_1^* < \cdots < y_n^* \right\}.$$

Being a supremum over w^*-lsc seminorms, $\| \cdot \|_p$ is a dual norm. We need to verify that $\| \cdot \|_p$ is UKK*. To this end, suppose that $(x_k^*) \subset B_{(Y^*, \|\cdot\|_p)}$ is an ε-separated sequence, and let $x^* = w^*\text{-}\lim x_k^*$. Since the original FDD is shrinking, without loss of generality, x^* are finitely supported, and $x_k^* = x^* + v_k^*$, where x^*, v_k^* are consecutively supported and $\|v_k^*\|_p \geq \frac{\varepsilon}{2}$. It is clear that $\|x_k^*\|_p^p \geq \|x^*\|_p^p + \|v_k^*\|_p^p$, and consequently $\|x^*\|_p \leq (1 - \varepsilon^p)^{\frac{1}{p}}$, which is the desired conclusion. As a last step, renorm X using the quotient norm of $(Y, \|\cdot\|_p)$, so that $X^* \hookrightarrow Y^*$ is a w^*-continuous isometry. The UKK* property clearly passes to X.

The main result, namely the proof of (i)$\Rightarrow$(iii), is technically easier in the case where X has a shrinking FDD (where it suffices to show that the dual to the given FDD has a $(p, 1)$-estimate). We will proceed with the proof in this special case, which will require the following lemmas.

Given an FDD $\{X_i\}_{i=1}^\infty$ and $\varepsilon > 0$, we define the *index*

$$\mathrm{SI}(\{X_i\}_{i=1}^\infty, \varepsilon) := \sup \left\{ k; \; \exists \varepsilon \leq |a_i| \leq 1, \; i = 1, 2, \ldots, k \right.$$

$$\left. \text{and normalized blocks } x_1 < x_2 < \cdots < x_k \text{ such that } \left\| \sum_{i=1}^k a_i x_i \right\| \leq 1 \right\}.$$

Lemma 2.49 (James; see [Fa˜01, Thm. 9.25]**).** *Let $\{X_i\}_{i=1}^\infty$ be a bimonotone FDD, $\mathrm{SI}(\{X_i\}_{i=1}^\infty, \frac{1}{4}) = n_0 < \infty$. Then $\{X_i\}_{i=1}^\infty$ satisfies the $(p, 1)$-estimate for $p = \log_2(4n_0 + 1)$. More precisely, for every $m \in \mathbb{N}$ and every finite block sequence, we have $\|\sum_{i=1}^m x_i\| \geq \frac{1}{2} \left(\sum_{i=1}^m \|x_i\|^p \right)^{\frac{1}{p}}$.*

Proof. Let $\{e_i\}_{i=1}^\infty$ be a normalized block basis of $\{X_i\}_{i=1}^\infty$. It suffices to prove that, for all $(a_i) \in S_{\ell_p^m}$, we have $\| \sum_{i=1}^m a_i e_i \| \geq \frac{1}{2}$. Assuming the contrary, choose a minimal m and $(a_i) \in S_{\ell_p^m}$ with $\| \sum_{i=1}^m a_i e_i \| < \frac{1}{2}$. Since $\{e_i\}_{i=1}^\infty$ is bimonotone, $|a_i| < \frac{1}{2}$. Choose the smallest n_1 satisfying $\sum_{i=1}^{n_1} |a_i|^p \geq (\frac{1}{2})^p$. Choose inductively a maximal finite increasing sequence of integers n_i, $i = 1, \ldots k$, so that n_{l+1} is the smallest index such that $\sum_{i=n_l+1}^{n_{l+1}} |a_i|^p \geq (\frac{1}{2})^p$. Then

$$\left(\sum_{i=n_l+1}^{n_{l+1}} |a_i|^p \right)^{\frac{1}{p}} \in \left[\frac{1}{2}, \frac{2^{\frac{1}{p}}}{2} \right] \quad \text{for } 0 \leq l < k.$$

From the maximality of k, we have $(1 - (\frac{1}{2})^p)^{\frac{1}{p}} \leq \frac{1}{2} 2^{\frac{1}{p}} k^{\frac{1}{p}}$, which implies that $k \geq \frac{1}{2}(2^p - 1) = 2n_0$. Set $x_l = \sum_{i=n_l+1}^{n_{l+1}} |a_i| e_i$. By the minimality of m and $\sum_{i=n_l+1}^{n_{l+1}} |a_i|^p \geq (\frac{1}{2})^p$, we have $\|x_l\| \geq \frac{1}{4}$. Thus $\mathrm{SI}((X_i), \frac{1}{4}) \geq k \geq 2n_0$. This is a contradiction. $\square$

Lemma 2.50. *Let X be a Banach space with a shrinking FDD $\{X_i\}_{i=1}^\infty$, and $\mathrm{Sz}(X) = \omega$. Then there exists a blocking $\{Z_i^*\}_{i=1}^\infty$ of $\{X_i^*\}_{i=1}^\infty$ and $p \in (0, 1)$ such that $\{Z_i^*\}_{i=1}^\infty$ has a $(p, 1)$-estimate for skipped blocks.*

Proof. We will assume without loss of generality that the FDD is bimonotone. By Lemma 2.42, given $\varepsilon > 0$, $(B_X^*)_\varepsilon^n = \emptyset$ for some $n \in \mathbb{N}$. We seek a blocking $\{Z_i\}_{i=1}^\infty$ of $\{X_i\}_{i=1}^\infty$ such that the dual FDD $\{Z_i^*\}_{i=1}^\infty$ has the *skipped SI property* from Lemma 2.49, i.e.,

$$\sup\left\{k;\ \exists\ \varepsilon \le |a_i| \le 1,\ i = 1, 2, \ldots, k,\ \text{and a normalized}\right.$$

$$\left.\text{skipped block basis } \{x_i\}_{i=1}^\infty \text{ such that } \left\|\sum_{i=1}^k a_i x_i\right\| \le 1\right\} < n_\varepsilon.$$

Suppose $(B_{X^*})_{\frac14}^n \ne \emptyset$ and $(B_{X^*})_{\frac14}^{n+1} = \emptyset$. We construct an increasing sequence $J_0 = \{j_k^0\}_{k=1}^\infty \subset \mathbb{N}$ of indices, so that the following is satisfied. Whenever $x \in (B_{X^*})_{\frac14}^n$, $y \in B_{X^*}$, supp $(x) \subset [1, j_k^0]$, supp $(y) \subset [j_{k+1}^0, \infty)$ and $\|y\| \ge \frac14$, we have $x + y \notin (B_{X^*})_{\frac14}^n$. We construct J_0 by induction. Having found the initial values up to j_k^0, suppose, by contradiction, that for every $j > j_k^0$ there exist x_j, y_j violating the conditions with supp $(x_j) \subset [1, j_k^0]$ and supp $(y_j) \subset [j, \infty)$. By compactness, without loss of generality, $x_j = x$ is a constant sequence. We have $y_j \overset{w^*}{\to} 0$, so if $x + y_j \in (B_{X^*})_{\frac14}^n$, then $x \in (B_{X^*})_{\frac14}^{n+1}$, a contradiction. Next we repeat the argument inductively (in l) n times, creating sequences $J_n \subset \cdots \subset J_l \subset \ldots J_0$ with the following properties. Whenever $x \in (B_{X^*})_{\frac14}^{n-l} \setminus (B_{X^*})_{\frac14}^{n-l+1}$, supp $(x) \subset [1, j_k^l]$, $y \in B_{X^*}$, supp $(y) \subset [j_{k+1}^l, \infty)$ and $\|y\| \ge \frac14$, then $x + y \notin (B_{X^*})_{\frac14}^{n-l}$.

We construct $J_l = \{j_k^l\}_{k=1}^\infty$ by induction. Having found the initial values up to j_k^l, suppose by contradiction that for every $j > j_k^l$ there exist x_j, y_j violating the conditions with supp $(x_j) \subset [1, j_k^l]$ and supp $(y_j) \subset [j, \infty)$. By compactness, without loss of generality $x_j = x$ is a constant sequence. Again, we have $y_j \overset{w^*}{\to} 0$, so if $x + y_j \in (B_{X^*})_{\frac14}^{n-l}$, then $x \in (B_{X^*})_{\frac14}^{n-l+1}$, a contradiction. Consider now the blocking $Z_k^* = \sum_{i=j_k^n}^{j_{k+1}^n - 1} X_i^*$, $Z_0^* = \sum_{i=1}^{j_1^n - 1} X_i^*$. By construction, we have that whenever $y = \sum_{i=0}^{n+1} y_i$ is a skipped sum with $\|y_i\| \ge \frac14$, we have $y \notin B_{X^*} = (B_{X^*})_{\frac14}^0$.

Indeed, denote $m_j = \max\{l; \sum_{i=0}^j y_i \in (B_{X^*})_{\frac14}^l\}$. Then, by construction, $m_{j+1} < m_j$ is decreasing, and since $m_0 \le n$, we have $m_{n+1} < 0$. The rest follows from Lemma 2.49. □

Lemma 2.51 (Johnson [John77]). *Let $\{F_n\}_{n=1}^\infty$ be a boundedly complete FDD of X that satisfies the skipped $(p, 1)$-estimate for some p. Then there exists a further blocking $\{G_n\}_{n=1}^\infty$ of $\{F_n\}_{n=1}^\infty$ that satisfies a $(p, 1)$-estimate.*

Proof. Without loss of generality, $\{F_n\}_{n=1}^\infty$ is bimonotone. For a fixed sequence $\varepsilon_i \searrow 0$, $\sum_{i=1}^\infty \varepsilon_i < \varepsilon$, we find an increasing sequence of integers $(n_i)_{i=1}^\infty$ so that whenever $x = \sum_{j=n_i+1}^{n_{i+1}} x_i$, $\|x\| \le 1$ and $x_i \in F_i$, there exists some $n_i + 1 \le k \le n_{i+1}$ for which $\|x_k\| < \varepsilon_i$. This is done by induction as follows. Having found the initial values $1 = n_0 < n_1 < \cdots < n_l$, we claim that there exists a large enough n_{l+1} that satisfies the property. Assuming the contrary, there exists for every $n > n_l$ a sequence $(x_j^n)_{n_l < j < n}$ such that $x_j^n \in F_j$, $\|x_j^n\| > \varepsilon_{l+1}$, and $\|\sum_{j=n_l+1}^n x_j^n\| \le 1$. By compactness, without loss of generality, there exist

$x_j = \lim_{n\to\infty} x_j^n$ and clearly $\|\sum_{j=n_l+1}^{n} x_j\| \leq 1$ for every $n > n_l + 1$. This contradicts the bounded completeness of $\{F_n\}_{n=1}^{\infty}$.

We put $G_i = \overline{\text{span}}\{F_{n_i+1}, \ldots, F_{n_{i+1}}\}$. To check the desired property, assume that $x \in S_X$, $x = \sum_{n=1}^{\infty} x_n$, $x_n \in F_n$. Put $z_i = \sum_{j=n_i+1}^{n_{i+1}} x_j \in G_i$. There exists some $n_i + 1 \leq k_i \leq n_{i+1}$ such that $\|x_{k_i}\| < \varepsilon_i$. Put $y_i = \sum_{i=k_i+1}^{k_{i+1}-1} x_i$. Clearly $(y_i)_{i=1}^{\infty}$ is a skipped block sequence of $\{F_n\}_{n=1}^{\infty}$, so we have

$$\left\| \sum_{i=1}^{\infty} y_i \right\| \geq A \left(\sum_{i=1}^{\infty} \|y_i\|^p \right)^{\frac{1}{p}}.$$

Since $\|z_i\| \leq \|y_i\| + \|x_{k_i}\| + \|y_{i+1}\|$, we obtain

$$\left(\sum_{i=1}^{\infty} \|z_i\|^p \right)^{\frac{1}{p}} \leq \left(\sum_{i=1}^{\infty} (\|y_i\| + \varepsilon_i + \|y_{i+1}\|)^p \right)^{\frac{1}{p}}$$

$$\leq \left(\sum_{i=1}^{\infty} 2(2\|y_i\|)^p \right)^{\frac{1}{p}} + \left(\sum_{i=1}^{\infty} |\varepsilon_i|^p \right)^{\frac{1}{p}}$$

$$\leq \frac{4}{A} \left\| \sum_{i=1}^{\infty} y_i \right\| + \varepsilon \leq \frac{4}{A} \left\| \sum_{i=1}^{\infty} z_i \right\| + \frac{4\varepsilon}{A} + \varepsilon = \frac{4}{A} + \frac{4\varepsilon}{A} + \varepsilon.$$

This finishes the proof. □

The proof of the special case of Theorem 2.48 where X has a shrinking FDD now follows by combining the lemmas above. Indeed, if $\{X_i\}_{i=1}^{\infty}$ is a shrinking FDD of X, then $\{X_i^*\}_{i=1}^{\infty}$ is a boundedly complete FDD of X^*. In order to reduce the general case to this situation, one can rely on the classical result from [DFJP74], according to which every separable Asplund space X (X is Asplund since it has $\text{Sz}(X) < \omega_1$) is a quotient of a separable Asplund space Y with a shrinking Schauder basis. Of course, a priori we do not know that $\text{Sz}(Y) = \omega$, so the main point of the proof in the general case is to renorm (nonequivalently) the space Y again to get this extra condition while preserving the continuity of the quotient mapping (or control the continuity of $X^* \hookrightarrow Y^*$, which is equivalent). □

Prior to the results of Knaust, Odell, and Schlumprecht, the equivalence of (ii) and (iii) in Theorem 2.48 for spaces with an FDD was established by Dilworth, Girardi, and Kutzarova [DGK95] based on the fundamental work of Prus [Prus83], [Prus89].

The Szlenk index of a Banach space X is interpreted as the so-called "oscillation index" of the identity map from (B_{X^*}, w^*) into $(X^*, \|\cdot\|)$ in [KeLou90].

2.5 Szlenk Index Applications to Universality

In this section, we are going to present some universality results in the non-separable setting. Under GCH, there exists a universal Banach space (of type $C(K)$) for every density by a result of Yesenin-Volpin. Let us point out (without proof) that some results due to Shelah imply that the nonexistence of a universal Banach space of density ω_1 is also consistent. We apply the Szlenk index approach to generalize the negative reflexive separable result to all densities.

The next theorem requires the assumption of the Generalized Continuum Hypothesis (GCH); i.e., $\tau^+ = 2^\tau$ for all infinite cardinal τ.

Theorem 2.52 (Yesenin-Volpin [Y-V49] (GCH)). *Let τ be an uncountable cardinal. Then there is a compact space K of weight τ such that every Banach space of density τ is isometrically isomorphic to a subspace of $C(K)$.*

Proof (Sketch). It is clear using the Hahn-Banach theorem that the *topological weight* (the minimal cardinality of the basis of the topology) $w((B_{X^*}, w^*))$ coincides with dens X. So the statement of the theorem follows from the next topological statement: *for every uncountable cardinal τ there exists a universal compact K of weight τ*; i.e., a compact with the property that for every compact C with $w(C) \leq \tau$, there exists a continuous surjection (quotient) $\phi : K \to C$. It is well-known (Alexandroff) that every compact C with $w(C) = \tau$ is a quotient of some zero-dimensional compact $\tilde{C}$ of the same weight; recall that a space is zero-dimensional if there is a basis of the topology consisting of clopen sets, which, in the case of compact spaces, coincides with totally disconnected, i.e., it does not contain connected subspaces having more than one point, see [Eng77, Thm. 6.2.6]. Let us indicate a short argument for this. Fix a norm-dense set $\{x_\gamma\}_{\gamma < \tau} \subset B_{C(K)}$. Then $t \to \{x_\gamma(t)\}_{\gamma < \tau}$ is a homeomorphic embedding of C into $[0,1]^\tau$. Since $[0,1]$ is a quotient of $\{0,1\}^\omega$ (use the Cantor function mapping from $[0,1]$ onto itself, which is locally constant outside the Cantor discontinuum), we have upon reindexing that $[0,1]^\tau$ is a quotient of $\{0,1\}^\tau$. The rest follows by the simple fact that $\{0,1\}^\tau$ is totally disconnected and so is each of its closed subsets. Thus it suffices to prove that there exists a zero-dimensional compact of weight τ universal (in the sense above) for all zero-dimensional compacts of weight τ. According to the fundamental Stone representation theorem (see, for example, [Wal74, p. 51]), for every Boolean algebra B there exists a zero-dimensional compact K (with card B = weight K), such that the Boolean algebra of clopen sets of K is isomorphic to B. Thus the last statement is equivalent to the following: there exists a universal Boolean algebra of cardinality τ, i.e., every Boolean algebra of weight at most τ is isomorphic to its subalgebra. To prove the theorem, it therefore suffices to construct a universal Boolean algebra for every cardinality $\tau > \omega$. We identify τ with the least ordinal of the same cardinality. Let $T = \{0,1\}^\tau$. Given $t = (a_i)_{i<\tau} \in T, \alpha < \tau$, define $t_\alpha = (a_i)_{i<\alpha}$. Given α, the set of all t_α is naturally identified with

$\{0,1\}^\alpha$. The Boolean algebra $B \subset 2^T$ consists of all subsets of T of the form $S = \{b \in T; \exists \alpha < \tau, U \subset \{0,1\}^\alpha, b_\alpha \in U\}$. It is easy to verify that this is a correct definition and that, using GCH, card $B = \tau$. Given any Boolean algebra H of cardinality τ, we need to find an embedding into B. By Stone's theorem, we may assume that H consists of all clopen sets of some zero-dimensional compact C. Suppose $\{(A_i^0, A_i^1)\}_{i<\tau}$ $(A_i^0 = C \setminus A_i^1)$ is a long sequence containing all pairs of complementary clopen sets in C, $A_0^0 = \emptyset$. We consider the set $K \subset \{0,1\}^\tau$, which consists of all sequences $(a_i)_{i<\tau}$ such that $\bigcap_{i<\tau} A_i^{a_i} \neq \emptyset$. This set is a compact embedded into the compact $\{0,1\}^\tau$. Indeed, if $(a_i)_{i<\tau} = f \notin K$, then $\bigcap_{i<\tau} A_i^{a_i} = \emptyset$. Since $A_i^{a_i}$ are closed subsets of a compact C, there exists a finite subset of indices $\{a_{i_1}, \ldots, a_{i_l}\}$ such that $\bigcap_{1 \leq j \leq l} A_{i_j}^{a_{i_j}} = \emptyset$. These indices determine the open neighborhood of f disjoint with K. Next we define a surjective mapping $\Phi : \{0,1\}^\tau \to K$ as follows. If $f \in K$, then $\Phi(f) = f$. If $f \notin K$, let $\alpha = \sup\{\beta; (\exists y \in K)(\forall i < \beta) \; y_i = f_i\}$. Since K is compact, there exists some $y \in K$ such that $y_i = f_i$ for all $i < \alpha$ and $y_\alpha \neq f_\alpha$. We define $\Phi(f) = y$, and, moreover, we choose our mapping so that $f_i = g_i, \forall i \leq \alpha$ implies that $\Phi(f) = \Phi(g)$. The embedding $e : H \to B$ is defined by $e(A_i^\varepsilon) = \{f; \Phi(f)(i) = \varepsilon\}$ for $\varepsilon = 0, 1$, $e(A_0^0) = \emptyset, e(A_0^1) = T$. It remains to verify that this definition is correct and represents a homomorphism. It is clear from the definition that Φ preserves $0, 1$ and the complements, so it remains to check that it also preserves $\wedge, \vee$ operations. This follows readily from the definition. $\qquad\square$

It is a classical result that, under CH, $\beta\omega \setminus \omega$ is a universal compact space of weight ω_1. On the other hand, Shelah has shown that it is consistent that there exists no universal compact of weight ω_1. For these results, see, e.g., [Bell00].

Let K be a scattered topological space. Recall the classical notion of *Cantor-Bendixon derivation*,

$$K' = K \setminus \{p; p \text{ is an isolated point in } K\},$$

its ordinal continuation

$$K^{(\alpha+1)} = K^{(\alpha)} \setminus \{p; p \text{ is an isolated point in } K^{(\alpha)}\},$$

and

$$K^{(\alpha)} = \bigcap_{\beta < \alpha} K^{(\beta)} \text{ for } \alpha \text{ a limit ordinal.}$$

We define $\chi(K) = \max\{\alpha; K^{(\alpha)} \neq \emptyset\}$, $n(K) = \text{card } K^{(\chi(K))} \in \mathbb{N}$. The *height* $\eta(K)$ of K is the least ordinal β for which the Cantor derivative $K^{(\beta)}$ is empty (i.e., $\eta(K) = \chi(K) + 1$).

Lemma 2.53. *Given an infinite cardinal τ, for every $\alpha < \tau^+$ there exists a strong Eberlein compact $K \subset c_0(\tau)$ with $\eta(K) \geq \alpha$.*

Proof. Recall that a *strong Eberlein compact* is a subset $K \subset c_0(\tau)$ consisting of $\{0,1\}$-valued finitely supported functions that is compact in the pointwise topology. It is easy to see that in such a compact there may not exist an infinite sequence $A_1 \subsetneq A_2 \subsetneq \ldots$ of finite subsets of τ such that $\chi_{A_i} \in K$. Consequently, K (hereditarily) has isolated points and is a scattered compact. Therefore the height $\eta(K)$ is well defined. Since K may be viewed as a subset of $\tau^{<\omega}$, it is clear that $|\eta(K)| < \tau^+$. To construct K with $\eta(K) \geq \alpha$ for all $\alpha < \tau^+$, we proceed by induction. For $\alpha = 1$, choose $K = \{0\} \cup \{\chi_t; t \in \tau\}$. Suppose we have constructed $K_\beta \subset c_0(\Gamma_\beta)$, $\eta(K_\beta) \geq \beta$ for all $\beta < \alpha < \tau^+$, where $|\Gamma_\beta| = \tau$ are pairwise disjoint index sets. Put $\Gamma = \bigcup \Gamma_\beta$, and $K = \{\chi_A; |A \setminus B| \leq 1$ for some $B \subset A, \chi_B \in K_\beta, \beta < \alpha\}$. It is standard to verify that K is a strong Eberlein compact. It follows that $|\Gamma| \leq |\tau \cdot \alpha| = \tau$ and $\eta(K) \geq \beta + 1$ for all $\beta < \alpha$, and so K satisfies $\eta(K) \geq \alpha$. $\qquad \square$

Theorem 2.54. *Let τ be an infinite cardinal. Then, for every $\tau \leq \alpha < \tau^+$, there exists a reflexive Banach space X of density τ and $\mathrm{Sz}(X) \geq \alpha$.*

Proof. Let $K \subset c_0(\tau)$ be a strong Eberlein compact with $\eta(K) \geq \alpha$. By the [DFJP74] factorization theorem, there exists a reflexive space X (without loss of generality of density τ) and a bounded linear and injective operator $T : X \to c_0(\tau)$, $K \subset T(B_X)$. Being reflexive, X is naturally a dual space, and so we have $\mathrm{Sz}(T^{-1}(K)) \leq \mathrm{Sz}(B_X)$. We have $\|f - g\| \geq 1$ whenever $f, g \in K$, $f \neq g$. Thus there exists some $\varepsilon > 0$ such that $\|T^{-1}(f) - T^{-1}(g)\| \geq 2\varepsilon$. It is now easy to verify that $T^{-1}(K^{(\beta)}) = (T^{-1}(K))_\varepsilon^\beta$, and so we have $\mathrm{Sz}(B_X) \geq \mathrm{Sz}(T^{-1}(K)) \geq \alpha$. $\qquad \square$

Theorem 2.55 ([HaLaMo]). *Given an infinite cardinal τ, there exists no Asplund space of density τ universal for all reflexive spaces of density τ.*

Proof. By Lemma 2.41, $\mathrm{Sz}(X) < \tau^+$ for every Asplund space of density τ. On the other hand, there exists a reflexive space Y of density τ and $\mathrm{Sz}(Y) > \alpha$ for every $\alpha < \tau^+$. Lemma 2.53 finishes the proof. $\qquad \square$

On the other hand, using the weak Szlenk index in a similar fashion, Argyros and Benyamini [ArBe87] proved that, *under GCH, there exists a WCG space X with* dens $X = \tau$ *universal for all WCG spaces of density τ if and only if* cof $\tau = \omega$.

Recently, Bell [Bell00] has proved that, under GCH, there exists a universal uniform Eberlein compact (UEC) (see Definition 6.29) for every weight, but on the other hand it is consistent that there does not exist a UEC of weight ω_1.

It is proved in [ArDo] and [DoFe] that given a countable ordinal α there is a Banach space Y_α with a separable dual such that every separable X with $\mathrm{Sz}(X) \leq \alpha$ isomorphically embeds into Y_α.

It is proved in [Todo95] that *there is no universal Corson compact of weight c* (see the definition of universal compact in the proof of Theorem 2.52), and also that *for every compact countably tight* (see Definition 3.31) *space K of* *weight c there is a first countable retractive Corson compact space S that is not a continuous image of any closed subspace of K* (a space S is *retractive* if every closed subset of S is a retract of S).

2.6 Classification of $C[0, \alpha]$ Spaces

As a main result, we are going to present characterizations of isomorphic classes of $C(K)$ spaces where K is countable due to Bessaga, Pełczyński, and Samuel. The structural theory of these spaces is a vast field. We refer to [JoLi01h, Chap. 36] for more information and references in this direction. Let us recall a well-known topological lemma. *Let K be a metrizable compact. Then there exists a unique decomposition $K = I \cup P$, where I is a countable set (of nonaccumulation points; i.e., points having a countable open neighborhood) and P is a perfect set. By Milyutin's fundamental theorem, $C(K) \cong C(L)$ whenever K, L are metrizable compacts with a nonempty perfect subset.* These are precisely $C(K)$ spaces with nonseparable dual, so in particular their Szlenk index is undefined. We proceed by classifying the remaining $C(K)$ spaces when K is a countable metrizable compact. It is clear from the Riesz representation theorem that for countable K, $C(K)^* = \ell_1(K) \cong \ell_1$ is separable, and by the Urysohn lemma, K is a 2-separated subset of $(C(K)^*, \|\cdot\|^*)$.

It is easy to see that, due to the norm separation of the points in K, the Szlenk derivation in $K \subset C(K)^*$ coincides with the Cantor derivation in K, so in particular $\chi(K) < \omega_1$. The pair $(\chi(K), n(K))$ completely topologically characterizes K.

Theorem 2.56 (Mazurkiewicz, Sierpiński). *Let K be a countable compact. Then K is homeomorphic to $[0, \omega^{(\chi(K))}.n(K)]$, the ordinal segment with the interval topology.*

Proof (Sketch). We omit the standard proof by transfinite induction that $\chi(\omega^\zeta \cdot \gamma) = \zeta$, $n(\omega^\zeta \cdot \gamma) = \gamma$. Note that K is zero-dimensional; i.e., it has a basis consisting of clopen sets. Indeed, for every $x \in K$ and a closed set $x \notin S \subset K$, choose a separating continuous function $f \geq 0$, $f(x) = 0$, $f(S) \geq 1$. Since K is countable, there exists a value $r \in (0, 1)$ such that $r \notin f(K)$, and so $f^{-1}[0, r)$ is a clopen neighborhood of x. We proceed with the proof by induction in α, n, $\chi(K) = \alpha, n(K) = n$, using the lexicographic order. The statement is obvious when $\alpha = 0$. Suppose it is true for all $\beta < \alpha, n \in \omega$, and α, k, where $k < n$. Suppose that $\chi(K) = \alpha$ and $n(K) = n$, and split $K = K_1 \cup K_2$, where K_i are disjoint clopen subsets of K with $0 < n(K_1), n(K_2) < n$. This is possible since the derivation operation is preserved for clopen subsets. By assumption, K_i are homeomorphic to closed initial intervals I_i of the ordinals, and so K is homeomorphic to the initial interval obtained by laying a copy of I_2 right after I_1 on the ordinal scale (denoted $I_1 \frown I_2$). This finishes the inductive step in k. Next assume the theorem is satisfied for all $\beta < \alpha, k \in \mathbb{N}$, and suppose $K^\alpha = \{p\}$. Choose a sequence $(p_n) \to p$ in the following way. If $\alpha = \beta + 1$, then $\{p_n\}_{n=1}^{\infty} = K^{\beta}$, and put $\alpha_n = \beta$. Otherwise, $p_n \in K^{\alpha_n} \setminus K^{\alpha_n+1}$, for an increasing sequence $(\alpha_n) \nearrow \alpha$. Choose a sequence of disjoint clopen sets S_n, $p_n \in S_n$, $\bigcup_{n=1}^{\infty} S_n = K \setminus \{p\}$. This is done in the following way. Denote by $(q_n)_{n=1}^{\infty}$ a sequence containing all points from $K \setminus (\{p\} \cup \{p_n\}_{n=1}^{\infty})$. We choose S_n to be clopen, $\{p_n\} = S_n \cap K^{\alpha_n}$, $q_k \in S_n$, $k \leq n$, if and only if $q_k \in K \setminus (K^{\alpha_n} \cup \bigcup_{i=1}^{n-1} S_i)$. Let I_n be the initial ordinal segment representing S_n. It is now clear that $I_1^{\frown} I_2^{\frown} \ldots$ with the last added compactifying point is homeomorphic to K. □

Lemma 2.57 (Bessaga and Pełczyński [BesPe60]). *Let α and β be two infinite ordinals so that $\omega \leq \alpha < \omega_1$ and $\alpha \leq \beta < \alpha^{\omega}$. Then $C[0, \alpha]$ is isomorphic to $C[0, \beta]$.*

Proof. Assume that $\alpha = \omega^{\alpha_1}$. Clearly, $C[0, \alpha] \cong C_0[0, \alpha]$ (by the isomorphism of all hyperplanes). Using the form of α, it is standard to choose a sequence of nonlimit ordinals $(\xi_n) \nearrow \alpha$ so that $[\xi_n, \xi_{n+1} - 1]$ forms a sequence of ordinal order types $(\eta_n)_{n=1}^{\infty}$, that repeats each of its elements infinitely many times. Clearly, $C_0[0, \alpha] \cong \sum(\oplus C[0, \eta_n])_{c_0}$. From the infinite repetitions of η_n, we have $C_0[0, \alpha\omega] \cong C_0[0, \alpha 2] \cong C_0[0, \alpha] \oplus C_0[0, \alpha] \cong C_0[0, \alpha]$. Similarly, $C_0[0, \alpha^2] \cong X \oplus Y$, where $X = \{f; f \upharpoonright (\alpha\xi, \alpha(\xi+1)] = \text{const.}, \xi < \alpha\}$, $Y = \{f; f(\alpha\xi) = 0, \xi < \alpha\}$. Now, $X \cong C_0[0, \alpha]$, and $Y \cong \sum(\oplus C_0[0, \alpha])_{c_0[0,\alpha]} \cong \sum(\oplus C_0[0, \alpha])_{c_0} \cong C_0[0, \alpha\omega] \cong C_0[0, \alpha]$. Thus $C_0[0, \alpha^2] \cong C_0[0, \alpha]$. Finally, by induction, we have $C[0, \alpha^{2^k}] \cong C[0, \alpha]$, and $\alpha^{2^k} \nearrow \alpha^{\omega}$. Using the initial interval projection, we see that for $\alpha < \beta < \gamma = \alpha^{2^k}$, $C[0, \alpha]$ is complemented in $C[0, \beta]$, and $C[0, \beta]$ is complemented in $C[0, \gamma]$. Using the Pełczyński decomposition method, together with the established facts, we see that $C[0, \alpha] \cong C[0, \gamma] \cong C[0, \beta]$. Let us pass to the case of general α. Recall that for every ordinal there exists a unique ordinal sum decomposition $\alpha = \omega^{\alpha_1} k_1 + \omega^{\alpha_2} k_2 + \cdots + \omega^{\alpha_n} k_n$, where $\alpha_1 > \alpha_2 > \cdots > \alpha_n$, $k_i \in \omega$. Thus $C[0, \alpha] \cong C[0, \omega^{\alpha_1}]$, and the result follows by a standard argument using the previous case. □

The isomorphic classification of $C(K)$ spaces below was established by Bessaga and Pełczyński. The additional Szlenk index characterization is a later result of Samuel.

Theorem 2.58 (Bessaga and Pełczyński [BesPe60], Samuel [Sam83]). *The scale of spaces $C[0, \omega^{\omega^{\alpha}}]$, where $\alpha \in [0, \omega_1)$, contains all isomorphic types of $C(K)$ with K countable. Moreover, $\mathrm{Sz}(C[0, \omega^{\omega^{\alpha}}]) = \omega^{\alpha+1}$, so the isomorphism class of $C(K)$ with K countable is determined by $\mathrm{Sz}(C(K))$.*

Proof. By the previous results, we have that every $C(K)$ is isomorphic to $C[0, \beta]$, where $\omega^{\omega^{\alpha}} \leq \beta < (\omega^{\omega^{\alpha}})^{\omega} = \omega^{\omega^{\alpha+1}}$, and so $C(K) \cong C[0, \omega^{\omega^{\alpha}}]$ for some $\alpha < \omega_1$. The fact that elements of the scale are mutually nonisomorphic follows from the next theorem. □

Theorem 2.59 (Samuel [Sam83]). *For every $0 \le \alpha < \omega_1$, $Sz(C[0, \omega^{\omega^{\alpha}}]) = \omega^{\alpha+1}$.*

Proof [HaLa]. We have noted before that, due to the norm 2-separation of the points in $K \subset B_{C(K)^*}$, the Szlenk ε-derivation, for all $\varepsilon < 1$, coincides with the Cantor derivation on the set $K \subset C(K)^*$. If $K = [0, \omega^{\omega^{\alpha}}]$, then $K_\varepsilon^{\omega^{\alpha}} = K^{(\omega^{\alpha})} \ne \emptyset$. By Theorem 2.43 and Theorem 2.56, we have $Sz(C(K)) \ge \omega^{\alpha+1}$. To prove the converse inequality, for a fixed $0 \le \alpha < \omega_1$, we denote $Z = \ell_1([1, \omega^{\omega^{\alpha}})) = C_0[0, \omega^{\omega^{\alpha}}]^*$. For $\gamma < \omega^{\omega^{\alpha}}$, we set $Z_\gamma = \ell_1([1, \gamma]) = C[0, \gamma]^*$ and $P_\gamma : Z \to Z_\gamma$ the canonical norm 1 projection. In the next statement, the Szlenk derived sets are meant, with the w^*-topologies coming from the respective preduals of Z and Z_γ.

Lemma 2.60. *Let $\alpha < \omega_1$, $\gamma < \omega^{\omega^{\alpha}}$, $\beta < \omega_1$, and $\varepsilon > 0$. If $z \in (B_Z)_{3\varepsilon}^\beta$ and $\|P_\gamma z\| > 1 - \varepsilon$, then $P_\gamma z \in (B_{Z_\gamma})_\varepsilon^\beta$.*

Proof. We will use a transfinite induction on β. The statement is trivially true for $\beta = 0$. Assume it is true for all $\mu < \beta$. If β is a limit ordinal, then clearly it is also true for β. So assume $\beta = \mu + 1$ and let $z \in B_Z$ such that $\|P_\gamma z\| > 1 - \varepsilon$ and $P_\gamma z \notin (B_{Z_\gamma})_\varepsilon^\beta$. We need to show that $z \notin (B_Z)_{3\varepsilon}^\beta$, so we may assume that $z \in (B_Z)_{3\varepsilon}^\mu$ and therefore that $P_\gamma z \in (B_{Z_\gamma})_\varepsilon^\mu$. Using all large enough coordinates of $P_\gamma z$, and suitable norming elements from the predual, we see that there is a w^*-open subset V of Z_γ containing $P_\gamma z$ such that $d = \text{diam}(V \cap (B_{Z_\gamma})_\varepsilon^\mu) < \varepsilon$. Using the Hahn-Banach theorem, we may choose V so that $V \cap (1 - \varepsilon)B_{Z_\gamma} = \emptyset$. Assume that

$$V = \bigcap_{i=1}^{n} \{x \in Z_\gamma, \ f_i(x) > \alpha_i\}, \text{ where } \alpha_i \in \mathbb{R} \text{ and } f_i \in C[0, \gamma].$$

Define functions $g_i \in C_0(\omega^{\omega^{\alpha}})$ by $g_i = f_i$ on $[1, \gamma]$ and $g_i = 0$ on $(\gamma, \omega^{\omega^{\alpha}})$, and put

$$U = \bigcap_{i=1}^{n} \{y \in Z, \ g_i(y) > \alpha_i\}.$$

It is clear that $z \in U \cap (B_Z)_{3\varepsilon}^\mu$. For any $y \in U \cap (B_Z)_{3\varepsilon}^\mu$, $P_\gamma y \in V$, so $\|P_\gamma y\| > 1 - \varepsilon$ and by the inductive hypothesis $P_\gamma y \in V \cap (B_{Z_\gamma})_\varepsilon^\mu$. Therefore, for all $y, y' \in U \cap (B_Z)_{3\varepsilon}^\mu$, $\|P_\gamma y - P_\gamma y'\| \le d < \varepsilon$. Since moreover $\|P_\gamma y\| > 1 - \varepsilon$ and $\|P_\gamma y'\| > 1 - \varepsilon$, we have that $\|y - y'\| \le d + 2\varepsilon < 3\varepsilon$. This shows that $z \notin (B_Z)_{3\varepsilon}^\beta$ and finishes our induction. $\square$

In order to conclude the proof of Theorem 2.59, it is enough to show that

$$\forall 0 \le \alpha < \omega_1 \ \forall \gamma < \omega^{\omega^{\alpha}} \ \forall \varepsilon > 0, \quad (B_{Z_\gamma})_\varepsilon^{\omega^{\alpha}} = \emptyset.$$

We proceed by transfinite induction on α. If $\alpha = 0$, then for any $\gamma < \omega$, Z_γ is finite-dimensional and therefore $s_\varepsilon(B_{Z_\gamma}) = \emptyset$, and the statement is true. It also passes easily to the limit ordinals. So assume now that it is true for α. Then Lemma 2.60 implies that

$$\forall \varepsilon > 0, \quad (B_Z)_\varepsilon^{\omega^\alpha} \subset \left(1 - \frac{\varepsilon}{3}\right) B_Z.$$

It now follows that

$$\forall \varepsilon > 0, \quad (B_Z)_\varepsilon^{\omega^{\alpha+1}} = \emptyset.$$

By Lemma 2.57, $C[0, \gamma] \cong C[0, \omega^{\omega^\alpha}] \cong C_0[0, \omega^{\omega^\alpha}]$ for any $\omega^{\omega^\alpha} \leq \gamma < \omega^{\omega^{\alpha+1}}$. So $(B_{Z_\gamma})_\varepsilon^{\omega^{\alpha+1}} = \emptyset$ for any $\varepsilon > 0$ and any $\gamma < \omega^{\omega^{\alpha+1}}$. This finishes the argument. $\qquad \square$

Recall that from the continuity of the ordinal exponential function we have $\omega_1 = \omega^{\omega_1}$. Thus $\omega_1.\omega = \omega^{\omega_1+1}$.

Proposition 2.61.

$$\mathrm{Sz}(C[0, \omega_1]) = \omega_1.\omega = \omega^{\omega_1+1}.$$

Proof. For any $\alpha < \omega_1$, $\mathrm{Sz}_1(C[0, \omega^\alpha]) > \alpha$ and $C[0, \omega^\alpha]$ embeds isometrically in $C[0, \omega_1]$, so $\mathrm{Sz}_1(C[0, \omega_1]) \geq \omega_1$. Since ω_1 is a limit ordinal, we obtain, using w^*-compactness, that $\mathrm{Sz}(C[0, \omega_1]) > \omega_1$. Then it follows from Theorem 2.43 that $\mathrm{Sz}(C[0, \omega_1]) \geq \omega_1.\omega$. On the other hand, the techniques of Lemma 2.60 yield similarly that $\mathrm{Sz}(C[0, \omega_1]) \leq \omega_1.\omega$. $\qquad \square$

Corollary 2.62. *For any* $\omega_1 \leq \alpha < \omega_1.\omega$,

$$\mathrm{Sz}(C[0, \alpha]) = \omega_1.\omega.$$

Proof. For any $\omega_1 \leq \alpha < \omega_1.\omega$, $C[0, \omega_1]$ embeds in $C[0, \alpha]$ and $C[0, \alpha]$ embeds in some finite sum $C[0, \omega_1] \oplus ... \oplus C[0, \omega_1]$. Then Propositions 2.46 and 2.61 imply that

$$\mathrm{Sz}(C[0, \omega_1]) = \omega_1.\omega = \mathrm{Sz}(C[0, \alpha]). \qquad \square$$

Unlike in the separable case, the Szlenk index does not distinguish the isomorphic classes for the nonseparable $C[0, \alpha]$ spaces.

Theorem 2.63 (Semadeni [Sema60]). *Given* $\omega_1 \leq \alpha < \beta < \omega_1.\omega$, $C[0, \alpha]$ *and* $C[0, \beta]$ *are isomorphic if and only if* $\omega_1.n \leq \alpha < \beta < \omega_1.(n+1)$ *for some integer number* n.

Proof (Sketch). On the one hand, $\omega_1.n \leq \alpha < \beta < \omega_1.(n+1)$ implies that $\alpha = \omega_1.n + \alpha'$, $\beta = \omega_1.n + \beta'$, where $\alpha', \beta' < \omega_1$. Thus α is homeomorphic with $\alpha' + \omega_1.n = \omega_1.n$, and similarly β is homeomorphic with $\omega_1.n$. On the other hand, for $X = C[0, \alpha]$, define $X_s \hookrightarrow C[0, \alpha]^{**} = \ell_\infty[0, \alpha]$ to be the linear subspace consisting of all w^*-sequentially continuous elements from X^{**}, i.e., $F \in X_s$ if and only if $\lim F(f_n) = 0$ for all w^*-null sequences $(f_n) \subset X^*$. It is easy to see that if $F \in \ell_\infty[0, \alpha]$ belongs to X_s, then F must be continuous at all points $\gamma \in \alpha$ with $\mathrm{cof}\, \gamma \leq \omega$. From this, using a standard argument, we obtain that $\omega_1.n \leq \alpha < \omega_1.(n+1)$ implies $\dim X_s/X = n$. The last number is, however, an isomorphic invariant of X, and this finishes the proof. $\qquad \square$

We include without proof the complete isomorphic characterization of $C[0, \alpha]$ spaces for ordinals α. This result can be proved using the ideas already present in the previous cases where $\alpha < \omega_1.\omega$, in particular the Szlenk index (Cantor derivation) and "long sequential continuity" properties of some subspaces in the bidual $C[0, \alpha]^{**} = \ell_\infty[0, \alpha]$.

Theorem 2.64 (Kislyakov [Kis75], Gulko and Oskin [GO75]). *Let $\xi < \eta$ be ordinals of the same cardinality and α be the least ordinal of this cardinality. Then $C[0, \xi] \cong C[0, \eta]$ is characterized as follows.*

If $\alpha = \omega$, or α is singular, or α is regular and $\xi, \eta \geq \alpha^2$, then $C[0, \xi] \cong C[0, \eta]$ if and only if $\xi < \eta < \xi^\omega$.

In the remaining case, where α is regular and $\xi = \alpha\tilde{\xi} + \gamma$, and $\eta = \alpha\tilde{\eta} + \delta$, where $\tilde{\xi}, \tilde{\eta} \leq \alpha$, $\gamma, \delta < \alpha$, then $C[0, \xi] \cong C[0, \eta]$ if and only if $\operatorname{card} \tilde{\eta} = \operatorname{card} \tilde{\xi}$.

2.7 Szlenk Index and Renormings

We are going to introduce the weak*-dentability index $\Delta(X)$ of a Banach space and show a result of Lancien claiming that spaces with a countable index have dual LUR renormings. We then proceed to prove a deep result of Bossard and Lancien, that shows the existence of a universal estimate $\Delta(X) \leq \Psi(\mathrm{Sz}(X))$ for spaces with a countable Szlenk index.

Definition 2.65. *Let X be an Asplund space and $B \subset X^*$ be a w^*-compact subset. Given $\varepsilon > 0$, put $\Delta_\varepsilon^0(B) = B$. Proceeding inductively, if α is an ordinal, put*

$$\Delta_\varepsilon^{\alpha+1}(B) = \Delta_\varepsilon^\alpha(B) \backslash \bigcup W \text{ for all } w^*\text{-open slices } W \text{ of } B_\alpha^\varepsilon \text{ with } \operatorname{diam} W < \varepsilon.$$

If α is a limit ordinal, put

$$\Delta_\varepsilon^\alpha(B) = \bigcap_{\beta < \alpha} \Delta_\varepsilon^\beta(B).$$

Clearly, $\Delta_\varepsilon^\alpha(B)$ are w^*-compact sets. Assume that α is the least ordinal so that $\Delta_\varepsilon^\alpha(B) = \emptyset$. Then we define $\Delta_\varepsilon(B) = \alpha$. We define the dentability index $\Delta(B) = \sup_{\varepsilon>0} \Delta_\varepsilon(B)$. In the case where $B = B_{X^*}$, we abuse the notation slightly by denoting $\Delta_\varepsilon(X) = \Delta_\varepsilon(B)$ and calling $\Delta(X) = \Delta(B)$ the w^*-dentability index of the space X.

The definition is correct for the same reason as in the Szlenk situation, and $\Delta(X)$ exists as an ordinal for every Asplund space X. One can prove similarly to the Szlenk index that for every separable Asplund space, $\Delta(X) = \omega^\alpha$ for some $\alpha < \omega_1$. Clearly, $\Delta(X) \geq \mathrm{Sz}(X)$ for every Asplund space.

Proposition 2.66 (Lancien [Lanc95]). *A Banach space X is superreflexive if and only if $\Delta(X) = \omega$.*

Proof. Every superreflexive space has an equivalent uniformly rotund renorming (see, e.g., [DGZ93a, Thm. 4.4.1]). It follows easily that $\Delta_\varepsilon(X) < \omega$ for every $\varepsilon > 0$, which yields one implication. To prove the converse, we show that there exists no ε-dyadic tree in B_{X^*}, $\{x_t\}_{t \in \bigcup_{l=1}^k \{0,1\}^l}$, with the root $x_\varnothing = 0$, provided that $k > \Delta_\varepsilon(X)$. This condition is known to characterize superreflexive spaces ([Jam72b]). Suppose the contrary. We have $x_t = \frac{x_{t \frown 0} + x_{t \frown 1}}{2}$ and $\|x_t - x_{t \frown 0}\| \geq \varepsilon$ for all t of length at most $k-1$. It is clear that every slice containing x_t must contain at least one of its followers $x_{t \frown 0}$, $x_{t \frown 1}$, so in particular it has diameter at least ε. It follows by a simple inductive argument that $x_t \in \Delta_\varepsilon^l(\{x_t\})$ whenever $|t| < k - l$. Thus $0 \in \Delta_\varepsilon^{k-1} \neq \emptyset$, a contradiction. $\square$

Theorem 2.67 (Lancien [Lanc93]). *Let X be an Asplund space, $\Delta(X) < \omega_1$. Then X has a dual LUR renorming* (see Definition 3.47).

Proof. Let $\Delta(X) = \beta < \omega_1$. Choose a doubly indexed set of positive numbers $\{a_{\alpha,k}\}_{\alpha < \beta, k \in \mathbb{N}}$ satisfying $\sum_{\alpha,k} a_{\alpha,k} < \frac{1}{2}$. Put $\psi_{\alpha,k}(f) = a_{\alpha,k}\mathrm{dist}(f, \Delta_{\frac{1}{k}}^\alpha(B_{X^*}))$ if $\Delta_{\frac{1}{k}}^\alpha(B_{X^*}) \neq \emptyset$ and identical to 0 otherwise. Note that since $\Delta_{\frac{1}{k}}^\alpha(B_{X^*})$ are w^*-compact and convex, $\psi_{\alpha,k}$ are w^*-lower semicontinuous. Thus

$$F(f) = \left(\|f\|^2 + \sum_{\alpha,k} \psi_{\alpha,k}^2(f) \right)^{\frac{1}{2}}$$

is an LUR convex function. Indeed, suppose that $2F(f)^2 + 2F(f_n)^2 - F(f + f_n)^2 \to 0$, for some $f, f_n \in B_{X^*}$, $\varepsilon > 0$. Choose k large enough so that $\frac{4}{k} < \varepsilon$. Consider the minimal $\alpha < \beta$ such that $f \notin \Delta_{\frac{1}{k}}^\alpha(B_{X^*})$. We have that $\alpha = \gamma + 1$ is a nonlimit ordinal. Clearly, $\psi_{\alpha,k}(f) > 0$, and by [DGZ93a, p. 42] we have also

$$\lim_{n \to \infty} \psi_{\alpha,k}(f_n) = \lim_{n \to \infty} \psi_{\alpha,k}\left(\frac{f + f_n}{2} \right) = \psi_{\alpha,k}(f).$$

Also, $\psi_{\gamma,k}(f) = 0$, and so by [DGZ93a, p. 42] we have also

$$\lim_{n \to \infty} \psi_{\gamma,k}(f_n) = \lim_{n \to \infty} \psi_{\gamma,k}\left(\frac{f + f_n}{2} \right) = 0.$$

Upon removing finitely many initial elements and making small perturbations (at most $\varepsilon/4$ in norm and tending to 0 in norm) of the rest of $\{f_n\}_{n=1}^\infty$, we can without loss of generality assume that in fact $\psi_{\gamma,k}(f_n) = 0$ for all $n \in \mathbb{N}$. Due to the convexity of $\Delta_{\frac{1}{k}}^\gamma(B_{X^*})$, we also obtain $\psi_{\gamma,k}(\frac{f+f_n}{2}) = 0$. For n large enough, $\psi_{\alpha,k}\left(\frac{f+f_n}{2} \right) > 0$ and $\psi_{\alpha,k}(f_n) > 0$ so we have that $f, f_n, \frac{f+f_n}{2} \in \Delta_{\frac{1}{k}}^\gamma \setminus \Delta_{\frac{1}{k}}^\alpha$. Therefore, there exists a w^*-slice of $\Delta_{\frac{1}{k}}^\gamma$ of diameter less than $\frac{\varepsilon}{4}$ containing $\frac{f+f_n}{2}$. Since such a slice necessarily contains one of the elements of f, f_n, we obtain that $\|f - f_n\| = 2\|f - \frac{f+f_n}{2}\| = 2\|f_n - \frac{f+f_n}{2}\| \leq \frac{\varepsilon}{2}$, for n large enough.

Thus $\lim_{n\to\infty}\|f-f_n\|=0$, so F is an LUR convex function. The Minkowski functional of the set $\{f\in X^*; F(f)\le 1\}$ defines an equivalent dual LUR renorming of X^*. $\square$

The following fundamental result of Bossard and Lancien (Bossard proved the result first under the assumption that X is separable) relates the Szlenk and w^*-dentability indices.

Theorem 2.68 (Bossard [Boss02], Lancien [Lanc96]). *There exists a function $\Psi:\omega_1\to\omega_1$ such that for every Asplund space X, $\mathrm{Sz}(X)<\omega_1$ implies $\Delta(X)\le\Psi(\mathrm{Sz}(X))$.*

Proof. Let $K=(B_{\ell_1^*},w^*)=(B_{\ell_\infty},\sigma(\ell_\infty,\ell_1))$. Since ℓ_1 is separable, K is a compact space with a complete metric ρ. Consider the metrizable compact space $\mathcal{K}=\{L; L\subset K \text{ is closed}\}$, with the Vietoris topology τ_V, whose subbasis consists of sets of the form $\mathcal{U}=\{L; L\subset U, \text{ for some open } U\subset K\}$ or $\mathcal{U}=\{L; L\cap U\ne\emptyset, \text{ for some open } U\subset K\}$ ([Eng77, p. 163]). The space $(\mathcal{K},\tau_V)$ is completely metrizable by the metric ρ introduced as $\rho(L,M)=\sup_{l\in L,m\in M}\{\rho-\mathrm{dist}(l,M),\rho-\mathrm{dist}(m,L)\}$ (in fact, the definition of this metric makes sense for any pair of nonempty subsets of K, and we will use it below also in the general setting). Given $\varepsilon>0$, denote by $s_\varepsilon:\mathcal{K}\to\mathcal{K}$ $d_\varepsilon:\mathcal{K}\to\mathcal{K}$ the functions $s_\varepsilon(L)=L_\varepsilon^1$, $d_\varepsilon(L)=\Delta_\varepsilon^1(L)$. We will also use the ordinal iterations s_ε^α, $d_\varepsilon^\alpha:\mathcal{K}\to\mathcal{K}$ defined for $\alpha<\omega_1$ as $s_\varepsilon^\alpha(L)=L_\varepsilon^\alpha$, $d_\varepsilon^\alpha(L)=\Delta_\varepsilon^\alpha(L)$.

Fact 2.69. *Let $\varepsilon>0$, $\alpha<\omega_1$. Then $s_\varepsilon^\alpha,d_\varepsilon^\alpha$ are Borel functions on $\mathcal{K}$.*

Proof. The fact that s_ε^α and d_ε^α are Borel functions follows by a standard induction in α from the case $\alpha=1$. If $\alpha=\beta+1$, then $s_\varepsilon^\alpha=s_\varepsilon\circ s_\varepsilon^\beta$, $d_\varepsilon^\alpha=d_\varepsilon\circ d_\varepsilon^\beta$, and the claim follows. For α a limit ordinal, $\alpha_n\nearrow\alpha$, we have $s_\varepsilon^\alpha=\lim_{\alpha_n\to\alpha}s_\varepsilon^{\alpha_n}$, $d_\varepsilon^\alpha=\lim_{\alpha_n\to\alpha}d_\varepsilon^{\alpha_n}$ pointwise on $\mathcal{K}$, so we are done again.

Now we present the argument for $s_\varepsilon(=s_\varepsilon^1)$. We need to show that $(s_\varepsilon)^{-1}(\mathcal{O})$ is a Borel set for every open $\mathcal{O}\subset\mathcal{K}$. It suffices to prove the statement for all elements of the subbasis of $\mathcal{K}$ consisting of sets of two kinds. Either $\mathcal{O}=\{L; L\subset U \text{ for some open } U\subset K\}$ or $\mathcal{O}=\{L; L\cap U\ne\emptyset, \text{ for some open } U\subset K\}$. Let us first observe that it is sufficient to prove the statement for the open sets of the first kind. Indeed, let $\mathcal{O}=\{L; L\cap U\ne\emptyset, \text{ for some open } U\subset K\}$. Let $V=K\setminus U$. As $V\subset K$ is closed, there exists a sequence $(V_n)_{n=1}^\infty$ of open subsets of K, $V\subset V_n$, $\rho(V,V_n)<\frac{1}{n}$. Denote $\mathcal{O}_n=\{L; L\subset V_n\}$, open subsets of $\mathcal{K}$ of the first kind. We have

$$(s_\varepsilon)^{-1}(\mathcal{O})=\bigcup_{n=1}^\infty(\mathcal{K}\setminus(s_\varepsilon)^{-1}(\mathcal{O}_n)).$$

This finishes the reduction to the first case. We continue with the proof for the open sets of the first kind. Denote by $\ell_1^\mathbb{Q}$ the set of all finitely supported vectors from ℓ_1 with rational coordinates. Consider a countable system $\{U_n\}_{n=1}^\infty$ of all open sets in K of the form $U = \{f; x_i(f) > r_i, i \in \{1, \ldots, k\}\}$ for all $k \in \mathbb{N}$, $x_i \in \ell_1^{\mathbb{Q}}$, and $r_i \in \mathbb{Q}$.

Let $O \subset K$ be open and $\mathcal{O} = \{L : L \subset O\}$ be the corresponding open set in $\mathcal{K}$, and suppose that $L \in s_\varepsilon^{-1}(\mathcal{O})$. Thus $L_\varepsilon^1 \subset O$, and, by a standard compactness argument, there exists a finite set $\{n_1, \ldots, n_k\} \subset \mathbb{N}$ such that $\|\cdot\|_\infty - \operatorname{diam}(U_{n_i} \cap L) \leq \varepsilon_i$ for some rational $\varepsilon_i < \varepsilon$, and $L \subset O \cup \bigcup_{i=1}^k L \cap U_{n_i}$. The first condition is characterized by $x(L \cap U_{n_i}) \leq \varepsilon_i$ for all $x \in \ell_1^{\mathbb{Q}}$, $\|x\|_1 \leq 1$, which is a Borel condition for $L \in \mathcal{K}$. The second condition is also Borel. Thus we obtain the Borel property of $s_\varepsilon^{-1}(\mathcal{O})$. $\qquad\square$

Denote $\mathcal{B}_\alpha = \{L \in \mathcal{K}; s_\varepsilon^\alpha(L) = \emptyset \text{ for all } \varepsilon > 0\}$. By Fact 2.69, this set is Borel in $\mathcal{K}$. Note, moreover, that every $L \in \mathcal{B}_\alpha$ is $\|\cdot\|_\infty$-separable. This can be seen as follows. Observe first that from the very definition, each set in $\mathcal{B}_\alpha$ is fragmented by the dual norm. Now use the following fact to prove $\|\cdot\|_\infty$-separability of the sets in $\mathcal{B}_\alpha$.

Fact 2.70. *Let Y be a separable Banach space and $L \subset Y^*$ be a w^*-compact subset. Then L is fragmented by the dual norm if and only if L is norm separable.*

Proof (Namioka). Let L be norm separable. Assume without loss of generality that $\emptyset \neq A \subset L$ is a w^*-closed set, and let $\varepsilon > 0$. Since A is norm separable, there is a sequence (a_i) in A such that $A \subset \bigcup_{i=1}^\infty B(a_i, \frac{\varepsilon}{2})$. As A in the w^*-topology is compact, it is a Baire space, and each $B(a_i, \frac{\varepsilon}{2})$ is w^*-closed. By the Baire category theorem, there is $i \in \mathbb{N}$ such that $A \cap B(a_i, \frac{\varepsilon}{2})$ contains a nonempty w^*-open set W in A. Clearly $\|\cdot\|$-$\operatorname{diam}(W) \leq \varepsilon$.

Now assume that L is fragmented by the dual norm of Y^*. Assume by contradiction that L is not norm separable. Then there is an uncountable subset $H \subset L$ and $\varepsilon > 0$ such that $\|u - v\| \geq \varepsilon$ whenever $u, v \in H$ and $u \neq v$. Since L in its weak* topology is separable and metrizable, by deleting countably many points, we may assume that each point of H is a limit point in the weak* topology. Since L is norm fragmented, there is a non empty weak*-relatively open subset U of H of norm diameter $\leq \frac{\varepsilon}{2}$. By the choice of H, U must then be a singleton $\{u\}$ for some $u \in H$. But this contradicts that u is a limit point of H. This contradiction shows that L is norm separable. $\quad\square$

Once the $\|\cdot\|_\infty$-separability of the sets in $\mathcal{B}_\alpha$ has been established, the w^*-dentability follows from the following remark and Proposition 2.72.

Remark 2.71. Let K be a $\|\cdot\|$-separable and w^*-compact subset of X^*. Then K is a boundary of $C := \overline{\operatorname{conv}}^{w^*}(K)$. From [Fa˜01, Thm. 3.46], it follows that $C = \overline{\operatorname{conv}}^{\|\cdot\|}(K)$, and so C is $\|\cdot\|$-separable and w^*-compact.

Let C be a w^*-compact convex subset of X^*. Define $\sigma_C : X \to \mathbb{R}$ by $\sigma_C(x) = \sup\{\langle x^*, x \rangle : x^* \in C\}$. Then it is easy to check that σ_C is a Lipschitz convex function on X, and it is well known that its convex conjugate function σ_C^* is the indicator function of C and, in particular, the domain of σ_C^* is C, and consequently, the range of the subdifferential mapping $\partial \sigma_C$ is a subset of C.

Proposition 2.72. *Let C be a norm-separable w^*-compact convex subset of X^*. Then σ_C is Fréchet differentiable on a dense G_δ subset of X. Moreover, every w^*-compact convex subset of C is w^*-dentable.*

Proof (Sketch). This is a consequence of [Tang99, Thm. 2] because σ_C^* has a separable domain. The Preiss-Zajíček theorem (see the proof of [Phel93, Thm. 2.11]) applies to the continuous convex functions whose subdifferentials have a norm separable range. Thus the proof of [Phel93, Theorem 2.12] shows that σ_C is Fréchet differentiable on a dense G_δ subset of X. Then Šmulyan's theorem as given in [DGZ93a, Lemma VIII.3.15] shows that a point $x_0 \in X$ of differentiability of σ_D (where D is a weak*-convex compact subset of C) strongly exposes its derivative f_0. □

To continue with the proof of Theorem 2.68, consider that, in particular, $\Delta_\varepsilon^{\omega_1}(L) = \emptyset$ for every $L \in \mathcal{B}_\alpha$, so $\Delta(L) < \omega_1$ for all $L \in \mathcal{B}_\alpha$. For a fixed $\varepsilon > 0$ define a partial ordering $\prec_\varepsilon$ on $\mathcal{B}_\alpha$ as follows: $L \prec_\varepsilon M$ iff $M \subset d_\varepsilon(L)$. Clearly, $\prec_\varepsilon$ is well founded. To see that $\prec_\varepsilon$ is analytic, consider a set $S \subset \mathcal{B}_\alpha^3$, $S = \{(A, B, C); C = d_\varepsilon(A)\} \cap \{(A, B, C); B \subseteq C\}$. S is clearly Borel, and $\prec_\varepsilon \subset \mathcal{B}_\alpha^2$ is a projection of S onto the first two coordinates, so it is analytic. By Theorem 2.6, we conclude that $\beta = \max_{\varepsilon > 0} \operatorname{rank}_{\prec_\varepsilon}(\mathcal{B}_\alpha) < \omega_1$. We claim that $\sup_{L \in \mathcal{B}_\alpha} \Delta(L) \leq \beta$, which implies the statement of the theorem for all separable Asplund spaces (since every separable Banach space X is a linear quotient of ℓ_1, and so $(B_{X^*}, w^*) \in \mathcal{K}$). Supposing the contrary, $\Delta_\varepsilon^\beta(L) \neq \emptyset$ for some $L \in \mathcal{B}_\alpha$, $\varepsilon > 0$. However, a standard inductive argument in β shows that $\Delta_\varepsilon^\beta(L) \neq \emptyset$ implies $\operatorname{rank}_{\prec_\varepsilon}(L) \geq \beta$, a contradiction. Indeed, if $\beta < \omega$, then the sequence $L \prec_\varepsilon \Delta_\varepsilon^1(L) \prec_\varepsilon \ldots \prec_\varepsilon \Delta_\varepsilon^\beta(L)$ witnesses that $\operatorname{rank}_{\prec_\varepsilon}(L) \geq \beta$. We omit the standard inductive argument for all $\beta < \omega_1$.

The general case where $\operatorname{Sz}(X) < \omega_1$ now follows using Theorem 2.44. □

Thus every Asplund space with $\operatorname{Sz}(X) < \omega_1$ has a dual LUR renorming by Theorem 2.67. Note that this condition is not necessary. Indeed, we have shown that there exists a reflexive space with $\operatorname{Sz}(X) = \alpha$ for every ordinal. By Troyanski's renorming theorem, every reflexive space has a (dual) LUR renorming.

Corollary 2.73 (Deville; (see, e.g., [DGZ93a, Thm. VII.4.8]]). *Let K be a scattered compact such that $\chi(K) < \omega_1$. Then $C(K)$ has an equivalent dual LUR renorming.*

Proof. By Exercise 2.6, $\operatorname{Sz}(C(K)) < \omega_1$. The result follows by the renorming in Theorem 2.67. □

This result is optimal in the sense that $C[0, \omega_1]$ has no dual LUR renorming (Talagrand [Tala86], see also [DGZ93a, p. 313]), and it is the space with

$\chi([0,\omega_1]) = \omega_1$. Recently, it was shown in [HaLaP] that $\Delta(C[0,\omega^{\omega^\alpha}]) = \omega^{\alpha+2}$ for $\alpha < \omega$, and $\Delta(C[0,\omega^{\omega^\alpha}]) = \mathrm{Sz}(C[0,\omega^{\omega^\alpha}]) = \omega^{\alpha+1}$ for $\omega \le \alpha < \omega_1$. This result indicated the possibility of a simple form for a function Ψ as in Theorem 2.68. This has recently been verified by M. Raja [Raja]. He proved that *for every Asplund Banach space X, $\Delta(X) \le \omega^{\mathrm{Sz}(X)}$.*

For more information on Szlenk's index and its applications, we refer to [Lanc06]. We only mention here in passing that it is proved in [GKL00] that a Banach space X is isomorphic to c_0 if it is (nonlinearly) Lipschitz isomorphic to c_0 and that the corresponding statement for a uniform homeomorphism is still an open problem. What is known [GKL01] is that X^* is isomorphic to ℓ_1 if X is uniformly homeomorphic to c_0.

2.8 Exercises

2.1. R. Grząślewicz [Gras81] has proved that for every $n \ge 1$ there exists a compact convex set Q in $\mathbb{R}^{n+2}$ such that every closed convex subset of the unit ball of $\mathbb{R}^n$ can be obtained as an intersection of Q with some n-dimensional affine subspace of $\mathbb{R}^{n+2}$. On the other hand, Grünbaum [Grun58] and Bessaga [Bess58] proved that there exists no n-dimensional normed space containing all two-dimensional normed spaces isometrically.

Hint. We indicate the main steps of Grünbaum's solution, which works only for $n = 3$. The general solution due to Bessaga is more involved. In fact, Bessaga has shown that there exists no n-dimensional symmetric convex body that is universal for all symmetric $2n + 2$-gons. The estimate is optimal.

Let K be a three-dimensional universal convex body and P_1, P_2, P_3 its different planar sections. (1) If P_1 and P_2 are smooth, P_3 is a polygon, and $x \in \partial P_1 \cap \partial P_2 \cap \partial P_3$, then x is not a vertex of P_3. (2) If P_1 is a polygon, P_2 is smooth and rotund, and $x \in \partial P_1 \cap \partial P_2$ is an internal point of an edge I of P_1, then $\partial P_3 \cap I \neq \emptyset$ implies that P_3 is not a polygon. (3) Using compactness, there exist sequences $(S_n^i)_{n=1}^\infty$, $i = 1, 2, 3$ of polygonal sections of K convergent to C_i, smooth, rotund, and with affinely nonequivalent sections. Let Q be a parallelogram section of K. Using (2) we see that $Q \cap C_i$ must consist of vertices of Q. Thus, for a pair of opposite vertices $v, -v$ of Q, there exist at lest two of C_i that contain them. This contradicts (1).

2.2. The original definition of Szlenk index, due to Szlenk, is based on the following derivation process. Let X be a separable Banach space. Given a closed bounded set $C \subset X^*$, $\varepsilon > 0$, we put

$$C_\varepsilon^1 = \{f \in C; \exists\{f_n\} \subset C, \exists\{x_n\} \subset B_X \text{ so that } f_n \xrightarrow{w^*} f, x_n \xrightarrow{w} 0, f_n(x_n) \ge \varepsilon\}.$$

We define higher derivations inductively and put $\sigma_\varepsilon(C) = \min\{\alpha; C_\varepsilon^\alpha = \emptyset\}$, $\sigma(C) = \sup \sigma_\varepsilon(C)$, $\sigma(X) = \sigma(B_{X^*})$. Show that $\mathrm{Sz}(X) = \sigma(X)$ for all spaces with a separable dual.

Hint. Rosenthal's ℓ_1 theorem. Note that $\sigma(\ell_1) = 1$, while $\mathrm{Sz}(\ell_1)$ is undefined.

2.3 (Bourgain [Bour79]). Let X be a separable Banach space containing an isomorphic copy of $C[0, \alpha]$ for all $\alpha < \omega_1$. Then X is universal for all separable Banach spaces.

Hint. Consider the compact set $\mathcal{K} = \{L \subset [0,1], L \text{ is compact}\}$ of all compact subsets with the Hausdorff metric, $\varepsilon > 0$, and $\mathcal{T} = \{T \in \mathcal{L}(C[0,1], X), \|T\| \leq 1\}$ equipped with the weak operator topology τ_w. The basis of this topology is generated by sets $\{T; x^*(T(f)) \in (\alpha, \beta) \text{ for some } \alpha < \beta, f \in C[0,1], x^* \in X^*\}$. Due to boundedness, we get that $(\mathcal{T}, \tau_w)$ is a Polish space. Let $\mathcal{C}_\varepsilon \subset \mathcal{K}$ be a set containing all compact $L \subset [0,1]$, such that for some $T \in \mathcal{T}$ we have $\|T(f)\|_X \geq \varepsilon \sup_{t \in L} |f(t)|$. The set of all pairs $(L, T) \in \mathcal{K} \times \mathcal{T}$ satisfying the last condition is closed. Being the range of projection of the last set, $\mathcal{C}_\varepsilon$ is analytic. To achieve a contradiction, we may assume that all elements from $\mathcal{C}_\varepsilon$ are scattered since the opposite would imply that $C(L) \hookrightarrow X$ for some L containing a Cantor discontinuum $\mathbb{D}$, which implies that $C[0,1] \cong C(\mathbb{D}) \hookrightarrow X$ and the conclusion sought follows. By our assumption, for $\varepsilon > 0$ small enough, $\mathcal{C}_\varepsilon$ contains scattered compacts of height arbitrarily close to ω_1. Note that the Cantor derivation process for elements of $\mathcal{C}_\varepsilon$ is Borel (it follows also from our result on Szlenk derivation, which coincides with the Cantor derivation for scattered compacts), and so the corresponding ordering $\prec$, $L \prec M$ if and only if $L \subset M^{(1)}$ is an analytic partial ordering. It remains to apply Theorem 2.6, which claims that $\mathrm{rank}_\prec(\mathcal{C}_\varepsilon) < \omega_1$, which is a contradiction.

2.4 (Lancien [Lanc96]). Suppose $Y \hookrightarrow X$, $\mathrm{Sz}(X) < \omega_1$, $\mathrm{Sz}(Y) < \omega_1$. Then $\mathrm{Sz}(X) \leq \mathrm{Sz}(X/Y)\,\mathrm{Sz}(Y)$.

Hint. Using separable reduction, assume that X is separable.

Step 1. Let $\varepsilon > 0$, $F = 3B_{Y^\perp}$ and $B = F + \frac{\varepsilon}{3}B_{X^*}$. Show that then $B_\varepsilon^{\omega \alpha} \subset F_{\frac{\varepsilon}{3}}^\alpha + \frac{\varepsilon}{3}B_{X^*}$ for all $\alpha < \omega_1$. Proceed by induction. In the nonlimit step $\alpha = \beta+1$, use (and prove) that $B_\varepsilon^{\omega\beta+k} \subset (F_{\frac{\varepsilon}{3}}^\beta \setminus \bigcup_{i=1}^k V_i) + \frac{\varepsilon}{3}B_{X^*}$ for all $k \in \mathbb{N}$, where $\{V_i\}_{i=1}^\infty$ is a basis in w^*-topology, consisting of sets of diameter at most $\frac{\varepsilon}{3}$, of the set $F_{\frac{\varepsilon}{3}}^\beta \setminus F_{\frac{\varepsilon}{3}}^{\beta+1}$.

Step 2. Let $Q : X^* \to X^*/Y^\perp$. Show that, for every $\alpha < \omega_1$, $Q((B_{X^*})_\varepsilon^{\gamma_\varepsilon \alpha}) \subset (B_{X^*/Y^\perp})_{\frac{\varepsilon}{4}}^\alpha$, where $\gamma_\varepsilon = \omega S_{\frac{\varepsilon}{3}}(F) = \omega\,\mathrm{Sz}_{\frac{\varepsilon}{9}}(X/Y)$. Deduce from previous steps that $\mathrm{Sz}_\varepsilon(X) \leq \omega\,\mathrm{Sz}_{\frac{\varepsilon}{9}}(X/Y)\,\mathrm{Sz}_{\frac{\varepsilon}{4}}(Y)$. Finally, use that $\mathrm{Sz}(Z) = \omega^\alpha$ for every separable Z.

2.5 (Lancien [Lanc95]). Let K be a compact space. Then the following are equivalent:

(i) $K^{(\omega)} = \emptyset$.
(ii) $\mathrm{Sz}(C(K)) = \omega$.
(iii) $C(K)$ admits an equivalent norm with property Lipschitz UKK*, i.e., δ is a linear function of ε in the definition of UKK*.

(iv) $C(K)$ is (nonlinear) Lipschitz isomorphic to $c_0(\Gamma)$.

Hint. (i)$\Rightarrow$(iii) We have that $K^{(n)} = \emptyset$ for some $n \in \mathbb{N}$, $C(K)^* = \ell_1(K)$. Consider the dual renorming of $\ell_1(K)$ given by the formula

$$|||f||| = \sum_{i=0}^{n-1} \sum_{t \in K^{(i)} \setminus K^{(i+1)}} \frac{1}{2^i} |f(t)|.$$

(iii)$\Rightarrow$(ii) is obvious. (ii)$\Rightarrow$(i) relies on the fact that if $x \in K^{(\alpha)}$, then $\delta_x \in (B_{(C(K))^*})_1^{(\alpha)}$. (i)$\Leftrightarrow$(iv) is a theorem in [JoLi01h, Chap. 41, Thm. 8.7].

2.6 (Lancien [Lanc96]). Let K be a scattered compact such that $K^{\omega^\alpha} \neq \emptyset$ and $K^{\omega^{\alpha+1}} = \emptyset$ for some $\alpha < \omega_1$. Then $\mathrm{Sz}(C(K)) = \omega^{\alpha+1}$.

Hint. Separable reduction.

2.7 (Prus [Prus89]). An example of a nonsuperreflexive space with a Schauder basis satisfying a (p, q)-estimate for $1 < p, q < \infty$.

Hint. Let $1 < q < p < \infty$, fix $n \in \mathbb{N}$, and set $m = [n^{\frac{q}{p-q}}] + 1$. Put $A_k = \{in + k; i = 0, \ldots, m - 1\}$, $k = 1, \ldots, n$. Define a norm on $E_n = \mathbb{R}^{mn}$ $\|x\| = \max\{\sum_{j \in A_k} |x_j|^q)^{\frac{1}{q}}; k = 1, \ldots, n\}$. Show that E_n has an (∞, q)-estimate. Next, the canonical basis $\{e_i\}_{i=1}^{mn}$ of $(E_n)_p$ satisfies a (p, q)-estimate. Consider the vectors $v_k = \sum_{i=1}^{m-1} m^{-\frac{1}{q}} e_{in+k}$. We claim that $\{v_k\}_{k=1}^n$ is equivalent to the unit basis of ℓ_∞^n. To this end, let $x = \sum_{i=1}^l x_i$ be a blocking $x_1 < \cdots < x_l$ of an element $x = \sum_{k=1}^n a_k v_k$. Distinguish the cases; for $N_1 = \{i; \mathrm{card}\ \mathrm{supp}\ x_i \cap A_k \leq 1; \forall k\}$, we get $(\sum_{i \in N_1} \|x_i\|^p)^{\frac{1}{p}} \leq \max |a_k|$, and for $i \notin N_1$, let p_i be the number of intervals $[jn + 1, (j + 1)n]$ having nonempty intersection with $\mathrm{supp}\ x_i$, and we get $\|x_i\| \leq ((\max_k |a_k| m^{-\frac{1}{q}})^q p_i)^{\frac{1}{q}}$. Since $\sum_i p_i \leq 2m$, we get in the end the estimate $(\sum_{i=1}^l \|x_i\|^p)^{\frac{1}{p}} \leq 3 \max_k |a_k|$. To finish, put $X = \bigoplus(\sum_{n=1}^\infty (E_n)_p)_{\ell_p}$. X has a (p, q)-estimate but is not superreflexive as it contains ℓ_∞^n uniformly.

2.8. Show that not every separable reflexive space admits an equivalent norm with the UKK* property.

Hint. Let $X_n = \ell_{\frac{n+1}{n}}; n = 1, 2, \ldots$. Then $X = (\sum X_n)_{\ell_2}$ is reflexive and has no equivalent UKK* norm. To see this, first we show that the canonical norm on X is not UKK*. If $\varepsilon > 0$ is given, denote by $\eta_\varepsilon(B_{X^*})$ the collection of such points f in B_{X^*} so that there is a sequence of points (f_n) in B_{X^*} such that $\|f_n - f\| \geq \varepsilon$ and $f_n \to f$ in the w^* topology. From the definition of UKK*, it follows that there is $\delta > 0$ so that $\eta_\varepsilon(B_{X^*}) \subset \delta B_{X^*}$. By iterating this procedure and using the homogeneity, we get that, for large n, the iterated $\eta_\varepsilon^{(n)}(B_{X^*}) = \emptyset$. Thus, for proving that the canonical norm of X is not UKK* it suffices to show the following *Claim. If* $Y = \ell_p$, *where* $\frac{1}{p} + \frac{1}{q} = 1$ *and*

$m \leq 2^p$, then $\eta_{\frac{1}{2}}^{(n)}(B_{Y^*}) \neq \emptyset$. In order to see the claim, first note that if e_k are the unit vectors in ℓ_p and $n_1 < n_2 < ...n_m$, then $\|\sum_{k=1}^{k=m} \frac{1}{2}e_k\| = (\frac{m}{2^p})^{\frac{1}{p}} \leq 1$. Thus $\sum_{k=1}^{k=m-1} \frac{1}{2}e_k$ is in $\eta_{\frac{1}{2}}(B_{Y^*})$, as it is a w^* limit of $\sum_{k=1}^{k=m-1} \frac{1}{2}e + \frac{1}{2}e_i$ when $i \to \infty$ since the latter points are in B_{Y^*} and have distance $\frac{1}{2}$ to $\sum_{k=1}^{k=m-1} \frac{1}{2}e_k$. By iterating, we get that any $\sum_{k=1}^{k=m-n} \frac{1}{2}e_k \in \eta_{\frac{1}{2}}^n(B_{Y^*})$ for each $n < m$. Since 0 is in the weak* closure of the collection of such elements, we get that O is in the w^*-closure of $\eta_1^n(B_{Y^*})$. This by the above means that the canonical norm of X is not UKK*. It cannot have such an equivalent norm either since if $B_1 \subset B_2$, then obviously $\eta_\varepsilon(B_1) \subset \eta_\varepsilon(B_2)$.

3

Review of Weak Topology and Renormings

In this chapter, we discuss some basic tools from nonseparable Banach space theory that will be used in subsequent chapters. The first part concentrates on some fundamental results concerning Mackey and weak topologies. For example, the first section presents some of Grothendieck's basic results on the dual Mackey topology on dual Banach spaces. The second section includes work of Odell, Rosenthal, Emmanuele, Valdivia, and others on the sequential agreement of dual Mackey and norm topologies in spaces that do not contain ℓ_1. In the third section, our attention turns to classical results of Dunford, Pettis, and Grothendieck on weak compactness in $L_1(\mu)$ spaces and in the duals of $C(K)$ spaces; this section ends with the Josefson-Nissenzweig theorem, which shows that, for all infinite-dimensional spaces X, S_{X^*} is weak*-sequentially dense in B_{X^*}. These results will be needed in Chapter 7.

A ubiquitous tool to deal with nonseparable Banach spaces consists in decomposing the space in an orthogonal-like way by means of a transfinite sequence of projections starting with one of separable range, progressively increasing their range and ending with the identity operator (what is called a *projectional resolution of the identity*). The pioneering trail in this direction (in the setting of weakly compactly generated (WCG) Banach spaces) was opened by Amir and Lindenstrauss, and since then it has become a general and important procedure—sometimes the only one—to produce a certain type of basis in the space (such as an M-basis) or an equivalent norm with particular differentiability or rotundity properties. Comprehensive treatments of these techniques can be found in [DGZ93a] and in [Fab97]. In the fourth section, we present material from this topic that is needed in subsequent chapters.

Most—if not all—projectional resolutions of the identity are the result of the fundamental notion of a projectional generator, a concept that can be traced back gradually to [JoZi74a], [Vas81], [Plic83], [Fab87] and [OrVa89], and was crystallized by Valdivia [Vald88]. Such a device appears naturally in many nonseparable spaces (in WCG spaces, the definition of a projectional generator is the expected one, and in the some of the more general classes of Banach spaces, such as weakly Lindelöf determined (WLD) spaces, special w^*-compactness properties of the dual unit ball of a Banach space ensure that a projectional generator exists). The underlying ideas are so interweaved that from the existence of an M-basis in a WLD space, a projectional generator—and consequently a projectional resolution of the identity—can be easily constructed, and in turn this allows us to construct equivalent norms with certain desirable properties on the space. The fifth section presents some renorming techniques; it does not attempt to cover all such techniques but focuses on what is essential for later chapters—with special attention given to properties of Day's norm on $\ell_\infty(\Gamma)$. We refer to [MOTV] for a survey on some recent advances in renorming theory that use covering principles and Nagata-Smirnov-like techniques.

The concluding sixth section presents a quantitative version of Krein's classical theorem on the weak compactness of the closed convex hull of a weakly compact set—a result that will be needed in Chapter 6.

3.1 The Dual Mackey Topology

In this section, we present basic material, mainly due to Grothendieck, on the dual Mackey topology in connection with dual limited sets and related notions.

A couple of vector spaces, E and F (over $\mathbb{R}$), form a *dual pair* when there is a bilinear form $\langle \cdot, \cdot \rangle$ from $E \times F$ into $\mathbb{R}$ that separates points; i.e., the mapping on E given by $e \to \langle e, f \rangle$ is injective for every $f \in F$, and the mapping on F given by $f \to \langle e, f \rangle$ is injective for every $e \in E$. A dual pair is denoted by $\langle E, F \rangle$. A set $B \subset E$ is *bounded* when $\sup_{b \in B} |\langle b, f \rangle|$ is finite for every $f \in F$. A set $D \subset E$ is *fundamental* in E when $\langle d, f \rangle = 0$ for all $d \in D$ implies $f = 0$. A family $\mathcal{M}$ of bounded subsets of E such that $\bigcup_{M \in \mathcal{M}} M$ is fundamental in E defines naturally a Hausdorff locally convex topology $\mathcal{T}_{\mathcal{M}}(F, E)$ on F: a basis of $\mathcal{T}_{\mathcal{M}}$-neighborhoods of 0 is given by $\{M^\circ; M \in \mathcal{S}\mathcal{M}\}$, where $M^\circ := \{f \in F; \langle m, f \rangle \geq -1 \text{ for all } m \in M\}$, is called the *polar* of M, and $\mathcal{S}\mathcal{M}$ is the *saturation* of the family $\mathcal{M}$, i.e., the family of sets $\Gamma(M)$ for $M \in \mathcal{M}$, their scalar multiples, and the finite intersections of these. Here $\Gamma(\cdot)$ denotes the *absolutely convex* (i.e., the convex and balanced) hull. We call $\mathcal{T}_{\mathcal{M}}(F, E)$ the *topology on F of uniform convergence on the sets M of $\mathcal{M}$ or on $\mathcal{M}$*. The situation is symmetric, so families in F define topologies in E. Suitable choices of fundamental families of bounded subsets of F give rise to different topologies on E. We collect some of them in the following definition.

Definition 3.1. *Let $\mathcal{F}$ be the family of all the bounded finite-dimensional subsets of F. The topology $\mathcal{T}_{\mathcal{F}}(E, F)$ on E is called the* weak topology on E *associated to the dual pair $\langle E, F \rangle$ and is denoted by $w(E, F)$ for short.*

Let $\mathcal{B}$ be the family of all the bounded subsets of F. The topology $\mathcal{T}_{\mathcal{B}}(E, F)$ on E is called the strong topology on E *associated to the dual pair $\langle E, F \rangle$, and is denoted by $\beta(E, F)$ for short.*

Let $\mathcal{ACWK}$ be the family of all the absolutely convex and $w(F, E)$-bounded subsets of F. The topology $\mathcal{T}_{\mathcal{ACWK}}(E, F)$ on E is called the Mackey topology on E associated to the dual pair $\langle E, F \rangle$, and is denoted by $\tau(E, F)$ for short.

A locally convex topology $\mathcal{T}$ on E is said to be *compatible* with the dual pair $\langle E, F \rangle$ if the topological dual of $(E, \mathcal{T})$ is F.

When we consider a locally convex space $(E, \mathcal{T})$, the couple formed by E and its topological dual space E^*, plus the natural bilinear form $\langle x, x^* \rangle := x^*(x)$, $x \in X$, $x^* \in X^*$, fits the preceding scheme. The original topology $\mathcal{T}$ on E is the topology $\mathcal{T}_{\mathcal{M}}(E, E^*)$ of uniform convergence on the family $\mathcal{M}$ of all the $\mathcal{T}$-equicontinuous subsets of E^*. The weak topology $w(E, E^*)$ on E is denoted sometimes by w and the topology $w(E^*, E)$ on E^* by w^*.

Theorem 3.2. *Given a dual pair $\langle E, F \rangle$, the Mackey topology $\tau(E, F)$ is the strongest locally convex topology on E compatible with the dual pair $\langle E, F \rangle$. In particular, the closures of convex subsets of E in the $\tau(E, F)$ topology and in the $w(E, F)$ topology coincide.*

For a proof, see, e.g., [Fa~01, Thm. 4.33].

Remark 3.3. It is a well-known fact that *every Fréchet locally convex space E (i.e., a complete metrizable locally convex space) has a topological dual E^* that, when endowed with the topology $\tau(E^*, E)$, is complete* (see, for example, [Ko69, §21.6.4]). This happens, in particular, if E is a Banach space.

Note also that, in the setting of Banach spaces, the Mackey topology $\tau(E^*, E)$ in E^* is the topology of uniform convergence on the family of all weakly compact sets in E because of Krein's theorem (see, e.g., [Fa~01, Thm. 3.58]).

A topological vector space $(E, \mathcal{T})$ is uniformizable since it is completely regular ([Fa~01, p. 96]): the family of vicinities is given by $\{N_U; \ U \in \mathcal{U}\}$, where $\mathcal{U}$ is the family of neighborhoods of 0 and $N_U := \{(x, y) \in E \times E; \ x - y \in U\}$, $U \in \mathcal{U}$. A set $P \subset E$ is called *precompact* if it is relatively compact in $(\widetilde{E}, \widetilde{\mathcal{T}})$, the completion of $(E, \mathcal{T})$. The family $\mathcal{PK}$ of all precompact subsets of a locally convex space $(E, \mathcal{T})$ is fundamental and saturated and coincides with the family of its totally bounded subsets. A set $P \subset E$ is *totally bounded* whenever P can be covered, for every vicinity V, by a finite number of V-*small* subsets of E (i.e., sets $S \subset E$ such that $S \times S \subset V$).

The following elementary proposition characterizes the precompact subsets of a locally convex space (see, for example, [Grot73, Chap. II, Prop. 34]).

Proposition 3.4. *Let $(E, \mathcal{T})$ be a locally convex space. A set $P \subset E$ is precompact in $(E, \mathcal{T})$ if and only if it is bounded and the uniformities induced by $w(E, E^*)$ and $\mathcal{T}$ agree on P.*

Proof. Every bounded set P is obviously precompact in $(E, w(E, E^*))$ and thus $\mathcal{T}$-precompact if both uniformities agree on P. On the other hand, if P is precompact and V is a neighborhood of 0, there exists a finite set F such that $P \subset F + V$, so P is bounded. In the completion $(\widetilde{E}, \widetilde{\mathcal{T}})$, the set P is relatively compact, so uniformities associated to $\widetilde{\mathcal{T}}$ and $w(\widetilde{E}, E^*)$ coincide on P. They induce on E the $\mathcal{T}$- and $w(E, E^*)$-uniformities, respectively.   □

A basic result is the following.

Theorem 3.5. *Let $(E, \mathcal{T})$ be a locally convex space. Then $\mathcal{T}_{\mathcal{PK}}(E^*, E)$ and $w(E^*, E)$ agree on every $\mathcal{T}$-equicontinuous subset of E^*.*

Proof. Every $\mathcal{T}$-equicontinuous subset of E^* is obviously $w(E^*, E)$-bounded (even more, it is $w(E^*, E)$-relatively compact). As a straightforward consequence of the Arzelà-Ascoli theorem, it is relatively compact in the topology of the uniform convergence on every precompact set in E; this gives the conclusion.   □

We recall also the following well-known lemma and its corollary.

Lemma 3.6. *Let u be a linear mapping from a locally convex space E into a locally convex space F. Then its restriction to an absolutely convex set $A \subset E$ is uniformly continuous if and only if it is continuous at 0.*

Proof. Given a neighborhood V of 0 in F, there exists a neighborhood U of 0 in E such that $u(U/2 \cap A) \subset V/2$. Take $x, y \in A$ such that $x - y \in U$. Then $x - y \in U \cap 2A$, so $(x - y)/2 \in U/2 \cap A$. We then get $u((x - y)/2) \subset V/2$; this proves the uniform continuity of u on A.   □

Corollary 3.7. *Suppose that two locally convex topologies $\mathcal{T}_1$ and $\mathcal{T}_2$ are given on a vector space E. If $\mathcal{T}_1$ and $\mathcal{T}_2$ coincide on an absolutely convex subset M of E (or just assume that the restrictions of $\mathcal{T}_1$ and $\mathcal{T}_2$ to M coincide at 0), then the uniformities induced on M by $\mathcal{T}_1$ and $\mathcal{T}_2$ are the same.*

The following basic result relates two dual pairs linked by a linear form and their topologies of the uniform convergence on precompact sets.

Theorem 3.8 ([Grot73] Chap. 2, Thm. 12). *Let $\langle E, E' \rangle$ and $\langle F, F' \rangle$ be two dual pairs, $\mathcal{E}$ a fundamental family of bounded subsets of E and $\mathcal{F}'$ a fundamental family of bounded subset of F', $u : E \to F$ a $w(E, E')$-$w(F, F')$-continuous linear mapping, and $u' : F' \to E'$ its adjoint mapping. Then the following are equivalent:*

(1) *u transforms the sets $A \in \mathcal{E}$ into precompact subsets of $\left(F, \mathcal{T}_{\mathcal{F}'}(F, F')\right)$.*
(1') *u' transforms the sets $B' \in \mathcal{F}'$ into precompact subsets of $\left(E', \mathcal{T}_{\mathcal{E}}(E', E)\right)$.*
(2) *The restriction of u to the sets $A \in \mathcal{E}$ is uniformly continuous when we equip E with the topology $w(E, E')$ and equip F with the topology $\mathcal{T}_{\mathcal{F}'}(F, F')$.*

(2′) *The restriction of u′ to the sets B′ ∈ 𝔉′ is uniformly continuous when we equip F′ with the topology w(F′, F) and equip E′ with the topology $\mathcal{T}_{\mathcal{E}}(E', E)$.*

(3) *The restriction of the function $(e, f') \to \langle ue, f' \rangle = \langle e, u'f' \rangle$ to the sets $A \times B'$, $A \in \mathcal{E}$, $B' \in \mathcal{F}'$, is uniformly continuous for the uniform structure of $w(E, E') \times w(F', F)$. It even suffices to suppose these restrictions are uniformly continuous for the uniform structure of $w(E, E') \times \beta(F', F)$ or, conversely, for the uniform structure of $\beta(E, E') \times w(F', F)$.*

  Furthermore, when all the sets $A \in \mathcal{E}$ (resp. $B' \in \mathcal{F}'$) are absolutely convex we can replace in Condition 2 (resp. 2′) the uniform continuity by continuity and even by continuity at 0.

  The preceding conditions imply that, for every $A \in \mathcal{E}$, $B' \in \mathcal{F}'$, the set $\{\langle u(e), f' \rangle ; e \in A, f' \in B'\}$ is bounded.

Proof. We equip E' with the topology $\mathcal{T}_{\mathcal{E}}(E', E)$ and equip F with the topology $\mathcal{T}_{\mathcal{F}'}(F, F')$.

(1) ⇒ (2′) (1) implies that $u' : \left(F', \mathcal{T}_{\mathcal{PK}}(F', F)\right) \to \left(E', \mathcal{T}_{\mathcal{E}}(E', E)\right)$ is uniformly continuous. $B' \in \mathcal{F}'$ is $\mathcal{T}_{\mathcal{F}'}(F, F')$-equicontinuous; hence $\mathcal{T}_{\mathcal{PK}}(F', F)$ and $w(F', F)$ agree on B' by Theorem 3.5, and we get (2′).

(2′) ⇒ (1′) $B' \in \mathcal{F}'$ is a $w(F', F)$-precompact; hence the set $u'(B')$ is $\mathcal{T}_{\mathcal{E}}(E', E)$-precompact.

By symmetry, we have (1′) ⇒ (2) and (2)⇒(1), so (1), (1′), (2), and (2′) are equivalent.

Given x_1, $x_2 \in A \in \mathcal{E}$ and y_1', $y_2' \in B' \in \mathcal{F}'$, we can write

$$\langle ux_1, y_1' \rangle - \langle ux_2, y_2' \rangle = \langle u(x_1 - x_2), y_1' \rangle + \langle ux_2, y_1' - y_2' \rangle,$$

and then (2) and (2′) imply (3).

The first weakened statement of (3) implies (2): given $\varepsilon > 0$, there exists a $w(E, E')$-vicinity U for $A \in \mathcal{E}$ such that

$$|\langle ux_1, y' \rangle - \langle ux_2, y' \rangle| < \varepsilon$$

for all $x_1, x_2 \in A$ such that $(x_1, x_2) \in U$ and $y' \in B' \in \mathcal{F}'$. This is (2). Analogously the second weakened statement of (3) implies (2′).

The last statement follows, as $A \times B'$ is precompact for the uniformity associated to $w(E, E') \times w(F', F)$.

The absolutely convex version is a consequence of Lemma 3.6. □

Definition 3.9. *Let $\langle E, F \rangle$ be a dual pair. A set $L \subset E$ is called F-limited if $\sup_{e \in L} |\langle e, f_n \rangle| \to 0$ whenever (f_n) is a sequence in F such that $f_n \xrightarrow{w(F, E)} 0$.*

Remark 3.10. It is simple to prove that, for a Banach space X, X-limited sets in X^* and X^*-limited sets in X are $\|\cdot\|$-bounded (see Exercise 3.11). A simple consequence of Theorem 3.5 is that every $\|\cdot\|$-relatively compact subset of X is X^*-limited.

The next result characterizes the $\tau(X^*, X)$-relatively compact subsets of the dual X^* of a Banach space X.

Theorem 3.11 (Grothendieck [Grot53]). *Let X be a Banach space. Then a bounded set in X^* is $\tau(X^*, X)$-relatively compact if and only if it is X-limited.*

Proof. If $K \subset X^*$ is $\tau(X^*, X)$-relatively compact, then K is $\tau(X^*, X)$-precompact. Let (x_n) be a weakly null sequence in X. Then $\{x_n; n \in \mathbb{N}\} \cup \{0\}$ is $w(X, X^*)$-compact, and thus $\tau(X^*, X)$-equicontinuous. It follows from Theorem 3.5 that $x_n \to 0$ in $\left(X, \mathcal{T}_{\mathcal{PK}}(X, X^*)\right)$, in particular uniformly on K, and so K is X-limited. On the other hand, put $i : X \to X$, the identity mapping, $\mathcal{ACWK}$ the family of all absolutely convex weakly compact subsets of X, and $\mathcal{L}$ the family of all X-limited sets in X^* instead of u, $\mathcal{E}$, and $\mathcal{T}'$, respectively, in Theorem 3.8. It is enough to prove that every $W \in \mathcal{ACWK}$ is $\mathcal{T}_{\mathcal{L}}(X, X^*)$-precompact; then the $\tau(X^*, X)$-completeness of X^* will give the result. According to Proposition 3.4 and Corollary 3.7, we need to prove that the two topologies $w(X, X^*)$ and $\mathcal{T}_{\mathcal{L}}(X, X^*)$ coincide on W. Let $A \subset W$ be a closed subset of $\left(W, \mathcal{T}_{\mathcal{L}}(X, X^*)\right)$. It is $w(X, X^*)$-relatively compact. Let $x \in \overline{A}^{w(X, X^*)}$. The angelicity of $w(X, X^*)$-compact sets in a Banach space (see, for example, [Fa~01, Thm. 4.50] and Definition 3.31) allows us to choose a sequence (a_n) in A that $w(X, X^*)$-converges to x. Then $(x - a_n)$ is $w(X, X^*)$-null, so it is $\mathcal{T}_{\mathcal{L}}(X, X^*)$-null. It follows that $x \in A$ and so A is $w(X, X^*)$-closed. □

3.2 Sequential Agreement of Topologies in X^*

In this section, we show that spaces not containing copies of ℓ_1 can be characterized by the sequential coincidence of the dual Mackey and norm topologies in their dual spaces.

Definition 3.12. *A Banach space X is said to have*

(S) *the* Schur property *whenever every w-null sequence in X is $\|\cdot\|$-null;*

(DP) *the* Dunford-Pettis property *whenever every w-null sequence in X^* is $\tau(X^*, X)$-null;*

(G) *the* Grothendieck property *if X^* has the same w and w^*-convergent sequences.*

We shall say also that, if this is the case, the space X is Schur *(resp.* Dunford-Pettis*) (resp.* Grothendieck*).*

Recall that $C(K)$ has the Dunford-Pettis property whenever K is a compact space ([Fa~01, p. 376]), $\ell_1(\Gamma)$ has the Schur property ([Fa~01, p. 146]), and $\ell_\infty(\Gamma)$ has the Grothendieck property for any set Γ (Theorem 7.18). Observe that a Banach space X has the Schur property if and only if B_{X^*} is an X-limited set (see Exercise 3.12).

Proposition 3.13. *A Banach space X has the Dunford-Pettis property if and only if $x_n^*(x_n) \to 0$ whenever (x_n) is a weakly null sequence in X and (x_n^*) is a weakly null sequence in X^*.*

Proof. Clearly, the condition about sequences is necessary. Assume now that this condition holds. Let (x_n^*) be a sequence in X^* such that $x_n^* \overset{w}{\to} 0$. Let K be a weakly compact set in X such that (x_n^*) does not converge to 0 uniformly on K. Then there are elements $x_n \in K$ and some $\varepsilon > 0$ with $|x_n^*(x_n)| \geq \varepsilon$ for each n. As K is weakly compact, by the Eberlein-Šmulyan theorem we may assume without loss of generality that $x_n \overset{w}{\to} x \in K$. Then $x_n^*(x_n - x) \to 0$ and $x_n^*(x) \to 0$, and thus $x_n^*(x_n) \to 0$, a contradiction. $\square$

Definition 3.14. *A bounded linear operator T from a Banach space X into a Banach space Y is called a* Dunford-Pettis operator *or a completely continuous operator if T maps weakly null sequences to norm null sequences.*

Theorem 3.15 (Odell, Rosenthal [Rose77]). *Let X be a Banach space. Then X does not contain an isomorphic copy of ℓ_1 if and only if every Dunford-Pettis operator from X into any Banach space is compact.*

Proof. Assume that X does not contain an isomorphic copy of ℓ_1 and T is an operator from X into Y that is not compact. Then, for some $\varepsilon > 0$, there is an ε-separated infinite family $y_n = Tx_n$ in $TB_X \subset Y$. By Rosenthal's ℓ_1 theorem, there is a subsequence (z_n) of (x_n) that is weakly Cauchy. Then $(z_{n+1} - z_n)$ is a sequence that is weakly null, and its image under T is ε-separated, contradicting the fact that T is a Dunford-Pettis operator.

On the other hand, assume that X contains an isomorphic copy of ℓ_1. We will find a noncompact Dunford-Pettis operator from X into $L_1[0,1]$ as follows.

Let (r_n) be a bounded sequence in $L_\infty[0,1]$ that is not norm compact in $L_1[0,1]$ (for example, the Rademacher functions). Recall that the *Rademacher functions* (r_n) on $[0,1]$ are defined as follows:

$r_1 = 1$ everywhere, r_2 is 1 on $[0, \frac{1}{2})$ and -1 on $[\frac{1}{2}, 1]$, r_3 is 1 on $[0, \frac{1}{4})$ and $[\frac{1}{2}, \frac{3}{4}]$ and -1 on $[\frac{1}{4}, \frac{1}{2})$ and $[\frac{3}{4}, 1]$, etc.

Let an operator T from ℓ_1 into $L_\infty[0,1]$ be defined as follows: $T(\sum a_n e_n) := \sum a_n r_n$, where e_n are the unit vectors in ℓ_1. As $L_\infty[0,1]$ is an injective space, extend T to an operator $\tilde{T}$ from X into $L_\infty[0,1]$. Let I be the identity map from $L_\infty[0,1]$ into $L_1[0,1]$. Then consider the operator from X into $L_1[0,1]$ defined by $\hat{T} = I\tilde{T}$. Since I is a weakly compact operator (Theorem 3.24), we have that $\hat{T}$ is a weakly compact operator from X into $L_1[0,1]$. The space $L_\infty[0,1]$ is isomorphic to ℓ_∞, so I is a Dunford-Pettis operator. Therefore $\hat{T}$ is also a Dunford-Pettis operator. As $\hat{T}(e_n) = s_n$ for each n, $\hat{T}$ is not norm compact. $\square$

Theorem 3.16 (Emmanuele [Emm86]). *Let X be a Banach space. Then X does not contain an isomorphic copy of ℓ_1 if and only if the $\tau(X^*, X)$ and norm compact sets in X^* coincide.*

Proof. Assume that X does not contain a copy of ℓ_1 and that K is a $\tau(X^*, X)$-compact set. Then K is X-limited by Theorem 3.11. Consider the restriction mapping $T : X \to B(K)$, where $B(K)$ denotes the space of bounded functions on K equipped with the supremum norm. By the definition of an X-limited set, T is a Dunford-Pettis operator, so by Theorem 3.15, T is a compact operator. Therefore T^* is compact, so $K = T^*(K) \subset T(B_{C(K)^*})$ is compact in the norm of X^*.

Assume now that the condition on the coincidence of compacts holds and that $T : X \to Y$ is a Dunford-Pettis operator. Then $T^*(B_{Y^*})$ is X-limited and thus $\tau(X^*, X)$-relatively compact (Theorem 3.11). From the assumption, it follows that $T^*(B_{Y^*})$ is $\| \cdot \|$-relatively compact, so T^* (and then T) is compact. Therefore X does not contain a copy of ℓ_1 by Theorem 3.15. $\square$

Theorem 3.17 (Ørno [Orn91], Valdivia [Vald93a]). *A Banach space X does not contain an isomorphic copy of ℓ_1 if and only if the two topologies $\tau(X^*, X)$ and $\| \cdot \|$ agree sequentially on X^*.*

Proof. The fact that the condition is necessary can be deduced from Theorem 3.16: if the family of all $\tau(X^*, X)$-compacts in X^* coincides with the family of all $\| \cdot \|$-compacts, then the two topologies $\tau(X^*, X)$ and $\| \cdot \|$ agree sequentially.

To prove that the condition suffices, suppose that Y is a subspace of X that is isomorphic to ℓ_1, and let $\{e_n\}_{n=1}^\infty$ be the image of the unit vector basis under some isomorphism from ℓ_1 onto Y. Define a bounded linear operator from Y into $L_\infty[0,1]$ by mapping e_n to the n-th Rademacher function r_n. By the injective property of $L_\infty[0,1]$, this operator extends to a bounded linear operator T from X into $L_\infty[0,1]$. Let r_n^* be the n-th Rademacher function in $L_1[0,1]$ considered as a subspace of $(L_\infty[0,1])^*$. Thus the sequence (r_n^*), being equivalent to an orthonormal sequence in a Hilbert space, converges weakly to zero. Since $L_\infty[0,1]$ has the Dunford-Pettis property (see, for example, [Dies75, p. 113]), (r_n^*) converges in the Mackey topology $\tau(X^*, X)$ to zero and a fortiori $(T^* r_n^*)$ converges in the Mackey topology $\tau(X^*, X)$ to zero. But $\langle e_n, T^* r_n^* \rangle = \langle r_n, r_n^* \rangle = 1$, so $(T^* r_n^*)$ does not converge to zero in norm. $\square$

Corollary 3.18. *Let X be a nonreflexive Banach space. Then S_{X^*} is dense in B_{X^*} in the Mackey topology $\tau(X^*, X)$; however, if X does not contain an isomorphic copy of ℓ_1, then S_{X^*} is sequentially closed in the Mackey topology $\tau(X^*, X)$.*

Proof. Since X is nonreflexive, given an absolutely convex weakly compact set W in X and given $\varepsilon > 0$, W does not contain εB_X and thus, by the separation theorem, there is $f_\varepsilon \in S_{X^*}$ such that $\sup_W |f_\varepsilon| < \varepsilon$. Therefore 0 is in the closure of S_{X^*} in the Mackey topology $\tau(X^*, X)$. Then the first part of the statement follows by an easy homogeneity argument (see [Fa˜01, p. 88]). The second part follows from Theorem 3.17. $\square$

3.3 Weak Compactness in $ca(\Sigma)$ and $L_1(\lambda)$

In this section, we shall review some classical results of Dunford, Pettis, and Grothendieck on weak compact sets in $L_1(\mu)$ spaces and in spaces dual to $C(K)$ spaces that will be used in further chapters. We also present the Josefson-Nissenzweig theorem.

Let Ω be a nonempty set and Σ be a σ-algebra of subsets of Ω. Let $ba(\Sigma)$ be the linear space of all bounded finitely additive scalar-valued measures on Σ. We define two equivalent norms, $\|\cdot\|_\infty$ and $\|\cdot\|_1$, on $ba(\Sigma)$, given respectively by the formulas $\|\mu\|_\infty := \sup\{|\mu(E)|; E \in \Sigma\}$ and $\|\mu\|_1 := |\mu|(\Omega)$, where $|\mu|(E) := \sup\left\{\sum_{i=1}^n |\mu(E_i)|; E_i \in \Sigma, E_i \subset E, i = 1, 2, \ldots, n, \ n \in \mathbb{N}\right\}$ is the variation of μ on $E \in \Sigma$. It is clear that $\|\mu\|_\infty \le \|\mu\|_1 \le 4\|\mu\|_\infty$ for $\mu \in ba(\Sigma)$. Equipped with either of those two norms, $ba(\Sigma)$ is a Banach space. Let $ca(\Sigma)$ be the closed subspace of $ba(\Sigma)$ of all countably additive bounded measures on Σ. In what follows, we are always going to use the term *measure* in the sense of *countably additive measure* unless we specify otherwise. Then $ca(\Sigma)$ is again a Banach space. Given a positive member λ of $ca(\Sigma)$ (i.e., an element $\lambda \in ca^+(\Sigma)$), let $L_1(\lambda)$ be the closed linear subspace of $ca(\Sigma)$ of all equivalence classes (modulo equality λ-almost everywhere) of scalar-valued measurable functions on Ω such that $|f|$ is Lebesgue integrable with respect to λ and equipped with the restriction of $\|\cdot\|_1$. The identification is as follows: to an element $f \in L_1(\lambda)$, we associate $\mu \in ca(\Sigma)$ given by $\mu(E) := \int_E f d\lambda$ (moreover, $|\mu|(E) = \int_E |f| d\lambda$) for every $E \in \Sigma$.

Theorem 3.19. *Let B be a bounded set in $L_1(\mu)$ for a finite measure μ. Then either B is weakly relatively compact or B contains a sequence equivalent to the unit vector basis of ℓ_1.*

Proof (Sketch of the main idea; we follow [JoLi01]). We will use the following lemma due to Grothendieck.

Lemma 3.20. *Let X be a Banach space and $A \subset X$ be a subset of X. If for every $\varepsilon > 0$ there is a weakly compact set $A_\varepsilon \subset X$ such that $A \subset A_\varepsilon + \varepsilon B_X$, then A is weakly relatively compact.*

Proof. Since A is bounded, it is enough to show that the weak*-closure of A in X^{**} is actually in X. Since A_ε and $\varepsilon B_{X^{**}}$ are both weak*-compact in X^{**}, we have $\overline{A_\varepsilon + \varepsilon B_X}^{w^*} = A_\varepsilon + \varepsilon B_{X^{**}}$. Moreover, X is closed in X^{**} in the norm topology, and thus $\overline{A}^{w^*} \subset \bigcap_\varepsilon (A_\varepsilon + \varepsilon B_{X^{**}}) \subset \bigcap_\varepsilon (X \cap \varepsilon B_{X^{**}}) = X$. $\qquad\square$

Assume that μ is a probability measure. For $k \in \mathbb{N}$, put

$$a_k(W) := \sup\{\|x \restriction \{x; |x| \ge k\}\|_1; x \in W\}.$$

If $a(W)_k \to 0$, then we use the fact that, for each k, we have

$$W \subset k B_{L_\infty(\mu)} + a_k(W) B_{L_1(\mu)}.$$

Since $B_{L_2(\mu)}$ is weakly compact and $B_{L_\infty(\mu)} \subset B_{L_2(\mu)}$, the weak relative compactness of W follows from Lemma 3.20.

If $\lim a_k(W) > 0$, put

$$c(W) := \sup \lim_n \{\|x_n \upharpoonright A_n\|_1\},$$

where the supremum is over all sequences (x_n) in W and $\{A_n\}$ of pairwise disjoint measurable sets. It follows from measure theory that $c(W) = a(W)$. Choose $x_n \in W$ and pairwise disjoint measurable sets A_n so that

$$0 < \|x_n \upharpoonright A_n\|_1 \to c(W).$$

If follows that $\{x_n \upharpoonright A_n\}$, and thus also $\{x_n\}$ is equivalent to the unit vector basis of ℓ_1 ([JoLi01, p. 18]).   □

The following is a well-known consequence of Theorem 3.19. A more general result will be presented in Theorem 6.38.

Corollary 3.21 (Steinhaus). *The space $L_1(\mu)$ is weakly sequentially complete for any finite measure μ.*

Proof. Let (x_n) be a weakly Cauchy sequence in $L_1(\mu)$. Then (x_n) is bounded as it is weakly bounded. Then (x_n) is either relatively compact or contains a subsequence equivalent to the unit vector basis of ℓ_1 (Theorem 3.19). In the first case, (x_n) is weakly relatively sequentially compact by the Eberlein-Šmulyan theorem, and so (x_n) has a weakly convergent subsequence. Thus (x_n) is weakly convergent, as it is weakly Cauchy. In the second case, (x_n) has a subsequence (x_{n_k}) equivalent to the unit vector basis of ℓ_1. The subsequence (x_{n_k}) is weakly Cauchy in ℓ_1 and thus norm convergent. Therefore, (x_n) is again weakly convergent, as it is weakly Cauchy.   □

The following result characterizes weakly convergent sequences in the space $ca(\Sigma)$ and, a fortiori, in $L_1(\lambda)$.

Theorem 3.22. *A sequence (μ_n) in $ca(\Sigma)$ converges weakly to some $\mu \in ca(\Sigma)$ if and only if for each $E \in \Sigma$, $\lim_n \mu_n(E) = \mu(E)$. In particular, if $\lambda \in ca^+(\Sigma)$, then a sequence (f_n) in $L_1(\lambda)$ converges weakly to some $f \in L_1(\lambda)$ if and only if, for each $E \in \Sigma$, $\lim_n \int_E f_n d\lambda = \int_E f d\lambda$. Moreover, $ca(\Sigma)$ (and thus $L_1(\lambda)$) is weakly sequentially complete.*

The following result characterizes weakly relatively compact subsets of $ca(\Sigma)$.

Theorem 3.23. *Let K be a subset of $ca(\Sigma)$. Then the following are equivalent.*

(i) *K is weakly relatively compact.*

(ii) K *is bounded and* uniformly countably additive *(i.e., given a decreasing sequence (E_n) in Σ such that $\bigcap_{n=1}^{\infty} E_n = \emptyset$ and given $\varepsilon > 0$, there exists $n_0 \in \mathbb{N}$ such that $|\mu(E_n)| \leq \varepsilon$ for all $n \geq n_0$ and $\mu \in K$).*

(iii) K *is bounded and there exists $\lambda \in ca^+(\Sigma)$ such that K is* uniformly λ-continuous *(i.e., given $\varepsilon > 0$, there exists $\delta > 0$ such that $E \in \Sigma$ and $\lambda(E) \leq \delta$ implies $|\mu(E)| \leq \varepsilon$ for all $\mu \in K$).*

Proof. For the proof, see [DuSch, Thm. IV.9.1]. □

A straightforward consequence of the Radon-Nikodym theorem and Theorems 3.22 and 3.23 is the following classical result.

Theorem 3.24 (Dunford-Pettis). *Let $\lambda \in ca^+(\Sigma)$, and let K be a subset of $L_1(\lambda)$. Then the following are equivalent:*

(i) K *is weakly relatively compact.*

(ii) K *is bounded, and the countably additive measures defined by members of K form a uniformly countably additive family (see Theorem 3.23).*

(iii) K *is bounded and given $\varepsilon > 0$, there exists $\delta > 0$ such that $E \in \Sigma$ and $\lambda(E) \leq \delta$ imply $\int_E |f|\,d\lambda \leq \varepsilon$ for all f in K (or we say K is* uniformly integrable*).*

Given a measure $\mu \in ba(\Sigma)$, the topology of convergence in measure on the linear space $\mathcal{M}(\Sigma, \mu)$ of all equivalence classes (modulo equality μ-almost everywhere) of Σ-measurable scalar-valued functions is a vector topology with a basis of neighborhoods of the origin given by $V(\varepsilon) := \{f \in \mathcal{M}; \mu\{t \in \Omega; |f(t)| \geq \varepsilon\} \leq \varepsilon\}$, $\varepsilon > 0$. $\mathcal{M}(\Sigma, \mu)$ endowed with the topology of convergence in measure is a metrizable topological vector space that is not locally convex in general (see [Fa~01, Chap. 4]).

Theorem 3.25 (Grothendieck). *Let $\lambda \in ca^+(\Sigma)$. Then every bounded sequence in $L_\infty(\lambda)$ that converges to some $f \in L_\infty(\lambda)$ in measure converges to f in $\tau(L_\infty(\lambda), L_1(\lambda))$. The converse is true if λ is finite. If this is the case, the topology $\tau(L_\infty(\lambda), L_1(\lambda))$ and the topology of the convergence in measure coincide on $B_{L_\infty(\lambda)}$ (and $\tau(L_\infty(\lambda), L_1(\lambda))$ restricted to $B_{L_\infty(\lambda)}$ is thus metrizable).*

Proof. Let (f_n) be a sequence in $B_{L_\infty(\lambda)}$ that converges to zero in measure. Let $\varepsilon > 0$ and W be a weakly compact set in $L_1(\lambda)$. Assume without loss of generality that $W \subset B_{L_1(\lambda)}$. By Theorem 3.24, there is $0 < \delta < \varepsilon$ such that $\sup_{w \in W} \int_B |w|\,d\lambda < \varepsilon$ whenever $\lambda(B) < \delta$.

Pick $n_0 \in \mathbb{N}$ so that for $n \geq n_0$ we have $\lambda(\{t \in \Omega; |f_n(t)| \geq \delta\}) \leq \delta$. Define $B_n := \{t \in \Omega; |f_n(t)| \geq \delta\}$ and $A_n := \Omega \setminus B_n$, $n \in \mathbb{N}$. Then, for every $w \in W$ and $n \geq n_0$, we have

$$\left| \int_{\Omega} f_n w d\lambda \right| \leq \left| \int_{A_n} f_n w d\lambda \right| + \left| \int_{B_n} f_n w d\lambda \right|$$

$$\leq \delta \int_{A_n} |w| d\lambda + \int_{B_n} |w| d\lambda$$

$$\leq \delta \int_{\Omega} |w| d\lambda + \int_{B_n} |w| d\lambda$$

$$\leq \delta + \varepsilon < 2\varepsilon.$$

This proves the first part.

Assume now that λ is finite. In this case, we identify $L_\infty(\lambda)$ with a subset of $L_1(\lambda)$. In this identification, $B_{L_\infty(\lambda)}$ is, again by Theorem 3.24, a weakly compact subset of $L_1(\lambda)$. Then a $\tau(L_\infty(\lambda), L_1(\lambda))$-null sequence (f_n) in $L_\infty(\lambda)$ converges to 0 uniformly on $B_{L_\infty(\lambda)}$, i.e., in $(L_1(\lambda), \|\cdot\|_1)$, in particular in measure.

To finish the proof, we need to ensure that the families of closed subsets of $B_{L_\infty(\lambda)}$ are the same in both the restriction to $B_{L_\infty(\lambda)}$ of the topologies of the convergence in measure and $\tau(L_\infty(\lambda), L_1(\lambda))$. That every $\tau(L_\infty(\lambda), L_1(\lambda))$-closed subset is closed for the convergence in measure is clear from the first part and the fact that this last topology is metrizable. Assume now that $A \subset B_{L_\infty(\lambda)}$ is closed in the topology of the convergence in measure. An element $f \in \overline{A}^{\tau(L_\infty(\lambda), L_1(\lambda))}$ is in $\overline{A}^{\|\cdot\|_1}$ from the previous argument, so there is a sequence (f_n) in A that $\|\cdot\|_1$-converges to f, in particular in measure. This concludes that $f \in A$. □

Let K be a compact topological space. As is well known ([DuSch, Thm. IV.6.3]), the Banach space $C(K)^*$ can be isometrically isomorphically identified with the space $rca(\mathcal{B})$ of all regular countably additive measures on the σ-algebra $\mathcal{B}$ of the Borel subsets of K. The following result characterizes the weakly relatively compact subsets of $C(K)^*$.

Theorem 3.26 (Grothendieck [Grot53]). *Let K be a compact space, and let M be a bounded set in $C(K)^*$. Then the following are equivalent:*

(i) *M is relatively compact in $[C(K)^*, w(C(K)^*, C(K)^{**})]$.*
(ii) *M is $C(K)$-limited.*
(iii) *M is relatively compact in $[C(K)^*, \tau(C(K)^*, C(K))]$.*
(iv) *Whenever (O_j) is a sequence of pairwise disjoint open sets in K, then $\mu(O_j) \to 0$ uniformly for $\mu \in M$.*

Proof. (i)⇒(ii) Let (f_n) be a weakly null sequence in $C(K)$. We shall prove that it converges to 0 uniformly on M. By Theorem 3.23, there exists $\lambda \in ca^+(\mathcal{B})$ such that M is λ-uniformly continuous. By Egoroff's theorem, (f_n) converges in λ-measure to 0. We will now use Theorem 3.25 to obtain that $(f_n) \to 0$ in $\tau(L_\infty(\lambda), L_1(\lambda))$. By the Riesz representation theorem, M can be viewed as a subset of $L_1(\lambda)$, and it is weakly compact there, so the conclusion follows.

(ii)$\Leftrightarrow$(iii) is a consequence of Theorem 3.11.

(ii)$\Rightarrow$(iv) If (iv) fails, there exist an $\varepsilon > 0$, a sequence (O_n) of pairwise disjoint open subsets of K, and a sequence (μ_n) in M such that $|\mu_n(O_n)| \geq \varepsilon$ for all n. Let $f_n \in C(K)$ be supported by O_n, $0 \leq f_n(x) \leq 1$ for all $x \in K$, and such that $\left|\int_K f_n d\mu_n\right| > \varepsilon$. Obviously $(f_n) \to 0$ pointwise, and hence weakly ([Fa~01, Chap. 12]), which is a contradiction with M being $C(K)$-limited.

Given a set $S \subset K$, we shall denote $S^c := K \setminus S$.

(iv)$\Rightarrow$(i) We will first prove that (iv) implies the following.

Claim: For every compact $L \subset K$ and for every $\varepsilon > 0$, there is an open neighborhood U of L such that $|\mu|(U \cap L^c) \leq \varepsilon$ for every $\mu \in M$.

Assume that the claim does not hold. We can then find a compact subset $L \subset K$ and some $\varepsilon > 0$ such that for every open neighborhood V of L there exists $\mu \in M$ such that $|\mu(V \cap L^c)| > \varepsilon/4$. An inductive procedure produces sequences (K_n) of compact subsets of K, open pairwise subsets (O_n) of K, and (μ_n) of elements in M such that $K_{n+1} \subset O_{n+1} \subset \overline{O}_{n+1} \subset V_n \cap L^c$ and $|\mu_n(O_n)| > \varepsilon/4$ for all $n \in \mathbb{N}$. The construction starts by choosing any open neighborhood V_0 of L in K and then a compact set $K_1 \subset V_0 \cap L^c$ and an open set O_1 in $V_0 \cap L^c$ together with an element $\mu_1 \in M$ such that $K_1 \subset O_1 \subset \overline{O}_1 \subset V_0 \cap L^c$, $|\mu_1(O_1)| \geq |\mu_1(K_1)| > \varepsilon/4$. Put $V_1 := (\overline{O}_1)^c$ and repeat the construction to obtain a compact set $K_2 \subset V_1 \cap L^c$ and an open set O_2 in $V_1 \cap L^c$ together with an element $\mu_2 \in M$ such that $K_2 \subset O_2 \subset \overline{O}_2 \subset V_1 \cap L^c$, $|\mu_2(O_1)| \geq |\mu_2(K_2)| > \varepsilon/4$. Proceed in this way. The sequence (O_n) violates (iv), so the claim follows from (iv).

We will show that the claim implies (i). This will finish the proof. By the Eberlein-Šmulyan theorem, it is enough to prove that M is w-relatively sequentially compact. Then let (μ_n) be a sequence in M, and consider the space $L_1(\lambda)$, where $\lambda := \sum(1/2^n)\mu_n \in ca^+(\mathcal{B})$. By the Radon-Nikodym theorem, we can identify μ_n with an element $f_n \in L_1(\lambda)$, $n \in \mathbb{N}$. The proof will be finished as soon as we can prove that $\{f_n; n \in \mathbb{N}\}$ is a w-relatively compact subset of $L_1(\lambda)$. By the Dunford-Pettis theorem (Theorem 3.24), we need to show that for every $\varepsilon > 0$ there exists $\delta > 0$ such that, for every open (due to the regularity of the measure) set U in K of λ-measure $\leq \delta$, we have

$$\int_U |f_n|\, d\lambda \leq \varepsilon \text{ for } n \in \mathbb{N}.$$

Arguing by contradiction and passing to a subsequence if necessary, we can assume that there exists $\varepsilon > 0$ and open sets U_n in K such that $\lambda(U_n) \leq 2^{-n}$ and

$$\int_{U_n} |f_n| d\lambda > \varepsilon.$$

Put

$$V_n = \bigcup_{i \geq n} U_i, \ n \in \mathbb{N}.$$

Then (V_n) is a decreasing sequence of open sets the λ-measure of which tends to zero and

$$\int_{V_n} |f_n| d\lambda > \varepsilon, \ n \in \mathbb{N}.$$

By taking complements, the claim reads: *for each open set $V \subset K$ and for every $\alpha > 0$, there is a compact $L \subset V$ such that $|\lambda|(V \cap L^c) < \alpha$. Then*

$$\int_{V \cap L^c} |f_n| d\lambda \le \alpha \quad \text{for all } n \in \mathbb{N}.$$

Thus for each n, let a compact $K_n \subset V_n$ be such that

$$\int_{V_n \cap (K_n)^c} |f_k| d\lambda \le 2^{-n-1} \varepsilon \quad \text{for all } k \in \mathbb{N}.$$

Put

$$K'_n := \bigcap_{1 \le i \le n} K_i \subset K_n \subset V_n.$$

Then we have

$$\int_{K'_n} |f_n| d\lambda \ge \int_{V_n} |f_n| d\lambda - \int_{V_n \cap (K'_n)^c} |f_n| d\lambda.$$

Moreover,

$$V_n \cap (K'_n)^c = \bigcup_{1 \le i \le n} (V_n \cap (K_i)^c) \subset \bigcup_{1 \le i \le n} (V_i \cap (K_i)^c).$$

Therefore

$$\int_{V_n \cap (K'_n)^c} |f_n| d\lambda \le \sum_{1 \le i \le n} \int_{V_i \cap (K_i)^c} |f_n| d\lambda \le \sum_{1 \le i \le n} 2^{-i-1} \varepsilon \le \frac{1}{2} \varepsilon,$$

and thus

$$\int_{K'_n} |f_n| d\mu \ge \varepsilon - \frac{1}{2} \varepsilon = \frac{1}{2} \varepsilon \text{ for all } n \in \mathbb{N}.$$

The sequence (K'_n) of nonempty compact sets is decreasing and has nonempty intersection L, which is a compact subset of K. We have $\mu(L) = 0$.

By the claim, there is an open neighborhood U of L so that

$$\int_{U \cap L^c} |f_n| d\lambda < \frac{1}{2} \varepsilon \text{ for all } n \in \mathbb{N}.$$

As $\mu(L) = 0$, we thus have

$$\int_U |f_n| d\lambda < \frac{1}{2} \varepsilon \text{ for all } n \in \mathbb{N}.$$

Since $L := \bigcap_n K'_n$, there is $n_0 \in \mathbb{N}$ so that $K'_{n_0} \subset U$. Recall that we have for each $n \in \mathbb{N}$

$$\int_{K'_n} |f_n| d\lambda \geq \frac{1}{2}\varepsilon;$$

then we get

$$\frac{1}{2}\varepsilon > \int_U |f_{n_0}| d\lambda \geq \frac{1}{2}\varepsilon,$$

a contradiction.   □

Theorem 3.27 (Josefson [Jos75], Nissenzweig [Niss75]). *Let X be an infinite-dimensional Banach space. Then there is a sequence in X^* that is weak* convergent but not norm convergent.*

By using a simple geometric homogeneity argument (see, e.g., [Fa~01, p. 88]), we get the following corollary.

Corollary 3.28. *Let X be an infinite-dimensional Banach space. Then the weak*-sequential closure of S_{X^*} equals B_{X^*}.*

Proof (Theorem 3.27, a main idea). Note that in $\ell_\infty[0,1]$ the Rademacher functions are equivalent to the unit vector basis of ℓ_1. Now the main strategy of the proof is through the following lemma.

Lemma 3.29 (Hagler and Johnson [HaJo77]). *Assume that X^* contains a copy of ℓ_1 and X does not contain a copy of ℓ_1. Then there is a weak*-null sequence in X^* that is not equivalent to the unit vector basis of ℓ_1.*

Proof (Sketch; we will follow [Dies75]). Assume that no weak*-null sequence in X^* is equivalent to the unit vector basis of ℓ_1. We will find a copy of ℓ_1 in X. Let (y_n^*) be a sequence in B_{X^*} that is equivalent to the unit vector basis of ℓ_1. Put

$$\delta := \sup_{x \in S_X} \limsup_n |y_n^*(x)|.$$

Since (y_n^*) is equivalent to the unit vector basis of ℓ_1, it is not norm convergent to 0 and thus, by our assumption, y_n^* does not weak*-converge to 0. Thus $\delta(y_n*) > 0$.

Let $\varepsilon > 0$ be given. There is $x_1 \in S_X$ and an infinite set $N_1 \in \mathbb{N}$ such that, for any $n \in N_1$,

$$y_n^*(x_1) < -\delta + \varepsilon.$$

Partition N_1 into two disjoint infinite subsets (m_k) and (n_k) of positive integers. The sequence $\frac{1}{2}(y_{n_k}^* - y_{m_k}^*)$ is a normalized block basis of ℓ_1, so there is $x_2 \in S_X$ and an infinite set of k's for which

$$\frac{1}{2}(y_{n_k}^* - y_{m_k}^*)(x_2) > \delta - \varepsilon.$$

As $(y_{n_k}^*)$ and $(y_{m_k}^*)$ are normalized ℓ_1 block bases, then for all but finitely many k, we get

$$\max\{|y^*_{n_k}(x_2)|, |y^*_{m_k}(x_2)|\} < \delta + \varepsilon.$$

Thus, for large enough k, we have

$$y^*_{n_k}(x_2) > \delta - 3\varepsilon \quad \text{and} \quad y^*_{m_k}(x_2) < -\delta + 3\varepsilon.$$

We put

$$N_2 = \{n_k; y^*_{n_k}(x_2) > \delta - \varepsilon\} \text{ and } N_3 = \{m_k; y^*_{m_k}(x_2) < -\delta + \varepsilon\}.$$

We can keep going; letting $\Omega_n = \{y^*_k; k \in N_n\}$, we get a tree of subsets of B_{X^*}. Furthermore, (x_n) has been selected so that if $2^{n-1} \leq k < 2^n$, then $(-1)^k x_n(y^*) \geq \delta - \varepsilon$ for all $y^* \in \Omega_k$. Thus (x_n) is a Rademacher-like sequence in X that produces, by the note at the beginning of the proof, a copy of ℓ_1 in X. □

We can now finish the proof of Theorem 3.27. Assume that X is a Banach space for which Theorem 3.27 does not hold. Then evidently X^* is Schur and as such contains a copy of ℓ_1 (Exercise 3.16). If X does not contain a copy of ℓ_1, we get a contradiction by Lemma 3.29. If X contains a copy of ℓ_1, then we get a contradiction by the proof of Theorem 3.17. □

3.4 Decompositions of Nonseparable Banach Spaces

This section presents the approach to projectional resolutions of the identity using the concept of a projectional generator; it also collects material on this topic that is needed in further chapters.

Definition 3.30. *A compact space K is called an* Eberlein *compact space if it is homeomorphic to a subset of $(c_0(\Gamma), w)$ for some set Γ. A compact space K is called a* Corson *compact space if K is homeomorphic to a subset C of $[-1,1]^\Gamma$, for some set Γ, such that each point in C has only a countable number of nonzero coordinates. A compact space K is called a* Valdivia *compact space if K is homeomorphic to a set V in $[-1,1]^\Gamma$, for some Γ, such that V contains a dense subset whose points have only a countable number of nonzero coordinates.*

Among the Corson compacta are all Eberlein compacta—in particular, all metrizable compacta—(see, e.g., [Fa~01, Example after Definition 11.14] and [Fa~01, Remark after Definition 12.44]). Due to the Amir-Lindenstrauss theorem ([AmLi68]; see, e.g., [Fa~01, Thm. 11.16]), dual balls of WCG spaces are Eberlein compact if endowed with their w^*-topologies.

Definition 3.31. *A topological space T is* angelic *if every relatively countably compact subset S of T is relatively compact and every point in the closure of S is the limit of a sequence in S. A topological space T is called a* Fréchet-Urysohn *space if for every subset S of T and for every point s in the closure* *of S there exists a sequence in S converging to s. A topological space T has* countable tightness *if for every subset S of T and for every point s in the closure of S there exists a countable set $C \subset S$ such that $s \in \overline{C}$.*

Note that, for compact spaces, the concepts of angelic and Fréchet-Urysohn spaces coincide. Note that all Corson compacta are angelic (see Proposition 5.27).

Definition 3.32. *A Banach space is* Vašák *(i.e., weakly countably determined) if there exists a countable collection $\{K_n\}$ of w^*-compact subsets of X^{**} such that for every $x \in X$ and $u \in X^{**} \setminus X$ there is n_0 for which $x \in K_{n_0}$ and $u \notin K_{n_0}$.*

A Banach space X is called weakly Lindelöf determined (WLD) *if (B_{X^*}, w^*) is a Corson compact.*

Every Vašák space is WLD (see Theorem 5.37 and 6.25).

For the construction of a projection in a Banach space, the following two simple lemmas are useful.

Lemma 3.33. *Let X be a Banach space and $\Delta \subset X$ a set such that $\overline{\Delta}$ is a linear subspace, and let $\nabla \subset X^*$ be a set that 1-norms Δ (i.e., $\|x\| = \sup_{b^* \in B_\nabla} |\langle x, b^* \rangle|$, where $B_\nabla := \{b^* \in \nabla; \|x^*\| = 1\}$). Then $\overline{\Delta} \oplus \nabla_\perp$ is a topological direct sum, and $\|P\| = 1$, where $P : \overline{\Delta} \oplus \nabla_\perp \to \overline{\Delta}$ is the canonical projection.*

Proof. Obviously $\overline{\Delta} \cap \nabla_\perp = \{0\}$, so $\overline{\Delta} \oplus \nabla_\perp$ is an algebraic direct sum. Moreover, given $x \in \overline{\Delta}$ and $y \in \nabla_\perp$,

$$\|P(x+y)\| = \|x\| = \sup_{b^* \in B_\nabla} |\langle x, b^* \rangle| = \sup_{b^* \in B_\nabla} |\langle x + y, b^* \rangle| \le \|x + y\|,$$

and hence $\|P\| = 1$ and the direct sum is topological (in particular, closed). □

The proof of the following lemma is easy, so we omit it and refer to [DGZ93a, Lemma VI.2.4].

Lemma 3.34. *Let X be a Banach space and Δ and ∇ be as in Lemma 3.33, with $\overline{\nabla}^{w^*}$ a linear subspace. Then $\overline{\Delta} \oplus \nabla_\perp = X$ if and only if $\Delta^\perp \cap \overline{\nabla}^{w^*} = \{0\}$.*

Definition 3.35. *Let X be a nonseparable Banach space, and let μ be the first ordinal with card $\mu = \operatorname{dens} X$. A* projectional resolution of the identity *(PRI, for short) on X is a family $\{P_\alpha : \omega \le \alpha \le \mu\}$ of linear projections on X such that $P_\omega \equiv 0$, P_μ is the identity mapping, and for all $\omega \le \alpha \le \mu$ the following hold:*

(i) $\|P_\alpha\| = 1$,
(ii) $\operatorname{dens} P_\alpha X \le \operatorname{card} \alpha$,
(iii) $P_\alpha P_\beta = P_\beta P_\alpha = P_\alpha$ *if* $\omega \le \beta \le \alpha$, *and*
(iv) $\bigcup_{\beta < \alpha} P_{\beta+1} X$ *is norm dense in $P_\alpha X$ if $\omega < \alpha$.*

Given a subset $S \subset X$ of a Banach space X, we denote $\mathrm{span}_{\mathbb{Q}}(S) := \{x; x = \sum_{i=1}^{n} a_i z_i, a_i \in \mathbb{Q}, z_i \in S, i = 1, 2, \ldots, n, \ n \in \mathbb{N}\}$. We say that $Y \subset X$ is $\mathbb{Q}$-*linear* if $\mathbb{Q}$-$\mathrm{span}(Y) = Y$.

Lemma 3.36. *Let X be a Banach space, $W \subset X^*$ be $\mathbb{Q}$-linear, and $\Phi : W \to 2^X$ and $\Psi : X \to 2^W$ be at most countably valued mappings. Suppose $A_0 \subset X$, $B_0 \subset W$, card (A_0), and card $(B_0) \leq \Gamma$ for some cardinal Γ. Then there exist $\mathbb{Q}$-linear sets A, B, $A_0 \subset A \subset X$, $B_0 \subset B \subset W$, such that card (A), card $(B) \leq \Gamma$ and $\Phi(B) \subset A$, $\Psi(A) \subset B$.*

Proof. We will construct by induction two sequences of sets $A_0 \subset A_1 \subset A_2 \cdots \subset X$, $B_0 \subset B_1 \subset B_2 \ldots W$ as follows. Having constructed $A_0 \ldots A_n$, $B_0 \ldots B_n$, we put

$$A_{n+1} := \mathbb{Q}\text{-span}(A_n \cup \Psi(B_n)),$$
$$B_{n+1} := \mathbb{Q}\text{-span}(B_n \cup \Phi(A_n)).$$

Finally, we set $A := \bigcup_{n=0}^{\infty} A_n$, $B := \bigcup_{n=0}^{\infty} B_n$. That A and B satisfy the required properties is obvious. $\qquad\square$

Lemma 3.37. *Let X, W, Φ, Ψ, A_ω, B_ω, card A_ω, and card $B_\omega \leq \omega$ be as in Lemma 3.36. Assume that μ is the first ordinal with cardinal dens $X > \omega$. Then there exist families $\{A_\alpha; \omega \leq \alpha \leq \mu\}$ and $\{B_\alpha; \omega \leq \alpha \leq \mu\}$ of $\mathbb{Q}$-linear subsets of X and W, respectively, such that for each $\omega \leq \alpha \leq \mu$ the following hold:*

(i) card $(A_\alpha) \leq \alpha$, card $(B_\alpha) \leq \alpha$, $\overline{A_\mu} = X$,
(ii) $\Phi(B_\alpha) \subset A_\alpha$, $\Psi(A_\alpha) \subset B_\alpha$,
(iii) $A_\alpha \subset A_\beta, B_\alpha \subset B_\beta$ *if* $\omega \leq \alpha < \beta \leq \mu$,
(iv) $A_\alpha = \bigcup_{\beta < \alpha} A_{\beta+1}$, $B_\alpha = \bigcup_{\beta < \alpha} B_{\beta+1}$ *if* $\omega < \alpha$.

Proof. Suppose that $(x_\alpha)_{\omega \leq \alpha < \mu}$ is a dense sequence in X. We denote $A_\omega := A$, $B_\omega := B$. We proceed by transfinite induction using Lemma 3.36, assuming that for some $\gamma \leq \mu$ and every α, $\omega \leq \alpha < \gamma$, we have already constructed the sets $A_\alpha \subset X$ and $B_\alpha \subset W$ satisfying the required properties. If γ is a limit ordinal, put $A_\gamma := \bigcup_{\alpha < \gamma} A_\alpha$ and $B_\gamma := \bigcup_{\alpha < \gamma} B_\alpha$. If γ is nonlimit, apply Lemma 3.36 to the pair $A_{\gamma-1} \cup \{x_{\gamma-1}\}$, $B_{\gamma-1}$, and the cardinality card γ, in order to obtain the sets A_γ and B_γ. $\qquad\square$

The following concept for constructing PRI's is due to Valdivia [Vald88]. This concept is now at the core of most of the results on decompositions of non-separable Banach spaces. For precedents, see, e.g., [JoZi74a], [Vas81], [Plic83], and [Fab87], and for further developments, see, e.g., [OrVa89], [Vald90a], [Vald91], and [Ori92].

Definition 3.38. *Let X be a Banach space and $W \subset X^*$ be a 1-norming $\mathbb{Q}$-linear subset. Let $\Phi : W \to 2^X$ be at most a countably valued mapping such that for every nonempty set $B \subset W$ with linear closure, $\Phi(B)^\perp \cap \overline{B}^{w^*} = \{0\}$. Then the couple (W, Φ) is called a* projective generator *(PG) on X. A projective generator is called* full *if $W = X^*$.*

Definition 3.39. *A Banach space X has the* separable complementation property (SCP) *if every separable subspace of X is contained in a separable complemented subspace Z of X. The space X has the* 1-separable complementation property (1-SCP) *if it has the SCP and, for all Z as above, the projection $P : X \to Z$ satisfies $\|P\| = 1$.*

The notion of SCP is quite useful, especially for constructing counterexamples. We will show in Theorem 3.42 that all spaces with a projectional generator (WCG, WLD, Plichko, duals to Asplund spaces) have 1-SCP.

Definition 3.40. *Let X be a Banach space. Let $A \subset X$ and $B \subset X^*$ be two nonempty sets. We say that A* countably supports B *(or that B is* countably supported by A) *if* card $\{a \in A; \langle a, b \rangle \neq 0\} \leq \aleph_0$ *for all $b \in B$.*

Definition 3.41. *Let X be a Banach space with a PRI $\{P_\alpha; \omega \leq \alpha \leq \mu\}$. We shall say that a set $G \subset X$ is* subordinated *to the given PRI (or that the PRI is* subordinated *to the set G) if $P_\alpha(x) \in \{0, x\}$ for all $\omega \leq \alpha \leq \mu$ and $x \in G$.*

Theorem 3.42. *Let X be a nonseparable Banach space X, and let μ be the first ordinal with cardinal* dens X. *If X has a projectional generator $(W, \tilde{\Phi})$, then X has the 1-SCP and admits a PRI $\{P_\alpha; \omega \leq \alpha \leq \mu\}$. Moreover, given a set $G \subset X$ that countably supports W, we may in addition assume that G is subordinated to $\{P_\alpha; \omega \leq \alpha \leq \mu\}$.*

Proof. Let (W, Φ) be the projectional generator on X defined by $\Phi(w) = \{x \in G; w(x) \neq 0\} \cup \tilde{\Phi}(w)$ for all $w \in W$. Choose a countable valued mapping $\Psi : X \to 2^W$ such that $\|x\| = \sup_{f \in \Psi(x) \cap B_{X^*}} f(x)$ for all $x \in X$. Given any separable $Y \hookrightarrow X$, choose a countable $\mathbb{Q}$-linear set A_ω such that $\overline{A_\omega} = Y$, and let $B_\omega := \emptyset$. Put $P_\omega := 0$. Now, applying Lemma 3.37, we get families $\{A_\alpha; \omega \leq \alpha \leq \mu\}$ and $\{B_\alpha; \omega \leq \alpha \leq \mu\}$ of subsets of X and W, respectively, with the properties listed therein. By Lemmas 3.33 and 3.34, there exist norm-1 projections $P_\alpha : X \to X$ such that $P_\alpha(X) = \overline{A_\alpha}$, $P_\alpha^{-1}(0) = B_\alpha^\perp$, and $P_\alpha^*(X^*) = \overline{B_\alpha}^{w^*}$. The properties of $\{A_\alpha\}$ and $\{B_\alpha\}$ listed in Lemma 3.37 yield that $\{P_\alpha; \omega \leq \alpha \leq \mu\}$ is a PRI on X. This implies the 1-SCP property, as $Y \hookrightarrow P_{\omega+1}X$. Next we claim that $G \subset A_\mu \cup \{0\}$. Assuming the opposite, choose $0 \neq x \in G \setminus A_\mu$. As $\Phi(B_\mu) \subset A_\mu$, we get $x \notin \Phi(B_\mu)$. It follows that $s(x) = 0$ for all $s \in B_\mu$, hence $x \in (B_\mu)_\perp = \{0\}$, a contradiction. This proves the claim. In order to prove the last statement, choose, for an element $x \in G$, a minimal α with $x \in A_\alpha$. By the definition of A_α, we know that $\alpha = \beta + 1$ for some β. Since $x \notin A_\beta$, we get $x \notin \Phi(B_\beta)$, and hence $x \in (B_\beta)_\perp$. It follows that $x \in \overline{A}_{\beta+1} \cap (B_\beta)_\perp = (P_{\beta+1} - P_\beta)X$, and the proof is complete. $\square$

Valdivia, in [Vald90b], strengthened Theorem 3.42 by proving that, *if X is a Banach space containing a linearly dense subset G with the property that the set $\Sigma(G)$ of all elements in X^* countably supported by G satisfies that $\Sigma(G) \cap B_{X^*}$ is w^*-dense in B_{X^*}, then X has a PRI subordinated to G.*

The following proposition provides a natural example of a PG for the class of WCG Banach spaces. As a consequence of it and Theorem 3.42, we obtain the Amir-Lindenstrauss theorem on the existence of a PRI in every WCG Banach space and the 1-SCP of those spaces.

Proposition 3.43. *If X is a* weakly compactly generated *Banach space and K is an absolutely convex weakly compact and linearly dense subset of X, then X admits a full projectional generator (X^*, Φ) such that $\Phi(x^*) \subset K$ for all $x^* \in X^*$. As a result, a PRI $\{P_\alpha; \omega \leq \alpha \leq \mu\}$ on X can be constructed in such a way that $P_\alpha(K) \subset K$ for all $\omega \leq \alpha \leq \mu$.*

Proof. Given $x^* \in X^*$, let $\Phi(x^*) \in K$ be such that $\langle \Phi(x^*), x^* \rangle = \sup |\langle K, x^* \rangle|$. We claim that (X^*, Φ) is a PG. In order to prove the claim, let $W \subset X^*$ be such that $\mathrm{span}_{\mathbb{Q}}(W) = W$. Let $x^* \in \Phi(W)^\perp \cap \overline{B}_W^{w^*}$. By the Mackey-Arens theorem (Theorem 3.2), $\overline{B}_W^{w^*} = \overline{B_W}^{\tau(X^*,X)}$. Note that $\overline{B_W}^{\tau(X^*,X)} \subset \overline{B_W}^{\mathcal{T}_K}$, where $\mathcal{T}_K$ is the (metrizable) topology on X^* of the uniform convergence on K. Let (x_n^*) be a sequence in B_W such that $x_n^* \xrightarrow{\mathcal{T}_K} x^*$. Fix $\varepsilon > 0$ and find $n_0 \in \mathbb{N}$ such that $\sup |\langle K, x^* - x_n^* \rangle| < \varepsilon$ for all $n \geq n_0$. Then, in particular, $\sup |\langle K, x_n^* \rangle| = |\langle \Phi(x_n^*), x_n^* \rangle| = |\langle \Phi(x_n^*), x^* - x_n^* \rangle| < \varepsilon$ for all $n \geq n_0$. This implies that $\sup |\langle K, x^* \rangle| < 2\varepsilon$. As $\varepsilon > 0$ is arbitrary, we get $x^*|_K \equiv 0$, and so $x^* = 0$. This proves the claim and the result. ☐

Theorem 3.44. *Let $(X, \|\cdot\|_0)$ be a Banach space with a full PG $\tilde{\Phi}$. Let $(M_n)_{n=1}^\infty$ be a sequence of absolutely convex closed and bounded subsets of X. Let $G \subset X$ be a set that countably supports X^*. Then there exists a PRI $\{P_\alpha; \omega \leq \alpha \leq \mu\}$ on X such that $P_\alpha(M_n) \subset M_n$ for all $n \in \mathbb{N}$ and G is subordinated to $(P_\alpha)_{\omega \leq \alpha \leq \mu}$.*

Proof. We follow ideas in [Fab97, Prop. 6.1.10]. The construction of the PRI mimics the proof of Theorem 3.42, adding some simple changes: for $n, m \in \mathbb{N}$, let $\|\cdot\|_{n,m}$ be the equivalent norm on X whose closed unit ball is $\overline{M_n + (1/m)B_{(X,\|\cdot\|_0)}}$; for $x \in X$, define $\Psi(x) \subset X^*$ such that $\Psi(x)$ is countable and, for $n, m \in \mathbb{N}$,

$$\|x\|_0 = \sup \langle x, \Psi(x) \cap B_{(X^*,\|\cdot\|_0)} \rangle, \quad \|x\|_{n,m} = \sup \langle x, \Psi(x) \cap B_{(X^*,\|\cdot\|_{n,m})} \rangle.$$

Then, as in the proof of Theorem 3.42, we obtain a PRI $\{P_\alpha; \omega \leq \alpha \leq \mu\}$ (where μ is the first ordinal with cardinal $\mathrm{dens}\,X$) with $\|P_\alpha\|_0 = 1$ and $\|P_\alpha\|_{n,m} = 1$ for all $\omega \leq \alpha \leq \mu$ and $n, m \in \mathbb{N}$. From this we obtain the additional properties recorded in the statement. Indeed, $P_\alpha(M_n + (1/m)B_{(X,\|\cdot\|_0)}) \subset \overline{M_n + (1/m)B_{(X,\|\cdot\|_0)}}$ for all $m, n \in \mathbb{N}$ and hence

$$P_\alpha(M_n) \subset \bigcap_{m \in \mathbb{N}} \overline{M_n + (1/m)B_{(X,\|\cdot\|_0)}} \subset M_n$$

for all $n \in \mathbb{N}$. To prove the last part of the statement, the same argument as in the proof of Theorem 3.42 can be used. ☐

The following notion is useful in dealing with transfinite induction processes. It will be investigated in more detail in Section 5.1, where some examples of $\mathcal{P}$-classes of Banach spaces are given.

Definition 3.45. *We say that a class $\mathcal{C}$ of Banach spaces is a $\mathcal{P}$-class if, for every $X \in \mathcal{C}$, there exists a PRI $\{P_\alpha; \omega \leq \alpha \leq \mu\}$ such that $(P_{\alpha+1} - P_\alpha)(X) \in \mathcal{C}$ for all $\alpha < \mu$, where μ is the first ordinal with cardinal $\operatorname{dens} X$.*

Theorem 3.46. *Let $\mathcal{C}$ be a $\mathcal{P}$-class of Banach spaces. Then, if $X \in \mathcal{C}$, there exists a family of projections $\{Q_\gamma; \omega \leq \gamma \leq \mu\}$ (where μ is the first ordinal with cardinal $\operatorname{dens} X$) such that, letting $R_\gamma := (Q_{\gamma+1} - Q_\gamma)/(\|Q_{\gamma+1}\| + \|Q_\gamma\|)$,*

  (i) *$Q_\gamma Q_\delta = Q_\delta Q_\gamma = Q_\gamma$ if $\omega \leq \gamma \leq \delta \leq \mu$.*
  (ii) *$Q_\omega(X)$ and $R_\gamma(X)$ are separable for $\omega \leq \gamma < \mu$.*
  (iii) *$Q_\mu = Id_X$.*
  (iv) *For every $x \in X$, $\{\|R_\gamma x\|; \gamma \in [\omega, \mu)\} \in c_0([\omega, \mu))$.*
  (v) *For every $x \in X$ and $\gamma \in [\omega, \mu)$, $Q_\gamma x \in \overline{\operatorname{span}}(\{R_\delta(x); \omega \leq \delta < \gamma\} \cup \{Q_\omega(x)\})$.*

Moreover, if $M \subset X$ countably supports X^*, the family $\{Q_\gamma; \omega \leq \gamma \leq \mu\}$ can be chosen such that $Q_\gamma(x) \in \{0, x\}$ for all $x \in M$ and all $\omega \leq \gamma \leq \mu$.

Proof. The construction is done by transfinite induction on $\operatorname{dens} X$. It is trivial if X is separable. If the result is true for every element of density less than some uncountable cardinal $\aleph$ in a given $\mathcal{P}$-class $\mathcal{C}$, and if $X \in \mathcal{C}$ has density $\aleph$, a family $\{Q_\gamma\}_{\omega \leq \gamma \leq \mu}$ of projections on X satisfying (i) to (v) exists, where μ is the first with cardinal $\aleph$. For every $\gamma \in [\omega, \mu)$, the set $(P_{\gamma+1} - P_\gamma)(M)$ countably supports $(P_{\gamma+1}^* - P_\gamma^*)(X^*)$. By the induction hypothesis, on $(P_{\gamma+1} - P_\gamma)(X)$ there exists a family $\{\pi_\beta^\gamma\}_{\omega \leq \beta \leq \mu_\gamma}$ of projections verifying (i) to (v) and subordinated to $(P_{\gamma+1} - P_\gamma)(M)$, where μ_γ is the first ordinal with cardinal $\operatorname{dens}(P_{\gamma+1} - P_\gamma)(X)$. For $\beta \in [\omega, \mu_\gamma)$ and $\gamma \in [\omega, \mu)$, set

$$Q_{\gamma,\beta} := P_\gamma + \pi_\beta^\gamma(P_{\gamma+1} - P_\gamma).$$

The family $\{Q_{\gamma,\beta}\}_{\beta \in [\omega, \mu_\gamma), \gamma \in [\omega, \mu)}$, ordered lexicographically, satisfies (i) to (v) and is subordinated to M. $\square$

The transfinite sequence of projections $\{Q_\alpha; \omega \leq \alpha \leq \mu\}$ constructed in Theorem 3.46 is called a *separable projectional resolution of the identity* (SPRI) (see, e.g., [Fab97, Def. 6. 26]). The reader should note that it is not, strictly speaking, a PRI, as the projections do not necessarily have norm 1 (in fact, the norms of the projections may form an unbounded set).

3.5 Some Renorming Techniques

This section presents a deeper study of Day's norm and the related renorming theorem of Troyanski that will be needed later in this book.

For the reader's convenience, we collect here some definitions.

Definition 3.47. *The norm* $\| \cdot \|$ *on a Banach space* X *is said to be:*

(i) Gâteaux differentiable ($\equiv$ smooth) (G) *if*

$$\lim_{t \to 0+} (1/t)(\|x + th\| + \|x - th\| - 2) = 0 \qquad (3.1)$$

for every $x \in S_X$ *and every* $h \in X$;

(ii) Fréchet differentiable (F) *if, for every* $x \in S_X$, *the limit in* (3.1) *is uniform in* $h \in S_X$;

(iii) uniformly Gâteaux differentiable (UG) *if, for every* $h \in S_X$, *the limit in* (3.1) *is uniform in* $x \in S_X$;

(iv) uniformly Fréchet differentiable (UF) *if the limit in* (3.1) *is uniform in* $x \in S_X$ *and* $h \in S_X$;

(v) rotund ($\equiv$ strictly convex) (R) *if* $x, y \in S_X$ *with* $\|x + y\| = 2$ *implies* $x = y$;

(vi) 2-rotund (2R) *if* $x_n \xrightarrow{\|\cdot\|} x$ *for some* $x \in X$ *whenever* $x_n \in S_X$ *are such that* $\lim_{m,n \to \infty} \|x_m + x_n\| = 2$;

(vii) *w*-2-rotund (*w*-2R) *if* $x_n \xrightarrow{w} x$ *for some* $x \in X$ *whenever* $x_n \in S_X$ *are such that* $\lim_{m,n \to \infty} \|x_m + x_n\| = 2$;

(viii) locally uniformly rotund (LUR) *if* $x_n \xrightarrow{\|\cdot\|} x$ *whenever* $x, x_n \in S_X$ *are such that* $\lim_n \|x + x_n\| = 2$; *and*

(ix) uniformly rotund (UR) *if* $\|x_n - y_n\| \to 0$ *whenever* $x_n, y_n \in S_X$ *are such that* $\lim_n \|x_n + y_n\| = 2$.

The dual norm $\| \cdot \|$ *in* X^* *is said to be*

(x) weakly*-uniformly rotund (W*UR) *if* $(x_n^* - y_n^*) \to 0$ *in the weak*-topology of* X^* *whenever* $x_n^*, y_n^* \in S_{X^*}$ *are such that* $\lim_n \|x_n^* + y_n^*\| = 2$.

For information on these concepts and renorming, we refer to [DGZ93a] and [JoLi01h, Chap. 18].

Theorem 3.48 (Troyanski [Troy71]). *Every Banach space with a strong M-basis has an equivalent LUR norm.*

Proof. Let $\{b_i; b_i^*\}_{i \in I}$ be a strong M-basis of a Banach space X. Assume, without loss of generality, that $\|b_i^*\| \leq 1$ for all $i \in I$. Define, for a finite set $A \subset I$ and some $x \in X$,

$$F_A(x)^2 := \sum_{i \in A} |\langle x, b_i^* \rangle|^2,$$

$$D_A(x) := \| \cdot \|\text{-dist}\,(x, \mathrm{span}\{b_i; i \in A\}).$$

We have $F_A(x) \leq \sqrt{\mathrm{card}\, A}\, \|x\|$ and $D_A(x) \leq \|x\|$ for all $x \in X$, so both F_A and D_A are continuous seminorms on X. Given positive integers l and n, let

$$G_{l,n}^2(x) := \sup\{l F_A^2(x) + D_A^2(x);\ \mathrm{card}\, A \leq n,\ A \subset I\}, \quad x \in X.$$

Then $G_{l,n}$ is a continuous seminorm with $G_{l,n}(x) \leq \sqrt{nl+1}\|x\|$ for all $x \in X$. Finally, define an equivalent norm $\|\|\cdot\|\|$ on X by

$$\|\|x\|\|^2 := \|x\|^2 + \sum_{l,n=1}^{\infty} \frac{1}{nl+1} \frac{1}{2^{l+n}} G_{l,n}^2(x), \ x \in X.$$

We shall prove that $\|\|\cdot\|\|$ is LUR.

For this, fix x and a sequence (x_k) in X such that

$$\|\|x + x_k\|\| \to 2, \ x \in S_{(X,\|\|\cdot\|\|)}, \ x_k \in S_{(X,\|\|\cdot\|\|)}, \ k \in \mathbb{N}. \tag{3.2}$$

We shall prove that $x_k \to x$ in norm. The strategy will be the following. First, we will check that $\{x_k; k \in \mathbb{N}\}$ is a $\|\cdot\|$-relatively compact subset of X by proving that $\{x_k; k \in \mathbb{N}\}$ is as close as we wish to a certain norm-compact subset of X. Second, we will see that $\langle x_k, b_i^* \rangle \to_k \langle x, b_i^* \rangle$ for every $i \in I$. Finally, the coincidence of the topologies $\|\cdot\|$ and $\sigma(X, \text{span}\{b_i^*; i \in I\})$ (i.e., the topology of the pointwise convergence on the set $\text{span}\{b_i^*; \ i \in I\}$) on the norm-compact set $\overline{\{x_k; k \in \mathbb{N}\} \cup \{x\}}^{\|\cdot\|}$ gives the result.

To accomplish it, let $\{i_1, i_2, \ldots\} := \{i \in I; \langle x, b_i^* \rangle \neq 0\}$, where $|\langle x, b_1^* \rangle| \geq |\langle x, b_2^* \rangle| \geq \ldots$. Fix $\varepsilon > 0$. The M-basis $\{b_i; b_i^*\}_{i \in I}$ is strong, so we can find a finite set $B := \{i_1, i_2, \ldots, i_m\} \subset \{i_n; n \in \mathbb{N}\}$ such that

$$D_B(x) < \varepsilon \text{ (this fixes } m \in \mathbb{N}\text{)}. \tag{3.3}$$

By enlarging B if necessary, we may always assume that $|\langle x, b_{i_m}^* \rangle| > |\langle x, b_{i_{m+1}}^* \rangle|$, and this gives $\delta > 0$ such that

$$\sup\{F_A^2(x); \ A \neq B, \text{card } A \leq m\} + \delta < F_B^2(x). \tag{3.4}$$

We can now find $l \in \mathbb{N}$ such that

$$l\delta - \|x\|^2 > 0. \tag{3.5}$$

By the definition of $G_{l,m}(x + x_k)$, we can find finite sets $A_k \subset I$ such that card $A_k \leq m$, for all $k \in \mathbb{N}$, and

$$(0 \leq) c_k := G_{l,m}^2(x + x_k) - lF_{A_k}^2(x + x_k) - D_{A_k}^2(x + x_k) \xrightarrow{k} 0. \tag{3.6}$$

We shall verify the following threefold *Claim*:

(∗) $A_k = B$ for big enough k.

(∗∗) $F_B(x_k) \to_k F_B(x)$ and $F_B(x + x_k) \to_k 2F_B(x)$.

(∗∗∗) $D_B(x_k) \to_k D_B(x)$.

Before proving this claim, let us note the following consequences.

(a) From (∗∗∗), and recalling inequality (3.3), there exists $k_0 = k_0(\varepsilon) \in \mathbb{N}$ such that $D_B(x_k) < 2\varepsilon$ for $k \geq k_0$; noticing that (x_k) is a bounded sequence, this provides a closed bounded subset S_ε of the finite-dimensional space span$\{b_i; i \in B\}$ (hence S_ε is a compact set) such that $\mathrm{dist}\,(x_k, S_\varepsilon) < 2\varepsilon$ for $k \geq k_0$. This gives another compact set in the finite-dimensional space span$\{x_1, \ldots, x_{k_0}, b_i; i \in B\}$ with the same property, and this holds for every $\varepsilon > 0$: *the set $\{x_k; k \in \mathbb{N}\}$ is thus a $\|\cdot\|$-relatively compact set in X.*

(b) Given $j \notin B$ and $k \geq k_0(\varepsilon)$, we have $\langle x_k, x_j^* \rangle = \langle x_k - y, x_j^* \rangle + \langle y, x_j^* \rangle = \langle x_k - y, x_j^* \rangle$ for every $y \in \mathrm{span}\{x_i; i \in B\}$. From $D_B(x_k) < 2\varepsilon$, we get, in particular, $|\langle x_k, x_j^* \rangle| < 2\varepsilon$ for $k \geq k_0$.

(c) To deal with the case $j \in B$, define a mapping T from span$\{b_i; i \in B\}$ onto $\ell_2(B)$ by $T(\sum_{i \in B} \alpha_i b_i) = (\alpha_i)_{i \in B}$, an isomorphism. From $(**)$, Fact 1 below, and the LUR property of ℓ_2, we get $T(x_k) \to_k T(x)$, which implies $\langle x_k, b_j^* \rangle \to_k \langle x, b_j^* \rangle$ for $j \in B$.

This proves that (x_k) is $\sigma(X, \mathrm{span}\{x_i^*; i \in \mathbb{N}\})$-convergent to x. Thus, it remains to prove the claim. Let us start by establishing $(*)$. It is convenient to recall two facts (see, e.g., [DGZ93a, Fact II.2.3]) that come from the convexity of a seminorm p and the nonnegativity of the expression $2p^2(x) + 2p^2(y) - p^2(x + y)$.

Fact 1. *For a sequence (x_k), the following are equivalent:*

(1) $\lim_k p(x_k) = p(x)$ *and* $\lim_k p(x + x_j) = 2p(x)$.
(2) $2p^2(x) + 2p^2(x_k) - p^2(x + x_k) \to_k 0$.

Fact 2. *Let (p_n) be a sequence of seminorms and (α_n) a sequence of positive numbers. Then, if the seminorm p defined by $p^2 := \sum_n \alpha_n p_n^2$ satisfies $2p^2(x) + 2p^2(x_k) - p^2(x + x_k) \to_k 0$ for a certain sequence (x_k), we have also $2p_n^2(x) + 2p_n^2(x_k) - p_n^2(x + x_k) \to_k 0$ for every $n \in \mathbb{N}$.*

Fix $k \in \mathbb{N}$ and assume that $A_k \neq B$. Then $F_{A_k}^2(x) + \delta < F_B^2(x)$. From this,

$$
\begin{aligned}
G_{l,m}^2(x) &\geq l F_B^2(x) + D_B^2(x) \\
&> l F_{A_k}^2(x) + l\delta + D_B^2(X) \geq l F_{A_k}^2(x) + l\delta - \|x\|^2 + D_{A_k}^2(X) \\
&= (l\delta - \|x\|^2) + \left(l F_{A_k}^2(x) + D_{A_k}^2(X) \right).
\end{aligned} \tag{3.7}
$$

So we have, for every $k \in \mathbb{N}$ such that $A_k \neq B$,

$$
\begin{aligned}
2G_{l,m}^2&(x) + 2G_{l,m}^2(x_k) - G_{l,m}^2(x + x_k) \\
&> 2\Big\{ (l\delta - \|x\|^2) + \left(l F_{A_k}^2(x) + D_{A_k}^2(x) \right) \Big\} + 2\Big\{ l F_{A_k}^2(x_k) + D_{A_k}^2(x_k) \Big\} \\
&\quad - \Big\{ l F_{A_k}^2(x + x_k) + D_{A_k}^2(x + x_k) + c_k \Big\} \\
&= 2(l\delta - \|x\|^2) + \Big\{ l\Big(2F_{A_k}^2(x) + 2F_{A_k}^2(x_k) - F_{A_k}^2(x + x_k) \Big) \Big\} \\
&\quad + \Big\{ l\Big(2D_{A_k}^2(x) + 2D_{A_k}^2(x_k) - D_{A_k}^2(x + x_k) \Big) \Big\} - c_k
\end{aligned}
$$

(here we used (3.7), the definition of $G_{l,m}^2(x_k)$ and (3.6), in that order). If k is big enough, we reach a contradiction, thanks to the following reasons:

- $2G_{l,m}^2(x) + 2G_{l,m}^2(x_k) - G_{l,m}^2(x + x_k) \to_k 0$ (recall (3.2) together with Facts 1 and 2 above),

- $2(l\delta - \|x\|^2) > 0$ (inequality (3.5)),
- the two expressions between square brackets are nonnegative, and
- $c_k \to 0$ when $k \to \infty$ (see (3.6)).

This proves claim $(*)$.

From now on, we may assume that $A_k = B$ for all $k \in \mathbb{N}$. We have

$$
\begin{aligned}
0 \stackrel{k}{\leftarrow} & \; 2G_{l,m}^2(x) + 2G_{l,m}^2(x_k) - G_{l,m}^2(x + x_k) \\
\geq & \; 2\Big(lF_B^2(x) + D_B^2(x)\Big) + 2\Big(lF_B^2(x_k) + D_B^2(x_k)\Big) \\
& -2\Big(lF_B^2(x + x_k) + D_B^2(x + x_k)\Big) \\
= & \; l\Big(2F_B^2(x) + 2F_B^2(x_k) - F_B^2(x + x_k)\Big) \\
& + \Big(2D_B^2(x) + 2D_B^2(x_k) - D_B^2(x + x_k)\Big) \geq 0.
\end{aligned}
$$

So, again from Facts 1 and 2 above, we get $(**)$ and $(***)$ and the claim is proved. This concludes the proof. $\qquad\square$

Note that the existence of an M-basis in a Banach space is not sufficient for having an LUR renorming; see [Fa~01, Thm. 6.45].

Definition 3.49. *Let Γ be a set. We will denote by $\ell_\infty^c(\Gamma)$ the closed subspace of $\ell_\infty(\Gamma)$ consisting of all vectors with only a countable number of nonzero coordinates. We will call a subspace S of $\ell_\infty(\Gamma)$ a Sokolov subspace if we can write $\Gamma = \bigcup_n \Gamma_n$ such that given $f_0 \in S$, given $\gamma_0 \in \Gamma$, and given $\varepsilon > 0$, there is $n \in \mathbb{N}$ such that $\gamma_0 \in \Gamma_n$ and $\{\gamma \in \Gamma_n; |f(\gamma)| > \varepsilon\}$ is finite.*

Proposition 3.50. *$S \subset \ell_\infty^c(\Gamma)$ for every Sokolov subspace S of $\ell_\infty(\Gamma)$.*

Proof. Indeed, observe that if $f \in S$ and $\varepsilon > 0$ are given, we get that Γ is covered by Γ_n such that $\{\gamma \in \Gamma_n; |\langle \gamma, f \rangle| \geq \varepsilon\}$ is finite. Thus $\{\gamma \in \Gamma; |\langle \gamma, f \rangle| \geq \varepsilon\}$ is countable. $\qquad\square$

Theorem 3.51. *Let Γ be a set. Then each Sokolov subspace S of $\ell_\infty(\Gamma)$ admits a norm that is* pointwise LUR, *i.e., for every $\gamma \in \Gamma$, $(f_n - f)(\gamma) \to 0$ whenever $f_n, f \in S_S$ are such that $\|f_n + f\| \to 2$.*

Before starting on the proof, we need to discuss some prerequisites. Let Γ be an infinite set. We recall that the *Day norm* $\| \cdot \|_\mathcal{D}$ on $\ell_\infty(\Gamma)$ is defined, for $u \in \ell_\infty(\Gamma)$, by

$$
\|u\|_\mathcal{D}^2 := \sup \Big\{ \sum_{j=1}^n 2^{-j} u(\gamma_j)^2;
$$
$$
n \in \mathbb{N}, \; \gamma_1, \ldots, \gamma_n \in \Gamma, \; \gamma_k \neq \gamma_l \text{ if } k \neq l \Big\}. \tag{3.8}
$$

It is easy to check that $\| \cdot \|_\mathcal{D}$ is an equivalent norm on $\ell_\infty(\Gamma)$.

The following elementary statement can be found in [Dies75, p. 95].

Lemma 3.52. *Let $(s_k)_{(k \in \mathbb{N})}$ and $(t_k)_{(k \in \mathbb{N})}$ be two nonincreasing sequences of non-negative numbers such that $s_k = t_k = 0$ for all large $k \in \mathbb{N}$. Let $\pi : \mathbb{N} \to \mathbb{N}$ be an injective surjection. Then*

$$\sum_{k=1}^{\infty} s_k(t_k - t_{\pi(k)}) \geq 0$$

and for every $K \in \mathbb{N}$ either $\pi\{1,\ldots,K\} = \{1,\ldots,K\}$ or

$$(s_K - s_{K+1})(t_K - t_{K+1}) \leq \sum_{k=1}^{\infty} s_k(t_k - t_{\pi(k)}).$$

Proposition 3.53. *Let Γ be an infinite set, let $u \in \ell_\infty(\Gamma)$, $\varepsilon > 0$, and assume that the set $\{\gamma \in \Gamma;\ |u(\gamma)| > \varepsilon\}$ is finite. Let $u_n \in \ell_\infty(\Gamma)$, $n \in \mathbb{N}$, be such that*

$$2\|u\|_{\mathcal{D}}^2 + 2\|u_n\|_{\mathcal{D}}^2 - \|u + u_n\|_{\mathcal{D}}^2 \to 0 \quad \text{as} \quad n \to \infty.$$

Then $\limsup_{n\to\infty} \|u - u_n\|_\infty \leq 3\varepsilon$.

Proof. The argument is an elaboration of that due to Rainwater (see, e.g., [Dies75, pp. 94–100]). Denote $A := \{\gamma \in \Gamma;\ |u(\gamma)| > \varepsilon\}$, and let $\{\alpha_1,\ldots,\alpha_K\}$ be an enumeration of A such that $|u(\alpha_1)| \geq |u(\alpha_2)| \geq \cdots \geq |u(\alpha_K)|\ (> \varepsilon)$. Denote

$$\Delta = \left(2^{-K} - 2^{-K-1}\right)\left(u(\alpha_K)^2 - \varepsilon^2\right);$$

this is a positive number. Fix an arbitrary $n \in \mathbb{N}$. We find a set $B_n \neq A$, $A \subset B_n \subset \Gamma$, such that

$$\|u + u_n\|_{\mathcal{D}}^2 - \tfrac{1}{n} < \big\|(u + u_n) \restriction B_n\big\|_{\mathcal{D}}^2.$$

Enumerate

$$B_n = \left\{\alpha_1^n,\ldots,\alpha_{K_n}^n\right\} = \left\{\beta_1^n,\ldots,\beta_{K_n}^n\right\}$$

in such a way that

$$|u(\alpha_1^n)| \geq |u(\alpha_2^n)| \geq \cdots \geq |u(\alpha_{K_n}^n)|,$$
$$|(u + u_n)(\beta_1^n)| \geq |(u + u_n)(\beta_2^n)| \geq \cdots \geq |(u + u_n)(\beta_{K_n}^n)|.$$

Then, of course, $\alpha_1^n = \alpha_1,\ldots,\alpha_K^n = \alpha_K$ and $K_n > K$. Note that

$$\sum_{k=1}^{K_n} 2^{-k} u(\alpha_k^n)^2 = \big\|u \restriction B_n\big\|_{\mathcal{D}}^2 \leq \|u\|_{\mathcal{D}}^2,$$

$$\sum_{k=1}^{K_n} 2^{-k} u_n(\beta_k^n)^2 \leq \big\|u_n \restriction B_n\big\|_{\mathcal{D}}^2 \leq \|u_n\|_{\mathcal{D}}^2,$$

and

$$\|u + u_n\|_{\mathcal{D}}^2 - \frac{1}{n} < \left\|(u + u_n) \restriction B_n\right\|_{\mathcal{D}}^2 = \sum_{k=1}^{K_n} 2^{-k}(u + u_n)(\beta_k^n)^2.$$

Let us estimate

$$2\|u\|_{\mathcal{D}}^2 + 2\|u_n\|_{\mathcal{D}}^2 - \|u + u_n\|_{\mathcal{D}}^2$$

$$> 2\|u \restriction B_n\|_{\mathcal{D}}^2 + 2\|u_n \restriction B_n\|_{\mathcal{D}}^2 - \|(u + u_n) \restriction B_n\|_{\mathcal{D}}^2 - \frac{1}{n}$$

$$\geq 2\sum_{k=1}^{K_n} 2^{-k}u(\alpha_k^n)^2 + 2\sum_{k=1}^{K_n} 2^{-k}u_n(\beta_k^n)^2 - \sum_{k=1}^{K_n} 2^{-k}(u + u_n)(\beta_k^n)^2 - \frac{1}{n}$$

$$= 2\sum_{k=1}^{K_n} 2^{-k}\left(u(\alpha_k^n)^2 - u(\beta_k^n)^2\right) + \sum_{k=1}^{K_n} 2^{-k}\left(u(\beta_k^n) - u_n(\beta_k^n)\right)^2 - \frac{1}{n} \geq -\frac{1}{n}.$$

Indeed, the first summand is nonnegative by Lemma 3.52. Hence, letting $n \to \infty$ here, we get

$$\sum_{k=1}^{K_n} 2^{-k}\left(u(\alpha_k^n)^2 - u(\beta_k^n)^2\right) \to 0 \quad \text{and} \quad u(\beta_k^n) - u_n(\beta_k^n) \to 0, \text{ for } k = 1, \dots, K.$$

Find $n_0 \in \mathbb{N}$ so large that, for all $n \in \mathbb{N}$ greater than n_0,

$$\sum_{k=1}^{K_n} 2^{-k}\left(u(\alpha_k^n)^2 - u(\beta_k^n)^2\right) < \Delta \text{ and } |u(\beta_k^n) - u_n(\beta_k^n)| < 3\varepsilon, \text{ for } k = 1, \dots, K.$$

$$(3.9)$$

Fix for awhile any such n. Let $\pi : \mathbb{N} \to \mathbb{N}$ be defined as

$$\pi(k) := \begin{cases} k \text{ if } k \in \mathbb{N} \text{ and } k > K, \\ j \text{ if } k \in \mathbb{N}, \ k \leq K_n, \text{ and } \beta_k^n = \alpha_j^n. \end{cases}$$

Clearly, π is an injective mapping from $\mathbb{N}$ onto $\mathbb{N}$. We claim that

$$\{\alpha_1^n, \dots, \alpha_K^n\} = \{\beta_1^n, \dots, \beta_K^n\},$$

that is, $\pi\{1, \dots, K\} = \{1, \dots, K\}$. Assume that this is not true. Putting $s_k = 2^{-k}$, $t_k = u(\alpha_k^n)^2$ for $k = 1, \dots, K_n$, and $s_k = t_k = 0$ for $k = K_n + 1, K_n + 2, \dots$, we get from Lemma 3.52 and (3.9)

$$(0 < \Delta \leq) \left(2^{-K} - 2^{-K-1}\right)\left(u(\alpha_K^n)^2 - u(\alpha_{K+1}^n)^2\right)$$

$$\leq \sum_{k=1}^{K_n} 2^{-k}\left(u(\alpha_k^n)^2 - u(\beta_k^n)^2\right) \ (< \Delta),$$

a contradiction. This proves the claim. For all $n > n_0$, we thus have that $\{\beta_1^n, \dots, \beta_K^n\} = A = \{\alpha_1, \dots, \alpha_K\}$ and thus

$$|(u_n - u)(\alpha_1)| < 3\varepsilon, \ldots, \ |(u_n - u)(\alpha_K)| < 3\varepsilon.$$

Now we are ready to prove that $\limsup_{n\to\infty} \|u_n - u\|_\infty \leq 3\varepsilon$. Assume the contrary. Then there is an infinite set $N \subset \mathbb{N}$ such that for every $n \in N$ there is $\gamma_n \in \Gamma$ so that $|(u_n - u)(\gamma_n)| > 3\varepsilon$. This immediately implies that $\gamma_n \notin A$ for all $n \in N$ with $n > n_0$. But for these n's we have

$$\sum_{k=1}^{K} 2^{-k} u_n(\alpha_k)^2 + 2^{-K-1} u_n(\gamma_n)^2 \leq \|u_n\|_{\mathcal{D}}{}^2,$$

and so

$$2^{-K-1} \limsup_{n\in N,\ n\to\infty} u_n(\gamma_n)^2 \leq \lim_{n\to\infty} \|u_n\|_{\mathcal{D}}{}^2 - \lim_{n\to\infty} \sum_{k=1}^{K} 2^{-k} u_n(\alpha_k)^2$$

$$= \|u\|_{\mathcal{D}}{}^2 - \sum_{k=1}^{K} 2^{-k} u(\alpha_k)^2 = \|u \restriction (\Gamma \backslash A)\|_{\mathcal{D}}{}^2 \leq \varepsilon^2 \sum_{k=K+1}^{\infty} 2^{-k} = \varepsilon^2 \cdot 2^{-K}.$$

Consequently, $\limsup_{n\in N,\ n\to\infty} |u_n(\gamma_n)| \leq \sqrt{2}\varepsilon < 2\varepsilon$, and thus

$$(3\varepsilon \leq) \ \limsup_{n\in N,\ n\to\infty} |(u_n - u)(\gamma_n)| < 2\varepsilon + \varepsilon = 3\varepsilon,$$

which is a contradiction. □

Proof of Theorem 3.51. For $n \in \mathbb{N}$, let the norm $\|\cdot\|_n$ on $\ell_\infty(\Gamma)$ be defined by

$$\|u\|_n := \|u \restriction \Gamma_n\|_{\mathcal{D}}.$$

Define an equivalent norm $\|\cdot\|$ on $\ell_\infty(\Gamma)$ by

$$\|u\|^2 := \sum_n \frac{1}{2^n} \|u\|_n^2 + \|u\|_\infty^2,$$

where $\|\cdot\|$ is the canonical norm on $\ell_\infty(\Gamma)$. If $2\|u_n\|^2 + 2\|u\|^2 - \|u_n + u\|^2 \to 0$, then the same holds for any norm $\|\cdot\|_n$. Given $u \in S$ and $\varepsilon > 0$, find n so that $u \in \Gamma_n$ and $\{\gamma \in \Gamma_n; |u(\gamma)| > \varepsilon\}$ is finite. Then it follows from Proposition 3.53 that

$$\limsup_{n\to\infty} |(u_n - u)(\gamma)| \leq 3\varepsilon. \qquad \square$$

In the following lemma, we will use a variant of Day's norm (defined in (3.8)), where the coefficient 2^{-j} is replaced by 4^{-j} for all $j \in \mathbb{N}$.

Lemma 3.54 ([HaJo04]). *Let Γ be an arbitrary set and let $\|\cdot\|$ be Day's norm on $c_0(\Gamma)$. Let $\{x_n\} \subset c_0(\Gamma)$ satisfy*

$$\lim_{m,n\to\infty} (2\|x_m\|^2 + 2\|x_n\|^2 - \|x_m + x_n\|^2) = 0.$$

Then $\{x_n\}$ has a weak cluster point $x \in c_0(\Gamma)$ if and only if $\lim_{n\to\infty} x_n = x$ (in the norm topology).

Proof. Let
$$\lim_{m,n\to\infty} (2\|x_m\|^2 + 2\|x_n\|^2 - \|x_m + x_n\|^2) = 0. \tag{3.10}$$

Every weak cluster point of (x_n) is a weak limit of some subsequence of (x_n) (indeed, $\overline{\{x_n\}}^w \subset c_0(\bigcup \operatorname{supp} x_n)$, which has a separable dual, and as $\overline{\{x_n\}}^w$ is bounded, it is metrizable).

Since obviously any subsequence of $\{x_n\}$ also satisfies (3.10), by the facts mentioned above, we may assume that $x_n \to x$ weakly and we have to find a subsequence of the $\{x_n\}$ norm convergent to x.

Let $\|\cdot\|_\infty$ denote the canonical norm on $c_0(\Gamma)$. Let $\{\alpha_k^n\}$ be the support of x_n enumerated so that $|x_n(\alpha_1^n)| \geq |x_n(\alpha_2^n)| \geq \dots$ and $\{\beta_k^{m,n}\}$ be the support of $(x_m + x_n)$ enumerated so that $|(x_m + x_n)(\beta_1^{m,n})| \geq |(x_m + x_n)(\beta_2^{m,n})| \geq \dots$. Note that we may and do assume that $\beta_k^{m,n} = \beta_k^{n,m}$, $k \in \mathbb{N}$.

From the definition of Day's norm,
$$\|x_n\|^2 = \sum_k 4^{-k} x_n^2(\alpha_k^n) \geq \sum_k 4^{-k} x_n^2(\gamma_k) \tag{3.11}$$

for any sequence $\{\gamma_k\} \subset \Gamma$. Hence

$$2\|x_m\|^2 + 2\|x_n\|^2 - \|x_m + x_n\|^2$$
$$= 2\sum 4^{-k} x_m^2(\alpha_k^m) + 2\sum 4^{-k} x_n^2(\alpha_k^n) - \sum 4^{-k}(x_m + x_n)^2(\beta_k^{m,n})$$
$$\geq 2\sum 4^{-k} x_m^2(\beta_k^{m,n}) + 2\sum 4^{-k} x_n^2(\beta_k^{m,n}) - \sum 4^{-k}(x_m + x_n)^2(\beta_k^{m,n})$$
$$= \sum 4^{-k} \left(x_m(\beta_k^{m,n}) - x_n(\beta_k^{m,n}) \right)^2 \geq 0. \tag{3.12}$$

As $2\|x_m\|^2 + 2\|x_n\|^2 - \|x_m + x_n\|^2 \geq \left(\|x_m\| - \|x_n\| \right)^2 \geq 0$, (3.10) implies that $\{\|x_n\|\}$ is Cauchy and hence $\{\|x_n\|_\infty\}$ is bounded. Therefore by passing to a suitable subsequence, we may assume that there is $z \in \ell_\infty$ such that $|x_n(\alpha_k^n)| \to z(k)$, $k \in \mathbb{N}$. Note that $z(1) \geq z(2) \geq \dots \geq 0$. The vector z represents the asymptotic "shape" of the vectors x_n.

We claim that $z \in c_0$. If this is not the case, then there is a $C > 0$ such that $z(k) > C$ for $k \in \mathbb{N}$. Then there is a finite $A \subset \Gamma$ such that $\|x \upharpoonright \Gamma \setminus A\|_\infty < \frac{C}{8}$. By (3.12) and (3.10), there is $m_0 \in \mathbb{N}$ such that

$$\sum_k 4^{-k} \left(x_m(\beta_k^{m,n}) - x_n(\beta_k^{m,n}) \right)^2 < 4^{-|A|-1} \frac{C^2}{16} \quad \text{for } m, n > m_0. \tag{3.13}$$

As $\left| x_n(\alpha_{|A|+1}^n) \right| \to z(|A|+1) > C$, there is $n_1 > m_0$ such that $\left| x_{n_1}(\alpha_{|A|+1}^{n_1}) \right| > C$. Thus we can choose $\gamma \in \Gamma \setminus A$ for which $|x_{n_1}(\gamma)| > C$. Next we find a finite $B \subset \Gamma$ such that

$$\|x_{n_1} \upharpoonright \Gamma \setminus B\|_\infty < \frac{C}{8}. \tag{3.14}$$

This implies that $\gamma \in B \setminus A$. Using the weak convergence of (x_n), we choose $n_2 > m_0$ such that $\|(x_{n_2} - x) \upharpoonright B\|_\infty < \frac{C}{8}$. Therefore, we have

$$\|x_{n_2}\restriction B \setminus A\|_\infty < \frac{C}{4} \tag{3.15}$$

and so $|x_{n_2}(\gamma)| < \frac{C}{4}$. Furthermore,

$$|x_{n_1}(\gamma) + x_{n_2}(\gamma)| > \frac{3}{4}C. \tag{3.16}$$

We find the smallest $k_0 \in \mathbb{N}$ for which $\beta_{k_0}^{n_1,n_2} \notin A$. It follows that $k_0 \le |A|+1$ and

$$\left|(x_{n_1} + x_{n_2})\left(\beta_{k_0}^{n_1,n_2}\right)\right| \ge \left|(x_{n_1} + x_{n_2})(\gamma)\right|. \tag{3.17}$$

Now either $\beta_{k_0}^{n_1,n_2} \in B \setminus A$ and we can use (3.17), (3.16), and (3.15) to obtain

$$\left|x_{n_1}(\beta_{k_0}^{n_1,n_2}) - x_{n_2}(\beta_{k_0}^{n_1,n_2})\right|$$
$$\ge \left|x_{n_1}(\beta_{k_0}^{n_1,n_2}) + x_{n_2}(\beta_{k_0}^{n_1,n_2})\right| - 2\left|x_{n_2}(\beta_{k_0}^{n_1,n_2})\right|$$
$$\ge \left|x_{n_1}(\gamma) + x_{n_2}(\gamma)\right| - 2\left|x_{n_2}(\beta_{k_0}^{n_1,n_2})\right| \ge \frac{3}{4}C - \frac{1}{2}C \ge \frac{C}{4},$$

or $\beta_{k_0}^{n_1,n_2} \in \Gamma \setminus (B \cup A)$ and we use (3.17), (3.16), and (3.14) instead to get the same conclusion. Finally,

$$\sum_k 4^{-k}\left(x_{n_1}(\beta_k^{n_1,n_2}) - x_{n_2}(\beta_k^{n_1,n_2})\right)^2$$
$$\ge 4^{-k_0}\left(x_{n_1}(\beta_{k_0}^{n_1,n_2}) - x_{n_2}(\beta_{k_0}^{n_1,n_2})\right)^2 \ge 4^{-|A|-1}\frac{C^2}{16},$$

which contradicts (3.13).

Now we stabilize the supports of the vectors x_n. By (3.12),

$$0 \le 2\sum 4^{-k}x_m^2(\alpha_k^m) + 2\sum 4^{-k}x_n^2(\alpha_k^n) - \sum 4^{-k}(x_m + x_n)^2(\beta_k^{m,n})$$
$$- \left(2\sum 4^{-k}x_m^2(\beta_k^{m,n}) + 2\sum 4^{-k}x_n^2(\beta_k^{m,n}) - \sum 4^{-k}(x_m + x_n)^2(\beta_k^{m,n})\right)$$
$$\le 2\|x_m\|^2 + 2\|x_n\|^2 - \|x_m + x_n\|^2,$$

which together with (3.11) and (3.10) gives

$$\lim_{m,n\to\infty}\left(\sum 4^{-k}x_n^2(\alpha_k^n) - \sum 4^{-k}x_n^2(\beta_k^{m,n})\right) = 0. \tag{3.18}$$

But, for every $j \in \mathbb{N}$,

$$\sum_{k=1}^\infty 4^{-k}x_n^2(\alpha_k^n) - \sum_{k=1}^\infty 4^{-k}x_n^2(\beta_k^{m,n})$$
$$= \sum_{k=1}^\infty \left(4^{-k} - 4^{-(k+1)}\right)\left(\sum_{i=1}^k x_n^2(\alpha_i^n) - \sum_{i=1}^k x_n^2(\beta_i^{m,n})\right) \tag{3.19}$$
$$\ge \left(4^{-j} - 4^{-(j+1)}\right)\left(x_n^2(\alpha_j^n) - x_n^2(\alpha_{j+1}^n)\right)$$

unless $\{\alpha_i^n;\ 1 \le i \le j\} = \{\beta_i^{m,n};\ 1 \le i \le j\}$.

Indeed, if $\{\alpha_i^n;\ 1 \le i \le j\} \ne \{\beta_i^{m,n};\ 1 \le i \le j\}$, then $x_n^2(\alpha_1^n) + x_n^2(\alpha_2^n) + \cdots + x_n^2(\alpha_{j-1}^n) + x_n^2(\alpha_{j+1}^n) \ge \sum_{i=1}^{j} x_n^2(\beta_i^{m,n})$.

If $z(1) = 0$, then easily $\|x_n\|_\infty \le |x_n(\alpha_1^n)| \to z(1) = 0$; otherwise, choose $0 < \varepsilon \le z(1)$. As $z \in c_0$, we can find $k_1 \in \mathbb{N}$ such that $z(k_1+1) < \varepsilon$ and $z(k_1) \ge \varepsilon$. Put $\delta = \frac{1}{3}\big(z(k_1) - z(k_1+1)\big)$. There is $n_3 \in \mathbb{N}$ such that $\big||x_n(\alpha_k^n)| - z(k)\big| < \min\{\delta, \varepsilon\}$ for $n > n_3$ and $1 \le k \le k_1 + 1$, and thus $|x_n(\alpha_{k_1}^n)| - |x_n(\alpha_{k_1+1}^n)| > \delta$ for $n > n_3$. By putting this fact together with (3.19) and (3.18), we obtain $m_1 > n_3$ such that $\{\alpha_k^n;\ 1 \le k \le k_1\} = \{\beta_k^{m,n};\ 1 \le k \le k_1\}$ for $m, n > m_1$. As $\{\alpha_k^m;\ 1 \le k \le k_1\} = \{\beta_k^{n,m};\ 1 \le k \le k_1\} = \{\beta_k^{m,n};\ 1 \le k \le k_1\} = \{\alpha_k^n;\ 1 \le k \le k_1\}$ for $m, n > m_1$, the sets $\{\alpha_k^n;\ 1 \le k \le k_1\}$ are equal for $n > m_1$ and we denote this set by E.

The definitions of E, n_3, and k_1 in the previous paragraph give $\|x_n \upharpoonright \Gamma \setminus E\|_\infty \le |x_n(\alpha_{k_1+1}^n)| < z(k_1 + 1) + \varepsilon < 2\varepsilon$ for $n > m_1$. This, together with the weak convergence, implies $\|x \upharpoonright \Gamma \setminus E\|_\infty < 2\varepsilon$ and so $\|(x - x_n) \upharpoonright \Gamma \setminus E\|_\infty \le \|x \upharpoonright \Gamma \setminus E\|_\infty + \|x_n \upharpoonright \Gamma \setminus E\|_\infty < 4\varepsilon$ for $n > m_1$. Finally, using weak convergence again, pick $n_0 > m_1$ such that $\|(x - x_n) \upharpoonright E\|_\infty < \varepsilon$ for $n > n_0$. Then $\|x - x_n\|_\infty \le \max\{\|(x - x_n) \upharpoonright E\|_\infty, \|(x - x_n) \upharpoonright \Gamma \setminus E\|_\infty\} < 4\varepsilon$, for $n > n_0$. $\qquad\square$

Let Γ be an infinite set. If $\beta \in \Gamma$, we define a canonical projection $\pi_\beta : \ell_\infty(\Gamma) \to \ell_\infty(\Gamma)$ by

$$\pi_\beta u(\gamma) := \begin{cases} u(\beta), & \text{if } \gamma = \beta, \\ 0, & \text{if } \gamma \in \Gamma \setminus \{\beta\}, \end{cases}$$

where $u \in \ell_\infty(\Gamma)$. We shall need the following easily provable facts.

Fact 3.55 ([Troy77]). *Let $u \in \ell_\infty(\Gamma)$ and $\beta \in \Gamma$ be such that $u(\beta) \ne 0$, and assume that $i := \mathrm{card}\{\gamma \in \Gamma;\ |u(\gamma)| \ge 2^{-1/2}|u(\beta)|\} < +\infty$. Then*

$$\|u\|_{\mathcal{D}}^2 \ge \|u - \pi_\beta u\|_{\mathcal{D}}^2 + 2^{-i-1} u(\beta)^2.$$

Fact 3.56 ([Troy77]). *Let $u, v \in B_{\ell_\infty(\Gamma)}$ and $\beta \in \Gamma$ be such that $u(\beta) + v(\beta) \ne 0$, and assume that $k := \mathrm{card}\{\gamma \in \Gamma;\ |u(\gamma) + v(\gamma)| \ge |u(\beta) + v(\beta)|\} < +\infty$. Then*

$$2\|u\|_{\mathcal{D}}^2 + 2\|v\|_{\mathcal{D}}^2 - \|u + v\|_{\mathcal{D}}^2 \ge 2^{-k-1}\big(u(\beta) - v(\beta)\big)^2.$$

Proposition 3.57 (Troyanski [Troy75]; see also [FGHZ03] and [FGMZ04]**).** *Let $\Gamma \ne \emptyset$ be a set, and consider a subspace (not necessarily closed) $Y \subset \ell_\infty(\Gamma)$. Assume that there exist $\varepsilon > 0$, and $i, k \in \mathbb{N}$ such that*

$$\forall u \in Y \cap B_{\ell_\infty(\Gamma)}, \quad \mathrm{card}\{\gamma \in \Gamma;\ |u(\gamma)| > \varepsilon\} < i,$$
$$\text{and} \quad \mathrm{card}\{\gamma \in \Gamma;\ |u(\gamma)| > 2^{-i-1}\varepsilon\} < k.$$

Let $u_n,\ v_n \in Y \cap B_{\ell_\infty(\Gamma)}$, $n \in \mathbb{N}$, be such that $2\|u_n\|_{\mathcal{D}}^2 + 2\|v_n\|_{\mathcal{D}}^2 - \|u_n + v_n\|_{\mathcal{D}}^2 \to 0$ as $n \to \infty$. Then $\limsup_{n \to \infty} \|u_n - v_n\|_{\ell_\infty(\Gamma)} \le 4\varepsilon$.

Proof. The argument is a refinement of the proof of [Troy75, Prop. 1]. Assume that the conclusion is false. Then, by passing to suitable subsequences, we may and do assume that $\|u_n - v_n\| > 4\varepsilon$ for all $n \in \mathbb{N}$. For every $n \in \mathbb{N}$, find $\gamma_n \in \Gamma$ so that $|u_n(\gamma_n) - v_n(\gamma_n)| > 4\varepsilon$. We shall first observe that $\limsup_{n\to\infty} |u_n(\gamma_n) + v_n(\gamma_n)| > 2^{-i}\varepsilon$. Assume this is not so. Then, for large enough $n \in \mathbb{N}$, we have $|u_n(\gamma_n) + v_n(\gamma_n)\rangle| \le 2^{-i}\varepsilon$ and so

$$2|u_n(\gamma_n)| \ge |u_n(\gamma_n) - v_n(\gamma_n)| - |u_n(\gamma_n) + v_n(\gamma_n)| > 4\varepsilon - 2^{-i}\varepsilon > 2\sqrt{2}\varepsilon,$$

and hence

$$\operatorname{card}\left\{\gamma \in \Gamma;\ |u_n(\gamma)| > 2^{-1/2}|u_n(\gamma_n)|\right\} \le \operatorname{card}\left\{\gamma \in \Gamma;\ |u_n(\gamma)| > \varepsilon\right\} < 2i,$$

and by Fact 3.55,

$$\|u_n\|_{\mathcal{D}}^2 \ge \|u_n - \pi_{\gamma_n}(u_n)\|_{\mathcal{D}}^2 + 2^{-2i-1}u_n(\gamma_n)^2 \ge \|u_n - \pi_{\gamma_n}(u_n)\|_{\mathcal{D}}^2 + 2^{-2i}\cdot\varepsilon^2.$$

Also, for large enough $n \in \mathbb{N}$, we have

$$\|(u_n + v_n)\|_{\mathcal{D}}^2 - \|(u_n + v_n) - \pi_{\gamma_n}(u_n + v_n)\|_{\mathcal{D}}^2$$
$$\le \frac{1}{2}\left(u_n(\gamma_n) + v_n(\gamma_n)\right)^2 \le 2^{-2i-1}\cdot\varepsilon^2.$$

Thus, by the above and the convexity,

$$2\|u_n\|_{\mathcal{D}}^2 + 2\|v_n\|_{\mathcal{D}}^2 - \|(u_n + v_n)\|_{\mathcal{D}}^2 \ge 2\|u_n - \pi_{\gamma_n}(u_n)\|_{\mathcal{D}}^2 + 2^{-2i+1}\cdot\varepsilon^2$$
$$+ 2\|v_n - \pi_{\gamma_n}(v_n)\|_{\mathcal{D}}^2 - \|(u_n + v_n) - \pi_{\gamma_n}(u_n + v_n)\|_{\mathcal{D}}^2$$
$$+ \|(u_n + v_n) - \pi_{\gamma_n}(u_n + v_n)\|_{\mathcal{D}}^2 - \|(u_n + v_n)\|_{\mathcal{D}}^2$$
$$\ge 2^{-2i+1}\cdot\varepsilon^2 - 2^{-2i-1}\cdot\varepsilon^2 > 0$$

for large enough $n \in \mathbb{N}$. But, for $n \to \infty$, the first term in the chain of inequalities above goes to 0, a contradiction. We have thus proved that

$$\limsup_{n\to\infty} |u_n(\gamma_n) + v_n(\gamma_n)| > 2^{-i}\varepsilon.$$

Then, for infinitely many $n \in \mathbb{N}$, we have from the assumptions

$$\operatorname{card}\left\{\gamma \in \Gamma;\ |u_n(\gamma) + v_n(\gamma)|\right.$$
$$\ge |u_n(\gamma_n) + v_n(\gamma_n)|\right\}$$
$$\le \operatorname{card}\left\{\gamma \in \Gamma;\ |u_n(\gamma) + v_n(\gamma)| > 2^{-i}\varepsilon\right\} < 2k.$$

Hence, by Fact 3.56,

$$0 = \lim_{n\to\infty}\left(2\|u_n\|_{\mathcal{D}}^2 + 2\|v_n\|_{\mathcal{D}}^2 - \|(u_n + v_n)\|_{\mathcal{D}}^2\right)$$
$$\ge 2^{-2k-1}\limsup_{n\to\infty}\left(u_n(\gamma_n) - v_n(\gamma_n)\right)^2 > 2^{-2k-1}16\varepsilon^2 \ (> 0),$$

a contradiction. Therefore $\limsup_{n\to\infty} \|u_n - v_n\|_{\ell_\infty(\Gamma)} \le 4\varepsilon$.   □

3.6 A Quantitative Version of Krein's Theorem

In this section, we present a quantitative version of Krein's classical theorem on the weak compactness of the closed convex hull of a weakly compact set.

Definition 3.58. *Let M be a subset of a Banach space X and let $\varepsilon \geq 0$. We say that M is ε-weakly relatively compact (ε-WRK) if it is bounded and* $\overline{M}^{w^*} \subset X + \varepsilon B_{X^{**}}$.

For $\varepsilon = 0$, we get the usual concept of weak relative compactness. The concept of ε-weakly relative compactness will be used to characterize SWCG spaces (Theorem 6.13). In this section, we shall discuss stability of this notion with respect to several operations.

Proposition 3.59. *Let X be a Banach space, $C \subset X$ a nonempty convex set, and $x^{**} \in \overline{C}^{w^*}$. Then*

$$\operatorname{dist}(x^{**}, C) \leq 2 \operatorname{dist}(x^{**}, X).$$

Proof. Take any δ such that $\operatorname{dist}(x^{**}, X) < \delta$, and find $x \in X$ such that $\|x^{**} - x\| < \delta$. Then $x \in \overline{C}^{w^*} + \delta B_{X^{**}} \subset \overline{C + \delta B_X}^{w^*}$. It follows that $x \in \overline{C + \delta B_X}^{\|\cdot\|}$. Therefore, given $\varepsilon > 0$, there exists $c \in C$ and $b \in B_X$ such that $\|x - c - \delta b\| < \varepsilon$; so $\|x - c\| \leq \varepsilon + \delta$. Finally, we get $\|x^{**} - c\| = \|x^{**} - x + x - c\| \leq 2\delta + \varepsilon$, and then $\operatorname{dist}(x^{**}, C) \leq 2\delta + \varepsilon$. As $\varepsilon > 0$ was arbitrary, $\operatorname{dist}(x^{**}, C) \leq 2\delta$. Therefore, $\operatorname{dist}(x^{**}, C) \leq 2 \operatorname{dist}(x^{**}, X)$. $\qquad\square$

Proposition 3.60. (i) *The kernel of every $x^{**} \in X^{**} \setminus X$ is a norming hyperplane of X^*.*
(ii) *If X is a subspace of a Banach space Z and $Y \subset X^*$ a norming subspace, then $q^{-1}(Y)$ is a norming subspace of Z^*, where $q : Z^* \to X^*$ is the canonical quotient mapping.*

Proof. (i) This is a particular case of Lemma 2.25.

(ii) We follow [FMZ05]. Assume, without loss of generality, that Y is a 1-norming subspace of X^* (extend the corresponding norm in X to an equivalent norm in Z; see, for example, [DGZ93a, Lemma II.8.1]). Take $z \in S_Z$. Suppose first that its distance to X is less than $1/4$. Choose $x \in X$ such that $\|z - x\| < 1/4$ and $y^* \in Y \cap B_{X^*}$ with $\langle x, y^* \rangle > \|x\| - 1/4$. Select $z^* \in q^{-1}(Y) \cap B_{Z^*}$ such that $q(z^*) = y^*$. It follows that the supremum of z on $q^{-1}(Y) \cap B_{Z^*}$ is greater than $1/4$. Second, if the distance from z to X is greater than or equal to $1/4$, choose $z^* \in S_{X^\perp}$ such that $\langle z, z^* \rangle = 1/4$. Since $X^\perp$ is contained in $q^{-1}(Y)$, the supremum of z on $q^{-1}(Y) \cap B_{Z^*}$ is greater than or equal to $1/4$. This proves that

$$\frac{1}{4} B_{Z^*} \subset \overline{q^{-1}(Y) \cap B_{Z^*}}^{w^*}.$$

Hence $q^{-1}(Y)$ is norming. $\qquad\square$

A useful estimate for $\mathrm{dist}(x^{**}, X)$, the distance in the norm from an element $x^{**} \in X^{**}$ to X, is given in the following proposition.

Proposition 3.61. *Let X be a Banach space. Consider $x^{**} \in X^{**}$ and denote its kernel by $Y \subset X^*$. Then*

$$\frac{1}{2}\|x^{**}\|_{\overline{B_Y}^{w^*}} \leq \mathrm{dist}(x^{**}, X) \leq 2\|x^{**}\|_{\overline{B_Y}^{w^*}}. \tag{3.20}$$

Proof. Fix any $x \in X$ and any $x^* \in \overline{B_Y}^{w^*}$. Then, for every $y^* \in B_Y$, we have

$$\langle x^{**}, x^* \rangle = \langle x^{**}, x^* - y^* \rangle$$
$$= \langle x^{**} - x, x^* - y^* \rangle + \langle x, x^* - y^* \rangle \leq 2\|x^{**} - x\| + \langle x, x^* - y^* \rangle.$$

Hence $\langle x^{**}, x^* \rangle \leq 2\|x^{**} - x\|$ for every $x \in X$ and every $x^* \in \overline{B_Y}^{w^*}$. Thus the left inequality is proved.

It remains to prove the right inequality in (3.20). We may assume that $\|x^{**}\| = 1$. Put $Y^{\perp} = \{u^{**} \in X^{**}; \langle u^{**}, y^* \rangle = 0 \ \forall y^* \in Y\}$. Using the canonical isometry between Y^* and $X^{**}/Y^{\perp}$, we get that $\|x\|_Y = \mathrm{dist}(x, Y^{\perp})$ for every $x \in X$. Then

$$\mathrm{dist}(S_X, Y^{\perp})\|x\| \leq \|x\|_Y \leq \|x\| \quad \text{for every} \quad x \in X. \tag{3.21}$$

The parallel hyperplane lemma (see, e.g., [Phel93, Lemma 6.10] or [Fa~01, Exer. 3.1]) gives that $\min\|x^{**} \pm x\| \leq 2\|x\|_Y$ for every $x \in S_X$. Thus

$$\mathrm{dist}(x^{**}, S_X) \leq 2\inf\{\|x\|_Y; \ x \in S_X\}$$
$$= 2\inf\{\mathrm{dist}(x, Y^{\perp}); \ x \in S_X\} = 2\,\mathrm{dist}(S_X, Y^{\perp}).$$

As $\mathrm{dist}(x^{**}, S_X) \geq \mathrm{dist}(x^{**}, X)$, the bidual form of (3.21) gives that

$$\mathrm{dist}(S_X, Y^{\perp}) \leq \|x^{**}\|_{\overline{B_Y}^{w^*}}.$$

Then the right-hand inequality in (3.20) follows. $\qquad\square$

Proposition 3.62. *Let $(Z, \|\cdot\|)$ be a Banach space, X a closed subspace of Z, $\varepsilon \geq 0$, and M an ε-WRK subset of Z. Then $M \cap X$ is a 4ε-WRK subset of X.*

Proof. Let $j : X \hookrightarrow Z$ be the inclusion mapping. Take any $x^{**} \in X^{**}$ belonging to the weak*-closure of the set $M \cap X$. We need to show that $\mathrm{dist}(x^{**}, X) \leq 4\varepsilon$. Let $Y \subset X^*$ be the kernel of x^{**}. Then, by Proposition 3.61, we have $\mathrm{dist}(x^{**}, X) \leq 2\|x^{**}\|_{\overline{B_Y}^*}$. Let $W \subset Z^*$ be the kernel of $j^{**}(x^{**})$. Using the Hahn-Banach theorem, we can easily check that $j^*\left(\overline{B_W}^{w^*}\right) = \overline{B_Y}^{w^*}$. Thus

$$\|x^{**}\|_{\overline{B_Y}^{w^*}} = \sup\left\langle x^{**}, \overline{B_Y}^{w^*}\right\rangle = \sup\left\langle j^{**}(x^{**}), \overline{B_W}^{w^*}\right\rangle$$

$$= \left\|j^{**}(x^{**})\right\|_{\overline{B_W}^{w^*}} \leq 2\mathrm{dist}\left(j^{**}(x^{**}), Z\right);$$

here we applied Proposition 3.61 in Z. Thus

$$\mathrm{dist}\left(x^{**}, X\right) \leq 4\ \mathrm{dist}\left(j^{**}(x^{**}), Z\right) \leq 4\varepsilon;$$

here we used the fact that

$$j^{**}(x^{**}) \in j^{**}(\overline{M \cap X}^{w^*}) \subset \overline{j^{**}(M \cap X)}^{w^*} \subset \overline{M}^{w^*} \subset Z + \varepsilon B_{Z^{**}}.$$

$\square$

Lemma 3.63. *Let $(X, \|\cdot\|)$ be a Banach space, $\varepsilon \geq 0$, and let $\|\|\cdot\|\|$ be an equivalent norm X whose unit ball is ε-weakly compact (with respect to $\|\cdot\|$). Let Y be a subspace of X^*, and consider $x^* \in \overline{B_Y}^{w^*}$. Then the $\|\|\cdot\|\|$-distance from x^* to Y is at most 2ε.*

Proof. Suppose that $\Delta = \|\|\cdot\|\|$-$\mathrm{dist}\,(x^*, Y) > 0$. Find $x^{**} \in X^{**}$ such that $\|\|x^{**}\|\| = 1$, $x^{**} \in Y^{\perp}$, and $\langle x^{**}, x^* \rangle = \Delta$. Recalling that $x^{**} \in \overline{B_{(X, \|\|\cdot\|\|)}}^{w^*}$ we get, from the assumptions, that $\|\cdot\|$-$\mathrm{dist}\,(x^{**}, X) \leq \varepsilon$. Then, by Proposition 3.61, $\|x^{**}\|_{\overline{B_Y}^{w^*}} \leq 2\varepsilon$. Therefore $\Delta = \langle x^{**}, x^* \rangle \leq 2\varepsilon$. $\square$

Theorem 3.64 ([FHMZ05]). *Let $(X, \|\cdot\|)$ be a Banach space. Let $M \subset X$ be a bounded subset of X. Assume that M is ε-WRK for some $\varepsilon > 0$. Then $\mathrm{conv}(M)$ is 2ε-WRK.*

Remark 3.65. Note that the constant 2ε in Theorem 3.64 is optimal (see [GHM04]). See also [Gr06], [CMR], [AnCaa], and [AnCab].

Before starting the proof, we prepare some material. Given a Banach space X and an element $x^{**} \in X^{**}$, the following function on (B_{X^*}, w^*) is introduced in [DGZ93a, III.2, p. 105]. $\hat{x}^{**} : B_{X^*} \to \mathbb{R}$ is the infimum of the real continuous functions on (B_{X^*}, w^*) that are greater than or equal to x^{**}. The following proposition gives two alternative descriptions of $\hat{x}^{**}$. The first is a standard result in general topology. The second is in [DGZ93a, III.2.3].

Proposition 3.66. *Let X be a Banach space. Then, given $x^{**} \in X^{**}$,*

(i)
$$\hat{x}^{**}(x_0^*) = \lim_{N \in \mathcal{N}(x_0^*)} \{\sup\langle x^{**}, N \rangle\}, \quad \forall x_0^* \in B_{X^*}, \qquad (3.22)$$

where $\mathcal{N}(x_0^*)$ denotes the filter of neighborhoods of x_0^* in (B_{X^*}, w^*), and

(ii)
$$\hat{x}^{**}(x_0^*) = \inf\{\langle x, x_0^* \rangle + \|x^{**} - x\|;\ x \in X\}, \quad \forall x_0^* \in B_{X^*}. \qquad (3.23)$$

Remark 3.67. In particular, it follows from (ii) that if $d := \mathrm{dist}(x^{**}, X)$ denotes the distance in the norm from x^{**} to X, then $\hat{x}^{**}(0) = d$. From (i) we then get that for every $N \in \mathcal{N}(0)$, $d \leq \sup\langle x^{**}, N \rangle$, and for every $\varepsilon > 0$, there exists $N_\varepsilon \in \mathcal{N}(0)$ such that $\sup\langle x^{**}, N_\varepsilon \rangle < d + \varepsilon$.

The use of double limits in the study of compactness is implicit in the approach of Eberlein [Eb47] and explicit in Grothendieck (see [Grot52]) and Pták ([Pt63]). The following concept is a quantitative version of the double-limit condition.

Definition 3.68. *Let M be a bounded set of a Banach space X, and let S be a bounded subset of X^*. We say that M ε-interchanges limits with S (and in this case we shall write $M\S\varepsilon\S S$) if for any two sequences (x_n) in M and (x_m^*) in S such that the limits*

$$\lim_n \lim_m \langle x_n, x_m^* \rangle, \quad \lim_m \lim_n \langle x_n, x_m^* \rangle$$

exist, then

$$|\lim_n \lim_m \langle x_n, x_m^* \rangle - \lim_m \lim_n \langle x_n, x_m^* \rangle| \leq \varepsilon.$$

The following is a quantitative version of Grothendieck's double-limit characterization of weak compactness.

Theorem 3.69 ([FHMZ05]). *Let M be a bounded subset of a Banach space X and let $\varepsilon \geq 0$ be some number. Then the following statements are valid.*

(i) *If M is ε-WRK, then $M\S 2\varepsilon\S B_{X^*}$.*
(ii) *If $M\S\varepsilon\S B_{X^*}$, then M is ε-WRK.*

Proof. (i) Let (x_n) and (x_m^*) be sequences in M and B_{X^*}, respectively, such that both limits

$$\lim_n \lim_m \langle x_n, x_m^* \rangle, \quad \lim_m \lim_n \langle x_n, x_m^* \rangle$$

exist. Let $x^{**} \in \overline{M}^{w^*}$ be a w^*-cluster point of (x_n). Then

$$\lim_n \langle x_n, x_m^* \rangle = \langle x^{**}, x_m^* \rangle \text{ for all } m.$$

Fix $\delta > 0$. By the assumption, we can find $x \in X$ such that $\|x^{**} - x\| \leq \varepsilon + \delta$. Choose a subsequence of (x_m^*) (denoted again by (x_m^*)) such that $\lim_m \langle x, x_m^* \rangle$ exists. Let $x^* \in X^*$ be a w^*-cluster point of (x_m^*). We get

$$\lim_m \langle x_n, x_m^* \rangle = \langle x_n, x^* \rangle \text{ for all } n,$$
$$\lim_n \lim_m \langle x_n, x_m^* \rangle = \lim_n \langle x_n, x^* \rangle = \langle x^{**}, x^* \rangle,$$

and then

$$|\lim_n \lim_m \langle x_n, x_m^* \rangle - \lim_m \lim_n \langle x_n, x_m^* \rangle| = |\lim_n \langle x_n, x^* \rangle - \lim_m \langle x^{**}, x_m^* \rangle|$$
$$= |\langle x^{**}, x^* \rangle - \lim_m \langle x^{**}, x_m^* \rangle| = |\lim_m \langle x^{**}, x^* - x_m^* \rangle|$$
$$\leq |\lim_m \langle x, x^* - x_m^* \rangle| + 2(\varepsilon + \delta) = 2(\varepsilon + \delta).$$

As $\delta > 0$ is arbitrary, we get the conclusion.

(ii) Now assume $M \S \varepsilon \S B_{X^*}$. Let $x^{**} \in \overline{M}^{w^*}$ and let $d := d(x^{**}, X)$. We shall define inductively two sequences, (x_n) in M and (x_m^*) in B_{X^*}. To begin, choose any $x_1 \in M$. Then define $N(x_1; 1) := \{x^* \in B_{X^*}; |\langle x_1, x^* \rangle| < 1\}$, a neighborhood of 0 in (B_{X^*}, w^*). By Remark 3.67, we can find $x_1^* \in N(x_1; 1)$ such that

$$d - 1 \le \langle x^{**}, x_1^* \rangle < d + 1.$$

Choose $x_2 \in M$ such that $|\langle x^{**} - x_2, x_1^* \rangle| < 1/2$. Define $N(x_1, x_2; 1/2) := \{x^* \in B_{X^*}; |\langle x_i, x^* \rangle| < 1/2, i = 1, 2\}$, a neighborhood of 0 in (B_{X^*}, w^*). Again by Remark 3.67, we can find $x_2^* \in N(x_1, x_2; 1/2)$ such that $d - 1/2 \le \langle x^{**}, x_2^* \rangle < d + 1/2$. Continue in this way. We get (x_n) and (x_m^*) such that

$$x_n \in M, \quad x_m^* \in B_{X^*}, \text{ for all } n, m,$$

$$|\langle x^{**} - x_n, x_m^* \rangle| < \frac{1}{n}, \quad m = 1, 2, \ldots, n-1,$$

$$|\langle x_n, x_m^* \rangle| < \frac{1}{m}, \quad n = 1, 2, \ldots, m,$$

$$d - \frac{1}{m} \le \langle x^{**}, x_m^* \rangle < d + \frac{1}{m}, \quad m = 1, 2, \ldots.$$

Then

$$\lim_n \langle x_n, x_m^* \rangle = \langle x^{**}, x_m^* \rangle \text{ for all } m,$$

$$\lim_m \lim_n \langle x_n, x_m^* \rangle = \lim_m \langle x^{**}, x_m^* \rangle = d,$$

$$\lim_m \langle x_n, x_m^* \rangle = 0 \text{ for all } n,$$

$$\lim_n \lim_m \langle x_n, x_m^* \rangle = 0,$$

so

$$\left| \lim_m \lim_n \langle x_n, x_m^* \rangle - \lim_n \lim_m \langle x_n, x_m^* \rangle \right| = d \le \varepsilon. \qquad \square$$

We need the following definitions:

$$C(\mathbb{N}) := \{\lambda \in ca^+(\mathbb{N}); \text{card supp}(\lambda) < \aleph_0, \lambda(\mathbb{N}) = 1\}.$$

Given $B \subset \mathbb{N}$, let

$$C(B) := \{\lambda \in C(\mathbb{N}); \text{ supp}(\lambda) \subset B\}.$$

Let $\mathcal{G}$ be a family of finite subsets of $\mathbb{N}$. Given $\gamma > 0$, let $C(B, \mathcal{G}, \gamma) := \{\lambda \in C(B); \lambda(G) < \gamma, \text{ for all } G \in \mathcal{G}\}$. Pták's combinatorial lemma (see, e.g., [Fa~01, p. 422]) reads as follows.

Lemma 3.70 (Pták[Pt63]). *The following two conditions on $\mathcal{G}$ are equivalent:*

(i) *There exists a strictly increasing sequence $A_1 \subset A_2 \subset \ldots$ of finite subsets of $\mathbb{N}$ and a sequence (G_n) in $\mathcal{G}$ with $A_n \subset G_n$ for all n.*
(ii) *There exists an infinite subset $B \subset \mathbb{N}$ and $\gamma > 0$ such that*

$$C(B, \mathcal{G}, \gamma) = \emptyset.$$

Theorem 3.71 ([FHMZ05]). *Let $(X, \|\cdot\|)$ be a Banach space. Let $M \subset X$ be a bounded subset of X. Assume that $M\S\varepsilon\S B_{X^*}$ for some $\varepsilon \geq 0$. Then $\mathrm{conv}(M)\S\varepsilon\S B_{X^*}$.*

Proof. Assume $\|x\| \leq \mu$ for all $x \in M$ and some $\mu > 0$. Choose $\varepsilon > 0$ and $0 < \beta < \varepsilon$. Now select $\delta > 0$ and $\gamma > 0$ such that $\beta + 2\gamma\mu < \varepsilon - \delta$. Suppose that there exists a sequence (x_n) in conv (M) and a sequence (x_m^*) in B_{X^*} such that

$$\varepsilon := |\lim_n \lim_m \langle x_n, x_m^* \rangle - \lim_m \lim_n \langle x_n, x_m^* \rangle| > 0.$$

Let $x_0^* \in B_{X^*}$ be a cluster point of (x_m^*) in (B_{X^*}, w^*). Let $T \subset M$ be a countable set such that $\{x_n;\ n \in \mathbb{N}\} \subset$ conv (T), and choose a subsequence (denoted again by (x_m^*)) such that $x_m^* \to x_0^*$ on the set T. Then, for some $\sigma \in \{-1, 1\}$,

$$\sigma(\lim_n \langle x_n, x_0^* \rangle - \lim_m \lim_n \langle x_n, x_m^* \rangle) = \varepsilon.$$

By suppressing a finite number of indices, we may assume

$$\sigma(\lim_n \langle x_n, x_0^* \rangle - \lim_n \langle x_n, x_m^* \rangle) = \sigma \lim_n \langle x_n, x_0^* - x_m^* \rangle > \varepsilon - \delta \text{ for all } m.$$

Define

$$\Gamma(t) := \{m \in \mathbb{N};\ |\langle t, x_0^* - x_m^* \rangle| \geq \beta\},\ t \in T.$$

Then $\Gamma(t)$ is a finite subset of $\mathbb{N}$ for each t. Let $\mathcal{G} := \{\Gamma(t);\ t \in T\}$. Assume $C(\mathbb{N}, \mathcal{G}, \gamma) \neq \emptyset$, and choose $\lambda \in C(\mathbb{N}, \mathcal{G}, \gamma)$. It follows that

$$\lambda(\Gamma(t)) < \gamma \text{ for all } t \in T.$$

Put $x^* := \sum_{k \in \mathbb{N}} \lambda(k)(x_0^* - x_k^*) \in 2B_{X^*}$. Given $t \in T$,

$$|\langle t, x^* \rangle| = \left| \sum_{k \in \mathbb{N}} \lambda(k)\langle t, x_0^* - x_k^* \rangle \right|$$

$$\leq \sum_{\Gamma(t)} \lambda(k)|\langle t, x_0^* - x_k^* \rangle| + \sum_{\mathbb{N}\backslash\Gamma(t)} \lambda(k)|\langle t, x_0^* - x_k^* \rangle| < 2\gamma\mu + \beta.$$

It follows that $|\langle x_n, x^* \rangle| \leq 2\gamma\mu + \beta$ for all n. Then

$$2\gamma\mu + \beta \geq \lim_n |\langle x_n, x^* \rangle| = \left| \sum_{k \in \mathbb{N}} \lambda(k) \lim_n \langle x_n, x_0^* - x_k^* \rangle \right|$$

$$= \sigma \sum_{k \in \mathbb{N}} \lambda(k) \lim_n \langle x_n, x_0^* - x_k^* \rangle > \varepsilon - \delta,$$

a contradiction.

Assume then that $C(\mathbb{N}, \mathcal{G}, \gamma) = \emptyset$. Then, by Lemma 3.70, we can find $A_p := \{m_i; i = 1, 2, \ldots, p\} \subset \mathbb{N}$ and $t_p \in T$ such that

$$A_p \subset \Gamma(t_p), \ \forall p \in \mathbb{N},$$

i.e., $|\langle t_p, x_0^* - x_{m_k}^* \rangle| \geq \beta, \ k = 1, 2, \ldots, p$. Choose a subsequence of (t_n) (denoted again by (t_n)) such that $\lim_n \langle t_n, x_0^* - x_{m_k}^* \rangle$ exists for every k. Then we get

$$\lim_n \lim_k \langle t_n, x_{m_k}^* \rangle = \lim_n \langle t_n, x_0^* \rangle,$$

$$|\lim_n \langle t_n, x_0^* \rangle - \lim_k \lim_n \langle t_n, x_{m_k}^* \rangle| = \lim_k \lim_n |\langle t_n, x_0^* - x_{m_k}^* \rangle| \geq \beta,$$

so

$$|\lim_n \lim_k \langle t_n, x_{m_k}^* \rangle - \lim_k \lim_n \langle t_n, x_{m_k}^* \rangle| \geq \beta. \tag{3.24}$$

As β satisfies $0 < \beta < \varepsilon$ and is otherwise arbitrary, we get the conclusion.

□

Proof of Theorem 3.64. It follows from Theorem 3.69 and Theorem 3.71.   □

For a separable space, the main result of this section (Theorem 3.64) has been proved by Rosenthal (unpublished). We thank Y. Benyamini for informing us about this. The optimality of the statement in Theorem 3.64 is proven in [GHM04]. For further results in the direction of ε-weak relative compactness, see, e.g., [Gr06], [CMR], [AnCaa], and [AnCab].

3.7 Exercises

3.1. Let μ be a finite measure. Assume that $T : L_1(\mu) \to X$ is an operator and X is reflexive. Then $TL_1(\mu)$ is separable.

Hint. T is weakly compact and $L_1(\mu)$ has the DP property, so T is completely continuous [Fa~01, p. 375]; i.e., it sends weakly convergent sequences onto norm-convergent sequences. If K is a weakly compact set generating $L_1(\mu)$, then TK is norm-compact. Thus $TL_1(\mu)$ is separable.

3.2. Let μ be a finite measure such that $L_1(\mu)$ is nonseparable. Show that there is no one-to-one operator T from $L_1(\mu)$ into a reflexive space.

Hint. Otherwise, $TL_1(\mu)$ is separable (see Exercise 3.1), so $L_1(\mu)^*$ is weak*-separable, a contradiction with the fact that $L_1(\mu)$ is WCG and nonseparable.

3.3. Let μ be a finite measure such that $L_1(\mu)$ is nonseparable. Does there exist a one-to-one operator from $L_1(\mu)$ into $\ell_1(\omega_1)$?

Hint. No. Otherwise, $T^*(\ell_1^*(\omega_1)$ is weak*-dense in $L_1^*(\mu)$, so $L_1^*(\mu)$ is weak*-separable.

3.4. Let μ be a finite measure. Is $\ell_1(\omega_1)$ isomorphic to a subspace of $L_1(\mu)$?

Hint. No since $L_1(\mu)$ has a Gâteaux differentiable equivalent norm and $\ell_1(\omega_1)$ does not (see [DGZ93a, Remark after Lemma II.5.4]).

3.5. Let μ be a finite measure. Does there exist a map from $L_1(\mu)$ onto a dense set in $\ell_1(\omega_1)$?

Hint. No. $\ell_1(\omega_1)$ is not WCG.

3.6. Does there exist a one-to-one operator from $c_0(\omega_1)$ into $\ell_1(\omega_1)$?

Hint. No. $T^*(\ell_1^*(\omega_1))$ is w^*-dense in $c_0^*(\omega_1)$, and so $c_0^*(\omega_1)$ is w^*-separable. This is false.

3.7. Does there exist an operator from $c_0(\omega_1)$ onto a dense set of $\ell_1(\omega_1)$?

Hint. No. $\ell_1(\omega_1)$ is not WCG.

3.8. Does there exist a one-to-one operator from $c_0(\omega_1)$ into a reflexive space?

Hint. No. Otherwise the dual operator is weakly compact and thus norm-compact (Schur), so $c_0(\omega_1)^*$ is weak*-separable, and this is false.

3.9. Does there exist an operator from $c_0(\omega_1)$ onto a dense set in a nonseparable reflexive space?

Hint. No. Otherwise, the dual operator is weakly compact and thus norm-compact; thus it maps nonseparable reflexive space in a one-to-one way into ℓ_1. So the second dual operator is w^*-w-continuous and maps (ℓ_∞, w^*) onto a dense set in the space, so the reflexive space would be separable.

3.10. Let μ be a finite measure. Is $\ell_4(\omega_1)$ isomorphic to a subspace of $L_4(\mu)$?

Hint. No. Let T be such isomorphism into. T^* is a quotient map from $L_{\frac{4}{3}}(\mu)$ onto $\ell_{\frac{4}{3}}(\omega_1)$. Let I be the canonical embedding of $L_2(\mu)$ into $L_{\frac{4}{3}}(\mu)$. Then T^*I maps $L_2(\mu) = \ell_2(\omega_1)$ into (a dense subspace of) $\ell_{\frac{4}{3}}(\omega_1)$, so it is compact, hence the range of this operator is separable and so $\ell_{\frac{4}{3}}(\omega_1)$ is separable, and this is false.

3.11. Given a dual pair $\langle E, F \rangle$, is an F-limited set $L \subset E$ necessarily $\beta(E, F)$-bounded?

Hint. Yes. Assume not; we can find a $w(F, E)$-bounded set $B \subset F$, a sequence (b_n) in B, and a sequence (l_n) in L such that $|\langle l_n, b_n \rangle| \geq n$ for all $n \in \mathbb{N}$. Then (b_n/n) is $w(F, E)$-null and does not converge to 0 uniformly on L. Note that, in particular, every X-limited set in the dual X^* of a Banach space X is $\| \cdot \|$-bounded, and every X^*-limited set in X is also $\| \cdot \|$-bounded.

3.12. Show that a Banach space X is Schur if and only if B_{X^*} is X-limited.

3.13. Show that no $C(K)$ space with a K infinite compact has the Schur property.

Hint. Let (O_n) be a pairwise disjoint sequence of nonempty open subsets of K. Choose $x_n \in O_n$ for all n. Let f_n be a $[0,1]$-valued continuous function such that $f_n(x) = 0$ for all x outside O_n and $f_n(x_n) = 1$. The sequence (f_n) is bounded and pointwise null, and hence weakly null, and $\|f_n\| = 1$ for all n.

3.14 (Howard [How73]). Let X be a Banach space. If $A \subset X^*$ is $\tau(X^*, X)$-relatively sequentially compact, then A is $\tau(X^*, X)$-relatively compact. The converse does not hold true.

Hint. Assume that A is not X-limited. This means that there is $\varepsilon > 0$, a weakly null sequence (x_n) in X and a sequence (a_n^*) in A such that $|\langle x_n, a_n^* \rangle| \geq \varepsilon$ for all $n \in \mathbb{N}$. Since $W := \{x_n\} \cup \{0\}$ is weakly compact and A is $\tau(X^*, X)$-relatively sequentially compact, there is a subsequence $(a_{n_k}^*)$ of (a_n^*) such that $a_{n_k}^* \to a^*$ for some $a^* \in X^*$ uniformly on W. Thus there exists m_0 such that $|\langle x_m, a^* \rangle| < \varepsilon/2$ for all $m \geq m_0$ and there exists k_0 such that $n_{k_0} \geq m_0$ and $|\langle x_p, a^* - a_{n_k}^* \rangle| < \varepsilon/2$ for all $p \in \mathbb{N}$ and all $k \geq k_0$. We get $|\langle x_{n_k}, a_{n_k}^* \rangle| < \varepsilon$ for all $k \geq k_0$, a contradiction.

To see that the converse does not hold, consider $X := \ell_1[0, 2\pi]$. This is a Schur space, so on its dual ball, the topology $\tau(X^*, X)$ coincides with the weak*-topology, so the dual ball is compact in the topology $\tau(X^*, X)$. However, it is not sequentially compact in the topology $\tau(X^*, X)$, as otherwise it would be sequentially compact in the weak*-topology, which is not the case. Indeed, the sequence $f_n \in B_{X^*}$ defined by $f_n(x) = \sin nx$ does not have a pointwise convergent subsequence. Indeed, if it had such a subsequence (f_{n_k}), then the sequence $g_k := f_{n_{k+1}} - f_{n_k}$ would pointwise converge to 0 on $[0, 2\pi]$, and yet $\int_0^{2\pi} g_k^2 \neq 2\pi$, contradicting Lebesgue's dominated convergence theorem.

3.15 (Kirk [Kirk73]). If K is a compact topological space and the Mackey topology $\tau(C(K)^*, C(K))$ on $K \subset B_{C(K)^*}$ agrees with the initial topology on K, then K is finite.

Hint. If K where $\tau(X^*, X)$-compact, it would be X-limited by Theorem 3.11. However, if K is infinite, it is always possible to find a weakly null sequence (f_n) in $C(K)$ such that $\|f_n\| = 1$ for all $n \in \mathbb{N}$ (see Exercise 3.13), a contradiction.

3.16. Show that every Schur space contains a copy of ℓ_1.

Hint. Rosenthal's ℓ_1 theorem.

3.17. Find a Schur space that is not isomorphic to any $\ell_1(\Gamma)$.

Hint. [Hag77b].

3.18. Show that ℓ_1 is not a quotient of $C[0,1]$.

Hint. $C[0,1]^*$ does not contain ℓ_∞, as it admits an LUR norm by Troyanski's result and ℓ_∞ does not (see [DGZ93a]).

3.19 (Schlüchtermann and Wheeler [ScWh88]). Show that X is a separable Schur space if and only if the following holds: there is a weakly compact subset $K \subset X$ and a sequence $\{x_n\}$ in X such that for every weakly compact set $L \subset X$ and for every $\varepsilon > 0$, $L \subset \{x_1, \ldots, x_n\} + K + \varepsilon B_X$ for some n.

Hint. First, every separable Schur space satisfies the condition above. Indeed, put $K = \{0\}$ and $\{x_n\}$ dense in X. Note that L is norm-compact. Let the condition hold. For each m, there is $n(m)$, so that $mK \subset \{x_1, \ldots, x_{m(n)}\} + K + \frac{\varepsilon}{2} B_X$. Select $m > 2$ so that $\frac{1}{m} K \subset \frac{\varepsilon}{2} B_X$. Then $K \subset \frac{1}{m} \{x_1, \ldots, x_{m(n)}\} + \varepsilon B_X$. Thus K is norm-compact. Thus X is a separable Schur space.

3.20. Show that any bounded linear operator from an Asplund space into $L_1(\mu)$ is weak-compact.

Hint. Rosenthal's ℓ_1 theorem (see, e.g., [LiTz77, Thm. e.2.5]) and the weak sequential completeness of L_1.

3.21. Show that any operator from Asplund space into $\ell_1(\Gamma)$ is norm-compact.

Hint. Use Exercise 3.20 and the Schur property of $\ell_1(\Gamma)$.

3.22. Show that the Mackey topology $\tau(X^*, X)$ on X^* coincides with the norm topology if and only if X is reflexive.

3.23. Show that X^* in the Mackey topology $\tau(X^*, X)$ is metrizable if and only if X is reflexive.

3.24. Show that B_{X^*} is compact in the Mackey topology $\tau(X^*, X)$ if and only if X is a Schur space.

Hint. Limited sets.

3.25. Prove that $C[0, \omega_1]$ does not have an unconditional Schauder basis.

Hint. Such a basis would be shrinking, as the space is Asplund. Thus the space would be WCG, a contradiction.

3.26. Prove that $C[0, \omega_1]^*$ is not weak*-separable.

Hint. $c_0[0, \omega_1]$ can be isomorphically embedded into $C[0, \omega_1]$.

3.27. Show that there is no nontrivial convergent sequence in $\beta\mathbb{N}$.

Hint. This is because then $s_n \to s$ in the weak topology of ℓ_∞^* (since ℓ_∞ is a Grothendieck space), and this is not the case.

3.28. Does there exist a c_0-saturated nonseparable Banach space that does not contain $c_0(\omega_1)$?

Hint. Yes, the space JL_0 of Johnson-Lindenstrauss; see [Zizl03].

4

Biorthogonal Systems in Nonseparable Spaces

The main theme of this chapter is the existence of biorthogonal systems in general nonseparable Banach spaces. An important role is played by the notion of long Schauder bases; the first section introduces this notion, which is a natural generalization of the usual Schauder basis. The first section also contains Plichko's improvement of the natural "exhaustion" argument that yields the existence of a bounded total biorthogonal system in every Banach space. The second section presents Plichko's characterization of spaces with a fundamental biorthogonal system as those spaces that admit a quotient of the same density with a long Schauder basis. In general, a total biorthogonal system as constructed in the first section has a cardinality that corresponds to the w^*-density of the dual space. Thus, such a system may be countable for certain nonseparable spaces (most notably all subspaces of ℓ_∞). It is therefore a priori unclear if every nonseparable subspace of ℓ_∞ contains an uncountable biorthogonal system. The third section singles out some natural classes of spaces that are obtained "constructively" (representable spaces)—and hence are well-behaved in this respect, as shown by results of Godefroy and Talagrand. However, the general question of the existence of uncountable biorthogonal systems in every nonseparable space is undecidable in ZFC.

In the fourth section, we present (under an additional axiom ♣) an example of a nonseparable subspace of ℓ_∞ that contains no uncountable biorthogonal system (the first such example was obtained by Kunen under the continuum hypothesis). On the other hand, recent work by Todorčević, using Martin's Maximum axiom, shows the existence of a fundamental biorthogonal system for every space with density ω_1, so in particular every nonseparable space contains an uncountable biorthogonal system. In the sixth and final section, we present a renorming, due to Godun, Lin, and Troyanski, of nonseparable subspaces of ℓ_∞ that excludes the existence of an Auerbach basis for these spaces.

4.1 Long Schauder Bases

We start by introducing long Schauder bases, a natural generalization of the usual Schauder bases in the nonseparable setting. We continue by showing the existence of bounded total biorthogonal systems in every Banach space (Plichko).

Definition 4.1. *Let Γ be an ordinal and $\{x_\gamma\}_{\gamma=0}^{\Gamma} := \{x_\gamma; 0 \leq \gamma < \Gamma\}$ be a transfinite sequence of vectors from X. We put $x = \sum_{\gamma=0}^{\Gamma} x_\gamma$ to be the sum of the series of the elements $\{x_\gamma; 0 \leq \gamma < \Gamma\}$ (and the series is called* convergent*) if there exists a continuous function $S : [1, \Gamma] \to X$, where $[1, \Gamma]$ is equipped with the order topology, such that*

$$S(1) = x_0, \ S(\Gamma) = x, \ S(\gamma + 1) = S(\gamma) + x_\gamma \ \text{for} \ \gamma < \Gamma.$$

One may easily check that for $\Gamma = \omega$ this definition coincides with the usual definition of convergence of a series.

Definition 4.2. *A transfinite sequence $\{e_\gamma\}_{\gamma=0}^{\Gamma}$ of vectors from a normed linear space X is called a* long *(or* transfinite*)* Schauder basis *if for every $x \in X$ there exists a unique transfinite sequence of scalars $\{a_\gamma\}_{\gamma=0}^{\Gamma}$ such that $x = \sum_{\gamma=0}^{\Gamma} a_\gamma e_\gamma$.*

If $\{e_\gamma\}_{\gamma=0}^{\Gamma}$ is a long Schauder basis of a normed linear space X, then the canonical projections $P_\alpha \colon X \to X$ are defined for $1 \leq \alpha < \Gamma$ by $P_\alpha \left(\sum_{\gamma=0}^{\Gamma} a_\gamma e_\gamma \right) := \sum_{\gamma=0}^{\alpha} a_\gamma e_\gamma$.

Lemma 4.3. *Let $\{e_\gamma\}_{\gamma=0}^{\Gamma}$ be a long Schauder basis of a normed linear space X. The canonical projections P_α satisfy*

(i) $\dim\big((P_{\alpha+1} - P_\alpha)(X)\big) = 1$, $\alpha < \Gamma$,
(ii) $P_\alpha P_\beta = P_\beta P_\alpha = P_{\min(\alpha,\beta)}$, *and*
(iii) $P_\alpha(x) = \lim_{\gamma \to \alpha} P_\gamma(x)$ *if α is a limit ordinal, and $\lim_{\alpha \to \Gamma} P_\alpha(x) = x \in X$ for every $x \in X$.*

Conversely, if bounded linear projections $\{P_\alpha\}_{\alpha=1}^{\Gamma}$ in a normed space X satisfy (i)–(iii), then P_α are the canonical projections associated with some long Schauder basis of X.

Proof. The set $\{e_\gamma\}_{\gamma=0}^{\Gamma}$ is linearly independent in X. Thus (i), (ii), and (iii) follow directly from the definition of a long Schauder basis.

Conversely, if bounded projections P_α satisfy (i)–(iii), put formally $P_0 = 0$, and choose a nonzero $e_\gamma \in P_{\gamma+1}(X) \cap \mathrm{Ker}(P_\gamma)$. Then

$$x = \lim_{\gamma \to \Gamma} \big(P_\gamma(x)\big) = \lim_{\gamma \to \Gamma} \big(P_\gamma(x) - P_0(x)\big)$$

$$\lim_{\gamma \to \Gamma} \sum_{\alpha=0}^{\gamma} \big(P_{\alpha+1}(x) - P_\alpha(x)\big) = \sum_{\alpha=0}^{\Gamma} a_\alpha e_\alpha$$

for some scalars a_α, as $\dim\left(P_{\alpha+1}(X)/P_\alpha(X)\right) = 1$. The uniqueness of a_α for $x \in X$ follows from the fact that if $x = \sum_{\gamma=0}^{\Gamma} b_\gamma e_\gamma$, then by the continuity of P_α we get $P_\alpha(x) = \sum_{\gamma=0}^{\alpha} b_\gamma e_\gamma$ and hence $b_\alpha e_\alpha = P_{\alpha+1}(x) - P_\alpha(x) = a_\alpha e_\alpha$. Thus $\{e_\alpha\}_{\alpha=0}^{\Gamma}$ is a long Schauder basis of X, and $\{P_\alpha\}_{\alpha=1}^{\Gamma}$ are projections associated with $\{e_\gamma\}_{\gamma=0}^{\Gamma}$. $\qquad\square$

Fact 4.4. *Let* $\{e_\gamma\}_{\gamma=0}^{\Gamma}$ *be a long Schauder basis of a normed linear space* X *with canonical projections* $\{P_\gamma\}_{\gamma=1}^{\Gamma}$. *If* $\sup_{\gamma<\Gamma} \|P_\gamma\| < \infty$ *(we say that* P_γ *are* uniformly bounded*), then* $\{e_\gamma\}_{\gamma=0}^{\Gamma}$ *is also a long Schauder basis of the completion* $\widetilde{X}$ *of* X.

Proof. First observe that the extensions $\widetilde{P}_\gamma : \widetilde{X} \to \widetilde{X}$ are uniquely determined by P_γ and $\|\widetilde{P}_\gamma\| = \|P_\gamma\|$. We will show that $\widetilde{P}_\gamma$ on $\widetilde{X}$ satisfy (i)–(iii) of Lemma 4.3. (i) and (ii) are extended from P_γ to $\widetilde{P}_\gamma$ by the continuity of P_γ. Since $\lim_{\gamma\to\alpha} \widetilde{P}_\gamma(x) = \widetilde{P}_\alpha(x)$ for all x in a dense subset $X \subset \widetilde{X}$ and $\widetilde{P}_\gamma$ are uniformly bounded, we have also $\lim_{\gamma\to\alpha} \widetilde{P}_\gamma(x) = \widetilde{P}_\alpha(x)$ in $\widetilde{X}$, so (iii) is also true. Since $e_{\gamma+1} \in P_{\gamma+1}(X) \cap \mathrm{Ker}(P_\gamma)$, we get $e_{\gamma+1} \in \widetilde{P}_{\gamma+1}(\widetilde{X}) \cap \mathrm{Ker}(\widetilde{P}_\gamma)$ for every $\gamma < \Gamma$. Therefore $\widetilde{P}_\gamma$ are canonical projections associated with the long Schauder basis $\{e_\gamma\}_{\gamma=0}^{\Gamma}$ of $\widetilde{X}$. $\qquad\square$

Lemma 4.5. *Let* $\{e_\gamma\}_{\gamma=0}^{\Gamma}$ *be a long Schauder basis of a Banach space* $(X, \|\cdot\|)$. *Define* $\|\|\cdot\|\|$ *on* X *by* $\|\|x\|\| := \sup_{\gamma<\Gamma} \|\sum_{\alpha=0}^{\gamma} a_\alpha e_\alpha\|$ *for* $x = \sum_{\alpha=0}^{\Gamma} a_\alpha e_\alpha$. *Then*

(i) $\|\|\cdot\|\|$ *is a norm on* X, $\{e_\gamma\}_{\gamma=0}^{\Gamma}$ *is a Schauder basis of* $(X, \|\|\cdot\|\|)$, *and* $\|\|P_\alpha\|\| = 1$; *and*

(ii) $\|\|\cdot\|\|$ *is an equivalent norm on* X.

Proof. (i) The triangle inequality and homogeneity of $\|\|\cdot\|\|$ are simple to check. Since for every $x \in X$ we have $\|x\| = \lim_{\gamma\to\Gamma} \|\sum_{\alpha=0}^{\gamma} a_\alpha e_\alpha\|$, we obtain that $\|\|x\|\| \geq \|x\|$ for every $x \in X$. This in particular means that $\|\|\cdot\|\|$ is a norm on the space X.

To show that $\{e_\gamma\}_{\gamma=0}^{\Gamma}$ is a long Schauder basis of $(X, \|\|\cdot\|\|)$, we use Lemma 4.3. Properties (i) and (ii) are straightforward. To check (iii), we note that for $x \in X$ we have

$$\|\|x - P_\beta(x)\|\| = \sup_\alpha \|P_\alpha(x) - P_\alpha P_\beta(x)\| = \sup_{\alpha\geq\beta} \|P_\alpha(x) - P_\beta(x)\| \to 0,$$

as $\beta \to \Gamma$. Finally, for $\beta < \Gamma$, we estimate

$$\|\|P_\beta\|\| = \sup_{\|\|x\|\|\leq 1} \|\|P_\beta(x)\|\| = \sup_{\|\|x\|\|\leq 1} \sup_\alpha \|P_\alpha P_\beta(x)\| = \sup_\alpha \sup_{\|\|x\|\|\leq 1} \|P_\alpha P_\beta(x)\|$$

$$= \sup_\alpha \left\{ \sup\{\|P_\alpha P_\beta(x)\|;\ x \text{ with } \sup_\gamma \|P_\gamma(x)\| \leq 1\} \right\} \leq 1.$$

(ii) We will show that $\|\|\cdot\|\|$ is a complete norm on X (i.e., that $\widetilde{X} \subset X$), where $\widetilde{X}$ is the completion of X in $\|\|\cdot\|\|$. By (i), we already know that $\{e_\gamma\}_{\gamma=0}^{\Gamma}$

is a Schauder basis of $\widetilde{X}$. Given $x \in \widetilde{X}$, there is a unique transfinite sequence of scalars a_γ such that $x = \sum_{\gamma=0}^{\Gamma} a_\gamma e_\gamma$, where the convergence is in the norm $\||\cdot\||$. Since $\||\cdot\|| \geq \|\cdot\|$ on X, we get that $\sum_{\gamma=0}^{\Gamma} a_\gamma e_\gamma$ is convergent to some $x' \in (X, \|\cdot\|)$. As shown in part (i), $\sum_{\gamma=0}^{\Gamma} a_\gamma e_\gamma$ then converges to x' in the norm $\||\cdot\||$. Thus $x = x' \in X$. This means that X is complete in $\||\cdot\||$. From the Banach open mapping principle, it follows that the formal identity map $I_X \colon (X, \||\cdot\||) \to (X, \|\cdot\|)$ is an isomorphism, which means that $\||\cdot\||$ is an equivalent norm on X. $\qquad\square$

Theorem 4.6. *Let* $\{e_\gamma\}_{\gamma=0}^{\Gamma}$ *be a long Schauder basis of a Banach space* X. *The associated canonical projections* $\{P_\alpha\}_{\alpha=1}^{\Gamma}$ *are uniformly bounded.*

Proof. Define $\||\cdot\||$ as in Lemma 4.5. Then $\||P_\alpha\|| \leq 1$ for every α, and since $\||\cdot\||$ is an equivalent norm, the result follows. $\qquad\square$

The value $\mathrm{bc}\{e_\gamma\}_{\gamma=0}^{\Gamma} = \sup_{\gamma < \alpha} \|P_\gamma\|$ is called the *basis constant* of $\{e_\gamma\}_{\gamma=0}^{\Gamma}$. Considering the vectors e_γ, we see that $\|P_\gamma\| \geq 1$; in particular, $\mathrm{bc}\{e_\gamma\}_{\gamma=0}^{\Gamma} \geq 1$. A long Schauder basis $\{e_\gamma\}_{\gamma=0}^{\Gamma}$ is called *normalized* if $\|e_\gamma\| = 1$ for every $1 \leq \gamma < \Gamma$. It is called *monotone* if $\mathrm{bc}\{e_\gamma\} = 1$; that is, its associated projections satisfy $\|P_\gamma\| = 1$ for every $1 \leq \gamma < \Gamma$. Let $\{e_\gamma\}_{\gamma=0}^{\Gamma}$ be a long Schauder basis of a Banach space X. For $0 \leq \alpha < \Gamma$ and $x = \sum_{\gamma=0}^{\Gamma} a_\gamma e_\gamma$, denote $f_\alpha(x) = a_\alpha$. Then $\|P_{\alpha+1}(x) - P_\alpha(x)\| = \|f_\alpha(x)e_\alpha\| = |f_\alpha(x)| \cdot \|e_\alpha\|$, and thus

$$\|f_\alpha\| = \sup_{x \in B_X} |f_\alpha(x)| = \|e_\alpha\|^{-1} \sup_{x \in B_X} \|f_\alpha(x)e_\alpha\| \leq 2\|e_\alpha\|^{-1} \sup_{1 \leq \gamma < \Gamma} \|P_\gamma\|.$$

Therefore $f_\alpha \in X^*$. The functionals $\{f_\gamma\}_{\gamma=0}^{\Gamma}$ are called the *associated biorthogonal functionals* (or *coordinate functionals*) to $\{e_\gamma\}_{\gamma=0}^{\Gamma}$ and $x = \sum_{\alpha=0}^{\Gamma} f_\alpha(x)e_\alpha$ for every $x \in X$. We will denote the biorthogonal functionals f_γ by e_γ^*. We have that $\{e_\gamma, e_\gamma^*\}_{\gamma=0}^{\Gamma}$ is a biorthogonal system, and we have just proved that $\|e_\gamma\|\,\|e_\gamma^*\| \leq 2\mathrm{bc}\{e_\gamma\}_{\gamma=0}^{\Gamma}$.

A standard transfinite induction argument yields the following result.

Fact 4.7. *Let* $\{e_\gamma\}_{\gamma=0}^{\Gamma}$ *be a long Schauder basis of* X, $x = \sum_{\gamma=0}^{\Gamma} \alpha_\gamma e_\gamma$. *Then there exists a reordering of all* γ *with* $\alpha_\gamma \neq 0$ *into a sequence* $\{\gamma_i\}_{i\in\mathbb{N}}$ *so that* $x = \sum_{i=1}^{\infty} \alpha_{\gamma_i} e_{\gamma_i}$. *In particular,* $\{e_\gamma; e_\gamma^*\}_{\gamma=0}^{\Gamma}$ *is a strong bounded M-basis.*

Proposition 4.8. *Let* Γ *be an ordinal. The transfinite sequence* $\{x_\gamma\}_{\gamma=1}^{\Gamma}$ *defined by* $x_\gamma = \chi_{[0,\gamma]}$ *is a long Schauder basis of* $C[0, \Gamma]$.

Proof (Sketch). This basis is a natural generalization of the classical summing basis of c_0. It is easy to verify that for any $\alpha_i \in \mathbb{R}$, x_{λ_i}, where $1 \leq i \leq n$, $\lambda_1 < \cdots < \lambda_n$, and $k < n$,

$$\left\| \sum_{i=1}^{k} \alpha_i x_{\lambda_i} \right\| = \max_{1 \leq l \leq k} \left| \sum_{i=l}^{k} \alpha_i \right| \leq 2\max_{1 \leq l \leq n} \left| \sum_{i=l}^{n} \alpha_i \right| = 2 \left\| \sum_{i=1}^{n} \alpha_i x_{\lambda_i} \right\|,$$

which proves that $\{x_\gamma\}_{\gamma=1}^\Gamma$ is a long basic sequence. To prove that it is a long Schauder basis, it remains to verify the density of finite linear combinations of $\{x_\gamma\}_{\gamma=1}^\Gamma$ in $C[0,\Gamma]$. We omit the details. $\square$

Example 4.9 (Plichko [Plic84a]). The (ω^2)-long Schauder basis $\{x_\gamma\}_{\gamma\leq\omega^2}$ of $C[\omega^2](\cong c_0)$ is not an ordinary Schauder basis under any rearrangement.

Proof. By contradiction. Let $\{x_n\}_{n=1}^\infty$ be a rearrangement of $\{x_\gamma\}_{\gamma\leq\omega^2}$. Choose a finite sequence $n_1 < n_2 < \cdots < n_k < m_1 < \cdots < m_k$ such that $x_{n_i}, x_{m_i} \in \{x_\gamma; \omega i \leq \gamma < \omega(i+1)\}$. Then $\left\|\sum_{i=1}^k x_{n_i}\right\| = k$, but $\left\|\sum_{i=1}^k x_{n_i} - \sum_{i=1}^k x_{m_i}\right\| = 1$, a contradiction with the uniform boundedness of the canonical projections of a Schauder basis. $\square$

The w^*-density of the dual X^* has some simple equivalent formulations.

Fact 4.10. *Let X be a Banach space and Ω a cardinal. The following are equivalent:*

(i) $\Omega = w^*$-dens X^*.
(ii) Ω *is minimal such that there exists a bounded injection* $T : X \to \ell_\infty(\Omega)$.
(iii) Ω *is minimal such that there exists* $\{f_\alpha\}_{\alpha<\Omega} \subset B_{X^*}$, *which separates the points of* X.

Proof. (i)$\Rightarrow$(ii) Let $\{f_\alpha\}_{\alpha<\Omega} \subset B_{X^*}$ be a set whose rational span is w^*-dense in X^*. Clearly, $T(x) = (f_\alpha(x))_{\alpha<\Omega}$ is the desired operator.

(ii)$\Rightarrow$(iii) is trivial using the system $\{T^*(e_\alpha)\}_{\alpha<\Omega}$, where e_α are the canonical biorthogonal functionals on $\ell_\infty(\Omega)$.

(iii)$\Rightarrow$(i) follows by taking the rational span of $\{f_\alpha\}_{\alpha<\Omega}$. $\square$

Corollary 4.11. *Let X be a Banach space with $\Omega = w^*$-dens $X^* > \omega$. Then X contains a monotone long Schauder basic sequence of length Ω. If $\Omega = \omega$ and $\varepsilon > 0$, then X contains a Schauder basic sequence with basis constant $1 + \varepsilon$.*

Proof. We proceed by induction using Mazur's technique. Having constructed an initial part $\{x_\alpha\}_{\alpha<\Gamma}$ of the Schauder basic sequence, together with auxiliary sets $S_\alpha \subset B_{X^*}$, card $(S_\alpha) \leq \omega + \alpha$, where $\Gamma < \Omega$ and necessarily card $\Gamma <$ card Ω, we choose a set $S_\Gamma \subset B_{X^*}$ with the following properties. $S_\alpha \subset S_\Gamma$ for all $\alpha < \Gamma$, card $(S_\Gamma) \leq \omega + \Gamma$, and for every $x \in \overline{\operatorname{span}}\{x_\alpha; \alpha < \Gamma\}$ there exists a sequence (f_n) in S_Γ such that $\|x\| = \sup_n f_n(x)$. It suffices now to choose $0 \neq x_\Gamma \in \bigcap_{f\in S_\Gamma} \operatorname{Ker} f$. Indeed, we can see that the canonical projections associated to $\{x_\alpha\}_{\alpha<\Omega}$ all have norm 1. If $\Omega = \omega$, instead of the sequence (f_n), choose a single element that almost norms x, as in [LiTz77, Thm. 1.e.5]. $\square$

Theorem 4.12 (Plichko [Plic80c]). *Let X be a Banach space. Denote $\Omega := w^*$-dens X^*. Then, for every $\varepsilon > 0$, X has a $4+\epsilon$ bounded total biorthogonal system $\{x_\alpha; f_\alpha\}_{\alpha<\Omega}$. Moreover, $\{x_\alpha\}_{\alpha<\Omega}$ forms a (long) Schauder basic sequence.*

Proof. First assume that $\Omega > \omega$. Then the proof is based on the following lemma. For a given Banach space X and a subspace $I : Y \hookrightarrow X$, we denote by $Q : X^*/Y^\perp \to Y^*$ the canonical isomorphism.

Lemma 4.13. *Let X be a Banach space, $\Omega = w^*\text{-}\mathrm{dens}\, X^*$. Let $I : Y \hookrightarrow X$ be a subspace with a monotone long Schauder basic sequence $\{y_\alpha\}_{\alpha<\Omega}$, and $\{h_\alpha\}_{\alpha<\Omega} \subset X^*$ be some chosen biorthogonal functionals. Then there exists a subset $G \subset Y^\perp$, card $G \le \Omega$, such that $Q\left(\overline{\mathrm{span}}^{w^*}\{G \cup \{h_\alpha\}_{\alpha<\Omega}\}\right) = Y^*$.*

Proof. We prove by induction that for every $\alpha \le \Omega$ there exists $G_\alpha \subset Y^\perp$ of cardinality bounded by Ω, so that $Q\left(\overline{\mathrm{span}}^{w^*}\{G_\alpha \cup H_\alpha\}\right) = P_\alpha^* Y^*$, where $H_\alpha = \{h_\beta\}_{\beta<\alpha}$ and P_α are the canonical projections associated to $\{y_\beta\}_{\beta<\Omega}$. For $\alpha = \Omega$, we thus obtain the statement of the lemma.

When $\alpha = 0$, $P_0 = 0$ and we may put $G_0 = \emptyset$. Having chosen G_β for every $\beta < \alpha$, we distinguish two cases. If α is a nonlimit ordinal, it suffices to put $G_\alpha = G_{\alpha-1}$. Suppose α is a limit ordinal. First observe that $\bigcup_{\beta<\alpha} B_{P_\beta^* Y^*}$ is w^*-dense in $B_{P_\alpha^* Y^*}$. Indeed, for every $y \in Y$, $h \in B_{P_\alpha^* Y^*}$,

$$h(y) = P_\alpha^*(h)(y) = h(P_\alpha(y)) = \lim_{\beta \to \alpha} h(P_\beta(y)) = \lim_{\beta \to \alpha} P_\beta^* h(y).$$

In every $B_{P_\beta^* Y^*}$, we choose a w^*-dense subset $\tilde{W}_\beta$ with card $\beta \le \Omega$, and let $W_\beta \subset B_{X^*}$ be a set of the same cardinality such that $Q(W_\beta) = \tilde{W}_\beta$. By inductive assumption, for every $h \in W_\beta$, $h = f + g_h$, where $g_h \in Y^\perp$ and $f \in \overline{\mathrm{span}}^{w^*}\{G_\beta \cup H_\beta\}$. We let $G_\alpha = \bigcup_{\beta<\alpha}\bigcup_{h \in W_\beta}\{g_h\} \cup \bigcup_{\beta<\alpha} G_\beta$. Clearly, we have $\overline{\bigcup_{\beta<\alpha} W_\beta}^{w^*} \subset \overline{\mathrm{span}}^{w^*}\{G_\alpha \cup \bigcup_{\beta<\alpha} H_\beta\}$. Due to boundedness, $\overline{\bigcup_{\beta<\alpha} W_\beta}^{w^*}$ is w^*-compact and so is its image under Q. By the inductive assumption, the image has to contain $B_{P_\alpha^* Y^*}$. Consequently, $Q\left(\overline{\mathrm{span}}^{w^*}\{G_\alpha \cup H_\alpha\}\right) = P_\alpha^* Y^*$. Since card $G_\alpha \le \Omega$, the statement of the lemma follows.   □

As an immediate corollary, we obtain the following.

Corollary 4.14. *Let X be a Banach space, $\Omega = w^*\text{-}\mathrm{dens}\, X^*$, $Y \hookrightarrow X$ be a subspace with a long Schauder basis $\{y_\alpha\}_{\alpha<\Omega}$ and $\{h_\alpha\}_{\alpha<\Omega} \subset X^*$ be some chosen biorthogonal functionals. Then there exists in $Y^\perp$ a subset H of cardinality Ω such that $\overline{\mathrm{span}}^{w^*}\{H \cup \bigcup_{\alpha<\Omega} h_\alpha\} = X^*$.*

Proof. Let S be a w^*-dense subset of X^* with card $S = \Omega$ and G be the set obtained in the lemma above. For every $s \in S$, we choose $h_s \in Y^\perp$ for which $s = f + h_s$, where $f \in \overline{\mathrm{span}}^{w^*}\{G_\alpha \cup H_\alpha\}$. The set $H = G \cup \bigcup_{s \in S}\{h_s\}$ satisfies the condition.   □

We proceed with the proof of Theorem 4.12. Let $\{x_\alpha; f_\alpha\}_{\alpha<\Omega}$ be a normalized long Schauder basic sequence in X with the projectional constant less than $1 + \frac{\varepsilon}{6}$. Thus $\|x_\alpha\| = 1$, $\|f_\alpha\| \le \|P_{\alpha+1}\| + \|P_\alpha\| \le 2 + \frac{\varepsilon}{3}$. Denote

$Y = \overline{\text{span}}\{x_\alpha; \alpha < \Omega\}$, and let $H = \{h_\alpha\}_{\alpha<\Omega} \subset Y^\perp$ be the set from Corollary 4.14.

Reindex the long basis as $\{x_\alpha^n, f_\alpha^n\}_{\alpha<\Omega, n\in\mathbb{N}}$. Let g_α be a w^*-cluster point of $\{f_\alpha^n\}_{n\in\mathbb{N}}$. Clearly, $g_\alpha \in Y^\perp$, and $\|g_\alpha\| \leq 2 + \frac{\varepsilon}{3}$. Let us form a set $\{\tilde{g}_\alpha\}_{\alpha<\Omega} = \{h_\alpha\}_{\alpha<\Omega} \cup \left\{\frac{1}{2+\frac{\varepsilon}{3}} g_\alpha\right\}_{\alpha<\Omega}$. We now set $\tilde{f}_\alpha^n = f_\alpha^n - g_\alpha + \frac{\varepsilon}{3}\tilde{g}_\alpha$. The system $\{x_\alpha^n, \tilde{f}_\alpha^n\}$ is biorthogonal, and $\|\tilde{f}_\alpha^n\| \leq 4 + \varepsilon$. We have that $\frac{\varepsilon}{3}\tilde{g}_\alpha$ is a w^*-cluster point of $\{\tilde{f}_\alpha^n\}_{n\in\mathbb{N}}$. Therefore $f_\alpha^n - g_\alpha \in \overline{\text{span}}^{w^*}\{\tilde{f}_\alpha^n; \alpha < \Omega, n \in \mathbb{N}\}$, but as $\frac{1}{2+\frac{\varepsilon}{3}} g_\alpha \in \{\tilde{g}_\alpha\}_{\alpha<\Omega}$, we also have $f_\alpha^n \in \overline{\text{span}}^{w^*}\{\tilde{f}_\alpha^n; \alpha < \Omega, n \in \mathbb{N}\}$, which finishes the proof in the case $\Omega > \omega$. The remaining case, $\Omega = \omega$, can be proved similarly using the standard Mazur technique of constructing Schauder basic sequences (see [Fa˜01, Thm. 6.14]). □

4.2 Fundamental Biorthogonal Systems

Spaces admitting a fundamental biorthogonal system are characterized as those admitting a quotient of the same density and having a long Schauder basis (Plichko). This important result is applied to various concrete spaces, such as $\ell_\infty(\Gamma)$ and some of their subspaces.

Theorem 4.15 (Plichko [Plic80b]). *The following statements about a Banach space X, $\Omega = \text{dens } X > \omega$, are equivalent:*

(i) *X has a fundamental biorthogonal system of cardinality Ω.*
(ii) *X has a quotient with a monotone Ω-long Schauder basis.*
(iii) *X has a quotient with a fundamental system of cardinality Ω.*
(iv) *X has a $(4 + \varepsilon)$-bounded fundamental biorthogonal system of cardinality Ω.*

Proof. To prove (i)⇒(ii), we need the following lemma.

Lemma 4.16. *Suppose that a Banach space X has a fundamental biorthogonal system $\{x_\alpha; f_\alpha\}_{\alpha<\Omega}$, $\Omega = \text{dens } X > \omega$. Then there exist:*
(a) *a partition of $[0, \Omega)$ into a well-ordered system of subsets I_α, $0 \leq \alpha < \Omega$;*
(b) *functionals $g_\alpha \in \{f_\gamma; \gamma \in I_\alpha\}$; and*
(c) *subsets Y_α of the unit sphere of $\text{span}\{x_\alpha\}_{\alpha<\Omega}$ with $\text{card } Y_\alpha \leq \omega + \alpha$,*
such that, for every $\alpha < \Omega$,
(1) *$\text{card } I_\alpha \leq \omega + \alpha$,*
(2) *$\|f\| = \sup\{f(y); y \in Y_\alpha\}$ whenever $f \in F_\alpha = \overline{\text{span}}^{w^*}\{f_\gamma; \gamma \in I_\beta, \beta < \alpha\}$,*
(3) *$Y_\beta \subset Y_\alpha$ for $\beta < \alpha$, and*
(4) *$g_\alpha \in Y_\alpha^\perp$.*

Proof. Let $g_0 = f_0, I_0 = \{0\}$, and $F_0 = Y_0 = \emptyset$. Proceeding by induction, suppose that we have constructed g_β, I_β, Y_β, and F_β from (a)–(c) satisfying (1)–(4) for all $\beta < \alpha$.

We have $F_\alpha = \overline{\mathrm{span}}^{w^*}\{f_\gamma; \gamma \in I_\beta, \beta < \alpha\}$. Let $Q : X \to X/(F_\alpha)_\perp$ be the quotient mapping. Then $Q^* : (X/(F_\alpha)_\perp)^* = F_\alpha$, so $\overline{\mathrm{span}}\{\hat{x}_\gamma; \gamma \in I_\beta, \beta < \alpha\} = (X/(F_\alpha)_\perp)$.

As $\{x_\gamma\}_{\gamma<\Omega}$ is fundamental, there exists a dense subset of cardinality at most α of unit vectors from $\mathrm{span}\{x_\gamma\}_{\gamma<\Omega}$, whose image under Q is dense in $B_{X/(F_\alpha)_\perp}$. This implies the existence of Y_α for which (2) and (3) are satisfied.

We choose $g_\alpha = f_{\xi(\alpha)}$, where $\xi(\alpha) = \min\left\{\gamma; \gamma \notin \bigcup_{y \in Y_\beta, \beta < \alpha} \mathrm{supp}\, y\right\}$, so that (4) will be satisfied.

Finally, we let $I_\alpha = \bigcup_{y \in Y_\alpha} \mathrm{supp}\, y \bigcup \mathrm{supp}\, g_\alpha \setminus \bigcup_{\beta<\alpha} I_\beta$, so that (1) will be satisfied. Our construction is set up to end after Ω steps, which finishes the proof. □

To finish the proof of the first implication, let $Z = \overline{\mathrm{span}}^{w^*}\{g_\alpha; \alpha < \Omega\}$, $g_\alpha = f_{\xi(\alpha)}$ be from Lemma 4.16. We have $Z = (X/Z_\perp)^*$, and so $\{\hat{x}_{\xi(\alpha)}, g_\alpha\}_{\alpha<\Omega}$ form an M-basis of $X/Z_\perp$. According to properties (2)–(4) of the preceding lemma, the canonical projections onto $\overline{\mathrm{span}}^{w^*}\{g_\beta; \beta < \alpha\}$ in the space Z have norm 1. Their adjoint projections share this property, showing that $\{\hat{x}_{\xi(\alpha)}\}_{\alpha<\Omega}$ is a monotone long Schauder basis.

(ii)$\Rightarrow$(iv) The following result is needed.

Lemma 4.17. *Let $\{x_\alpha; f_\alpha\}_{\alpha<\Omega}$ be a C-bounded biorthogonal system in X, $\|f_\alpha\| = 1$, $1 \le \|x_\alpha\| \le C$. Assume that there exists a reindexing of $\{x_\alpha; f_\alpha\}_{\alpha<\Omega}$ into $\{x_n^\alpha; f_n^\alpha\}_{n\in\mathbb{N}, \alpha<\Omega}$ such that none of the sequences $\{x_n^\alpha\}_{n\in\mathbb{N}}$ is equivalent to the canonical basis of ℓ_1. Given any $\{y_\alpha\}_{\alpha<\Omega}$ with $y_\alpha \in \bigcap_{\gamma<\Omega} \mathrm{Ker}\, f_\gamma \in X$, for all $\alpha < \Omega$, there exists a $(C+\varepsilon)$-bounded biorthogonal system $\{\tilde{x}_\alpha; f_\alpha\}_{\alpha<\Omega}$ with $\overline{\mathrm{span}}\{x_\alpha, y_\alpha; \alpha < \Omega\} \subset \overline{\mathrm{span}}\{\tilde{x}_\alpha; \alpha < \Omega\}$.*

Proof. By assumption, there exist (without loss of generality positive) scalars $\{a_\alpha^n\}_{\alpha<\Omega, n\in\mathbb{N}}$ such that $\sum_{n=1}^\infty a_\alpha^n x_\alpha^n = z_\alpha$ and $\sum_{n=1}^\infty |a_\alpha| = \infty$. We put $\tilde{x}_\alpha^n = x_\alpha^n + \frac{\varepsilon}{\|y_\alpha\|} y_\alpha$. It is clear that $\|\tilde{x}_\alpha^n\| \le \|x_\alpha^n\| + \varepsilon \le C + \varepsilon$. We claim that $\{\tilde{x}_n^\alpha; f_n^\alpha\}_{n\in\mathbb{N}, \alpha<\Omega}$ is the system sought. We have

$$\lim_{N\to\infty} \frac{\sum_{n=1}^N a_\alpha^n \tilde{x}_\alpha^n}{\sum_{n=1}^N a_\alpha^n} = \lim_{N\to\infty} \frac{z_\alpha}{\sum_{n=1}^N a_\alpha^n} + \frac{\varepsilon}{\|y_\alpha\|} y_\alpha = \frac{\varepsilon}{\|y_\alpha\|} y_\alpha.$$

This implies the claim. □

We continue the proof of the implication. Let $E = X/Y$ be a quotient with a long Schauder basis, $\{x_\alpha; f_\alpha\}_{\alpha<\Omega}$. Let us split this space into separable subspaces $E_\alpha = (P_{\omega(\alpha+1)} - P_{\omega\alpha})E$, $\alpha < \Omega$. Then each space E_α admits a $1+\varepsilon$ bounded fundamental biorthogonal system $\{e_\alpha^n; h_\alpha^n\}_{n\in\mathbb{N}}$, $\|e_\alpha^n\| = 1$, $\|h_\alpha^n\| < 1 + \varepsilon$, such that $\{e_\alpha^n\}_{n\in\mathbb{N}}$ is not equivalent to the canonical basis of ℓ_1.

To see this, apply Lemma 1.25 to $X = E_\alpha$ to obtain a biorthogonal system $\{z_m; w_m\}_{m=1}^\infty$ satisfying (i) to (iv). Choose $\{z_m'\}_{m=1}^\infty \subset B_X \cap \{w_m; m \in \mathbb{N}\}_\perp$ such that $\overline{\mathrm{span}}\{z_m, z_m'; m \in \mathbb{N}\} = X$.

Reindex the biorthogonal system $\{z_m; w_m\}_{m=1}^{\infty}$ using (iii) as $\{z_m^l; w_m^l\}_{l,m\in\mathbb{N}}$ so that $\{z_m^l\}_{l=1}^{\infty}$ contains arbitrarily long finite sequences 2-equivalent to the ℓ_2-basis. Form a new biorthogonal system $\{e_\alpha^n; h_\alpha^n\}_{n\in\mathbb{N}} \sim \{z_m^l+\varepsilon z_m'; w_m^l\}_{l,m\in\mathbb{N}}$. Using the proof of Lemma 4.17, we see that the new biorthogonal system is fundamental and not equivalent to the basis of ℓ_1.

We define $g_\alpha^n(x) := h_\alpha^n\big((P_{\omega(\alpha+1)} - P_{\omega\alpha})(x)\big)$ for $x \in E$. Then the system $\{e_\alpha^n; g_\alpha^n\}_{\alpha<\Omega,n\in\mathbb{N}}$ is easily verified to be biorthogonal, and moreover $\|g_\alpha^n\| \leq 2(1+\varepsilon)$. It remains to lift this biorthogonal system to the original space. This is the content of the following result.

Lemma 4.18 (Godun [Godu83b]). *Let $\{\hat{z}_\alpha^n; \psi_\alpha^n\}_{\alpha<\Omega,n\in\mathbb{N}}$ be a C-bounded fundamental biorthogonal system in the quotient X/Y such that $\psi_\alpha^n \in Y^\perp \hookrightarrow X^*$ has the following properties:*

a) $\|z_\alpha^n\| \leq C$, $\|\psi_i^n\| = 1$.
b) $\text{dens}\, X = \Omega$.
c) *For every $\alpha < \Omega$, the sequence $\{\hat{z}_\alpha^n\}_{n=1}^{\infty}$ is not equivalent to the unit basis of ℓ_1.*

Let $Q: X \to X/Y$ be the canonical quotient mapping. Then, for every $\varepsilon > 0$, $\{\hat{x}_\alpha^n; \psi_\alpha^n\}_{\alpha<\Omega,n\in\mathbb{N}}$ admits a lifting to a fundamental biorthogonal system $\{x_\alpha^n; \psi_\alpha^n\}_{\alpha<\Omega,n\in\mathbb{N}}$, $Q(x_\alpha^n) = \hat{x}_\alpha^n$, in X, such that

$$\|x_\alpha^n\| \leq 2C + \varepsilon.$$

Proof. Choose $z_\alpha^n \in X$ so that $\hat{x}_\alpha^n = Q(z_\alpha^n)$ and $\|z_\alpha^n\| \leq C + \frac{\varepsilon}{4}$. We claim that for each $\alpha < \Omega$ there exists a sequence $\{y_\alpha^n\}_{n\in\mathbb{N}} \subset Y$ such that $\{z_\alpha^n - y_\alpha^n\}_{n\in\mathbb{N}}$ is not equivalent to the unit vector basis of ℓ_1 and $\sup_{1\leq n<\infty} \|z_\alpha^n - y_\alpha^n\| \leq 2C + \frac{\varepsilon}{2}$. Fix $\alpha < \Omega$. Since $\{\hat{x}_\alpha^n\}_{n\in\mathbb{N}}$ is not equivalent to the basis of ℓ_1, there exists a sequence of scalars (without loss of generality positive) $\{b_\alpha^n\}_{n\in\mathbb{N}}$ such that $\sum_{n=1}^{\infty} b_\alpha^n \hat{x}_\alpha^n$ converges and $\sum_{n=1}^{\infty} b_\alpha^n = \infty$. We may without loss of generality assume that $\sup_{p<\infty} \sum_{n=1}^{p} \|b_\alpha^n \hat{x}_\alpha^n\| < \frac{1}{4}$. Using a simple argument, there exists a sequence of indexes $\{n_k\}_{k=1}^{\infty}$, $n_1 = 1$, such that

$$\left\| \sum_{n=n_k}^{n_{k+1}-1} b_\alpha^n \hat{x}_\alpha^n \right\| \leq \frac{1}{2^k} \quad \text{and} \quad \sum_{n=n_k}^{n_{k+1}-1} b_\alpha^n = d_k \geq 1.$$

For each $k \in \mathbb{N}$, choose $g_\alpha^k \in Y$ such that $\left\| \sum_{n=n_k}^{n_{k+1}-1} b_\alpha^n z_\alpha^n - g_\alpha^k \right\| \leq \frac{1+\frac{\varepsilon}{4}}{2^k}$, and let $y_\alpha^n = \frac{1}{d_k} g_\alpha^k$ for all $n \in \{n_k, n_{k+1}-1\}$. Again, $\{z_\alpha^n - y_\alpha^n\}_{n=1}^{\infty}$ is not equivalent to the basis of ℓ_1 since we have $\sum_{n=1}^{\infty} b_\alpha^n = \infty$, but

$$\left\| \sum_{n=1}^{\infty} b_\alpha^n(z_\alpha^n - y_\alpha^n) \right\| \leq \sum_{k=1}^{\infty} \left\| \sum_{n=n_k}^{n_{k+1}-1} b_\alpha^n z_\alpha^n - \frac{\sum_{n=n_k}^{n_{k+1}-1} b_\alpha^n g_\alpha^k}{d_k} \right\| \leq 1 + \frac{\varepsilon}{4} \sum_{k=1}^{\infty} \frac{1}{2^k}.$$

Take a dense subset $\{\tilde{y}_\alpha\}_{\alpha<\Omega}$ of B_Y and put $x_\alpha^n = z_\alpha^n - y_\alpha^n - \frac{\varepsilon}{4}\tilde{y}_\alpha$. We have $Q(x_\alpha^n) = \hat{x}_\alpha^n$, and the biorthogonal system $\{x_\alpha^n; \psi_\alpha^n\}_{\alpha<\Omega,n\in\mathbb{N}}$ has the required properties. $\qquad\square$

Combining Lemmas 4.17 and 4.18 finishes the proof of the implication. The remaining implications follow easily.   □

Corollary 4.19 (Davis and Johnson [DaJo73a]). *Assume X has a weakly Lindelöf determined quotient of the same density. Then X admits a fundamental biorthogonal system.*

Proof. The separable case follows from Theorem 1.22. Nonseparable WLD spaces Y have an M-basis of cardinality dens Y (Theorem 5.37).   □

In the proof of the following corollary, property C is used. Recall that a closed convex subset M of a Banach space X is said to have *property C* if for every family $\mathcal{A}$ of closed convex subsets of M with empty intersection there is a countable subfamily $\mathcal{B}$ of $\mathcal{A}$ with empty intersection. We say that X has *property C* if the set X has property C (see, for example, [Fa˜01, Def. 12.36]).

Corollary 4.20. *The space JL_2 (resp. JL_0) defined by Johnson and Lindenstrauss [JoLi74] (see, e.g., [JoLi01h, Chap. 41]) has a fundamental biorthogonal system of cardinality c, although it contains no nonseparable subspace with an M-basis.*

Proof (Sketch). Recall that $(JL_2)^*$ is w^*-separable [JoLi74]. Also, $JL_2/c_0 \cong \ell_2(c)$, so JL_2 has a fundamental biorthogonal system by Corollary 4.19. The nonexistence of an M-basis for nonseparable subspaces follows by using property C. Indeed, property C is a three-space property (see, e.g., [Fa˜01, Thm. 12.37]). Thus, JL_2 (together with all subspaces) has property C. But property C is equivalent to WLD (see Definition 3.32) under the assumption of existence of an M-basis (Theorem 5.37). If $Y \hookrightarrow JL_2$ is WLD, then Y is DENS (see Definition 5.39 and the proof of Proposition 5.40), so $\omega = w^*\text{-dens } JL_2^* \geq w^*\text{-dens } (JL_2^*/Y^\perp) = w^*\text{-dens } Y^* = \text{dens } Y$, a contradiction.   □

Let Γ be a nonempty set. We call a family $\mathcal{C} \subset 2^\Gamma$ *uniformly independent* if for any distinct sets $X_1, \ldots, X_n, Y_1, \ldots, Y_m$ in $\mathcal{C}$,

$$\text{card} \left(\bigcap_{i=1}^n X_i \cap \bigcap_{i=1}^m (\Gamma \setminus Y_i) \right) = \text{card } \Gamma.$$

Lemma 4.21 (Pospíšil, see [Je78]). *For every infinite cardinal Γ, there exists a uniformly independent family $\mathcal{C} \subset 2^\Gamma$, card $\mathcal{C} = 2^\Gamma$.*

Proof. Consider the set $P = \{(F, \mathcal{F}); \mathcal{F} = \{F_1, \ldots, F_k\}, \emptyset \neq F, F_i \in \Gamma^{<\omega}\}$. Since card $P = \text{card } \Gamma$, it suffices to find the independent family in 2^P. For each $U \subset \Gamma$, we let $X_U = \{(F, \mathcal{F}) \in P : F \cap U \in \mathcal{F}\}$ and let $\mathcal{C} = \{X_U; U \subset \Gamma\}$. Trivially, $X_U \neq X_V$ for $U \neq V$. It remains to prove that $\mathcal{C}$ is uniformly independent. Let $U_1, \ldots, U_n, V_1, \ldots, V_m$ be distinct subsets of Γ. For every $1 \leq i \leq n, 1 \leq j \leq m$, choose $\gamma_{ij} \in (U_i \setminus V_j) \cup (V_j \setminus U_i) \neq \emptyset$. For every finite $F \subset \Gamma$ containing $\{\gamma_{ij}; i \leq n, j \leq m\}$, let $\mathcal{F} = \{F \cap U_i; i \leq n\}$. We have $(F, \mathcal{F}) \in X_{U_i}, i \leq n$ but $(F, \mathcal{F}) \notin X_{V_j}, j \leq m$, verifying the claim.   □

Theorem 4.22 (Rosenthal [Rose68b]). *The space $\ell_\infty(\Gamma)^*$ contains an iso-morphic copy of $\ell_2(2^\Gamma)$. Consequently, $\ell_2(2^\Gamma)$ is a quotient of $\ell_\infty(\Gamma)$.*

Proof. First, we construct a functional $\phi \in \ell_\infty^*(\Gamma)$, $\|\phi\| = 1$, with the property that for every distinct sets $U_1, \ldots, U_n, V_1, \ldots, V_m$ in $\mathcal{C}$ (the independent family from Lemma 4.21 with card $\mathcal{C} = 2^\Gamma$), we have

$$\phi\left(\prod_{i=1}^n \chi_{U_i} \cdot \prod_{i=1}^m \chi_{\Gamma\setminus V_i}\right) = 2^{-n-m}.$$

Let us use this formula as a definition of a norm-1 linear functional for a linear subspace of $\ell_\infty(\Gamma)$ generated by

$$\left\{\prod_{i=1}^n \chi_{U_i} \cdot \prod_{i=1}^m \chi_{\Gamma\setminus V_i}; U_1, \ldots, U_n, V_1, \ldots, V_m \in \mathcal{C}, n, m \in \mathbb{N}_0\right\}.$$

Using the fact that $\mathcal{C}$ is a uniformly independent family, it is straightforward to check that this definition is correct, and so, by the Hahn-Banach theorem there exists an extension of this functional to the whole $\ell_\infty(\Gamma)$, which preserves the defining formula. Since $\ell_\infty(\Gamma) \cong C(\beta\Gamma)$, where $\beta\Gamma$ is the Čech-Stone compactification of Γ, $\phi \in rca(\beta\Gamma)$ is a Radon measure. Note that passing to the variation $|\phi| \in rca^+(\beta\Gamma)$ does not change the defining relations, so we may without loss of generality assume that ϕ is in fact a probability measure on $\beta\Gamma$. Next, let us define $\phi_U \in B_{\ell_\infty(\Gamma)^*}, U \in \mathcal{C}$ by the formula $\phi_U(f) = \phi((\chi_U - \chi_{\Gamma\setminus U})f)$ for $f \in \ell_\infty(\Gamma)$. We claim that $\{\phi_U; U \in \mathcal{C}\}$ is equivalent to the canonical unit basis of $\ell_2(2^\Gamma)$. Suppose $a_i \in \mathbb{R}, U_i \in \mathcal{C}, 1 \leq i \leq n$ are given. We use the notation $U_i^1 = U_i, U_i^{-1} = \Gamma \setminus U_i$. We have

$$\left\|\sum_{i=1}^n a_i\phi_{U_i}\right\| = \sup_{f \in B_{\ell_\infty}} \sum_{i=1}^n a_i\phi_{U_i}(f)$$

$$= \sup_{f \in B_{\ell_\infty}} \sum_{\varepsilon_1,\ldots,\varepsilon_n \in \{1,-1\}} \left(\sum_{i=1}^n \varepsilon_i a_i\right) \phi\left(f \cdot \prod_{i=1}^n \phi_{U_i^{\varepsilon_i}}\right)$$

$$\sum_{\varepsilon_1,\ldots,\varepsilon_n \in \{1,-1\}} \left|\sum_{i=1}^n \varepsilon_i a_i\right| \frac{1}{2^n} = \int_0^1 \left|\sum_{i=1}^n a_i r_i(t)\right| dt,$$

where $\{r_i(t)\}_{i=1}^\infty$ are the Rademacher functions on $[0, 1]$. By the Khintchine inequality (see, e.g., [Fa~01, Lemma 6.29]), there exist $A_1, B_1 \in \mathbb{R}$ such that for all $a_i \in \mathbb{R}$

$$A_1 \left(\sum_{i=1}^n a_i^2\right)^{\frac{1}{2}} \leq \int_0^1 \left|\sum_{i=1}^n a_i r_i(t)\right| dt \leq B_1 \left(\sum_{i=1}^n a_i^2\right)^{\frac{1}{2}}.$$

The consequent part follows from Lemma 5.11. $\qquad\square$

Corollary 4.23. *Let Γ be an infinite set. Then $\ell_\infty(\Gamma)$ has a fundamental biorthogonal system.*

However, the following theorem holds.

Theorem 4.24 (Godun [Godu84]). *There exists a nonseparable subspace X of ℓ_∞ with a fundamental biorthogonal system, but no fundamental biorthogonal system in X can be extended to a fundamental biorthogonal system in ℓ_∞.*

Proof. Let X be the space JL_0 of Johnson and Lindenstrauss [JoLi74]; it has a fundamental biorthogonal system (see Corollary 4.20). By contradiction, assume that $\{x_\gamma; x_\gamma^*\}_{\gamma \in \Gamma}$ is a fundamental biorthogonal system in ℓ_∞ (we can always assume that $\|x_\gamma\| = 1$ for all γ) that extends a fundamental biorthogonal system $\{x_\beta; x_\beta^*\}_{\beta \in \Gamma_1}$ in X. Since Γ is uncountable, there is a sequence $\{x_n; x_n^*\}_{n \in \mathbb{N}}$ such that $\sup_n \|x_n^*\| < \infty$, so $x_n^* \xrightarrow{w^*} 0$ and then $x_n^* \xrightarrow{w} 0$ by the Grothendieck property of ℓ_∞. We claim that some subsequence of (x_n) is equivalent to the canonical basis of ℓ_1, which is impossible since JL_0 is Asplund. If the claim fails, by Rosenthal's ℓ_1 theorem, there is a weakly Cauchy subsequence (x_{n_k}) of (x_n). Then $(x_{n_{k+1}} - x_{n_k}) \xrightarrow{w} 0$ in ℓ_∞. By the Dunford-Pettis property of ℓ_∞ (see, e.g., [Fa~01, Thm. 11.36]), we have $-1 = \langle x_{n_{k+1}} - x_{n_k}, x_{n_k}^* \rangle \to_k 0$, a contradiction. $\qquad\square$

Corollary 4.25. *Let $X \hookrightarrow \ell_\infty$ be a subspace with ℓ_∞/X reflexive. Then X has a fundamental biorthogonal system.*

Proof. Since ℓ_∞/X is reflexive, we deduce from [LiTz77, p. 111] that $\ell_\infty \hookrightarrow X$. Using the complementability of ℓ_∞ in all overspaces, together with Corollary 4.23 and Corollary 4.19 we see that X has a reflexive quotient of density character c and so X admits a fundamental biorthogonal system of cardinality c. $\qquad\square$

We will see below that (under some set-theoretical assumptions) there exist nonseparable subspaces of ℓ_∞ without a fundamental system (even without an uncountable biorthogonal system).

Theorem 4.26 (Godun and Kadets [GoKa80], Plichko [Plic80d]). *$\ell_\infty^c(\Gamma)$ has a fundamental biorthogonal system if and only if card $\Gamma \leq c$.*

Proof. First assume that card $\Gamma \leq c$. By Theorem 4.22, $\ell_2(c)$ is a quotient of ℓ_∞ (and thus also of $\ell_\infty^c(\Gamma)$). By Corollary 4.19, $\ell_\infty^c(\Gamma)$ has a fundamental biorthogonal system. In order to prove the converse, suppose card $(\Gamma) > c$.

For a given $\phi \in \ell_\infty^c(\Gamma)^*$, there exists a countable set Γ_ϕ such that $\phi(x) = 0$ for every x, supp $(x) \subset \Gamma \setminus \Gamma_\phi$. Indeed, let us assume the contrary; i.e., for every countable $S \subset \Gamma$, there exists $x_S \in B_X$, supp $(x_S) \subset \Gamma \setminus S$, such that $\phi(x_S) \neq 0$.

We arrive at a long sequence $\{x_\alpha\}_{\alpha<\omega_1}$ of nonzero elements with pairwise disjoint supports for which $\phi(x_\alpha) \neq 0$. There exists $\varepsilon > 0$ such that for uncountably many α, $\phi(x_\alpha) > \varepsilon$, which is a contradiction with the boundedness of ϕ.

We proceed with the proof by contradiction, assuming that $\ell_\infty^c(\Gamma)$ has a fundamental biorthogonal system. By Theorem 4.15, there exists a quotient map $Q : \ell_\infty^c(\Gamma) \to X$, where X has a long Schauder basis $\{x_\alpha, f_\alpha\}_{\alpha<\Gamma}$, $\|f_\alpha\| = 1$. Put $\phi_\alpha = Q^* f_\alpha$, $\Gamma_\alpha = \Gamma_{\phi_\alpha}$. There exists $\delta > 0$ such that card $\{\alpha; \|\phi_\alpha\| > \delta\} > c$. Without loss of generality, we may assume this to be true for every $\alpha < \Gamma$. By Fact 4.7, the operator $T(x) = (\phi_\alpha(x))_{\alpha<\Gamma}$ is bounded and into $c_0(\Gamma)$. Thus every $x \in X$ lies in the kernel of all except at most countably many functionals ϕ_α. So for every countable set $S \subset \Gamma$, there exist $L_S \subset \Gamma$, card $(\Gamma \setminus L_S) \leq c$, such that $\ell_\infty(S) \subset \bigcap_{\alpha \in L_S} \operatorname{Ker} \phi_\alpha$. Let us construct by induction sequences $\{x_n\}_{n=1}^\infty$, $x_n \in B_{\ell_\infty^c(\Gamma)}$, $\{\psi_n\}_{n=1}^\infty$, $\psi_n \in \{\phi_\alpha\}_{\alpha<\Lambda}$, and $S_n \subset \Gamma$, S_n countable and pairwise disjoint, so that:

1. $\Gamma \setminus S_n \subset \operatorname{Ker} \psi_n$;
2. supp $(x_n) \subset S_n$ and $\psi_n(x_n) > \delta$.

Put $\psi_1 = \phi_1$ and $S_1 = \Gamma_1$, and choose x_1 using that $\|\phi_1\| > \delta$. Having constructed these sequences up to n, the inductive step consists of choosing $\psi_{n+1} = \phi_\beta \in L_{\bigcup_{i=1}^n S_i}$, $S_{n+1} = \Gamma_\beta \setminus \bigcup_{i=1}^n S_i$, and x_{n+1} exists due to the assumption $\|\phi_\beta\| > \delta$. Observe now that $x_0 = \sum_{n=1}^\infty x_n \in B_{\ell_\infty^c(\Gamma)}$, and $T(x_0)$ contains infinitely many coordinates larger than δ, which is a contradiction. $\qquad\square$

4.3 Uncountable Biorthogonal Systems in ZFC

As we will see later, the existence of an uncountable biorthogonal system in a nonseparable Banach space (in particular, a subspace of ℓ_∞) is undecidable in ZFC. In this section, we focus on the absolute (ZFC) results, singling out some natural classes of spaces well behaved in this respect. In particular, representable subspaces of ℓ_∞, introduced by Godefroy and Talagrand, and certain $C(K)$ spaces lead to positive results in this direction.

A weaker version of biorthogonality is easy to obtain in every nonseparable Banach space.

Fact 4.27. *Let X be a nonseparable Banach space, $\varepsilon > 0$. Then there exist long sequences $\{x_\alpha\}_{\alpha<\omega_1} \subset B_X$, $\{f_\alpha\}_{\alpha<\omega_1} \subset (1+\varepsilon)B_{X^*}$ satisfying*

$$f_\beta(x_\alpha) = 0 \text{ whenever } \beta > \alpha, f_\alpha(x_\alpha) = 1.$$

Moreover, if dens $X = \omega_1$, *we obtain in addition that $\tilde{T}(x) = (f_\alpha(x))_{\alpha<\omega_1}$ maps X into $\ell_\infty^c(\omega_1)$.*

Proof. Since X is nonseparable, there exists an ordinal sequence $\{X_\alpha\}_{\alpha<\omega_1}$ of nested separable subspaces $X_\alpha \subsetneq X_\beta \subset X$, for $\beta > \alpha$, such that dim$X_{\alpha+1}/$

$X_\alpha = 1$ and $X_\beta = \overline{\text{span}}\{\bigcup X_\alpha; \alpha < \beta\}$ for every limit ordinal $\beta < \omega_1$. Moreover, if dens $X = \omega_1$, we may assume that $X = \overline{\text{span}}\{\bigcup X_\alpha; \alpha < \omega_1\}$. Note that the last condition implies, in particular, that for every $x \in X$, $x \in X_\alpha$ for some $\alpha < \omega_1$. To finish the proof, it suffices to choose $f_\alpha \in (1+\varepsilon)B_{X^*}$, $X_\alpha \subset \text{Ker} f_\alpha \not\subset X_{\alpha+1}$, and a suitable $x_\alpha \in B_X \cap X_{\alpha+1}$ with $f_\alpha(x_\alpha) = 1$. $\quad\square$

Proposition 4.28 (Finet and Godefroy [FiGo89]). *Let X be a Banach space and $Y \hookrightarrow X$ be such that $\Omega = w^*\text{-dens}\,(X/Y)^*$. Then X contains a biorthogonal system of cardinality Ω. If, in addition, X has property C, $\{x_\alpha; f_\alpha\}_{\alpha<\Gamma}$ is an uncountable biorthogonal system in X, and $Y = \bigcap_{\alpha\in\Gamma} \text{Ker} f_\alpha$, then we have $w^*\text{-dens}\,(X/Y)^* = \Gamma$.*

Proof. By Corollary 4.11, X/Y contains a biorthogonal system $\{\tilde{x}_\alpha; \tilde{f}_\alpha\}_{\alpha<\Omega}$. Let $Q : X \to X/Y$ be the quotient mapping. It suffices to choose $\{x_\alpha; f_\alpha\}_{\alpha<\Omega}$, where $x_\alpha \in Q^{-1}\tilde{x}_\alpha$, $f_\alpha = \tilde{f}_\alpha \in Y^\perp \subset X^*$.

Let us now assume that X has property C. By [Fa~01, Thm. 12.41] this is equivalent to the following condition. Let $A \subset B_{X^*}$, $f \in \overline{A}^{w^*}$; then there exists a sequence $\{f_i\}_{i=1}^\infty \subset A$ such that $f \in \overline{\text{conv}}^{w^*}\{f_i\}_{i=1}^\infty$. Using the Banach-Dieudonné theorem, $\overline{Z}^{w^*} = \bigcup_{n\in\mathbb{N}} n\overline{B_Z}^{w^*}$ whenever $Z \hookrightarrow X^*$. Consequently, if $y \in \overline{Z}^{w^*}$, then there exists a sequence $\{z_i\}_{i=1}^\infty \subset Z$ such that $y \in \overline{\text{conv}}^{w^*}\{z_i\}_{i=1}^\infty$. To finish the proof, we need to show that

$$w^*\text{-dens}\,Y^\perp = w^*\text{-dens}\,\overline{\text{span}}^{w^*}\{f_\alpha; \alpha < \Gamma\} \geq \Gamma.$$

Let $\{g_\beta\}_{\beta\in A}$ be a w^*-dense subset of $Y^\perp$. For every β, there exists a countable $I_\beta \subset \Omega$ such that $g_\beta \in \overline{\text{conv}}^{w^*}\{f_\alpha\}_{\alpha\in I_\beta}$. We have then

$$Y^\perp = \overline{\{g_\beta\}_{\beta\in A}}^{w^*} = \overline{\text{conv}}^{w^*}\{f_\alpha\}_{\alpha\in\bigcup_{\beta\in A} I_\beta}.$$

Since $f_\alpha \in Y^\perp$ for every α, we have $\bigcup_{\beta\in A} I_\beta = \Gamma$, and so we have card $A = \Gamma$. $\quad\square$

Corollary 4.29 (Finet and Godefroy [FiGo89]). *Let X be a Banach space with property C, $\Omega = w^*\text{-dens}\,X^*$. Then X contains no subspace $Y \hookrightarrow X$ with a total biorthogonal system $\{x_\alpha; f_\alpha\}_{\alpha<\Gamma}$, where $\Gamma > \Omega$.*

Proof. Assume $Y \hookrightarrow X$ has a total biorthogonal system $\{x_\alpha; f_\alpha\}_{\alpha<\Gamma}$. Since Y has property C, and using Proposition 4.28, $w^*\text{-dens}\,Y^* \leq \Gamma$. Set $Z := \bigcap_{\alpha<\Gamma} \text{Ker} f_\alpha \hookrightarrow Y = \{0\}$. By the previous theorem, $w^*\text{-dens}(Y/Z)^* = \Gamma$, so $\Gamma \leq \Omega$ as stated. $\quad\square$

This result applies to $C(K)$ spaces, where K is a separable and nonmetrizable Rosenthal compact, or X^{**}, for every separable Banach space X, X^* nonseparable and $\ell_1 \not\hookrightarrow X$ (e.g., James tree space JT) ([God80]). In particular, let K be the two arrow space, $K = [0,1] \times \{0,1\}$, equipped with the lexicographic order. Then $C(K)$ contains no uncountable M-basic system, but it contains a biorthogonal system of cardinality c, namely $\{f_x; \mu_x\}_{x\in[0,1)}$, where $f_x = 1_{(x,1]}$, $\mu_x(f) = \lim_{t\to x^+} f(t) - f(x)$.

In the framework of $C(K)$ spaces, a result analogous to Plichko's theorem is the following result of Todorčević, which completes the previous weaker result of Lazar.

Definition 4.30. *A topological space T is said to have* property CCC *(the* countable chain condition *or the* Suslin property*) if T does not contain any uncountable family of nonempty open pairwise disjoint sets.*

The following result should be compared with Corollary 4.11.

Theorem 4.31 (Lazar [Laza81], Todorčević [Todo06]). *Let K be a compact set containing a nonseparable subset. Then $C(K)$ contains an uncountable biorthogonal system.*

Proof. By using the Tietze theorem on extension of continuous functions on compact spaces, we may without loss of generality assume that K is nonseparable. Moreover, we may assume that it has the countable chain condition (CCC) property (see Definition 4.30), since otherwise by Theorem 7.22, $C(K)$ contains a copy of $c_0(\omega_1)$.

We seek a system of points $x_\alpha \neq y_\alpha \in K$ and continuous functions $f_\alpha \in C(K)$, $\alpha < \omega_1$, such that:

(i) $f_\alpha(x_\alpha) = 1$, $f_\alpha(y_\alpha) = 0$,
(ii) $f_\alpha(x_\beta) = f_\alpha(y_\beta) = 0$ for $\beta < \alpha$,
(iii) $f_\alpha(x_\beta) = f_\alpha(x_\beta)$ for $\alpha < \beta$.

Once the system has been constructed, we see that $\{f_\alpha; \delta_{x_\alpha} - \delta_{y_\beta}\}_{\alpha<\omega_1}$ is the sought uncountable biorthogonal system in $C(K)$.

We construct the system by transfinite induction. Suppose we have constructed $x_\alpha, y_\alpha, f_\alpha$ for all $\alpha < \beta < \omega_1$. The inductive step follows from the following claim.

Claim. There exist $x \neq y$ in $K \setminus \overline{\{x_\alpha, y_\alpha\ \alpha < \beta\}}$ such that $f_\alpha(x) = f_\alpha(y)$ for all $\alpha < \beta$.

Proof of the claim. Proceeding by contradiction, we may assume that for every open set $U \subset K$, $U \cap \overline{\{x_\alpha, y_\alpha; \alpha < \beta\}} = \emptyset$, $f_\alpha, \alpha < \beta$ is a separating family of functions. Thus U is metrizable and separable. Choose a maximal pairwise disjoint family $\{U_\tau\}_{\tau\in T}$ of such open subsets of K. By the CCC property, T is countable. Clearly, $\bigcup_{\tau\in T} U_\tau \cup \{x_\alpha, y_\alpha; \alpha < \beta\}$ is a separable and dense subset of K, and the proof of the claim is finished. $\square$

This also finishes the proof of Theorem 4.31. $\square$

Definition 4.32. *We say that a Banach space X is a* representable *space if X is isomorphic to a Banach space $Y \hookrightarrow \ell_\infty$ that is analytic in the topology of pointwise convergence σ_p of ℓ_∞.*

Among representable spaces are all duals of separable Banach spaces, $C(K)$, where K is a separable Rosenthal compact [God80] and $\ell_1(c)$. Since $\ell_\infty = \ell_1^*$, the topological space (B_{ℓ_∞}, w^*) is a metrizable compact, and it is homeomorphic to $(B_{\ell_\infty}, \sigma_p)$. Also, $\ell_\infty = \bigcup_{n=1}^\infty n B_{\ell_\infty}$. Thus, in the definition above, we can replace σ_p by the w^*-topology of ℓ_∞ (coming from the predual ℓ_1).

Theorem 4.33 (Godefroy and Talagrand [GoTa82]).

Let X be a representable space. Then:

(i) *The space X is nonseparable if and only if X contains a biorthogonal system $\{x_\alpha; f_\alpha\}_{\alpha < c}$.*

(ii) *The space X contains an uncountable M-basic set if and only if X contains $\ell_1(c)$.*

Proof. Without loss of generality, $X \hookrightarrow \ell_\infty$ is an analytic subspace. Let $\{e_i\}_{i=1}^\infty$ be the canonical basis of ℓ_1, where $\ell_1 \hookrightarrow \ell_\infty^*$ is a 1-norming linear subspace. Denote the inclusion $I : X \hookrightarrow \ell_\infty$. Then $D = I^*(\ell_1) \subset X^*$ is a 1-norming subspace, so we have in particular $\overline{B_D}^{w^*} = B_{X^*}$. Suppose that X is nonseparable. In order to prove (i), we will construct a biorthogonal system $\{x_\alpha; f_\alpha\}_{\alpha \in 2^{\mathbb{N}}}$ in X, X^*, with the additional property that $(\{x_\alpha\}_{\alpha \in 2^{\mathbb{N}}}, \sigma_p)$ is homeomorphic to the Cantor set $\{0,1\}^{\mathbb{N}}$. For $\varepsilon > 0$, choose a system $\{y_\alpha; g_\alpha\}_{\alpha < \omega_1}$ from Fact 4.27 such that $y_\alpha \in B_X$, $g_\alpha \in (1+\varepsilon)B_{X^*}$, $g_\beta(y_\alpha) = 0$ for $\beta > \alpha$ and $g_\alpha(y_\alpha) = 1$. The Baire space $\mathcal{N} = \mathbb{N}^{\mathbb{N}}$ is a Polish space, completely metrizable by a metric ρ. Since (X, σ_p) is analytic, there exists a continuous surjection $\phi : \mathcal{N} \to (X, \sigma_p)$. Choose $\sigma_\alpha \in \mathcal{N}$ such that $\phi(\sigma_\alpha) = y_\alpha$, $\alpha < \omega_1$, and denote $\Theta = \{\sigma_\alpha\}_{\alpha < \omega_1}$. We may without loss of generality assume that every σ_α is a condensation point of Θ, i.e., $\forall \delta > 0$, card $\{\gamma; \rho(\sigma_\alpha, \sigma_\gamma) < \delta\} > \omega$. A set $U \subset \mathcal{N}$ will be called *special* if it is open and $U \cap \Theta \neq \emptyset$. We are going to construct by induction in n a (so-called *pre-Haar*) system of special sets $\{B_s^n\}_{n \in \mathbb{N}, s \in \{0,1\}^n}$ and functions $f_s^n \in (1+\varepsilon)B_D$ satisfying

(1) ρ-diam $B_s^n < \frac{1}{n}, B_s^n \cap B_r^n = \emptyset$ for $s \neq r$, $\overline{B_{s \frown 0}^{n+1}} \cup \overline{B_{s \frown 1}^{n+1}} \subset B_s^n$,

(2) $\qquad f_s^n \circ \phi(B_s^n) \geq 1 - \varepsilon, |f_s^n \circ \phi(B_r^n)| \leq \frac{1}{n}$ for $s \neq r$.

By assumption, $g_2 \circ \phi(\sigma_1) = 0$, $g_2 \circ \phi(\sigma_2) = 1$. As D is 1-norming, by a version of Helly's result ([Fa~01, Exer. 3.36]), there is $f_0^1 \in (1+\varepsilon)B_D$ satisfying $f_0^1 \circ \phi(\sigma_1) = 0$ and $f_0^1 \circ \phi(\sigma_2) = 1$. Since $f_0^1 \circ \phi$ is continuous, there exist open and disjoint neighborhoods D_i^0 of σ_i, $i = 1, 2$, $\rho - \text{diam} D_i^0 < 1$, and $|f_0^1 \circ \phi(D_1^0)| < \varepsilon$, $f_0^1 \circ \phi(D_2^0) > 1 - \varepsilon$. Using that both σ_1 and σ_2 are accumulation points, and the main property of $\{y_\alpha, g_\alpha\}_{\alpha < \omega_1}$, there exists a pair of points $\sigma_\beta \in D_1^0$, $\sigma_\alpha \in D_2^0$, $\beta > \alpha$, such that $g_\beta \circ \phi(\sigma_\alpha) = 0$ and $g_\beta \circ \phi(\sigma_\beta) = 1$. Using Helly's theorem, there exists $f_1^1 \in (1+\varepsilon)B_D$ satisfying $f_1^1 \circ \phi(\sigma_\alpha) = 0$ $f_1^1 \circ \phi(\sigma_\beta) = 1$. To finish the first step, choose special neighborhoods of $\sigma_\alpha \in B_0^1$, $\overline{B_0^1} \subset D_2^0$, $\sigma_\beta \in B_1^1$, $\overline{B_1^1} \subset D_1^0$ that satisfy condition 2. This completes the first step.

The inductive step from n to $n+1$. Define a function

$$F(\{O\}) := \min\{\alpha; \sigma_\alpha \in O\}$$

for all special sets, and extend it to a function

$$F(\{O_1,\ldots,O_k\}) := \max\{F(\{O_i\}); \ 1 \le i \le k\}$$

defined for all finite tuples of special sets.

In order to find $\{B_s^{n+1}\}_{s \in \{0,1\}^{n+1}}$, we use a finite inductive argument of its own. Choose first any system D_s^0, $s \in \{0,1\}^{n+1}$, of pairwise disjoint special sets with ρ-diameter $D_s^0 < 1/(n+1)$ that satisfies the inclusions $D_{t^\frown 0}^0, D_{t^\frown 1}^0 \subset B_t^n$, for all $t \in \{0,1\}^n$. To simplify the notation, we use a lower index k, $1 \le k \le 2^{n+1}$, in place of $s \in \{0,1\}^{n+1}$. The induction proper (in i) consists of constructing a system $\{D_k^i\}_{1 \le i, k \le 2^{n+1}}$ of special sets and functions $f^i \in (1+\varepsilon)B_D$ with the following properties:

1. $D_k^{i+1} \subset D_k^i$ for $i \ge 0$.

2. $f^i(D_i^i) > 1 - \varepsilon, |f^i(D_l^i)| < \dfrac{1}{n+1}$ whenever $i \ne l$.

Let us describe the inductive step from i to $i+1$ of this construction (we omit the case $i = 1$, which is similar). Choose $\alpha > F(\{D_k^i\}_{k=1}^{2^{n+1}})$ and a set $\{\alpha_k\}_{k=1}^{2^{n+1}}$ such that $\sigma_{\alpha_k} \in D_k^i$ and $\alpha = \alpha_{i+1} = \max\{\alpha_k\}_{k=1}^{2^{n+1}}$. Then we have $g_\alpha \circ \phi(\sigma_{\alpha_k}) = \delta_{i+1}^k$. As before, using Helly's theorem, there exist $f^{i+1} \in (1+\varepsilon)B_D$ and open disjoint neighborhoods of $\sigma_{\alpha_k} \in D_k^{i+1} \subset \overline{D_k^{i+1}} \subset D_k^i$ such that condition 2 is satisfied. Once the systems $\{D_k^i\}_{1 \le i, k \le 2^{n+1}}$ and $f^i \in (1+\varepsilon)B_D$ have been constructed, it suffices to use the (properly reindexed using s instead of k again) systems consisting of $B_k^{n+1} = \bigcap_{i=1}^{2^{n+1}} D_k^i$ and $f_k^{n+1} = f^k$ in the former inductive step. To finish the proof of (i), we choose $f_\alpha \in \overline{\{f_{\alpha \upharpoonright n}^n\}_{n \in \mathbb{N}}}^{w^*}$ and $x_\alpha = \phi(\bigcap B_{\alpha \upharpoonright n}^n)$ for every $\alpha \in \{0,1\}^{\mathbb{N}}$.

The proof of (ii) rests on deeper results. Assuming $\ell_1(c) \not\hookrightarrow X$, by Theorem 7.36, $\beta\mathbb{N} \not\subset (B_{X^*}, w^*)$. Since B_D is dense in (B_{X^*}, w^*) and consists of continuous functions on (X, τ_p), which is analytic, by [BFT78], (B_{X^*}, w^*) is an angelic space (see Definition 3.31) consisting of Baire-one functions on (X, τ_p). Consequently, X has property C and the conclusion follows by Corollary 4.29 above. $\square$

Corollary 4.34 (Stegall [Steg75], Fabian and Godefroy [FaGo88]). *Every nonseparable dual space X^* contains an uncountable biorthogonal system.*

Proof. If X is separable, let $\{x_n\}_{n=1}^\infty$ be a norm-dense sequence in B_X. Then $T : \ell_1 \to X$, $T(e_n) = x_n$, is a linear quotient map with $\|T\| = 1$. Thus $T^* : (X^*, w^*) \to (\ell_\infty, \tau_p)$ is w^*-w^*-continuous and $(B_{T^*(X^*)}, w^*)$ is w^*-compact, so $T(X^*)$ is a w^*-K_σ set.

If X is nonseparable and Y is a separable subspace of X such that Y^* is nonseparable, then there exists an uncountable biorthogonal system in $Y^* = X^*/Y^\perp$ that can be easily pulled back to X^*. In the remaining case, when X is a nonseparable Asplund space, the dual space belongs to a $\mathcal{P}$-class ([DGZ93a, Thm. VI.3.4]), so we may apply Theorem 5.1.   □

4.4 Nonexistence of Uncountable Biorthogonal Systems

The first example of a nonseparable Banach space that admits no uncountable biorthogonal system was obtained by Kunen (under CH). Kunen's result was unpublished and appeared only later in Negrepontis' survey [Negr84]. The first published example was by Shelah [Shel85] (under a stronger axiom, $\diamondsuit$). We have chosen to present a modification of yet another construction, due to Ostaszewski (under ♣) [Ost76], of a hereditarily separable scattered nonmetrizable locally compact space. As K is separable, the resulting nonseparable $C(K)$ space embeds isometrically into ℓ_∞. For completeness, let us mention that, in ZFC, $\diamondsuit \Leftrightarrow ♣ + \text{CH}$, and ♣ is independent of CH.

Denote by $\Lambda = \{\alpha < \omega_1; \alpha \text{ is a limit ordinal}\}$, and let $\iota : [0, \omega_1) \to \Lambda$ be an order isomorphism. We will use the notation $\lambda_\alpha = \iota(\alpha) \in \Lambda$.

We say that s is an *ω-sequence* if $S = \{\alpha_k\}_{k=1}^\infty$ is an (ordinary) increasing sequence of ordinals.

Definition 4.35. ♣ *denotes an additional set-theoretical axiom, consistent in ZFC, which claims that there exists a transfinite sequence of countable sets $\{s_\alpha\}_{\omega \le \alpha < \omega_1}$ such that*

(i) *s_α is an ω-sequence convergent to λ_α for each $\omega \le \alpha < \omega_1$, and*
(ii) *for every uncountable subset $S \subset \omega_1$, $s_\alpha \subset S$ for some $\alpha < \omega_1$.*

The intuitive meaning of this axiom is that countable subsets s_α of ω_1 "predict" all (i.e., uncountable) subsets of ω_1. This naturally provides an ideal tool for the construction of hereditarily separable spaces.

Theorem 4.36. (♣) *There exists a scattered, non-Lindelöf, locally compact space K such that K^n is hereditarily separable for all $n \in \mathbb{N}$.*

Proof. The proof consists of constructing a new topology τ on $K = [0, \omega_1)$ stronger than the canonical interval topology. Denote

$$K_<^n = \{(\alpha_1, \ldots, \alpha_n); \alpha_1 < \cdots < \alpha_n\} \subset K^n.$$

We first claim that, in order to prove that (K^n, τ^n) are hereditarily separable for all $n \in \mathbb{N}$, it suffices to show that $(K_<^n, \tau^n)$ are hereditarily separable for all n. Define a set of all ordered partitions of $\{1, \ldots, n\}$,

$$\mathcal{A} := \left\{ (A_1, \ldots, A_l); A_i \ne \emptyset, A_i \cap A_j = \emptyset, \bigcup_{i=1}^l A_i = \{1, \ldots, n\} \right\}.$$

Split K^n into finitely many disjoint subsets

$$S_{(A_1,\ldots,A_l)} = \{(\alpha_1,\ldots,\alpha_n); \alpha_i = \alpha_j \text{ iff } i,j \in A_p,$$
$$\alpha_i < \alpha_j \text{ iff } i \in A_p, j \in A_q, p < q\}.$$

Given any uncountable $M \subset K^n$, note that $M \cap S_{(A_1,\ldots,A_l)}$ is homeomorphic to a subset of $K^l_<$, and $M = \bigcup_{t\in A} M \cap S_t$, so the claim follows.

Before we start the construction of τ, we need some preliminaries. These are needed in order to obtain hereditary separability for all the finite powers of K. Let $M \subset [0,\omega_1)$ be infinite. Then there exists an increasing sequence $\{x_n\}_{n=1}^\infty \subset M$. This follows easily by choosing $x_1 = \min M$, $x_{n+1} = \min M \setminus \{x_1,\ldots,x_n\}$. Similarly, if $M \subset [0,\omega_1]$ is uncountable, it contains an uncountable increasing transfinite sequence. For every $n \in \mathbb{N}$, there exists a bijection $B_n : [0,\omega_1) \to [0,\omega_1)^n$, so by ♣ there exist systems $\{\tilde{X}^n_\alpha\}_{\alpha<\omega_1}$, $\tilde{X}^n_\alpha = B_n(s_\alpha)$ of countable subsets of $[0,\omega_1)^n$ such that for every uncountable $X \subset [0,\omega_1)^n$ there exists $\alpha < \omega_1$ for which $\tilde{X}^n_\alpha \subset X$. For every $\alpha < \omega_1$, choose some (if it exists) $X^n_\alpha \subset \tilde{X}^n_\alpha \cap K^n_<$ such that $X^n_\alpha = \{(x^{n,\alpha}_{1,i},\ldots,x^{n,\alpha}_{n,i})\}_{i=1}^\infty$, where $x^{n,\alpha}_{n,i} < x^{n,\alpha}_{1,i+1}$ and $\lim_{i\to\infty} x^{n,\alpha}_{k,i} \to \lambda^n_\alpha$ is independent of k. For notational convenience, without loss of generality, assume that X^n_α is defined for every $\alpha < \omega_1$. Observe that since s_α is a sequence convergent to λ_α, card $\{\alpha; \lambda^n_\alpha \le \beta\} \le \omega$ for every $\beta < \omega_1$. For each $\alpha < \omega_1$, we select a countable system of ω-sequences s^k_α, $k \in \mathbb{N}$ with the following properties:

(i) $\bigcup_{k=1}^\infty s^k_\alpha$ is an ω-sequence convergent to λ_α.
(ii) $s^k_\alpha \cap s^l_\alpha = \emptyset$ whenever $k \ne l$.
(iii) Given any $\beta < \omega_1$, if for some n, α, $\lambda_\beta = \lambda^n_\alpha$, and $k_1 < \cdots < k_n \in \mathbb{N}$ are arbitrary, then the set $\{i; x^{n,\alpha}_{j,i} \in s^{k_j}_\alpha$ for $1 \le j \le n\}$ is infinite.

This is done by a standard argument based on the fact that, for every $\beta < \omega_1$, we have card $\{(n,\alpha); \lambda^n_\alpha = \lambda_\beta\} \le \omega$. We will now define by transfinite induction a family of topologies τ_α on $[0,\lambda_\alpha)$ in such a way that $\bigcup_{\alpha<\omega_1} \tau_\alpha$ generates the desired topology on $K = [0,\omega_1)$. The topologies τ_α will be defined in such a way that the following conditions are satisfied:

(1) If $\beta < \alpha$, then id $: ([0,\lambda_\beta),\tau_\beta) \to ([0,\lambda_\alpha),\tau_\alpha)$ is a homeomorphic embedding.
(2) $\gamma < \lambda_\alpha$ implies $[0,\gamma) \in \tau_\alpha$.
(3) If $\lambda_\beta = \lambda^n_\alpha$, $\gamma > \beta$, and $\lambda_\beta \le \gamma_1 < \cdots < \gamma_n < \lambda_\gamma$, then $(\gamma_1,\ldots,\gamma_n) \in \overline{(X^n_\alpha)}$ in the topology of $([0,\lambda_\gamma)^n, \tau^n_\gamma)$.
(4) τ_α is a metrizable, by ρ_α, locally compact topology for all $\alpha < \omega_1$.

We begin the induction by taking τ_0 to be the discrete topology on $[0,\lambda_0)$, where $\lambda_0 = \omega$.

The inductive step. If α is a limit ordinal, we set τ_α to be a topology on $[0,\lambda_\alpha)$ generated by $\bigcup_{\gamma<\alpha} \tau_\gamma$. The validity of (1)–(3) is clear (condition (3) is considered subject to $\beta < \gamma \le \alpha$). The topological space $[0,\lambda_\alpha)$ has a countable weight and it is regular, and thus it is metrizable ([Eng77, p. 179]), which verifies condition (4). In order to define τ_α for $\alpha = \beta+1$, we will use the system $\{s_\beta^k\}_{k=1}^\infty$ of ω-sequences convergent to λ_β, $s_\beta^k = \{s_n^k\}_{n=1}^\infty$. Besides the open sets described by conditions (1) and (2), we will add the neighborhood bases of the countably many points $\{p_k\}_{k=1}^\infty$, $p_k = \lambda_\beta + k - 1$, described below. Using condition (i) and the inductive assumptions, we can choose a system of pairwise disjoint clopen sets $\{U_n^k\}_{k,n\in\mathbb{N}}$, $s_n^k \in U_n^k$, $\rho_\beta - \text{diam} U_n^k < \frac{1}{2^{k+n}}$. The neighborhood base of p_k will consist of sets $\bigcup_{i=n}^\infty U_i^k \cup \{p_k\}$ for all $n \in \mathbb{N}$. Properties (1) and (2) are again clear. To check property (4), note that the local compactness of $[0, \lambda_\alpha)$ is clear, and metrizability follows from the countable weight (e.g., [Eng77, p. 179]). It remains to check (3). We will treat the main case when $\beta + 1 = \alpha = \gamma$; the other cases follow rather easily. So, let $\lambda_\beta = \lambda_\xi^n$, and $\lambda_\beta \leq \gamma_1 < \cdots < \gamma_n < \lambda_\alpha$. Condition (iii) implies immediately that in every neighborhood of $(\gamma_1, \ldots, \gamma_n) \in [0, \lambda_\alpha)^n$ there exists a point from $\{(x_{1,i}^{n,\xi}, \ldots, x_{n,i}^{n,\xi})\}_{i=1}^\infty$, which yields the conclusion. It remains to prove that $K_<^n$ is hereditarily separable. We proceed by induction in n. The case $n = 1$ is straightforward. Indeed, let $M \subset [0, \omega_1)$ be uncountable. By $\clubsuit$, there exists some $s_\alpha \subset M$. By property (3) of the construction, $M \cap [0, \lambda_\alpha)$ is a dense countable subset of M. In the inductive step, let $M \subset K_<^n$ be uncountable. We have two possibilities. Either there exists $\{(t_1^\alpha, \ldots, t_n^\alpha)\}_{\alpha < \omega_1} \subset M$ for which $\lim_{\alpha \to \omega_1} t_1^\alpha = \omega_1$ or it does not. If it does, then there exists $M \supset M_0 = \{(t_1^\alpha, \ldots, t_n^\alpha)\}_{\alpha < \omega_1}$, where $t_n^\alpha < t_1^\beta$ for $\alpha < \beta < \omega_1$. We already know that $\tilde{X}_\alpha^n \subset M_0$ for some α. From the form of M_0, we see that $\tilde{X}_\alpha^n$ has a subset $X_\alpha^n = \{(x_{1,i}^{n,\alpha}, \ldots, x_{n,i}^{n,\alpha})\}_{i=1}^\infty$, where $x_{n,i}^{n,\alpha} < x_{1,i+1}^{n,\alpha}$, and $\lim_{i \to \infty} x_{k,i}^{n,\alpha} \to \lambda_\alpha^n$ for every k. By property (3) of our construction, we obtain that $X_\alpha^n \subset M$ is dense in $M_{\lambda_\alpha^n} = \{(\gamma_1, \ldots, \gamma_n) \in M; \gamma_1 > \lambda_\alpha^n\}$. However, $M \backslash M_{\lambda_\alpha^n}$ is a countable union of sets homeomorphic to subsets of $K_<^{n-1}$ (according to the value of their first coordinate less than λ_α^n), and the conclusion follows in this case by induction. In the remaining case, there exists $\sup\{t_1; (t_1, \ldots, t_n) \in M\} = \gamma < \omega_1$. As before, M can be viewed as a countable union of sets homeomorphic to subsets of $K_<^{n-1}$, and the proof is finished. $\qquad\square$

Lemma 4.37. *Let $\mathfrak{X}$ be a topological space with $\mathfrak{X}^n$ hereditarily separable for every $n \in \mathbb{N}$. Then $\mathfrak{X}^\omega$ is hereditarily separable. Moreover, if $\mathfrak{X}$ is a scattered compact, then $(C(\mathfrak{X})^*, w^*)$ is hereditarily separable.*

Proof. Let $S \subset \mathfrak{X}^\omega$. For every finite $F \subset \omega$, we choose a countable set $D_F \subset S$ such that $\pi_F(D_F)$ is dense in $\pi_F(S)$. The set $D = \cup\{D_F; F \subset \omega, \text{card } F < \omega\}$ is countable and dense in S. To prove the last statement of the lemma, consider a mapping

$$T : \mathfrak{X}^\omega \times (B_{\ell_1}, \|\cdot\|_1) \to (B_{C(\mathfrak{X})^*}, w^*),$$

defined as $T((x_i)_{i\in\omega}, (a_i)_{i\in\omega}) = \sum a_i \delta_{x_i}$. T is clearly a continuous and onto mapping. The space $\mathfrak{X}^\omega \times (B_{\ell_1}, \|\cdot\|_1)$ is also hereditarily separable, being a product of a hereditarily separable and a separable metric space. Thus $(B_{C(\mathfrak{X})^*}, w^*)$ is hereditarily separable, and so is $(C(\mathfrak{X})^*, w^*)$. $\qquad\square$

Theorem 4.38 (Zenor [Zen80], Velichko [Ve81]; see also [Negr84]). *Let* $\mathcal{X}$ *be a topological space with* $\mathcal{X}^\omega$ *hereditarily separable. Then* $(C(\mathcal{X}), \sigma_p)$ *is hereditarily Lindelöf.*

Proof. We let $\Phi : \mathcal{X}^\omega \times C(\mathcal{X}) \to \mathbb{R}^\omega$ be defined by $\Phi(x, f) = (f(x_n))$, where $x = (x_n) \in \mathcal{X}^\omega, f \in C(\mathcal{X})$. For an open set U in $\mathbb{R}^\omega$ and $x \in \mathcal{X}^\omega$, we set

$$N(x, U) = \{f \in C(\mathcal{X}); \Phi(x, f) \in U\}.$$

We fix a countable basis $\mathcal{B}$ or $\mathbb{R}^\omega$ consisting of basic open sets in the Cartesian topology. Then the family

$$\mathcal{P} = \{N(x, U); x \in \mathcal{X}^\omega, U \in \mathcal{B}\}$$

is a basis for $(C(\mathcal{X}), \sigma_p)$. It is enough to prove that for any $\mathcal{C} \subset \mathcal{P}$ there is a countable family $\mathcal{D} \subset \mathcal{C}$ with $\bigcup \mathcal{D} = \bigcup \mathcal{C}$. For $U \in \mathcal{B}$ we set

$$Y_U := \{x \in \mathcal{X}^\omega; N(x, U) \in \mathcal{C}\} \subset \mathcal{X}^\omega.$$

By assumption, Y_U is separable, and hence there is a countable set D_U dense in Y_U. We put

$$\mathcal{D} := \{N(x, U); x \in D_U, U \in \mathcal{B}\}.$$

It follows that $\mathcal{D}$ is countable, $\mathcal{D} \subset \mathcal{C}$, and $\bigcup \mathcal{D} = \bigcup \mathcal{C}$. □

Definition 4.39. *A family* $\{x_\alpha\}_{\alpha < \omega_1}$ *in a topological space* $\mathcal{X}$ *is called* right-separated *if* $x_\alpha \notin \overline{\{x_\beta\}_{\beta > \alpha}}$ *for all* $\alpha < \omega_1$.

Proposition 4.40. *A topological space* $\mathcal{X}$ *is hereditarily Lindelöf if and only if it contains no right-separated family.*

Proof. If $\mathcal{Y} = \{x_\alpha\}_{\alpha < \omega_1}$ is right-separated, then setting $U_\alpha = \mathcal{Y} \setminus \overline{\{x_\beta\}_{\beta < \alpha}}$ for $\alpha < \omega_1$ we obtain an open cover of $\mathcal{Y}$ without any countable subcover. Suppose now that there is $\mathcal{Y} \subset \mathcal{X}$ that is not Lindelöf. Thus there exists an open cover $\mathcal{U}$ of $\mathcal{Y}$ without any countable subcover. We choose inductively $x_\alpha \in \mathcal{Y}$ and $U_\alpha \in \mathcal{U}$ such that $x_\alpha \in U_\alpha \setminus \bigcup_{\gamma < \alpha} U_\gamma$, $\mathcal{Y} \subset \bigcup_{\alpha < \omega_1} U_\alpha$. Since $\mathcal{U}$ has no countable subcover, the index set of all α is uncountable. The family $\{x_\alpha\}_{\alpha < \omega_1}$ is right-separated. □

Theorem 4.41. (♣) *There exists a scattered nonmetrizable compact space* L *such that* $C(L)^*$ *is hereditarily separable in the* w^**-topology and* $C(L)$ *is hereditarily Lindelöf in the weak topology. Moreover, for every* $\{f_\alpha\}_{\alpha < \omega_1} \subset C(L)$, *there exists* $\beta < \omega_1$ *with* $f_\beta \in \overline{\mathrm{conv}}\,\{f_\alpha\}_{\beta < \alpha < \omega_1}$. *In particular,* $C(L)$ *is a nonseparable Banach space without an uncountable biorthogonal system.*

Proof. Let L be a one-point compactification of the locally compact space K constructed in Theorem 4.36. It is clear that L is a scattered compact space such that L^n is hereditarily separable for all $n \in \mathbb{N}$. By Lemma 4.37, $(C(L)^*, w^*)$ is hereditarily separable. We claim that K, and therefore also

L, is nonmetrizable. Assume, by contradiction, that K is metrizable. Since K is separable, it is Lindelöf ([Eng77, p. 177]). As K is locally countable, we obtain that K is countable, which is a contradiction. By Theorem 4.38, $(C(L), \sigma_p)$ is hereditarily Lindelöf. Since the σ_p and the weak topology coincide for bounded subsets of $C(L)$, L scattered, $(C(L), w)$ is hereditarily Lindelöf. By Proposition 4.40, there is no weakly right-separated family in $C(L)$. Since the weak and norm convex closures coincide in Banach spaces, we obtain in particular that for every $\{f_\alpha\}_{\alpha < \omega_1}$ there exists $\beta < \omega_1$ with $f_\beta \in \overline{\{f_\alpha\}}^w_{\alpha > \beta} \subset \overline{\mathrm{conv}}\,^w \{f_\alpha\}_{\alpha > \beta} \subset \overline{\mathrm{conv}}\,\{f_\alpha\}_{\alpha > \beta}$.   □

4.5 Fundamental Systems under Martin's Axiom

In this section, we will show how Martin's axiom MA_{ω_1} (resp. Martin's maximum MM) can be used to construct fundamental biorthogonal systems in every Banach space of density ω_1 and with a countably tight dual ball (resp. every Banach space of density ω_1). This result is in sharp contrast with Theorem 4.41. However, there is no contradiction since Martin's axiom MA_{ω_1} and ♣ (or CH) are well known to be mutually exclusive additional axioms of ZFC. MA_{ω_1} can be shown to be equivalent to a statement generalizing the classical Baire theorem. However, this translation is less useful for applications. Instead, it pays off to view MA_{ω_1} as a way to construct "ideal" objects from their "finite approximations". Let $\mathcal{P}$ be a partially ordered set. We say that $p, q \in \mathcal{P}$ are *compatible* if there exists $r \in \mathcal{P}$ for which $r \geq p, r \geq q$. We say in this case that r *extends* both p and q. We say that $\mathcal{P}$ satisfies the *countable chain condition* (CCC), if for every uncountable $R \subset \mathcal{P}$ there exist $p, q \in R$ that are compatible. A subset $D \subset \mathcal{P}$ is called *dense* if for every $p \in \mathcal{P}$ there exists $q \in D$ that extends p; i.e., $q \geq p$. Finally, we say that $\mathcal{F} \subset \mathcal{P}$ is a *filter* if any $p, q \in \mathcal{F}$ are compatible and if $p \leq q \in \mathcal{F}$ implies $p \in \mathcal{F}$. (Let us point out here that in set theory it is usual to use the reverse inequalities in order to define filters on partially ordered sets.)

Definition 4.42. *Martin's axiom* MA_{ω_1} *is an additional axiom of set theory consistent in* ZFC *that claims the following.*

Let $\mathcal{P}$ *be a CCC partially ordered set and* $\{D_\alpha\}_{\alpha < \omega_1}$ *be a system of dense subsets of* $\mathcal{P}$. *Then there exists a filter* $\mathcal{F} \subset \mathcal{P}$ *satisfying* $\mathcal{F} \cap D_\alpha \neq \emptyset$ *for every* $\alpha < \omega_1$.

In the topological context, CCC (see Definition 4.30) means that a compact K does not contain a pairwise disjoint uncountable system of open sets. We can interpret $\mathcal{P}$ as a system of all open subsets of K, and $U, V \in \mathcal{P}$ satisfy $V \leq U$ if and only if $U \subseteq V$. Under this interpretation, it is rather easy to observe that we have the following fact.

Fact 4.43 (MA_{ω_1}). *Let* K *be a CCC compact topological space and* $\{U_\alpha\}_{\alpha < \omega_1}$ *be a system of open and dense subsets of* K. *Then* $\bigcap_{\alpha < \omega_1} U_\alpha$ *is dense in* K.

In ZFC, one can prove that the last statement, in turn, implies the validity of MA_{ω_1}, and thus they are ZFC-equivalent. From here it is also clear that MA_{ω_1} contradicts the continuum hypothesis (i.e., $2^{\aleph_0} = \aleph_1$), just applying to the unit interval $[0, 1]$ and the system of complements of singletons. Another interpretation is designing $\mathcal{P}$ as a CCC system of finite sets (of "properties") whose "union" is an "ideal" object we are trying to construct. Of course, the "union" is done not over the whole $\mathcal{P}$ but only over the filter $\mathcal{F}$ because we need that all finite ingredients of the union be compatible. So, philosophically, MA_{ω_1} arranges for us the compatibility of a large subsystem of partial "properties". For completeness, let us state the well-known basic result needed in the sequel, whose proof can be found, e.g., in [Fa~01, Exer. 12.28].

Lemma 4.44 (root lemma). *Let $\mathcal{A}$ be an uncountable system of finite sets. Then there exists a finite set Δ and an uncountable $\mathcal{B} \subset \mathcal{A}$, so that $A \cap B = \Delta$ for every distinct $A, B \in \mathcal{B}$.*

Theorem 4.45 (Todorčević [Todo06]). (MA_{ω_1}). *Let X be a Banach space of dens $X = \omega_1$, and suppose that there exists a bounded linear operator $T : X \to c_0(\omega_1)$ with nonseparable range. Then X has a fundamental biorthogonal system.*

Proof. By a simple argument, we may without loss of generality assume that there exists $\{f_\gamma\}_{\gamma<\omega_1} \subset S_{X^*}$ such that $T(x) = (f_\gamma(x))_{\gamma<\omega_1}$. The proof of the next claim is the simplest example of an argument involving MA_{ω_1} in that it involves no additional parameters. The same scheme of proof will be used repeatedly later.

Claim. There exists an uncountable $\Gamma \subset \omega_1$ and a dense linear subspace $Y \hookrightarrow X$, such that $\sum_{\gamma \in \Gamma} |f_\gamma(y)| < \infty$ for all $y \in Y$.

Proof of the claim. Denote by $\mathcal{P} = \{(D_p, \Gamma_p)\}$ a system of all pairs $p = (D_p, \Gamma_p)$, where D_p is a finite subset of X, Γ_p is a finite subset of ω_1, and such that $\sum_{\gamma \in \Gamma_p} |f_\gamma(x)| < 1$ for every $x \in D_p$. We partially order $\mathcal{P}$ by letting $p \leq q$ if
$$D_p \subseteq D_q, \quad \Gamma_p \subseteq \Gamma_q.$$

We are now going to prove that $\mathcal{P}$ satisfies the CCC property. Let $\{p_\gamma\}_{\gamma<\omega_1}$ be any subset of $\mathcal{P}$, $p_\gamma = (D_\gamma, \Gamma_\gamma)$. Our objective is to prove that there exist $\alpha < \beta < \omega_1$ such that p_α, p_β are comparable; i.e., they have common extension $r \in \mathcal{P}$, $r \geq p_\alpha, p_\beta$.

By the root lemma, we may without loss of generality assume that there exist finite sets $D \subset X$, $\Delta \subset [0, \omega_1)$ such that, for every $\alpha < \beta < \omega_1$, we have $D_\alpha \cap D_\beta = D$, $\Gamma_\alpha \cap \Gamma_\beta = \Delta$.

By passing to a subsequence, using also the fact that $\{\alpha; f_\alpha(x) \neq 0\}$ is countable for every $x \in X$, we may without loss of generality assume that $\{p_\gamma\}_{\gamma<\omega_1}$ has the following additional properties. There exist $k \in \mathbb{N}$, $\varepsilon > 0$, such that for every $\alpha < \beta$ we have

$$\Gamma_\alpha \setminus \Delta = \{\gamma_\alpha^1, \gamma_\alpha^2, \ldots, \gamma_\alpha^k\}, \text{ where } \gamma_\alpha^1 < \gamma_\alpha^2 < \cdots < \gamma_\alpha^k,$$

$$\gamma_\alpha^k < \gamma_\beta^1,$$

$$f_\gamma(x) = 0 \text{ for every } x \in D_\alpha, \gamma \in \Gamma_\beta \setminus \Delta,$$

$$\sum_{\gamma \in \Gamma_\alpha} |f_\gamma(x)| < 1 - \varepsilon \text{ for every } x \in D_\alpha.$$

Let us now consider p_ω. Since the range of T lies in $c_0(\omega_1)$, there exists $\alpha < \omega$ for which

$$\sum_{i=1}^k |f_{\gamma_\alpha^i}(x)| < \varepsilon \text{ for every } x \in D_\omega.$$

This implies that $p_{\alpha,\omega} = (D_\alpha \cup D_\omega, \Gamma_\alpha \cup \Gamma_\omega)$ extends both p_α and p_ω. The CCC is established.

Let $\alpha < \omega_1$, $x \in X$. We set

$$\mathcal{P}_\alpha = \{p \in \mathcal{P}; \max \Gamma_p > \alpha\}, \mathcal{P}_x = \{p \in \mathcal{P}; tx \in D_p \text{ for some } t > 0\}.$$

Let us see that the open sets $\mathcal{P}_\alpha$, $\mathcal{P}_x$ are dense in $\mathcal{P}$ for every α, x. Given any $p \in \mathcal{P}$, there exists $\beta > \alpha$, $\max \Gamma_p$, such that $f_\beta(x) = 0$ for all $x \in D_p$. Thus $p \leq (D_p, \Gamma_p \cup \{\beta\}) \in \mathcal{P}_\alpha$. Similarly, choosing $t > 0$ small enough, so that $t \sum_{\gamma \in \Gamma_p} |f_\gamma(x)| < 1$, we have $p \leq (D_p \cup \{tx\}, \Gamma_p) \in \mathcal{P}_x$.

Let $S \subset X$, $|S| = \omega_1$ be a dense subset. Invoking MA_{ω_1}, we obtain a filter $\mathcal{F} \subset \mathcal{P}$, satisfying $\mathcal{F} \cap \mathcal{P}_\alpha \neq \emptyset$, $\mathcal{F} \cap \mathcal{P}_x \neq \emptyset$ for any choice $\alpha < \omega_1$, $x \in S$. Now let

$$\Gamma = \bigcup_{p \in \mathcal{F}} \Gamma_p, \ D = \bigcup_{p \in \mathcal{F}} D_p.$$

It is clear that $\sum_{\gamma \in \Gamma} |f_\gamma(x)| < \infty$ for every $x \in Y = \text{span } S$. $\qquad\qquad \square$

The proof of the theorem now continues along the lines of the Johnson-Rosenthal result [JoRo72]. The main difference lies in the indexing. For $x \in Y$, put $\Gamma_\varepsilon(x) \subset \omega_1$ to be a finite set such that $\sum_{\gamma \notin \Gamma_\varepsilon(x)} |f_\gamma(x)| < \varepsilon$. Let $\mathcal{P} = \{(D_p, \Gamma_p, \varepsilon_p)\}$ be a system of all triples $p = \{(D_p, \Gamma_p, \varepsilon_p)\}$, where D_p is a finite subset of Y, Γ_p is a finite subset of ω_1, and a rational $\varepsilon_p \in (0, 1)$, satisfying that

for every $f^* \in \overline{\text{span}}^{w^*}\{f_\gamma; \gamma \in \Gamma_p\}$, $\|f^*\| = 1$, there exists $x \in D_p$, $\|x\| = 1$, such that $|f^*(e) - e(x)| \leq \frac{\varepsilon_p}{3}\|e\|$ for all $e \in \overline{\text{span}}\{f_\gamma; \gamma \in \Gamma_p\}$

We partially order $\mathcal{P}$ by letting $p \leq q$ if

$$D_p \subseteq D_q, \ \Gamma_p \subseteq \Gamma_q, \ \varepsilon_p \geq \varepsilon_q,$$

$$\Gamma_{\frac{\varepsilon_p}{3}}(x) \cap (\Gamma_q \setminus \Gamma_p) = \emptyset \text{ for every } x \in D_p.$$

We are now going to prove that $\mathcal{P}$ satisfies the CCC property. Let $\{p_\gamma\}_{\gamma < \omega_1}$ be any subset of $\mathcal{P}$, $p_\gamma = (D_\gamma, \Gamma_\gamma, \varepsilon_\gamma)$. Our objective is to prove that there exist $\alpha < \beta < \omega_1$ such that p_α, p_β have a common extension $r \in \mathcal{P}$, $r \geq p_\alpha, p_\beta$. By the root lemma, we may without loss of generality assume that there exist finite sets $D \subset Y$, $\Delta \subset [0, \omega_1)$ and a rational $\varepsilon \in (0, 1)$ such that, for every $\alpha < \beta < \omega_1$, we have $D_\alpha \cap D_\beta = D$, $\Gamma_\alpha \cap \Gamma_\beta = \Delta$, $\varepsilon_\alpha = \varepsilon$. By passing to a subsequence, using also the fact that $\{\alpha; f_\alpha(x) \neq 0\}$ is countable for every $x \in Y$, we may without loss of generality assume that $\{p_\gamma\}_{\gamma < \omega_1}$ has the following additional properties. There exists a $k \in \mathbb{N}$ such that, for every $\alpha < \beta$, we have

$$\Gamma_\alpha \setminus \Delta = \{\gamma_\alpha^1, \gamma_\alpha^2, \dots, \gamma_\alpha^k\}, \text{ where } \gamma_\alpha^1 < \gamma_\alpha^2 < \cdots < \gamma_\alpha^k,$$

$$\gamma_\alpha^k < \gamma_\beta^1,$$

$$f_\gamma(x) = 0 \text{ for every } x \in D_\alpha, \gamma \in \Gamma_\beta \setminus \Delta.$$

Let us now consider p_ω. Since, for every $x \in Y$, $T(x)$ lies in $\ell_1(\omega_1)$, there exists $\alpha < \omega$ for which

$$\Gamma_{\frac{\varepsilon}{3}}(x) \cap (\Gamma_\alpha \setminus \Delta) = \emptyset \text{ for every } x \in D_\omega.$$

Applying the local reflexivity principle (see, e.g., [Fa~01, Thm. 9.15]) together with the compactness of finite-dimensional unit balls, we have the existence of a finite set $D_{\alpha,\omega} \subset Y$, $D_\alpha \cup D_\omega \subset D_{\alpha,\omega}$, and such that $p_{\alpha,\omega} = (D_{\alpha,\omega}, \Gamma_p \cup \Gamma_\omega, \varepsilon) \in \mathcal{P}$. We have that $p_{\alpha,\omega}$ extends both p_α and p_ω. The CCC is established.

Let $\alpha < \omega_1$, $x \in Y$, $\varepsilon \in (0, 1)$ be rational. We set

$$\mathcal{P}_\alpha = \{p \in \mathcal{P}; \max \Gamma_p > \alpha\}, \mathcal{P}_x = \{p \in \mathcal{P}; x \in D_p\}, \mathcal{P}_\delta = \{p \in \mathcal{P}; \varepsilon_p \leq \delta\}.$$

Let $p \in \mathcal{P}$. Choose $\gamma > \alpha$ large enough so that $f_\gamma(x) = 0$ for all $x \in D_p$. Applying the local reflexivity principle again, together with the compactness of finite-dimensional unit balls, we have the existence of a finite $\tilde{D}_p \supseteq D_p$ such that $p \leq (\tilde{D}_p, \Gamma_p \cup \{\gamma\}, \varepsilon) \in \mathcal{P}_\alpha$. Thus $\mathcal{P}_\alpha$ is open and dense in $\mathcal{P}$. Similarly, $p \leq (D_p \cup \{x\}, \Gamma_p, \varepsilon) \in \mathcal{P}_x$, so $\mathcal{P}_x$ is open and dense in $\mathcal{P}$. Once again, given $\delta > 0$, there is a finite set $\tilde{D}_p \supseteq D_p$ such that $p \leq (\tilde{D}_p, \Gamma_p, \min\{\varepsilon_p, \delta\}) \in \mathcal{P}_\alpha$. Thus $\mathcal{P}_\delta$ is open and dense in $\mathcal{P}$. Invoking MA_{ω_1}, we obtain a filter $\mathcal{F} \subset \mathcal{P}$ satisfying $\mathcal{F} \cap \mathcal{P}_\alpha \neq \emptyset$, $\mathcal{F} \cap \mathcal{P}_x \neq \emptyset$, and $\mathcal{F} \cap \mathcal{P}_\varepsilon \neq \emptyset$ for any choice $\alpha < \omega_1$, $x \in S$, and a rational $\varepsilon \in (0, 1)$. Now let $\tilde{\Gamma} = \bigcup_{p \in \mathcal{F}} \Gamma_p$. Then $\tilde{\Gamma}$ is uncountable. Using a standard countable exhaustion argument, there exists an increasing subsequence $\{\xi_\alpha\}_{\alpha < \omega_1} = \Gamma \subset \tilde{\Gamma} \subset [0, \omega_1)$ having the property that

$$(\forall \gamma \in \Gamma)(\forall \text{ finite } A \subset \Gamma, \ \max A < \gamma)(\forall \ \varepsilon > 0),$$

$$(\exists p \in \mathcal{F} : \varepsilon_p < \varepsilon, A \subset \Gamma_p \subset [0, \gamma) \subset [0, \omega_1)).$$

Note that using the filter property it follows that for every $p \in \mathcal{F}$ we have that $\forall x \in D_p, \Gamma_{\frac{\varepsilon_p}{3}}(x) \cap \Gamma \subset \Gamma_p$. Let us see that the long sequence $\{f_\gamma\}_{\gamma \in \Gamma} \subset B_{X^*}$ forms a long monotone Schauder basic sequence. To this end, it suffices to prove for every finitely supported $f = \sum_{i=1}^m a_i f_{\gamma_i}$, $\|f\| = 1$,

$g = \sum_{i=1}^{n} a_i f_{\gamma_i}$, where $n \geq m$, $\gamma_1 < \gamma_2 < \cdots < \gamma_n \in \Gamma$, that the inequality $\|g\| \geq (1 - \varepsilon)$ holds for every $\varepsilon > 0$. Indeed, the last statement is equivalent to all canonical projections having norm 1. Let us choose $p \in \mathcal{F}$ with the properties $\{\gamma_1, \ldots, \gamma_m\} \subset \Gamma_p$, $\Gamma_p \cap [\gamma_{m+1}, \omega_1) = \emptyset$, $\varepsilon_p < \varepsilon$. For some $x \in D_p$, $\|x\| = 1$, we have $|f(x)| \geq 1 - \frac{\varepsilon_p}{3}$. By the filter property, there exists $q \geq p$ satisfying $\{\gamma_{n+1}, \ldots, \gamma_n\} \subset \Gamma_q$. Thus

$$|g(x)| = \left| f(x) + \sum_{i=m+1}^{n} a_i f_{\gamma_i}(x) \right| \geq 1 - \frac{\varepsilon_p}{3} - \sum_{i=m+1}^{n} |a_i| \geq 1 - \frac{2\varepsilon_p}{3}$$

due to the fact that $\gamma_i \notin \Gamma_{\frac{\varepsilon_p}{3}}(x)$ for $i > m$.

Since $I : \overline{\operatorname{span}}\{f_\gamma; \gamma \in \Gamma\} \hookrightarrow X^*$, we have a natural (restricted) bounded operator $I^* : X \to \overline{\operatorname{span}}^{w^*}\{f_\gamma; \gamma \in \Gamma\}$. Note that $I^*(x) = \sum_{\alpha=0}^{\omega_1} f_{\xi_\alpha}(x) f_{\xi_\alpha}^*$ is an absolutely convergent series for every $x \in Y$. It follows that $I^*(X) \subset \overline{\operatorname{span}}\{f_\gamma^*; \gamma \in \Gamma\}$. Given a finite set $A \subset \Gamma$, denote by P_A the natural projection onto coordinates from A in the space $\overline{\operatorname{span}}\{f_\gamma^*; \gamma \in \Gamma\}$.

Next we claim that for every finitely supported $g^* \in S_{\overline{\operatorname{span}}\{f_\alpha^*; \alpha \in \Gamma\}}$ and $\varepsilon > 0$, there exists $x \in B_Y$ such that $\|I^*(x) - g^*\| < \varepsilon$. Choose $p \in \mathcal{F}$ for which $\operatorname{supp}(g^*) \subset \Gamma_p$, $\varepsilon_p < \varepsilon$. There exists $x \in D_p$, $\|x\| = 1$, which satisfies

$$\|(I^*(x) - g^*) \restriction \overline{\operatorname{span}}\{f_\gamma; \gamma \in \Gamma_p\}\|$$
$$= \|(P_{\Gamma_p} \circ I^*(x) - g^*) \restriction \overline{\operatorname{span}}\{f_\gamma; \gamma \in \Gamma_p\}\| \leq \frac{\varepsilon_p}{3}.$$

As $p \in \mathcal{F}$, we have $\Gamma_{\frac{\varepsilon_p}{3}}(x) \cap \Gamma \subset \Gamma_p$, so $\sum_{\gamma \in \Gamma \setminus \Gamma_p} |f_\gamma(x)| \leq \frac{\varepsilon_p}{3}$. Thus $\|I^*(x) - P_{\Gamma_p} \circ I^*(x)\| \leq \frac{\varepsilon_p}{3}$, and

$$\|I^*(x) - g^*\| \leq \|I^*(x) - P_{\Gamma_p} \circ I^*(x)\| + \|P_{\Gamma_p} \circ I^*(x) - g^*\| \leq \frac{2\varepsilon_p}{3}.$$

This proves the claim. From the claim we obtain immediately that I^* is a quotient operator. In particular, X has a quotient with a ω_1-long Schauder basis. By Plichko's theorem (Theorem 4.15), we conclude that X has a fundamental biorthogonal system. $\square$

Theorem 4.46 (Todorčević [Todo06] (MA_{ω_1})). *Let X be a Banach space of* dens $X = \omega_1$ *such that (B_{X^*}, w^*) is countably tight. Then there exists a bounded linear operator $T : X \to c_0(\omega_1)$ with nonseparable range; in particular, X has a fundamental biorthogonal system.*

Proof. By the basic Fact 4.27, there exists a sequence $\{f_\alpha\}_{\alpha < \omega_1} \subset B_{X^*}$ such that the operator $\tilde{T}(x) = (f_\alpha(x))$ maps X into $\ell_\infty^c(\omega_1)$, and, moreover, for every $\alpha < \omega_1$, there exists some $x \in B_X$ such that $f_\alpha(x) \geq \frac{1}{2}$. Our objective now will be to select an uncountable subsequence $\Gamma \subset \omega_1$, so that $T(x) = (f_\alpha(x))_{\alpha \in \Gamma}$ maps X into $c_0(\Gamma)$. The stated property of $\tilde{T}$ ensures that T has a nonseparable range for every choice of uncountable Γ. Let $\mathcal{P} = \{(D_p, \Gamma_p, \varepsilon_p)\}$

be a system of all triples, where D_p is a finite subset of X, Γ_p is a finite subset of ω_1, and $\varepsilon_p \in (0,1)$ is rational. We partially order $\mathcal{P}$ by letting $p \leq q$ if

$$D_p \subseteq D_q, \ \Gamma_p \subseteq \Gamma_q, \ \varepsilon_p \geq \varepsilon_q,$$

$$|f_\gamma(x)| < \varepsilon_p \text{ for every } x \in D_p, \ \gamma \in \Gamma_q \setminus \Gamma_p.$$

We are now going to prove that $\mathcal{P}$ satisfies the CCC property. Let $\{p_\gamma\}_{\gamma < \omega_1}$ be any subset of $\mathcal{P}$, $p_\gamma = (D_\gamma, \Gamma_\gamma, \varepsilon_\gamma)$. Our objective is to prove that there exist $\alpha < \beta < \omega_1$ such that p_α, p_β have a common extension $r \in \mathcal{P}$, $r \geq p_\alpha, p_\beta$. By the root lemma, we may without loss of generality assume that there exist finite sets $D \subset X$, $\Delta \subset [0, \omega_1)$ and a rational $\varepsilon \in (0,1)$ such that, for every $\alpha < \beta < \omega_1$, we have $D_\alpha \cap D_\beta = D$, $\Gamma_\alpha \cap \Gamma_\beta = \Delta$, $\varepsilon_\alpha = \varepsilon$. By passing to a subsequence, using also the fact that $\{\alpha; f_\alpha(x) \neq 0\}$ is countable for every $x \in X$, we may without loss of generality assume that $\{p_\gamma\}_{\gamma < \omega_1}$ has the following additional properties. There exists a $k \in \mathbb{N}$ such that for every $\alpha < \beta$ we have

$$\Gamma_\alpha \setminus \Delta = \{\gamma_\alpha^1, \gamma_\alpha^2, \ldots, \gamma_\alpha^k\}, \text{ where } \gamma_\alpha^1 < \gamma_\alpha^2 < \cdots < \gamma_\alpha^k,$$

$$\gamma_\alpha^k < \gamma_\beta^1,$$

$$f_\gamma(x) = 0 \text{ for every } x \in D_\alpha, \gamma \in \Gamma_\beta \setminus \Delta.$$

The last property immediately implies that $p_{\alpha,\beta} = (D_\alpha \cup D_\beta, \Gamma_\alpha \cup \Gamma_\beta, \varepsilon)$ extends p_α. In order to verify that $\mathcal{P}$ has CCC, it is therefore enough to find $\alpha < \beta$ for which $p_{\alpha,\beta}$ also extends p_β. We will prove the last statement by contradiction, assuming that, for every $\alpha < \beta < \omega_1$, we have that $|f_\gamma(x)| \geq \varepsilon$ for some $x \in D_\beta$ and $\gamma \in \Gamma_\alpha \setminus \Delta$. Define for $\alpha < \omega_1$ and $\delta > 0$

$$F_\alpha = (f_{\gamma_\alpha^1}, f_{\gamma_\alpha^2}, \ldots, f_{\gamma_\alpha^k}) \in (B_{X^*})^k,$$

$$A_\alpha^\delta = \{(h_1, \ldots, h_k) \in (B_{X^*})^k; |h_i(x)| \geq \delta \text{ for some } x \in D_\alpha, 1 \leq i \leq k\},$$

$$B_\alpha^\delta = \{(h_1, \ldots, h_k) \in (B_{X^*})^k; |h_i(x)| > \delta \text{ for some } x \in D_\alpha, 1 \leq i \leq k\}.$$

Clearly, A_α^δ is closed in $(B_{X^*}, w^*)^k$, while B_α^δ is open. By our assumption, we have $F_\alpha \in A_\beta^\varepsilon$ and $F_\beta \notin B_\alpha^{\frac{\varepsilon}{2}}$ whenever $\alpha < \beta$. Consequently,

$$\overline{\{F_\alpha\}_{\alpha \leq \gamma}}^{w^*} \cap \overline{\{F_\alpha\}_{\alpha > \gamma}}^{w^*} = \emptyset$$

for every $\gamma < \omega_1$. Let $z \in \bigcap_{\gamma < \omega_1} \overline{\{F_\alpha\}_{\alpha > \gamma}}^{w^*} \neq \emptyset$. $(B_{X^*}, w^*)^k$ is countably tight as a product of countably tight spaces [Juh83, p. 113] and so there exists $\gamma < \omega_1$ such that $z \in \overline{\{F_\alpha\}_{\alpha \leq \gamma}}^{w^*}$. This is a contradiction.

In order to finish the proof of the theorem, we need to check that certain subsets of $\mathcal{P}$ are dense. Let $\alpha < \omega_1$, $x \in X$, and $\varepsilon \in (0,1)$ be rational. We set

$$\mathcal{P}_\alpha = \{p \in \mathcal{P}; \max \Gamma_p > \alpha\}, \mathcal{P}_x = \{p \in \mathcal{P}; x \in D_p\}, \mathcal{P}_\varepsilon = \{p \in \mathcal{P}; \varepsilon_p \leq \varepsilon\}.$$

It is easy to verify that for any $p \in \mathcal{P}$ and $\beta > \alpha$ such that $f_\beta(D_p \cup \{x\}) = 0$, $p \leq (D_p \cup \{x\}, \Gamma_p \cup \{\beta\}, \min\{\varepsilon_p, \varepsilon\}) \in \mathcal{P}_\alpha \cap \mathcal{P}_x \cap \mathcal{P}_\varepsilon$. Consequently, $\mathcal{P}_\alpha$, $\mathcal{P}_x$, and $\mathcal{P}_\varepsilon$ are open and dense in $\mathcal{P}$. Let $S \subset X$ be such that $\overline{S} = X$, $|S| = \omega_1$. Invoking MA_{ω_1}, we obtain a filter $\mathcal{F} \subset \mathcal{P}$ satisfying $\mathcal{F} \cap \mathcal{P}_\alpha \cap \mathcal{P}_x \cap \mathcal{P}_\varepsilon \neq \emptyset$ for any $\alpha < \omega_1$, $x \in S$, $\varepsilon \in (0,1) \cap Q$. Now let $\Gamma = \bigcup_{p \in \mathcal{F}} \Gamma_p$. It is standard to verify that $(f_\gamma(x))_{\gamma \in \Gamma} \in c_0(\Gamma)$ for every $x \in X$.

To finish the proof, use Theorem 4.45. $\qquad \square$

In order to prove the next theorem, a stronger axiom than MA_{ω_1} is needed. The so-called Martin's Maximum (MM) is provably the strongest version of Martin's axiom consistent with ZFC. In particular, we have $MM \Rightarrow MA_{\omega_1}$. We refer to the original article [FMS88] for the precise statement and some consequences of this principle and to [Todo06] for the proof of the next theorem, as well as for an enlightening discussion regarding the content of MM.

Theorem 4.47 (Todorčević [Todo06] (MM)). *Let X be a Banach space with* dens $X = \omega_1$. *If (B_{X^*}, w^*) is not countably tight, then there exists a bounded linear operator $T : X \to c_0(\omega_1)$ with a nonseparable range.*

Theorem 4.48 (Todorčević [Todo06] (MM)). *Let X be a Banach space satisfying* dens $X = \omega_1$. *Then X has a fundamental biorthogonal system. In particular, every nonseparable Banach space contains an uncountable biorthogonal system.*

Proof. If (B_{X^*}, w^*) is countably tight, the result follows from Theorem 4.45 and Theorem 4.46, relying on MA_{ω_1}. If (B_{X^*}, w^*) is not countably tight, then it follows from Theorem 4.47 and Theorem 4.45, relying on MM. $\qquad \square$

4.6 Uncountable Auerbach Bases

As mentioned earlier in this book, the existence of an Auerbach basis for every separable Banach space is an open problem. In this section, we are going to present a renorming of nonseparable spaces with a w^*-separable dual ball, due to Godun, Lin, and Troyanski, for which no Auerbach basis exists.

Theorem 4.49 (Godun, Lin, and Troyanski [GLT93]). *Let X be a nonseparable Banach space such that (B_{X^*}, w^*) is separable. Then X admits an equivalent norm $||| \cdot |||$ such that $(X, ||| \cdot |||)$ does not have an Auerbach basis.*

In order to prove this theorem, we shall need some preliminary results. The Faber-Schauder basis of $(C([0,1]), \|\cdot\|_\infty)$ related to a dense sequence (t_n) of distinct points in $[0,1]$ is better described by the associated sequence of canonical projections $P_n : C([0,1]) \to C([0,1])$ onto the subspace generated by the first n elements of the basis. Precisely, $P_n(f)$ is a piecewise linear function whose value at t_i coincides with $f(t_i)$, $i = 1, 2, \ldots, n$. We say that the corresponding basis is an *interpolating basis with nodes* (t_n).

This can be generalized to the case of a metrizable compact space K and a dense sequence (t_n) in K, giving an interpolating basis of $(C(K), \|\cdot\|_\infty)$ with nodes (t_n) (V. I. Gurarii [Gura66]; see [Sema82, p. 96]).

Lemma 4.50. *Let K be a metrizable compact space. Then, for every dense sequence (t_n) in K, there exists an equivalent norm $\|\cdot\|_0$ on $(C(K), \|\cdot\|_\infty)$ with the property that if (f_n) is a sequence in $C(K)$ such that $\|f_n\|_0 = 1$, $n \in \mathbb{N}$, and $\lim_n f_n(t_m) = 0$ for all $m \in \mathbb{N}$, then $\lim_n \mathrm{dist}_{\|\cdot\|_0}(f_n, G) = 1$ for every finite-dimensional subspace G of $C(K)$.*

Proof. Let (t_n) be a dense sequence in K. From [Gura66], there exists in the space $(C(K), \|\cdot\|_\infty)$ an interpolating Schauder basis $\{e_n, e_n^*\}_{n=1}^\infty$ with nodes at (t_n). Let $R_n := \mathrm{Id}_{C(K)} - P_n$, where $P_n : C(K) \to \mathrm{span}\{x_i\}_{i=1}^n$ is the canonical projection associated to the basis, $n \in \mathbb{N}$, and $\mathrm{Id}_{C(K)}$ is the identity mapping on $C(K)$. Define an equivalent norm on $C(K)$ by

$$\|f\|_0 := \sup_n |R_n(f)|_\infty, \quad f \in C(K).$$

Obviously, $\|R_n\|_0 = 1$, $n \in \mathbb{N}$, and $\lim_n \|R_n(f)\|_0 = 0$ for every $f \in C(K)$. We shall prove that this norm satisfies the conclusion of the theorem.

Take a sequence (f_n) in $C(K)$ such that $\|f_n\|_0 = 1$ for all $n \in \mathbb{N}$, and $\lim_n f_n(t_m) = 0$ for all $m \in \mathbb{N}$.

First of all, observe that $\lim_n \mathrm{dist}_{\|\cdot\|_0}(f_n, F_m) = 0$, where

$$F_m := \{f \in C(K); f(t_i) = 0, \; i = 1, 2, \ldots, m\}.$$

This follows from the fact that

$$\mathrm{dist}_{\|\cdot\|_0}(f_n, F_m) = \|q_m(f_n)\|_0 = \sup\{\langle f_n, \mu\rangle; \mu \in S_{F^\perp} \subset C(K)^*\},$$

where $q_m : X \to X/F_m$ is the canonical quotient mapping, and from the obvious fact that $F_m^\perp = \mathrm{span}\{\delta_{t_i}\}_{i=1}^m$.

Now fix a finite-dimensional subspace G of $C(K)$. To conclude the proof, it will suffice to show that, for any subsequence (f_{n_k}) of (f_n), there exists another subsequence $(f_{n_{k_j}})$ such that $\lim_j \mathrm{dist}_{\|\cdot\|_0}(f_{n_{k_j}}, G) = 1$. To avoid excessive indexing, call an arbitrary subsequence of (f_n) again (f_n). As G is finite-dimensional, we can choose $g_n \in G$ such that $\|f_n - g_n\|_0 = \mathrm{dist}_{\|\cdot\|_0}(f_n, G)$, $n \in \mathbb{N}$, and, moreover, $\lim_n \|g_n - g\|_0 = 0$ for some $g \in G$. Fix $\varepsilon > 0$ and choose $m \in \mathbb{N}$ such that $\|R_m(g)\|_0 < \varepsilon$. By the previous observation, we can find $N \in \mathbb{N}$ such that, for all $n \geq N$, $\mathrm{dist}_{\|\cdot\|_0}(f_n, F_m) < \varepsilon$ and $\mathrm{dist}_{\|\cdot\|_0}(f_n, G) > \|f_n - g\|_0 - \varepsilon$. Choose $h_n \in F_m$ such that $\|f_n - h_n\|_0 < \varepsilon$. Since $R_m(h_n) = h_n$, $n \in \mathbb{N}$, we have, for all $n > N$,

$$\begin{aligned}
\mathrm{dist}_{\|\cdot\|_0}(f_n, G) &> \|f_n - g\|_0 - \varepsilon \geq \|R_m(f_n - g)\|_0 - \varepsilon \\
&\geq \|R_m(f_n)\|_0 - \|R_m(g)\|_0 - \varepsilon \\
&\geq \|R_m(h_n)\|_0 - \|R_m(f_n - h_n)\|_0 - 2\varepsilon \\
&> \|h_n\|_0 - 3\varepsilon \geq \|f_n\|_0 - \|h_n - f_n\|_0 - 3\varepsilon > 1 - 4\varepsilon. \qquad \square
\end{aligned}$$

Corollary 4.51. *Let X be a separable Banach space and let F be a norming subspace of X^*. Then X admits an equivalent norm $\|\cdot\|_0$ such that, if (x_n) is a sequence in X, $\|x_n\|_0 = 1$, $n \in \mathbb{N}$, and $\lim_n \langle x_n, f \rangle = 0$ for every $f \in F$, then $\lim_n \operatorname{dist}_{\|\cdot\|_0}(x_n, G) = 1$ for every finite-dimensional subspace $G \subset X$.*

Proof. By renorming, we may assume that F is 1-norming, so $\overline{B_F}^{w^*} = B_{X^*}$. Choose a dense subset $\{f_n; n \in \mathbb{N}\}$ of (B_F, w^*); then it is also dense in (B_{X^*}, w^*). Let $\|\cdot\|_0$ be an equivalent norm on $(C((B_{X^*}, w^*)), \|\cdot\|_\infty)$ as given by Lemma 4.50. Then X is linearly isometric to a subspace of the space $(C((B_{X^*}, w^*)), \|\cdot\|_\infty)$, and $\|\cdot\|_0$ induces on X an equivalent norm (denoted again $\|\cdot\|_0$) with the required properties. □

The following lemma elaborates on the previous corollary. This time we need the separability of the norming subspace.

Lemma 4.52. *Let X be a separable Banach space and let $F \subset X^*$ be a separable norming subspace. Then X admits an equivalent norm $\||\cdot\||$ such that $\operatorname{dist}_{\||\cdot\||}(x^*, F) < \||x^*\||$ for all $x^* \in X^* \setminus \{0\}$.*

Proof. Let $\|\cdot\|$ be a norm in X with the property in Corollary 4.51 (we do not need separability of F for this). Let $\{X_n\}_{n \in \mathbb{N}}$ be an increasing family of finite-dimensional subspaces of X such that $\bigcup_{n \in \mathbb{N}} X_n$ is dense in X. For $x^* \in X^*$, define

$$\||x^*\||_0 := \sum_{n=1}^{\infty} 2^{-n} \operatorname{dist}_{\|\cdot\|}(x^*, X_n^\perp)$$

and

$$\||x^*\|| := \|x^*\| + \||x^*\||_0.$$

Clearly, $\||\cdot\||$ is an equivalent norm on X^*; precisely, $\|x^*\| \leq \||x^*\|| \leq 2\|x^*\|$ for all $x^* \in X^*$. It is easy to see that $\||\cdot\||$ is w^*-lower semicontinuous, and hence it is a dual norm. Denote again by $\||\cdot\||$ the norm in X such that its dual norm in X^* is $\||\cdot\||$. Certainly, $\||x^{**}\|| \leq \|x^{**}\|$ for all $x^{**} \in X^{**}$. We shall prove that $\||x^{**}\|| \geq \|x^{**}\|$ for $x^{**} \in F^\perp$, so $\||\cdot\||$ and $\|\cdot\|$ will coincide on $F^\perp$. Fix $x^{**} \in F^\perp \subset X^{**}$, $\||x^{**}\|| = 1$. We can find a sequence (y_n^*) in $B_{(X^*, \|\cdot\|)}$ such that $\lim_n \langle x^{**}, y_n^* \rangle = \|x^{**}\|$. As F is $\||\cdot\||$-separable, it is possible to choose a sequence (x_n) in X such that it satisfies the following four conditions:

$$\||x_n\|| = 1, \ n \in \mathbb{N}, \tag{4.1}$$

$$(\|x_n\|) \text{ converges to some } \alpha \neq 0 \text{ when } n \to \infty, \tag{4.2}$$

$$\lim_n \langle x_n, y_m^* \rangle = \langle x^{**}, y_m^* \rangle \text{ for all } m \in \mathbb{N}, \tag{4.3}$$

$$\lim_n \langle x_n, f \rangle = \langle x^{**}, f \rangle \text{ for all } f \in F. \tag{4.4}$$

Fix $\varepsilon > 0$ and find $m \in \mathbb{N}$ such that $2^{-m} < \varepsilon$. Observe that, for $x^* \in X_m^\perp$, $\||x^*\||_0 = \sum_{k=m+1}^{\infty} 2^{-k} \operatorname{dist}_{\|\cdot\|}(x^*, X_k^\perp) \leq 2^{-m} \|x^*\| < \varepsilon\|x^*\|$. From Corollary

4.51 we have $\lim_n \mathrm{dist}_{\|\cdot\|}\left(\frac{x_n}{\|x_n\|}, X_m\right) = 1$, so for $n \in \mathbb{N}$ we can find $x_n^* \in X_m^\perp$ such that $\|x_n^*\| = 1$ and $\lim_n \langle \frac{x_n}{\|x_n\|}, x_n^* \rangle = 1$. It follows from (4.2) that $\lim_n \langle x_n, x_n^* \rangle = \alpha$. We get

$$1 = \||x^{**}|\| = \||x_n|\| \geq \left\langle x_n, \frac{x_n^*}{\||x_n^*|\|} \right\rangle = \frac{\langle x_n, x_n^* \rangle}{\|x_n^*\| + \||x_n^*|\|_0}$$

$$\geq \frac{\langle x_n, x_n^* \rangle}{(1+\varepsilon)\|x_n^*\|} = \frac{1}{1+\varepsilon}\langle x_n, x_n^* \rangle \xrightarrow{n} \frac{1}{1+\varepsilon}\alpha. \qquad (4.5)$$

Now, given $\delta > 0$, there exists by (4.3) some $n \in \mathbb{N}$ such that $\|x^{**}\| - \langle x^{**}, y_n^* \rangle < \delta$. We can then find $k_0 \in \mathbb{N}$ such that $|\langle x_k, y_n^* \rangle - \langle x^{**}, y_n^* \rangle| < \delta$ for all $k \geq k_0$. We get $\|x^{**}\| \leq \langle x_k, y_n^* \rangle + 2\delta \leq \|x_k\| + 2\delta$ for all $k \geq k_0$, so $\|x^{**}\| \leq \alpha + 2\delta$. As $\delta > 0$ is arbitrary, we get $\|x^{**}\| \leq \alpha$, so, from (4.5), $1 \geq 1/(1+\varepsilon)\|x^{**}\|$ and finally, as $\varepsilon > 0$ is arbitrary, $\|x^{**}\| \leq 1 = \||x^{**}|\|$.

Let $x^* \neq 0$ in X^*. Since $\bigcup_{n=1}^\infty X_n$ is dense in X, there exists $n \in \mathbb{N}$ such that $\mathrm{dist}_{\|\cdot\|}(x^*, X_n^\perp) > 0$. This implies $\|x^*\| < \||x^*|\|$. Then

$$\mathrm{dist}_{\||\cdot|\|}(x^*, F) = \sup_{x^{**} \in F^\perp, \, \||x^{**}|\|=1} |\langle x^{**}, x^* \rangle| = \sup_{x^{**} \in F^\perp, \, \|x^{**}\|=1} |\langle x^{**}, x^* \rangle|$$

$$= \mathrm{dist}_{\|\cdot\|}(x^*, F) \leq \|x^*\| < \||x^*|\|. \qquad \square$$

Proof of Theorem 4.49. . Let $F \subset X^*$ be the separable closed subspace of X^* spanned by a countable w^*-dense subset of B_{X^*}. Choose a separable subspace $Y \subset X$ that 1-norms F. Considered as a subspace of F^*, Y is a separable 1-norming subspace. By Lemma 4.52, F admits an equivalent norm $\|| \cdot |\|$ such that $\mathrm{dist}_{\||\cdot|\|}(e^*, Y) < \||e^*|\|$ for all $0 \neq e^* \in F^*$. The norm $\|| \cdot |\|$ on F^* induces an equivalent norm $\|| \cdot |\|$ on X with the property that $\mathrm{dist}_{\||\cdot|\|}(x, Y) < \||x|\|$ for all $0 \neq x \in X$. Suppose that X has an Auerbach basis $(x_i)_{i \in I}$. Then I is uncountable. Since Y is separable, there is a countable subset $J \subset I$ such that $Y \subset \overline{\mathrm{span}}\{x_j\}_{j \in J}$. Then, for any $i \in I \setminus J$,

$$\mathrm{dist}_{\||\cdot|\|}(x_i, \overline{\mathrm{span}}\{x_j\}_{j \neq i})$$
$$\leq \mathrm{dist}_{\||\cdot|\|}(x_i, \overline{\mathrm{span}}\{x_j\}_{j \in J}) \leq \mathrm{dist}_{\||\cdot|\|}(x_i, Y) < \||x_i|\|.$$

This contradicts the fact that $(x_i)_{i \in I}$ is an Auerbach basis. $\qquad \square$

Remark 4.53. The usual basis of $\ell_1(c)$ is an Auerbach basis, but $\ell_1(c) \subset C[0,1]^* \subset \ell_\infty$, and hence $(B_{\ell_1^*(c)}, w^*)$ is separable; thus $\ell_1(c)$ can be renormed so that it has no Auerbach basis.

4.7 Exercises

4.1. Show that the cardinality of any fundamental minimal system in a Banach space coincides with its density.

4.2. Let $\{x_\gamma; x_\gamma^*\}_{\gamma \in \Gamma}$ and $\{y_\delta; y_\delta^*\}_{\delta \in \Delta}$ be two total bounded fundamental system in $X \times X^*$, where X is a Banach space. Is necessarily card $\Gamma = $ card Δ?

Hint. No. $X := \ell_1(\omega_1) \subset C[0,1]^*$, so X^* is w^*-separable. By Theorem 4.12, X has a countable total (and bounded) biorthogonal system. On the other hand, $\{e_\gamma; e_\gamma^*\}_{0 \le \gamma < \omega_1}$ is also a total (and bounded) biorthogonal system in $X \times X^*$, where $\{e_\gamma\}_{0 \le \gamma < \omega_1}$ is the canonical basis of $\ell_1(\omega_1)$.

4.3. Show that if K is zero-dimensional and compact, then $C(K)$ is not indecomposable.

Hint. Assume $C(K)$ is indecomposable and K is zero-dimensional. Then there must be at least two nonisolated points in K. Thus there is a clopen $A \subset K$ such that A and $K \setminus A$ are both infinite. Then consider the restriction to A and to $K \setminus A$.

4.4. A Banach space X is called *hereditarily indecomposable* if there is no subspace Y of X that can be split into $Y = W \oplus Z$, where W and Z are both infinite-dimensional. The first example of a hereditarily indecomposable Banach spaces was constructed by Gowers and Maurey (see [JoLi01h, Chap. 29]). Argyros and Tolias in [ArTo04] have constructed examples of nonseparable hereditarily indecomposable spaces. Show that if X^* is not w^*-separable, then X cannot be hereditarily indecomposable.

Hint. X has a subspace with a long Schauder basis.

4.5. Show that if the density of X is strictly greater than c, then X contains a subspace that has an uncountable transfinite Schauder basis and thus is not hereditarily indecomposable.

Hint. Check the w^*-density of X^*.

4.6. Let X have a fundamental biorthogonal system. Is there an injection $T : X \to c_0(\Gamma)$?

Hint. No. Consider $\ell_\infty(\Gamma)$ for Γ uncountable. This space does not have a rotund renorming.

4.7. Let $T : X \to c_0(\Gamma)$ be injective. Show that there exists $\tilde{T} : X \to c_0(\Gamma)$ that is injective and has a dense range.

Hint. $\overline{T(X)}$ is a WCG Asplund space, so it has a shrinking M-basis (Theorem 6.3), and then consider the associated injection into $c_0(\Gamma)$.

4.8. Show that there exists a normed (noncomplete, and this is crucial) nonseparable subspace X of $c_0(\Gamma)$ and $\{f_i\}_{i=1}^\infty \subset X^*$ that distinguishes points of X. In particular, X contains no copy of the unit basis of $c_0(\Gamma)$.

Hint. Use $I : \ell_1(c) \to c_0(\Gamma)$.

4.9. Let $T : \ell_\infty^c(\Gamma) \to X$ be an operator with dense range, dens $X > c$. Then ℓ_∞ is isomorphic to a subspace of X, or, more precisely, T is an isomorphism when restricted to some subspace of $\ell_\infty^c(\Gamma)$ isomorphic to ℓ_∞.

Hint. See the proof of Theorem 4.26.

4.10. Is $\ell_2(\omega_1)$ isomorphic to a subspace of ℓ_∞?

Hint. No. Check the w^*-separability of the dual spaces.

5

Markushevich Bases

In this chapter, we investigate spaces with M-bases and their relationship with several closely connected notions, such as projectional resolutions of identity, the separable complementation property, and projectional generators. Renorming theory also plays an important role here, as spaces with an M-basis admit an equivalent rotund norm, while spaces with a strong M-basis even have an LUR renorming thanks to the fundamental result of Troyanski.

The first section treats the existence issue; namely a space has a (strong) M-basis if it belongs to the $\mathcal{P}$ class. In the second section, it is shown that an M-basis can be linearly perturbed to become a bounded M-basis; this is a result of Plichko. The third and fourth sections treat various aspects of weakly Lindelöf determined (WLD) spaces, a class admitting many equivalent descriptions—in particular, as spaces whose dual ball is Corson and also as spaces admitting a weakly Lindelöf M-basis. The fifth section shows the impact of the additional axioms to ZFC on the structure of $C(K)$ spaces, where K is a Corson compactum. The last two sections examine extensions of M-bases and quasicomplements. Among other things, it is shown that WLD spaces admit extensions of M-bases from subspaces, which implies that every subspace of a WLD space is quasicomplemented. The seventh section also contains Rosenthal's theory of quasicomplements in ℓ_∞ spaces.

5.1 Existence of Markushevich Bases

The main general method for constructing an M-basis is based on the technique of PRI's. As we will see (Theorem 5.1), this method actually leads to a strong M-basis, a nontrivial strengthening of an M-basis (see Definition 1.32). The existence of PRI's alone does not guarantee the existence of M-bases. Take, e.g., $X = \ell_2(2^{2^\Gamma}) \oplus \ell_\infty(\Gamma)$, card $\Gamma = c$. This space has a PRI $\{P_\alpha \ \omega \leq \alpha \leq 2^{2^\Gamma}\}$ such that $(P_{\alpha+1} - P_\alpha)(X)$ has dimension 1 for all $\alpha < \Gamma$, while $(P_{\Gamma+1} - P_\Gamma)(X) = \ell_\infty(\Gamma)$. The space X cannot have an M-basis since it does not have a rotund renorming [DGZ93a, Cor. II.7.13]. Another example of a Banach space without an M-basis is given at the end of this section. In order to obtain a positive result, one has to put some additional conditions on the building blocks $(P_{\alpha+1} - P_\alpha)(X)$ of the space X. For example, assuming that all $(P_{\alpha+1} - P_\alpha)(X)$ already have an M-basis, the existence of an M-basis on the whole X follows easily. This leads to the useful notion of a $\mathcal{P}$-class introduced in Definition 3.45.

There are several sorts of spaces that are known to form $\mathcal{P}$-classes [DGZ93a]. These are duals of Asplund spaces, WCG (see Proposition 3.43, and Theorem 3.42), Vašák (i.e., weakly countably determined, see Definition 3.32 and Section 6.2), WLD (see Definition 3.32, Theorem 5.36 and Theorem 3.42), Plichko spaces (see Definition 5.46 and Theorem 5.63), and $C(K)$ spaces, where K is a Valdivia compact (see Section 5.4).

Theorem 5.1. *Let $\mathcal{C}$ be a $\mathcal{P}$-class of Banach spaces. Then every $X \in \mathcal{C}$ has a strong M-basis.*

Proof. This is obtained by transfinite induction on the density of X: for separable spaces, this follows from Theorem 1.36. Assuming that the result holds for all spaces with density less than $\mu := \mathrm{dens}\, X$, the existence of a PRI $\{P_\alpha; \omega \leq \alpha \leq \mu\}$ on X and strong M-bases $\{x_\gamma^\alpha; f_\gamma^\alpha\}_{\gamma \in \Gamma_\alpha}$ in $(P_{\alpha+1} - P_\alpha)X \times \left((P_{\alpha+1} - P_\alpha)X\right)^*$ gives, as it is standard to check, a strong M-basis in $X \times X^*$ given by $\{x_\gamma^\alpha; (P_{\alpha+1}^* - P_\alpha^*)f_\gamma^\alpha\}_{\gamma \in \Gamma_\alpha, \omega \leq \alpha < \mu}$.   □

Corollary 5.2. *Let X belong to one of the following classes of spaces: duals to Asplund spaces, Vašák, WLD, Plichko, or $C(K)$, where K is a Valdivia compact. Then X has a strong M-basis.*

Valdivia, in [Vald90b], proved that *if X is a Banach space containing a linearly dense subset G with the property that the set $\Sigma(G)$ of all elements in X^* countably supported by G satisfies that $\Sigma(G) \cap B_{X^*}$ is w^*-dense in B_{X^*}, then there exists a PRI on X and a subordinated M-basis $\{x_\gamma; x_\gamma^*\}_{\gamma \in \Gamma}$ in $X \times X^*$ such that* $\mathrm{span}\{x_\gamma; \gamma \in \Gamma\} = \mathrm{span}\, G$ *and* $\Sigma(\{x_\gamma; \gamma \in \Gamma\}) = \Sigma(G)$.

Theorem 5.3. *If X admits an M-basis of cardinality Γ, then X linearly injects into $c_0(\Gamma)$.*

Proof. Let $\{x_\gamma; f_\gamma\}_{\gamma \in \Gamma}$ be an M-basis such that $\|f_\gamma\| = 1$ for all $\gamma \in \Gamma$. The operator $T : X \to \ell_\infty(\Gamma)$, $T(x) = (f_\gamma(x))_{\gamma \in \Gamma}$ has norm 1. Moreover, $T(\mathrm{span}\{x_\gamma\}_{\gamma \in \Gamma}) \hookrightarrow c_0(\Gamma) \hookrightarrow \ell_\infty(\Gamma)$, and $c_0(\Gamma)$ is closed in $\ell_\infty(\Gamma)$.   □

Corollary 5.4. *If X has an M-basis, then X admits a rotund norm. In particular, $\ell_\infty(\Gamma)$, for Γ uncountable, is not a subspace of a space with an M-basis. However, ℓ_∞ is a complemented subspace of a Banach space with an M-basis.*

Proof. It is well known that $c_0(\Gamma)$ admits a rotund renorming and that Banach spaces that inject into spaces with a rotund norm themselves admit a rotund renorming. The second statement follows from the first one and the fact that $\ell_\infty(\Gamma)$, under every equivalent renorming, contains a subspace isometric to $(\ell_\infty, \|\cdot\|_\infty)$, which is not a rotund space. Details on these well-known renorming results can be found in [DGZ93a, Chap. 2]. The second part, for example, is in [Fa~01, Thm. 6.45]. □

We proved in Theorem 3.48 that *every Banach space with a strong M-basis has an equivalent LUR renorming.* We note in passing that the Ciesielski-Pol $C(K)$ space admits an LUR renorming [DGZ93a, Thm. VII.4.8]; however, it has no injection into any $c_0(\Gamma)$ [DGZ93a, Thm. VI.8.8.3]. Hence, by Theorem 5.3, it has no M-basis.

Proposition 5.5 (Plichko). *There exists a Banach space admitting an M-basis but no LUR renorming; in particular, no strong M-basis.*

Proof. By [Fa~01, Thm. 6.45], there exists a Banach space Z with an M-basis containing a (complemented) copy of ℓ_∞. As ℓ_∞ does not admit an LUR renorming [DGZ93a, Thm. II.7.10], neither does Z. By Theorem 3.48, Z has no strong M-basis. □

The following proposition gives a property of Banach spaces with density character ω_1 having a PRI. We shall prove a more precise result in Proposition 5.48. See also Remark 5.47.

Proposition 5.6. *Let X be a Banach space, dens $X = \omega_1$. If X admits a PRI, then X has a strong M-basis and 1-SCP.*

Proof. Let $\{P_\alpha; \omega \leq \alpha < \omega_1\}$ be the PRI of X. Then $X_\alpha = (P_{\alpha+1} - P_\alpha)(X)$, $\alpha < \omega_1$, are separable complemented subspaces of X. The existence of a strong M-basis follows by the argument in Theorem 5.1. The 1-SCP property follows from the simple fact that $X = \bigcup_{\alpha < \omega_1} X_\alpha$. □

Theorem 5.7 (Plichko and Yost [PlYo01]). *There exists an (RNP) Banach space X with dens $X = \omega_1$ and strong M-basis but failing the SCP. In particular, X does not have PRI under any equivalent renorming.*

Proof. In order to construct X, we will use the following lemma.

Lemma 5.8. *There exists a collection $\{N_\alpha\}_{\alpha < \omega_1}$ of infinite subsets of $\mathbb{N}$ such that:*

(i) *$N_\alpha \cap N_\beta$ is finite whenever $\alpha \neq \beta$.*
(ii) *If $A \subset [0, \omega_1)$ is uncountable and $\{\beta_1, \ldots, \beta_k\} \subset [0, \omega_1) \setminus A$, then there exists an infinite $B \subset A$ and $j \in \mathbb{N}$ such that $j \in \bigcap_{\alpha \in B} N_\alpha \setminus \bigcup_{i=1}^k N_{\beta_i}$.*
(iii) *Given any distinct $\alpha_1, \ldots, \alpha_n \in [\omega, \omega_1)$, there exist distinct $\gamma_1, \ldots, \gamma_n \in [0, \omega)$ such that the sets in the collection $\{N_{\alpha_k} \setminus N_{\gamma_k}\}_{1 \leq k \leq n} \cup \{N_{\gamma_k} \setminus N_{\alpha_k}\}_{1 \leq k \leq n}$ are pairwise disjoint.*

Proof. Let $D = \{0,1\}^{<\omega}$ be the full dyadic tree and $\phi : \mathbb{N} \to D$ a bijection. The set of all branches of D can be canonically identified with the Cantor discontinuum $\mathbb{D} := \{0,1\}^{\omega}$. Choose a set of branches $\{B_\alpha\}_{0 \leq \alpha < \omega_1}$ that corresponds to a subset without isolated points in $\mathbb{D}$. (To achieve this, it suffices to start from an arbitrary subset of cardinality ω_1 in $\mathbb{D}$ and discard the countable subset of its isolated points). In the language of the tree D, this condition implies that for every B_α and a node $t \in B_\alpha$, there exists $B_\beta \neq B_\alpha$ with $t \in B_\beta$. We claim that $\{N_\alpha\}_{\alpha < \omega_1}$, $N_\alpha = \phi^{-1}(B_\alpha)$ is the system sought. It is clear that (i) holds. To check (ii), let $A \subset [0, \omega_1)$ be uncountable and $\{\beta_1, \ldots, \beta_k\} \subset [0, \omega_1) \setminus A$. It is easy to see that there exists a node $t \in \bigcap_{\alpha \in B} B_\alpha \setminus \bigcup_{i=1}^{k} B_{\beta_i}$ for some infinite set $B \subset A$. Thus $j = \phi^{-1}(t)$ can be used to prove (ii). To see (iii), it suffices to find nodes $t_j \in B_{\beta_j} \setminus \bigcup_{i=1, i \neq j}^{k} B_{\beta_i}$ and use properties of A that guarantee the existence of $B_{\gamma_i} \neq B_{\beta_i}$, $t_j \in B_{\gamma_i}$. $\qquad\square$

Let $y_\alpha = \chi_{N_\alpha} \in \ell_\infty$, and let $\{e_\alpha\}_{0 \leq \alpha < \omega_1}$ be the canonical basis of $\ell_1(\omega_1)$. Put $T : \ell_1(\omega_1) \to \ell_\infty$, $T(e_\alpha) = y_\alpha$. For each $n \in \mathbb{N}$, we define an equivalent norm on $\ell_1(\omega_1)$,

$$\|x\|_n = \max \left\{ \frac{1}{n} \|x\|_1, \|Tx\|_\infty \right\}.$$

Put $X_n = (\ell_1(\omega_1), \|\cdot\|_n)$ and $L_n = (\ell_1(\omega), \|\cdot\|_n) \hookrightarrow X_n$. The following lemma is the main step of the construction.

Lemma 5.9. *Let Z be an arbitrary Banach space and U be a separable complemented subspace of $X_n \oplus_1 Z$ containing L_n. Then the norm of any projection of $X_n \oplus_1 Z$ onto U is at least $\frac{1}{4}(n-3)$.*

Proof. Suppose $V \oplus U = X_n \oplus_1 Z$, $P : X_n \oplus_1 Z \to X_n$ is the canonical projection. As U is separable, there exist $\alpha_0 < \omega_1$ such that $P(U) \subset \overline{\text{span}}\{e_\alpha; 0 \leq \alpha < \alpha_0\}$. We have $\alpha_0 \geq \omega$ as $L_n \subset P(U) \cap X_n$ and $U \hookrightarrow \overline{\text{span}}\{e_\alpha; 0 \leq \alpha < \alpha_0\} + Z$. Thus we can find a sequence $\{w_m\}_{m=1}^{\infty} \subset \text{span}\{e_\alpha\}_{0 \leq \alpha < \alpha_0} + Z$ such that $U \subset \overline{\{w_m\}_{m=1}^{\infty}} \subset U + \frac{1}{n} B_{X_n}$. For any $\alpha > \alpha_0$, there exists an $m \in \mathbb{N}$ such that $d(e_\alpha + w_m, V) < \frac{1}{n}$. Consequently, there exists an $m \in \mathbb{N}$ and an uncountable $A \subset (\omega, \omega_1)$ such that $d(e_\alpha + w_m, V) < \frac{1}{n}$ whenever $\alpha \in A$. Suppose $w = w_m = \sum_{i=1}^{k} \lambda_i e_{\beta_i} + z$, $\lambda_i \in \mathbb{R}$, $\beta_i \in [0, \alpha_0)$, and $z \in Z$. Let $B \subset A$ and $j \in \bigcap_{\alpha \in B} N_\alpha \setminus \bigcup_{i=1}^{k} N_{\beta_i}$ be given by Lemma 5.8. Choose $\{\alpha_1, \ldots, \alpha_n\} \subset B$ and $\gamma_1, \ldots, \gamma_n$ by applying (iii) in Lemma 5.8. Let us now define elements $u = \sum_{i=1}^{n} e_{\gamma_i} \in \ell_1(\omega_1)$ and $v = \sum_{i=1}^{n} y_{\gamma_i} \in \ell_\infty$. Clearly, $Tu = v$, $\|u\|_1 = n$. By (iii), $\|\sum_{i=1}^{n} y_{\alpha_i} - v\|_\infty = 1$. Letting $x = \sum_{i=1}^{n} e_{\alpha_i} + nw$, we have $x - nw - u \in X_n$. We have

$$\|x\| = \left\| \sum_{i=1}^{n} e_{\alpha_i} + n \left(\sum_{i=1}^{k} \lambda_i e_{\beta_i} + z \right) \right\| = \left\| \left\| \sum_{i=1}^{n} e_{\alpha_i} + n \sum_{i=1}^{k} \lambda_i e_{\beta_i} \right\| \right\|_n + n\|z\|$$

$$\geq \left\| T \left(\sum_{i=1}^{n} e_{\alpha_i} + n \sum_{i=1}^{k} \lambda_i e_{\beta_i} \right) \right\|_{\infty} = \left\| \sum_{i=1}^{n} y_{\alpha_i} + n \sum_{i=1}^{k} \lambda_i y_{\beta_i} \right\|_{\infty}$$

$$\geq \left(\sum_{i=1}^{n} y_{\alpha_i} + n \sum_{i=1}^{k} \lambda_i y_{\beta_i} \right)(j) = \sum_{i=1}^{n} y_{\alpha_i}(j) = n$$

and

$$\mathrm{dist}(x, U) \leq \|\|x - nw - u\|\|_n + \mathrm{dist}(nw - u, U)$$

$$= \max \left\{ \frac{1}{n} \|x - nw - u\|_1, \|T(x - nw - u)\|_{\infty} \right\} + \mathrm{dist}(nw, U)$$

$$\leq \max \left\{ \frac{1}{n} \left\| \sum_{i=1}^{n} e_{\alpha_i} - u \right\|_1, \left\| \sum_{i=1}^{n} y_{\alpha_i} - v \right\|_{\infty} \right\} + 1 = 3.$$

By a similar argument, we obtain $\mathrm{dist}(x, V) \leq \sum_{i=1}^{n} \mathrm{dist}(e_{\alpha_i} + w, V) < 1$. To finish the proof, pick $x_U \in U$ and $x_V \in V$ so that $\|x - x_V\| < 1$ and $\|x - x_U\| - 3 < \varepsilon$ is arbitrarily small. We have $\|x_U + x_V\| < 4$ and $\|x_U\| \geq \|x\| - \|x - x_U\| \geq n - 3 - \varepsilon$. Thus the norm of the projection of $X_n \oplus_1 Z$ onto U along V is bounded below by $\frac{n-3}{4}$. □

To finish the proof of the theorem, it suffices to put $X := \sum_{n=1}^{\infty} \oplus_1 X_n$ and $L := \sum_{n=1}^{\infty} \oplus_1 L_n \hookrightarrow X$, a separable subspace. By Lemma 5.9, there exist no separable overspaces of L in X complemented in X. This proves the nonexistence of PRI under any equivalent renorming. On the other hand, putting together the canonical bases of the X_n's gives rise to a strong M-basis in X. □

We close this section by showing that, unlike in the separable case, non-separable Banach spaces may lack M-bases. An example, based on the next theorem, is provided by the space $\ell_{\infty}(\Gamma)$, where Γ is an infinite set (see Theorem 5.12).

Theorem 5.10 (Johnson [John70a]). *A Grothendieck Banach space with an M-basis is reflexive.*

In the proof, we shall need the following lemma.

Lemma 5.11 (Rosenthal [Rose69a]). *Let X be a Banach space. Then every reflexive subspace Y of X^* is w^*-closed in X^*. Consequently, Y^* is a quotient of X.*

Proof. The unit ball B_Y of Y is w-compact and hence w^*-compact in X^*, and so it is w^*-closed in X^*. It is enough to apply now the Banach-Dieudonné theorem (see, e.g., [Fa~01, Thm. 4.4]). Consequently, $X/Y_{\perp}$ is isomorphic to Y^*. □

Proof of Theorem 5.10. Assume that a Grothendieck Banach space X has an M-basis $\{x_\gamma; x_\gamma^*\}_{\gamma \in \Gamma}$. Let $Y := \overline{\text{span}}^{\|\cdot\|}\{x_\gamma^*\}_{\gamma \in \Gamma}$.

We claim that Y *is reflexive.* In order to prove the claim it will be enough, by the Eberlein-Šmulyan theorem, to show that B_Y is weakly sequentially compact. Each $y \in Y$ is *countably supported on* $\{x_\gamma; \gamma \in \Gamma\}$, i.e., $\{\gamma \in \Gamma; \langle x_\gamma, y \rangle \neq 0\}$ is countable. Let (y_n) be a sequence in B_Y. We can then find a countable subset $N \subset \Gamma$ such that $\langle x_\gamma, y_n \rangle = 0$ for all $\gamma \in \Gamma \setminus N$ and for all $n \in \mathbb{N}$. By a diagonal procedure, we can extract a subsequence of (y_n) (say (y_{n_k})) such that $(\langle x_\gamma, y_{n_k} \rangle)_k$ converges for every $\gamma \in N$. On $\{x_\gamma; \gamma \in \Gamma \setminus N\}$ we know that each y_{n_k} vanishes. Then (y_{n_k}) is w^*-convergent to some $x^* \in B_{X^*}$. By the Grothendieck property of X, (y_{n_k}) w-converges to $x^* (\in B_Y)$ and the claim is proved.

Now use Lemma 5.11 to conclude that Y is w^*-closed. It is also w^*-dense, hence $Y = X^*$ and then X^* (and thus X) is reflexive.   □

Theorem 5.12 (Johnson [John70a]). *Let Γ be an infinite set. Then the space $\ell_\infty(\Gamma)$ does not admit any M-basis.*

Proof. $\ell_\infty(\Gamma)$ is a Grothendieck space (see Theorem 7.18) and is not reflexive. The use Theorem 5.10.   □

5.2 M-bases with Additional Properties

The main result in this section, due to Plichko, is that a space with an M-basis also contains a bounded M-basis.

Theorem 5.13 (Plichko [Plic83]). *Suppose that a nonseparable Banach space X admits an M-basis, and let $\varepsilon > 0$. Then X admits a $\left(2(1+\sqrt{2})^2 + \varepsilon\right)$-bounded M-basis whose vectors and functionals preserve the original spans.*

Proof. Denote $\Omega = \text{dens } X > \omega$, $J = [0, \Omega)$, and let $\{x_i; f_i\}_{i \in J}$ be an M-basis of X. For $I \subset J$, put $X_I = \overline{\text{span}}\{x_j; j \in I\}$, $X^I = \overline{\text{span}}\{x_j; j \notin I\}$. Let $F_I = (X^I)^\perp$, $F^I = (X_I)^\perp$.

Lemma 5.14. *For every infinite $I \subset J$, there is a subset $I \subset I' \subset J$, with card $I' = $ card I, such that $d(S_{F_I}, F^{I'}) = 1$.*

Proof. Let $Q : X \to X/X^I$ be the quotient map. Since $F_I = (X/X^I)^*$, there is a subset $\hat{A} \subset S_{X/X^I}$ with card $\hat{A} = $ card I and such that $\sup\{f(\hat{y}); \hat{y} \in \hat{A}\} = \|f\|$ for every $f \in F_I$. For every $\hat{y} \in \hat{A}$, let $\{y_n\}_{n=1}^\infty \subset X$ be a sequence satisfying $Q(y_n) = \hat{y}$ with $\|y_n\| \to 1$. Then the set $A = \{y_n/\|y_n\|; \hat{y} \in \hat{A}\}$ has the cardinality of I and $\sup\{f(y); y \in A\} = \|f\|$ for every $f \in F_I$. Let $I' = I \cup \{i \in J; \exists y \in A, f_i(y) \neq 0\}$. Since supp (y) is countable for every $y \in A$, we have card $I' = $ card I.   □

Lemma 5.15. *There exists a collection* $\{J_\beta\}_{\omega \le \beta < \Omega}$, $J_\beta \subset J$ *for all* β, *such that* $J = \bigcup_{\beta=\omega}^{\Omega} J_\beta$ *and*

(1) $J_\beta \subset J_\gamma$ *for* $\beta < \gamma$.
(2) card $J_\beta = \beta$.
(3) $J_{\beta+1} \setminus J_\beta \ne \emptyset$.
(4) $d(S_{F_{J_\beta}}, F^{J_{\beta+1}}) = 1$.

Proof. Proceed by transfinite induction. Let $J_\omega = \omega$. Having constructed J_γ for all $\gamma < \beta$, with β a limit ordinal, we let $J_\beta = \bigcup_{\gamma < \beta} J_\gamma$. If β is a nonlimit ordinal, then card $J_{\beta-1} < \Omega$. Choose an element $i \in J \setminus J_{\beta-1}$ and let $J_\beta = (J_{\beta-1} \cup \{i\} \cup \{\beta - 1\})'$ using Lemma 5.14. $\square$

Lemma 5.16. *Let* $\{x_i; f_i\}_{i \in J}$ *be an M-basis of a nonseparable Banach space* X *and let* $\varepsilon > 0$. *Then* X *has an M-basis* $\{x_i'; f_i'\}_{i \in J}$ *for which*

$$\mathrm{span}\{x_i\}_{i \in J} = \mathrm{span}\{x_i'\}_{i \in J}, \mathrm{span}\{f_i\}_{i \in J} = \mathrm{span}\{f_i'\}_{i \in J},$$

and there exists a subset $I \subset J$, card $I = $ card J, *such that* $\sup_{j \in I} \|x_j'\|.\|f_j'\| < 2 + \varepsilon$.

Proof. Without loss of generality, $\|f_i\| = 1$. Let $\{J_\beta\}_{\beta=\omega}^{\Omega}$ be sets from Lemma 5.15. For each ordinal of the form $\gamma = \delta + 3n - 2$, where δ is a limit ordinal and $n > 0$, we choose an element $h_\gamma = f_{i_\gamma}$, $i_\gamma \in J_{\gamma+1} \setminus J_\gamma$. The subspaces $F_{J_{\gamma-1}}$ and $F^{J_{\gamma+2}}$ are w^*-closed, and clearly $F_{J_{\gamma-1}} \cap F^{J_{\gamma+2}} = \{0\}$. Thus $G_\gamma = F_{J_{\gamma-1}} + F^{J_{\gamma+2}} = F_{J_{\gamma-1}} \oplus F^{J_{\gamma+2}}$ is also a topological sum. Then $G_\gamma = \bigcup_{n=1}^{\infty} n(B_{F_{J_{\gamma-1}}} + B_{F^{J_{\gamma+2}}})$, which is a union of w^*-compact convex sets; so by the Banach-Dieudonné theorem, G_γ is w^*-closed. Let $g = g_1 + g_2$, $g_1 \in F_{J_{\gamma-1}}, g_2 \in F^{J_{\gamma+2}}$. By part (4) of Lemma 5.15, $\|h_\gamma - g\| = \|h_\gamma - g_1 - g_2\| \ge \|h_\gamma - g_1\|$, and similarly $\|h_\gamma - g\| \ge \|g_1\|$. Considering separately the cases $\|g_1\| > \frac{1}{2}$ or the opposite, we get that $\|h_\gamma - g\| \ge \frac{1}{2}$. Accordingly, $d(h_\gamma, G_\gamma) \ge \frac{1}{2}$. Canonically, $X_{J_{\gamma+2} \setminus J_{\gamma-1}} = (G_\gamma)_\perp = G_\gamma^\perp \cap X$, so $X_{J_{\gamma+2} \setminus J_{\gamma-1}}^* = X^*/G_\gamma$. By the Hahn-Banach theorem, there exists $y_\gamma \in G_\gamma^\perp \subset X^{**}$ with $\|y_\gamma\| < 2 + \varepsilon$ and such that $h_\gamma(y_\gamma) = 1$. Since $\mathrm{span}\{x_i; i \in J_{\gamma+2} \setminus J_{\gamma-1}\}$ is dense in $G_\gamma^\perp \cap X$, by Goldstine's theorem there is an element $z_\gamma \in \mathrm{span}\{x_i; i \in J_{\gamma+2} \setminus J_{\gamma-1}\}$ such that $h_\gamma(z_\gamma) = 1$ and $\|z_\gamma\| < 2 + \varepsilon$. We let

$$x_i' = x_i \text{ and } f_i' = f_i - f_i(x_\gamma')f_\gamma' \text{ for all } i \in J_{\gamma+2} \setminus (J_{\gamma-1} \cup \{i_\gamma\})$$

and put $x_{i_\gamma}' = z_\gamma$, $f_{i_\gamma}' = f_{i_\gamma} = h_\gamma$. Finally, we put $I = \{i_\gamma; \gamma = \delta + 3n - 2\}$. It is now standard to check that the constructed system $\{x_i'; f_i'\}_{i \in J}$, together with I, satisfies the desired conditions. $\square$

Lemma 5.17 (Olevskii; Ovsepian and Pełczyński [OvPe75]). *Let* X *be a Banach space,* $n \in \mathbb{N}$, *and* $x_0, \dots, x_{2^n-1} \in X$, $h_0, \dots, h_{2^n-1} \in X^*$ *be a biorthogonal sequence. There exists a unitary matrix* $(a_{k,j}^n)_{0 \le k, j < 2^n}$ *such that if*

$$e_k := \sum_{j=0}^{2^n-1} a_{k,j}^n x_j, f_k = \sum_{j=0}^{2^n-1} a_{k,j}^n h_j, \ for \ k = 0, \dots, 2^n - 1,$$

then

(1) $\max_{0 \le p < 2^n} \|e_p\| < (1 + \sqrt{2}) \max_{1 \le j < 2^n} \|x_j\| + 2^{-\frac{n}{2}} \|x_0\|$.
(2) $\max_{0 \le p < 2^n} \|f_p\| < (1 + \sqrt{2}) \max_{1 \le j < 2^n} \|h_j\| + 2^{-\frac{n}{2}} \|h_0\|$.
(3) $f_p(e_q) = \delta_q^p$.
(4) $\mathrm{span}\{e_p; 0 \le p < 2^n\} = \mathrm{span}\{x_p; 0 \le p < 2^n\}$, and $\mathrm{span}\{h_p; 0 \le p < 2^n\} = \mathrm{span}\{f_p; 0 \le p < 2^n\}$.

Proof. Conditions (3) and (4) are satisfied for every unitary matrix, so we need to construct one having additionally properties (1) and (2). To this end, we choose a matrix that transforms the usual basis of $\ell_2^{2^n}$ onto its Haar basis. We put

$$a_{k,0}^n := 2^{-\frac{n}{2}} \ \text{for} \ 0 \le k < 2^n,$$

$$a_{k,2^s+r}^n := \begin{cases} 2^{\frac{(s-n)}{2}} & \text{for } 2^{n-s-1}2r \le k < 2^{n-s-1}(2r+1), \\ -2^{\frac{(s-n)}{2}} & \text{for } 2^{n-s-1}(2r+1) \le k < 2^{n-s-1}(2r+2), \\ 0 & \text{for } k < 2^{n-s-1}2r \text{ and for } k \ge 2^{n-s-1}(2r+2), \end{cases}$$

where $s = 0, \dots, n-1$, $r = 0, \dots, 2^s - 1$. We have

$$\sum_{j=1}^{2^n-1} |a_{k,j}^n| = \sum_{s=0}^{n-1} 2^{-\frac{n-s}{2}} < 1 + \sqrt{2} \quad \text{for } 0 \le k < 2^n.$$

Checking the properties is standard. $\qquad \square$

Lemma 5.18. *Let $\{x_i; h_i\}_{i \in J}$ be a biorthogonal system in X, $\varepsilon > 0$. If there exists $I \subset J$, card $I = $ card J, and $M < \infty$, such that $\sup_{i \in I} \|x_i\| . \|h_i\| = M$, then there exists a biorthogonal system $\{e_i; f_i\}_{i \in J}$ in X such that*

$$\mathrm{span}\{e_i\}_{i \in J} = \mathrm{span}\{x_i\}_{i \in J}, \mathrm{span}\{f_i\}_{i \in J} = \mathrm{span}\{h_i\}_{i \in J}, \ and$$

$$\sup_{i \in J} \|e_i\| . \|f_i\| \le M(1 + \sqrt{2})^2 + \varepsilon.$$

Proof. Denote $\Gamma = $ card J, and assume without loss of generality that $\|x_i\| = 1$ for all $i \in J$. Split $J = \bigcup_{\gamma \in \Gamma} J_\gamma$, where card $J_\gamma = 2^{n_\gamma}$, $J_\gamma = \{i_0^\gamma, \dots, i_{2^{n_\gamma}-1}^\gamma\}$, and so that

$$J_\gamma \setminus I = \{i_0^\gamma\} \ \text{and}$$

$$2^{-\frac{n_\gamma}{2}} \max\{\|x_{i_0^\gamma}\|, \|h_{i_0^\gamma}\|\} < \frac{\varepsilon}{3(1 + \sqrt{2})^2 M}.$$

Applying Lemma 5.17 to the finite set of vectors and functionals $\{x_i, h_i\}_{i \in J_\gamma}$ yields a biorthogonal system $\{e_i; f_i\}_{i \in J_\gamma}$ with the properties

$$\mathrm{span}\{e_i\}_{i \in J_\gamma} = \mathrm{span}\{x_i\}_{i \in J_\gamma}, \quad \mathrm{span}\{f_i\}_{i \in J_\gamma} = \mathrm{span}\{h_i\}_{i \in J_\gamma},$$

and $\max_{i \in J_\gamma} \|e_i\| . \|f_i\| \le M(1+\sqrt{2})^2 + \varepsilon$. To finish the proof, it suffices to use the system $\{e_i; f_i\}_{i \in J}$. $\qquad \square$

Combining the previous lemmas finishes the proof of Theorem 5.13. Note the additional fact that the new M-basis of X has the same linear spans of both the vectors and the functionals of the original basis. □

Plichko [Plic79] has improved Theorem 5.13 for the class of WCG spaces, showing that the boundedness constant can be chosen to be $2 + \varepsilon$. This is an almost optimal result in light of the next example.

Example 5.19 (Plichko [Plic86a]). There exists a WCG space X of dens $X = c$ that has no λ-bounded M-basis for $\lambda < 2$.

Proof. We let $X = c_0[0,1] + C[0,1] \hookrightarrow \ell_\infty[0,1]$ with the canonical norm inherited from $\ell_\infty[0,1]$. Since the decomposition of any element $x \in X$ into $x = y + z$, $y \in c_0[0,1]$, $z \in C[0,1]$ is unique, $X \cong c_0[0,1] \oplus C[0,1]$, so X is a WCG space. Let $\{x_\alpha; f_\alpha\}_{\alpha \in \Gamma}$ be an M-basis of X. Let $\{g_n\}_{n=1}^\infty$ be w^*-dense in $c_0[0,1]^\perp = C[0,1]^*$. Since X is in particular WLD, by Theorem 5.37, for every $f \in X^*$, card $\{\alpha; f(x_\alpha) \neq 0\} \leq \omega$. Hence there exists a countable set $J \subset \Gamma$ such that $\bigcup_{n \in \mathbb{N}} \{\alpha; g_n(x_\alpha) \neq 0\} \subset J$. Thus $C[0,1] \subset \overline{\text{span}}\{x_j; j \in J\}$. If $\alpha \notin J$, then $x_\alpha \in c_0[0,1]$, $f_\alpha \in C[0,1]^\perp$, and we have

$$\|x_\alpha\| \|f_\alpha\| = \frac{\|x_\alpha\|}{\text{dist}(x_\alpha, (f_\alpha)_\perp)} \geq \frac{\|x_\alpha\|}{\text{dist}(x_\alpha, C[0,1])}.$$

One checks easily that, for every $x \in c_0[0,1]$, $\text{dist}(x, C[0,1]) \leq \frac{\|x\|}{2}$. Thus $\|x_\alpha\| \|f_\alpha\| \geq 2$. □

Recall that every separable or reflexive Banach space has a 1-norming M-basis (under every renorming). In the separable case, it follows from Theorem 1.22, and in the reflexive case this follows from the fact that every M-basis is 1-norming by reflexivity.

Proposition 5.20 (Godefroy [Gode95]). *Let $X \cong Y \oplus Z$, where Y is separable and Z is reflexive. Then X has a $1/4$-norming M-basis (under every renorming).*

Proof. By the 1-SCP property of WCG spaces, we may without loss of generality assume that Y is 1-complemented. Both Y and Z have 1-norming M-bases $\{x_n; f_n\}_{n \in \mathbb{N}}$, $\{y_\gamma; f_\gamma\}_{\gamma \in \Gamma}$. We claim that $\{(x_n; f_n)\}_{n \in \mathbb{N}} \cup \{(y_\gamma; f_\gamma)\}_{\gamma \in \Gamma}$, where the functionals are extended naturally by zero on the respective complemented subspace, is a $1/4$-norming M-basis. Note that the extended functionals satisfy $\|f_n\| = 1$, $\|f_\gamma\| \leq 2$. The rest follows from the fact that for $y \in Y, z \in Z$ we have $\sup\{\|y\|, \|z\|\} \geq \frac{\|y+z\|}{2}$. □

These observations, together with the fact that every WCG space admits an M-basis $\{x_\gamma; f_\gamma\}_{\gamma \in \Gamma}$ such that the range of $T(f) = (f(x_\gamma))_{\gamma \in \Gamma}$ is contained in $c_0(\Gamma)$, lead to the natural question of whether every WCG space admits a norming M-basis. This problem is still open. The following three results provide some insight into this problem.

Theorem 5.21 (Valdivia [Vald94]). *Let X be a Banach space such that the compact space (B_{X^*}, w^*) has countable tightness. Assume that $\{x_\gamma^*; x_\gamma^{**}\}_{\gamma \in \Gamma}$ is an M-basis in $X^* \times X^{**}$ such that $X \subset \overline{\operatorname{span}}\{x_\gamma^{**}; \gamma \in \Gamma\}$. Then X is Asplund and WCG.*

Proof (Sketch). The space X^* can be identified with $Z^*/X^\perp$. Let $\Psi : Z^* \to Z^*/X^\perp$ be the canonical quotient mapping. Every $x^* \in X^*$ extends (uniquely) to a continuous linear mapping $e(x^*) : Z \to \mathbb{R}$, and so X^* can also be identified to the subspace $e(X^*) \subset Z^*$. The mapping $\eta : Z^* \to \mathbb{R}^\Gamma$ given by $\eta(z^*) := (\langle x_\gamma^{**}, z^* \rangle)_{\gamma \in \Gamma}$, for $z^* \in Z^*$, is one-to-one and $w(z^*, Z)$-$\mathcal{T}_p$-continuous, where $\mathcal{T}_p$ denotes the product topology in $\mathbb{R}^\Gamma$. By the bipolar theorem, $B_{e(X^*)}$ is dense in $(B_{Z^*}, w(Z^*, Z))$. Obviously, $\eta(B_{e(X^*)}) \subset \Sigma(\Gamma)$, where $\Sigma(\Gamma)$ denotes the subspace of $\mathbb{R}^\Gamma$ of vectors with countable support, so $K := (B_{Z^*}, w(Z^*, Z))$ is a Valdivia compact. Let $\mathcal{T}_\Sigma$ be the topology on $C(K)$ of the pointwise convergence on the set $K \cap \{z^* \in Z^*; \eta(z^*) \in \Sigma(\Gamma)\}$. It is plain that Z is a closed subspace of $(C(K), \mathcal{T}_\Sigma)$. We claim that X is a closed subspace of $(Z, \mathcal{T}_\Sigma)$. Indeed, let $z \in Z \setminus X$. There exists $u^* \in B_{Z^*}$ such that $u^* \upharpoonright X \equiv 0$ and $\varepsilon := \langle z, u^* \rangle \neq 0$. Put $M := \{z^* \in B_{Z^*}; \eta(M) \subset \Sigma(\Gamma), \langle z, z^* \rangle \geq \varepsilon/2\}$. Then $u^* \in \overline{M}^{w(Z^*, Z)}$ and $\Psi(u^*) = 0 \in \overline{\Psi(M)}^{w(X^*, X)}$. By the countable tightness of $(B_{X^*}, w(X^*, X))$, there exists a countable set $P \subset M$ such that $0 \in \overline{\Psi(P)}^{w(x^*, X)}$. Then there exists $w^* \in \overline{P}^{w(Z^*, Z)}$ such that $\Psi(w^*) = 0$, We get $w^* \upharpoonright X \equiv 0$, $\langle z, w^* \rangle \geq \varepsilon/2$ and $\eta(w^*) \in \Sigma(\Gamma)$, and this proves the claim. In particular, X is closed in $(C(K), \mathcal{T}_\Sigma)$. The space $C(K)$ has a PRI $\{P_\alpha; \omega \leq \alpha \leq \mu\}$ (see [DGZ93a, Thm. VI.7.6]) with pointwise continuous projections P_α. A standard procedure gives an M-basis $\{v_j; v_j^*\}_{j \in J}$ in $X \times X^*$ such that $\{v_j; j \in J\}$ countably supports X^*. Now, the fact that $\{x_\gamma^*\}_{\gamma \in \Gamma}$ countably supports X gives, again by a standard technique, another M-basis $\{y_\delta; y_\delta^*\}_{\delta \in \Delta}$ in $X \times X^*$ with $\operatorname{span}\{y_\delta; \delta \in \Delta\} = \operatorname{span}\{v_j; j \in J\}$ and $\operatorname{span}\{y_\delta^*; \delta \in \Delta\} = \operatorname{span}\{x_\gamma^*; \gamma \in \Gamma\}$. This proves that the M-basis $\{y_\delta; y_\delta^*\}_{\delta \in \Delta}$ is shrinking and, by Theorem 6.3, the space X is Asplund and WCG.   □

Using Theorem 5.21, an example of a compact scattered space K and an equivalent norm $\||\cdot\||$ on $C(K)$ such that $(C^*(K), \||\cdot\||)$ has no 1-norming M-basis is provided in [Vald94]. The techniques in Theorem 5.21 give also the following result [Vald94]: *Let X be a Banach space such that X^* is a nonseparable Vašák space. Then there exist a Banach space Y and a real number $0 < \delta < 1$ such that Y^* is isomorphic to $X^* \oplus l_1$ and it does not admit any α-norming Markushevich basis for $\delta \leq \alpha \leq 1$.*

Theorem 5.22 (Troyanski, see, e.g., [VWZ94]**).** *Let $X = JL_2$ be the space of Lindenstrauss and Johnson (see Corollary 4.20). Then X has an equivalent LUR norm $\|\cdot\|$. The dual space X^* under the dual norm $\|\cdot\|$ is a WCG space that does not admit any 1-norming M-basis.*

Proof (Sketch). Assume that $\{x_\alpha^*; x_\alpha^{**}\}$ is a 1-norming M-basis in $X^* \times X^{**}$ in this norm. Put $Y := \overline{\operatorname{span}}(\{x_\alpha^{**}\})$. Given $x \in S_X$, get $x^* \in S_{X^*}$ such that

$\langle x, x^* \rangle = 1$. As Y is 1-norming, choose $y_n \in S_Y$ such that $\langle y_n, x^* \rangle \to 1$. We have $\|y_n + x\| \to 2$ and thus $\|y_n - x\| \to 0$, as the norm is LUR. Thus $X \subset Y$. The space Y admits a separable PRI $\{P_\alpha; \omega \le \alpha \le \mu\}$, as it has a 1-norming basis. Therefore, by an easy exhaustion argument, $c_0 \subset P_\alpha(Y)$ for some α (as $c_0 \subset X$). Thus c_0 is complemented in X by the Sobczyk theorem, which is not the case. $\qquad \square$

Theorem 5.22 has been improved by Godefroy as follows.

Theorem 5.23 (Godefroy [Gode95]). *Let X be a WCG space of* dens $X \ge$ *c. Then, for some $\lambda < 1$, there exists an equivalent renorming of $(\ell_1 \oplus X, \|\cdot\|)$ that admits no λ-norming M-basis.*

Proof (Sketch). The main ingredient of the proof is a result due to Finet [Fin89] that implies, in particular, that for every WCG space Y of density at least c, there exists a $\xi(Y) < 1$ and a renorming of Y such that no proper subspace of Y^* is $\xi(Y)$-norming. Let $\|\cdot\|$ be a norm on $Y := \ell_1 \oplus X$ with the mentioned properties ($\xi = \xi(Y) < 1$). Let us assume by contradiction that there is a ξ-norming M-basis $\{x_\gamma; f_\gamma\}_{\gamma \in \Gamma}$ in $\ell_1 \oplus X \times (\ell_1 \oplus X)^*$ with $\|f_\gamma\| = 1$ for all $\gamma \in \Gamma$. The operator $T : \ell_1 \oplus X \to c_0(\Gamma)$ defined by $Tx = (f_\gamma(x))_{\gamma \in \Gamma}$ has a dual $T^* : \ell_1(\Gamma) \to (\ell_1 \oplus X)^*$ with span$\{f_\gamma\}_{\gamma \in \Gamma} \subset T^*(\ell_1(\Gamma))$. Thus $\overline{T^*(\ell_1(\Gamma))} = (\ell_1 \oplus X)^*$ by the nonexistence of a proper ξ-norming subspace. Thus T^{**} is one-to-one and thus, by [DGZ93a, VI.5.4], $\ell_1 \oplus X$ has to be an Asplund space, which is a contradiction. $\qquad \square$

Definition 5.24. *An M-basis $\{x_\gamma; f_\gamma\}_{\gamma \in \Gamma}$ is* countably λ-norming *for some $\lambda > 0$ if $Y = \{f \in X^*; \text{card}\ \{\gamma \in \Gamma; f(x_\gamma) \ne 0\} \le \omega\}$ is a λ-norming subspace. We say that the basis is* countably norming *if it is countably λ-norming for some $\lambda > 0$.*

Theorem 5.25 (Alexandrov and Plichko [AlPl]). *The space $C[0, \omega_1]$ has a strong and countably norming M-basis, but it has no norming M-basis.*

Proof. For convenience, we work in the space $C_0[0, \omega_1] \cong C[0, \omega_1]$. Put $x_\gamma = \chi_{[0,\gamma]}$, $f_\gamma = \delta_\gamma - \delta_{\gamma+1}$, $\gamma < \omega_1$. It is standard to check that $\{x_\gamma; f_\gamma\}_{\gamma < \omega_1}$ is a strong M-basis. To check that it is countably 1/2-norming, set $S = \{f \in \ell_1[0, \omega_1); \sum_{\alpha=0}^{\omega_1} f(\alpha) = 0\}$. We have $f(x_\gamma) = 0$ whenever $\gamma > \max \text{supp}\ f$. As every element x has a countable support supp x, there exists a norming functional $f = \delta_\alpha - \delta_\beta \in S$, $\|f\| = 2$, where $|x(\alpha)| = \|x\|$ and $\beta \notin \text{supp}\ x$. Note that $S \ne \overline{\text{span}}\{f_\gamma; \gamma < \omega_1\}$. Indeed, $\delta_0 - \delta_\omega \notin \overline{\text{span}}\{f_\gamma; \gamma < \omega_1\}$. This finishes the first part. For the second part, assume by contradiction that there exists some $\{x_\gamma; f_\gamma\}_{\gamma < \omega_1}$, $\|f_\gamma\| = 1$, that is a norming M-basis of $C_0[0, \omega_1]$.

Claim. For every $\alpha < \omega_1$, card $\{\gamma; f_\gamma(\alpha) \ne 0\} \le \omega$.

Proof. We proceed by induction in α. Suppose that the claim has been proven for all $\beta < \alpha$. If α is nonlimit, $\chi_{\{\alpha\}} \in C_0[0, \omega_1]$. Since the operator $T(x) :=$

$(f_\gamma(x))_{\gamma<\omega_1}$ has a range contained in $c_0[0,\omega_1)$, we have card $\{\gamma; f_\gamma(\chi_{\{\alpha\}}) \neq 0\} \leq \omega$ as claimed. For a limit ordinal $\alpha < \omega_1$, let $G_\alpha := \{\gamma; f_\gamma(\alpha) \neq 0\}$. If card $G_\alpha > \omega$, then using the inductive hypothesis, we may without loss of generality assume that $f_\gamma(\beta) = 0$ for all $\beta < \alpha$ and $\gamma \in G_\alpha$. Again, we have reached a contradiction since $f_\gamma(\chi_{[0,\alpha]}) = f_\gamma(\alpha) \neq 0$ for uncountably many $\gamma \in G_\alpha$. $\square$

It follows that for every $\alpha < \omega_1$ there exists $\phi(\alpha) \in [0,\omega_1)$ such that whenever supp $f_\gamma \cap [0,\alpha] \neq \emptyset$, then supp $f_\gamma \subset [0,\phi(\alpha)]$. Using the notation $\phi^n(\alpha) = \phi \circ \cdots \circ \phi(\alpha)$, we put $\Phi(\alpha) = \lim_{n\to\infty}\phi^n(\alpha)$. Clearly, whenever supp $f_\gamma \cap [0,\Phi(\alpha)) \neq \emptyset$, then supp $f_\gamma \subset [0,\Phi(\alpha))$. Thus there is a transfinite sequence of consecutive intervals $\{I_\alpha\}_{\alpha<\omega_1} = [m_\alpha, m_{\alpha+1})$ of $[0,\omega_1)$ such that supp $f_\gamma \cap I_\alpha \neq \emptyset$ implies supp $f_\gamma \subset I_\alpha$. Next observe that card $\{\gamma; \sum_\alpha f_\gamma(\alpha) \neq 0\} \leq \omega$ since otherwise $T([0,\beta]) \notin c_0[0,\omega_1)$ for some large enough β, a contradiction. So there exists $\beta < \omega_1$ such that $\xi \geq \beta$ and supp $f_\gamma \subset I_\xi$ implies that $\sum_{\alpha\in I_\xi} f_\gamma(\alpha) = 0$. Put $y_i = \chi_{(m_{\beta+i}, m_{\beta+i+1}]} \in C_0[0,\omega_1)$, $i \in \mathbb{N}$. Consider an element

$$x = \sum_{i=1}^n \frac{i}{n}y_i + \sum_{i=n+1}^{2n-1} \frac{2n-i}{n}y_i \in C_0[0,\omega_1), \; \|x\| = 1.$$

Let $f \in \text{span}\{f_\gamma\}_{\gamma<\omega_1}$, $\|f\| = 1$. We have $f = \sum_{j=1}^m h_j$ where supp $h_j \subset I_{\alpha_j}$ for some $\alpha_j < \omega_1$. Thus

$$h_j(y_i) = h_j(\chi_{I_{\beta+i}} - \chi_{\{m_{\beta+i}\}} + \chi_{\{m_{\beta+i+1}\}}) = h_j(m_{\beta+i+1}) - h_j(m_{\beta+i}).$$

In particular, $h_j(y_i) = h_j(m_{\beta+i+1})$ if and only if $\beta+i+1 = \alpha_j$, and $h_j(y_i) = -h_j(m_{\beta+i})$ if and only if $\beta+i = \alpha_j$ and is 0 otherwise. Thus

$$|f(x)| = \left| \sum_{j=1}^m \left(\sum_{i=1}^n \frac{i}{n}h_j(y_i) + \sum_{i=n+1}^{2n-1} \frac{2n-i}{n}h_j(y_i) \right) \right| \leq \frac{1}{n}\sum_{i=1}^{2n} |f(m_{\beta+i})| \leq \frac{1}{n}.$$

Since n can be chosen arbitrary, span$\{f_\gamma\}_{\gamma<\omega_1}$ is not norming. $\square$

5.3 Σ-subsets of Compact Spaces

We are going to prove some topological results to be used in the theory of WLD spaces (which will be investigated in the next section). One of the main results here is a theorem of Deville and Godefroy that characterizes Valdivia compacta that are not Corson as those containing a homeomorphic copy of $[0,\omega_1]$.

For a nonempty set Γ, the space $\mathbb{R}^\Gamma$ will always be endowed with its product topology.

Definition 5.26. *Let Γ be a nonempty set. For $x \in \mathbb{R}^\Gamma$, we denote* $\mathrm{supp}(x) = \{\gamma \in \Gamma; x(\gamma) \neq 0\}$. *We put* $\Sigma(\Gamma) = \{x \in \mathbb{R}^\Gamma; \mathrm{card}\,(\mathrm{supp}(x)) \leq \omega\}$. *If $I \subset \Gamma$, we define the projection* $P_I(x(\gamma)) = \chi_I \cdot x(\gamma)$.

Recall that a subset S of a topological space T is *countably closed* if S contains the closure of every countable set $N \subset S$.

Proposition 5.27. $\Sigma(\Gamma) \subset \mathbb{R}^\Gamma$ *is countably closed and Fréchet-Urysohn. In particular, every Corson compact is angelic.*

Proof. The first statement is clear. To prove the Fréchet-Urysohn property, let S be a subset of $\Sigma(\Gamma)$ and let $m_0 \in \overline{S}$. To every $x \in \Sigma(\Gamma)$, let us assign an infinite sequence $\Gamma \supset \{\gamma_i(x)\}_{i=1}^\infty \supset \mathrm{supp}\,(x)$. We construct by induction a sequence $m_n \in S$ such that

$$|m_n(\gamma_k(m_l)) - m_0(\gamma_k(m_l))| < \frac{1}{n} \text{ for } 0 \leq l < n, 1 \leq k \leq n.$$

It is easy to check that $m_n \to m_0$. $\qquad\square$

Definition 5.28. *Let K be a compact space and $A \subset K$. We say that A is a Σ-subset of K if there is a homeomorphic injection $\phi : K \to \mathbb{R}^\Gamma$ for some Γ such that $\phi(A) = \phi(K) \cap \Sigma(\Gamma)$.*

In particular (see Definition 3.30), a compact space is a Corson compact if and only if it is a Σ-subset of itself, and it is a Valdivia compact if and only if it has a dense Σ-subset.

Lemma 5.29 (Kalenda [Kal00a]). *Let K be a Valdivia compact, and let $A \subset K$ be a dense Σ-subset of K. Then the following assertions hold:*

(1) *A is countably closed in K.*
(2) *A is an angelic space. In particular, if $x \in K$ is a G_δ point (i.e., $\{x\} = \bigcap_{n=1}^\infty O_n$, where O_n are open sets in K), then $x \in A$.*
(3) *If $G \subset K$ is a G_δ set, then $G \cap A$ is dense in G. In particular, a closed and G_δ subset of a Valdivia compact is again a Valdivia compact.*
(4) *If K has a dense subset G consisting of G_δ-points, then A is the unique Σ-subset of K.*

Proof. Parts (1) and (2) follow from Proposition 5.27.

(3) Let $G = \bigcap_{n=1}^\infty U_n$, where U_n are open and dense sets $U_n \subset K$. Let $x \in G$ and W be an open neighborhood of x. We will show that $W \cap A \cap G \neq \emptyset$. To this end, we construct by induction open sets V_n, $n \in \mathbb{N}$, such that

$$V_1 \subset W, x \in V_n, n \in \mathbb{N},$$

$$\overline{V_{n+1}} \subset V_n \cap U_1 \cap \cdots \cap U_n, n \in \mathbb{N}.$$

As A is dense in K, we have $V_n \cap A \neq \emptyset$ for every n. Moreover, $\overline{V_{n+1}} \subset V_n$ and A is countably compact, and hence $A \cap \bigcap_{n \in \mathbb{N}} V_n \neq \emptyset$. By the construction we have $\bigcap_{n \in \mathbb{N}} V_n \subset G \cap W$, which completes the proof.

(4) Let G be a dense set of G_δ points. We claim that A is the unique dense Σ-subset of K. Let A' be another dense Σ-subset of K. Both A' and A are dense and countably compact; from (2) it follows that $G \subset A \cap A'$. Thus $A' \cap A$ is dense (and a Σ-subset) in K. Let $x \in A$. Then $x \in \overline{A' \cap A}$, and since $\Sigma(\Gamma)$ is Fréchet-Urysohn (see Proposition 5.27), there is a sequence $x_n \in A' \cap A$ such that $x_n \to x$. From (1), A' is sequentially closed, so we have $x \in A'$. Therefore $A \subset A'$. By interchanging the roles of A and A', we get $A' = A$.   □

Proposition 5.30 (Valdivia [Vald97]). *The space $[0, \omega_1]$ is a Valdivia compact but not a Corson compact. The quotient space L obtained from $[0, \omega_1]$ by identifying points ω and ω_1 is not a Valdivia compact.*

Proof. The homeomorphic embedding $h : [0, \omega_1] \to \mathbb{R}^{\omega_1}$, defined as $h(\alpha) = \chi_{[0,\alpha]}$, shows that $[0, \omega_1)$ is a dense Σ-subset of $[0, \omega_1]$. On the other hand, ω_1 is not a limit of a sequence from $[0, \omega_1)$, so this space is not angelic and thus not Corson (see Proposition 5.27).

To show that L is not a Valdivia compact, assume by contradiction that there is a dense Σ-subset A of L. As A must contain all isolated points and is sequentially closed, we get that $A = L$. The collated point $p = \{\omega, \omega_1\}$ lies in A since it is a limit of a sequence $n \to \omega$, $n < \omega$. On the other hand, $p \in \overline{(\omega, \omega_1)} \subset L$, and by the angelicity of A there should exist a convergent sequence to it, which is a contradiction.   □

Theorem 5.31 (Deville and Godefroy [DeGo93]). *The following are equivalent for a Valdivia compact K.*

(i) *K is Corson.*
(ii) *K does not contain a homeomorphic copy of $[0, \omega_1]$.*

Proof. (i) $\Rightarrow$ (ii) Every Corson compact is angelic (see Proposition 5.27).
  To prove (ii) $\Rightarrow$ (i), we will use Fact 5.32 and Lemma 5.33.

Fact 5.32. *Let $K \subset [0, 1]^I$ be a Valdivia compact and $A = \Sigma(I) \cap K$ be dense in K. If $J \subset I$, then there is $\tilde{J} \supseteq J$ and $\mathrm{card}\, \tilde{J} = \mathrm{card}\, J$, such that $P_{\tilde{J}} K = K$.*

Proof. We construct, by a standard countable exhaustion procedure (see, e.g., [DGZ93a, Lemma VI.7.5]), a subset $\tilde{J}$ such that $P_{\tilde{J}}(A) \subset A$. Since $\overline{A} = K$, the conclusion follows.   □

Lemma 5.33. *Let (H, τ) be a compact space and $\{\phi_\alpha\}_{\alpha \leq \Omega}$ be a sequence of maps from H to H such that:*

(i) *$\phi_\alpha \phi_\beta = \phi_\beta \phi_\alpha = \phi_\alpha$ for $0 \leq \alpha \leq \beta \leq \Omega$.*
(ii) *For every $x \in H$, the map $E_x : \alpha \to \phi_\alpha(x)$ is continuous from $[0, \Omega]$ to (H, τ).*

Then, for every $x \in H$, $E_x([0, \Omega])$ is homeomorphic to an ordinal interval.

Proof. We introduce a well-ordering $\leq$ on the compact $E_x([0, \Omega])$ by the condition $a \leq b$ if and only if $\min E_x^{-1}(a) \leq \min E_x^{-1}(b)$. Thus $(E_x([0, \Omega]), \leq)$ is order-isomorphic to an ordinal interval, so it can be equipped with the interval topology τ'. It is clear that $E_x^{-1}(F)$ is closed for every closed $F \subset (E_x([0, \Omega]), \tau')$. Thus $E_x(E_x^{-1}(F)) = F$ and so F is τ closed as well. Therefore the two compact topologies τ and τ' coincide. □

Returning to the proof of Theorem 5.31, $A = K \cap \Sigma(I)$ is dense in K and there exists $x \in K \setminus A$. Let $J \subset \text{supp } x$, $\text{card } J = \omega_1$. We can write $J = \{i_\alpha; 0 \leq \alpha < \omega_1\}$. We now construct inductively a transfinite sequence $(I_\alpha)_{0 \leq \alpha \leq \omega_1}$ of subsets of I such that

(a) I_α is countable whenever $\alpha < \omega_1$;
(b) if $\alpha < \beta$, then $I_\alpha \subset I_\beta$;
(c) $i_\alpha \in I_{\alpha+1}$;
(d) $P_{I_\alpha}(K) \subset K$ for every $\alpha \in [0, \omega_1]$; and
(e) if α is a limit ordinal, then $I_\alpha = \bigcup_{\beta < \alpha} I_\beta$.

To pass from α to $\alpha+1$, note that $x \in U_\alpha = \{y \in K; y(i_\alpha) \neq 0\}$, so we can pick $x_\alpha \in U_\alpha \cap A$. To obtain $I_{\alpha+1}$, we apply Fact 5.32 to $I_\alpha \cup \text{supp } x_\alpha$. Once this construction is performed, apply Lemma 5.33 to K and $\phi_\alpha = P_{I_\alpha}$ to see that $E_x([0, \omega_1])$ is homeomorphic to $[0, \eta]$ for some ordinal η. By conditions (a) and (c), $E_x([0, \omega_1])$ is uncountable, and therefore K contains a subset homeomorphic to $[0, \omega_1]$. □

5.4 WLD Banach Spaces and Plichko Spaces

In this section, we investigate the class of WLD spaces. These spaces can be characterized in many ways, including the following: as spaces whose dual balls are Corson compacta in the weak* topology; as spaces with a full projectional generator; as spaces with a weakly Lindelöf M-basis; or as spaces whose dual balls are Valdivia compact in the weak* topology under every renorming. The larger class of Plichko spaces is introduced.

Definition 5.34. *Let X be a Banach space. We say that $S \subset X^*$ is a Σ-subspace of X^* if there is a linear one-to-one w^*-continuous mapping $T : X^* \to \mathbb{R}^\Gamma$ for some set Γ such that $T(S) \subset \Sigma(\Gamma)$.*

Lemma 5.35. *Let X be a Banach space with an M-basis $\{x_\gamma; x_\gamma^*\}_{\gamma \in \Gamma}$ and a dual unit ball that has w^*-countable tightness. Then $\{x_\gamma; \gamma \in \Gamma\}$ countably supports X^*; i.e., $\text{card}\{\gamma \in \Gamma; \langle x_\gamma, x^* \rangle \neq 0\} \leq \aleph_0$ for every $x^* \in X^*$.*

Proof. Let $S := \{x^* \in X^*; \text{card}\{\gamma : \gamma \in \Gamma, \langle x_\gamma, x^* \rangle \neq 0\} \leq \aleph_0\}$, a linear subspace of X^* containing $\{x_\gamma^*; \gamma \in \Gamma\}$. By the countable tightness of (B_{X^*}, w^*) and the Banach-Dieudonné theorem, S is w^*-closed, so $S = X^*$ and then $\{x_\gamma; \gamma \in \Gamma\}$ countably supports X^*. □

Theorem 5.36. *Every WLD Banach space has a full PG.*

Proof. (B_{X^*}, w^*) is Corson and hence, for some nonempty Γ, it is a subspace of $(\Sigma(\Gamma), \mathcal{T}_p)$ (see Definition 5.26), where $\mathcal{T}_p$ is the topology of pointwise convergence. Given $\gamma \in \Gamma$, let $\pi_\gamma : \Sigma(\Gamma) \to \mathbb{R}$ be the γ-th coordinate mapping, an element in $C((B_{X^*}, w^*))$. In this last space, the algebra generated by the elements in X and the constant functions is $\|\cdot\|$-dense, so there exists a countable set $X_\gamma \subset X$ such that π_γ is in the $\|\cdot\|$-closure of the algebra $\mathcal{A}(X_\gamma, I_{\text{const}})$ generated by X_γ and the constant function I_{const} on (B_{X^*}, w^*). Define

$$\Phi : X^* \to 2^X \text{ as } \begin{cases} \Phi(0) := \{0\}, \\ \Phi(x^*) := \{0\}, & \text{if } x^* \notin B_{X^*}, \\ \Phi(x^*) := \bigcup_{\pi_\gamma(x^*) \neq \pi_\gamma(0)} X_\gamma, & \text{if } x^* \in B_{X^*},\ x^* \neq 0. \end{cases} \quad (5.1)$$

We claim that (X^*, Φ) is a PG. To prove the claim, take $W \subset X^*$ such that $\text{span}_{\mathbb{Q}} W = W$. Let $x^* \in \Phi(W)^\perp \cap \overline{B_W}^{w^*}$. Assume $x^* \neq 0$. Then there exists $\gamma \in \Gamma$ such that $\pi_\gamma(x^*) \neq \pi_\gamma(0)$. As $x^* \in \overline{B_W}^{w^*}$, there exists $w^* \in B_W$ such that $\pi_\gamma(w^*) \neq \pi_\gamma(0)$. Then $X_\gamma \subset \Phi(w^*)$. As $x^* \in \Phi(B)^\perp$ and $\Phi(w^*) \subset \Phi(W)$, we have $\langle X_\gamma, x^* \rangle = 0$. Now, every element of $\mathcal{A}(X_\gamma, I_{\text{const}})$ is of the form $f := a_0 + \sum_i a_i \prod_j x_{i,j}^{n_j}$, where a_0, a_i are constant functions and $x_{i,j} \in X_\gamma$. It follows that $f(x^*) = a_0 = f(0)$. Then, as $\pi_\gamma \in \overline{\mathcal{A}}^{\|\cdot\|}(X_\gamma, I_{\text{const}})$, we get $\pi_\gamma(x^*) = \pi_\gamma(0)$, a contradiction. $\qquad\square$

Theorem 5.37 (see, e.g., [VWZ94]). *For a Banach space X, the following statements are equivalent:*

(i) *X is WLD.*
(ii) *There exists an M-basis $\{x_\gamma; x_\gamma^*\}_{\gamma \in \Gamma}$ in $X \times X^*$ that countably supports X^*; i.e., $\text{card}\{\gamma; \gamma \in \Gamma, \langle x_\gamma, x^* \rangle \neq 0\} \leq \aleph_0$ for every $x^* \in X^*$.*
(iii) *X^* is a Σ-subspace of itself.*
(iv) *X is weakly Lindelöf and admits an M-basis.*
(v) *X has property C and admits an M-basis.*

Moreover, if X is WLD, then every M-basis in X has the property stated in (ii).

Proof. (i)$\Rightarrow$(ii) If X is WLD, then it has a (full) PG (see Theorem 5.36), so X has a PRI (Theorem 3.42). As every complemented subspace of a WLD is again WLD, the class of WLD spaces is a $\mathcal{P}$-class and then X has a (strong) M-basis (Theorem 5.1). Let $\{x_\gamma; x_\gamma^*\}_{\gamma \in \Gamma}$ be any M-basis in X. Nor use Lemma 5.35 to conclude (ii).

(ii)$\Rightarrow$(iii) Use the evaluation map $X^* \to \mathbb{R}^\Gamma$ given by $x^* \mapsto (\langle x_\gamma, x^* \rangle)_\gamma$.

(iii)$\Rightarrow$(i) Use the restriction to B_{X^*} of the map T in Definition 5.34.

(i)$\Rightarrow$(iv) We only need to show that X is weakly Lindelöf. We use the following result, noting that X has a PG.

Lemma 5.38 (Orihuela [Ori92]). *Let X be a Banach space. Assume that for every map ϕ from X into finite subsets of X^*, X admits a norm-1 projection P such that $P(X)$ is separable and for some countable dense set $A \subset P(X)$, $P^*(f) = f$ for all $f \in \phi(A)$. Then X is weakly Lindelöf.*

Proof. Consider X in its weak topology. Assume that $\{V_\alpha\}_{\alpha \in \Gamma}$ is an open cover of X. For every $x \in X$, let $r_x > 0$ be the supremum of all positive numbers r such that $B_X^Q(x, r)$, the open ball of radius r centered at x, lies in some V_α. Choose $V_x \in \{V_\alpha\}$ so that $B_X^Q(x, \frac{r_x}{2}) \subset V_x$ and assume that V_x is formed by the intersection of half-spaces given by a finite set K_x of functionals. Define $\phi(x) = K_x$.

By our assumption, there is a projection P of X onto $P(X)$ and an appropriate set $A \subset P(X)$ constructed for the map ϕ. We claim that $\{V_z;\ z \in A\}$ is a cover of X, which will complete the proof.

Choose any $x \in X$ and let $B_X^Q(P(x), r) \subset V_{P(x)}$. Find $z \in A$ such that $z \in B_X^Q(P(x), \frac{1}{10}r)$. Then $r_z > \frac{9}{10}r$ and thus $B_X^Q(z, \frac{2}{5}r) \subset V_z$. Hence $P(x) \in V_z$. As $\phi(z) \subset P^*(X^*)$, it follows that $x \in V_z$. Indeed, if, say, $|f(Px - z)| < \varepsilon$ for all $f \in \phi(z)$, then $|f(x - z)| = |P^*(f)(x - z)| = |f(P(x) - P(z))| = |f(P(x) - z)| < \varepsilon$. This shows that $\{V_z\}$ is a cover of X. □

We resume the proof of Theorem 5.37.

(iv)$\Rightarrow$(v) is trivial.

(v)$\Rightarrow$(i) Use the same argument as in (i)$\Rightarrow$(ii), replacing the angelicity by the dual characterization of property C (see, e.g., [Fa˜01, Thm. 12.41]).

The final statement is contained in the proof of (i)$\Rightarrow$(ii). □

In Theorem 5.44, Proposition 5.45, and Theorem 5.51, we will add several other conditions that characterize WLD spaces.

Definition 5.39. *A Banach space X is DENS if $\operatorname{dens} X = w^*\text{-}\operatorname{dens} X^*$.*

Note that, in general, we have $\operatorname{dens} X \geq w^*\text{-}\operatorname{dens} X^*$.

Proposition 5.40. *A WLD Banach space X is DENS.*

Proof. Let $\{x_\gamma; x_\gamma^*\}_{\gamma \in \Gamma}$ be an M-basis in $X \times X^*$. It countably supports X^* (see Theorem 5.37). Let D be a w^*-dense subset of X^* with $\operatorname{card} D = w^*\text{-}\operatorname{dens} X^*$. Given $x^* \in X^*$, let $\operatorname{supp} x^* := \{\gamma \in \Gamma; \langle x_\gamma, x^* \rangle \neq 0\}$. Then $S := \{x_\gamma; \gamma \in \operatorname{supp} d^*,\ d^* \in D\}$ is fundamental in X, and $\operatorname{card} S = \operatorname{card} D$, so we have $\operatorname{dens} X \leq w^*\text{-}\operatorname{dens} X^* \leq \operatorname{dens} X$ and X is DENS. To show that S is fundamental, assume on the contrary that there exists $0 \neq x^* \in X^*$ such that $\langle s, x^* \rangle = 0$ for all $s \in S$. We can find $\gamma \in \Gamma$ such that $\langle x_\gamma, x^* \rangle \neq 0$. Find $d^* \in D$ such that $\langle x_\gamma, d^* \rangle \neq 0$. Then $x_\gamma \in S$ and $\langle x_\gamma, x^* \rangle \neq 0$, a contradiction. □

Proposition 5.41. *If a Banach space $(X, \|\cdot\|)$ has a full PG, then every subspace Y of X has a PRI.*

Proof. Let Φ be a full PG for X. From Lemmas 3.33 and 3.34 together with Lemma 3.36 for mappings, Φ and $\Psi : X \to 2^{X^*}$, where for every $x \in X$, $\Psi(x)$ is a countable 1-norming (for x) subset of S_{X^*}, we see that there exists a 1-complemented subspace Z of X such that $Y \subset Z \subset X$ and $\operatorname{dens} Y = \operatorname{dens} Z$. Obviously, every complemented subspace of X has a full PG, so we may assume from the beginning that $\operatorname{dens} Y = \operatorname{dens} X$. The fact that a PRI $\{P_\alpha; \omega \le \alpha \le \mu\}$ can be defined on X with the additional property that $P_\alpha(B_Y) \subset B_Y$ (and then $P_\alpha(Y) \subset Y$) for all α (see Theorem 3.44) concludes the proof, since then $\{P_\alpha \restriction Y; \omega \le \alpha \le \mu\}$ is the PRI on Y sought. $\qquad\square$

Corollary 5.42. *The class of subspaces of WLD spaces is a $\mathcal{P}$-class (in particular, every subspace of a WLD Banach space has a (strong) M-basis).*

Corollary 5.43. *Every subspace of a WLD Banach space is WLD. In particular, the class of WLD Banach spaces is a $\mathcal{P}$-class.*

Proof. Let X be a WLD Banach space and let Y be a subspace of X. By Corollary 5.42, Y has an M-basis. It is simple to prove that the continuous image $f(K)$ of a compact angelic space K is itself angelic. Indeed, take any set $B \subset f(K)$ and $y \in \overline{B}$. We claim that there is a sequence in B converging to y. If $y \in B$, there is nothing to prove. If not, take a minimal compact $A \subset K$ so that $f(A) = \overline{B}$ and $x \in A$ such that $f(x) = y$. Set $A_0 \subset A$ such that $f(A_0) = B$. We will prove that $x \in \overline{A_0}$. Otherwise, $\overline{A_0}$ is a proper subset of A with $f(\overline{A_0}) = \overline{B}$, a contradiction with the minimality of A. By the angelicity of K, there exists a sequence (x_n) in A_0 converging to x, and so $(f(x_n))$ is a sequence in B converging to y. In particular, (B_{Y^*}, w^*) is angelic. It is now enough to apply Lemma 5.35 and the equivalence (i)$\Leftrightarrow$(ii) in Theorem 5.37. $\qquad\square$

Theorem 5.44 ([GoMo]). *Let X be a Banach space. Then the following are equivalent.*

(i) *X is WLD.*

(ii) *X has a full PG.*

Proof. (i)$\Rightarrow$(ii) has been proved in Theorem 5.36.

(ii)$\Rightarrow$(i) We shall prove this implication by induction on $\operatorname{dens} X$. The separable case is clear. Assume that the theorem has been proved for all Banach spaces with a full projectional generator of density less than μ for some uncountable ordinal μ. Let X be a Banach space of density μ with a full projectional generator Φ. Then X has a projectional resolution of the identity $\{P_\alpha; \omega \le \alpha \le \mu\}$. Given $\omega \le \alpha < \mu$, $(P_{\alpha+1} - P_\alpha)X$ has, as is easy to prove, a full projectional generator. From the induction hypothesis, it is WLD, so it has an M-basis (countably supported by $(P_{\alpha+1}^* - P_\alpha^*)X^*$); see Theorem 5.37. By a standard argument (see, for example, [Fab97, Prop. 6.2.4]), we can glue together all those bases in one single M-basis $\{x_\gamma; x_\gamma^*\}_{\gamma \in \Gamma}$ in $X \times X^*$. We shall prove that this M-basis (in fact, any M-basis in $X \times X^*$) is countably supported by X^*. It will follow then, from Theorem 5.37, that X is WLD.

So select any M-basis $\{x_\gamma; x_\gamma^*\}_{\gamma \in \Gamma}$ in $X \times X^*$. Given $x^* \in X^*$, put $\operatorname{supp} x^* :=$ $\{\gamma \in \Gamma; \langle x_\gamma, x^* \rangle \neq 0\}$. Let $S := \{x^* \in X^* : \operatorname{card} \operatorname{supp} x^* \leq \aleph_0\}$. Then S is a linear subspace of X^*, and S is w^*-dense in X^*, as it contains $\{x_\gamma; \gamma \in \Gamma\}$. We claim that $S \cap B_{X^*}$ is w^*-closed. Once this claim is proved, the Banach-Dieudonné theorem will conclude that S is w^*-closed, and hence $S = X^*$ and so $\{x_\gamma; x_\gamma^*\}_{\gamma \in \Gamma}$ will countably support X^*. Then let $x_0^* \in \overline{S \cap B_{X^*}}^{w^*}$. To prove the claim, choose any $x_1^* \in S \cap B_{X^*}$. Let $\{r_n : n \in \mathbb{N}\}$ be an enumeration of the rational numbers. Given $n \in \mathbb{N}$, put $\mathbb{N}_n := \{1, 2, \ldots, n\}$ and $P_n := (\mathbb{N}_n)^{\mathbb{N}_n}$. Set $\Phi\left(r_1 \frac{x_0^* - x_1^*}{2}\right) := \{x_1^1, x_2^1, \ldots\}$ and choose $x_2^* \in S \cap B_{X^*}$ such that

$$|\langle x_k^1, x_0^* - x_2^* \rangle| < 1/2, \ k = 1, 2.$$

Set

$$\Phi\left(\sum_{i=1}^2 r_{\pi(i)} \frac{x_0^* - x_i^*}{2}\right) := \{x_k^\pi : k = 1, 2, \ldots\}, \quad \pi \in P_2,$$

and choose $x_3^* \in S \cap B_{X^*}$ such that

$$|\langle x_k^\pi, x_0^* - x_3^* \rangle| < 1/3, \ \pi \in P_j, \ j = 1, 2, \ k = 1, 2, 3.$$

Continue in this way to get a sequence $(x_n^*)_{n=1}^\infty$ in $S \cap B_{X^*}$. Let $y^* \in B_{X^*}$ be a w^*-cluster point of the sequence $(x_n^*)_{n=1}^\infty$. Then

$$\frac{x_0^* - x_n^*}{2} \quad w^*\text{-clusters to} \quad \frac{x_0^* - y^*}{2} \in \overline{W}^{w^*},$$

where $W := \operatorname{span}_{\mathbb{Q}} \{\frac{x_0^* - x_n^*}{2} : n \in \mathbb{N}\}$. Choose an arbitrary $x \in \Phi(W)$, say $x \in \Phi\left(\sum_{i=1}^n r_{\pi(i)} \frac{x_0^* - x_i^*}{2}\right)$, for some $n \in \mathbb{N}$ and for some $\pi \in P_n$, so $x = x_k^\pi$ for some $k \in \mathbb{N}$. Now, given $m_0 \geq \max\{k, n-1\}$ and $m \geq m_0$, we get $|\langle x, x_0^* - x_m^* \rangle| < \frac{1}{m}$, hence $|\langle x, x_0^* - y^* \rangle| = 0$. It follows that $x_0^* - y^* = 0$ from the definition of projectional generator. Each x_n^* has countable support on Γ, and so does $y^* (= x_0^*)$ and $x_0^* \in S$. $\qquad \square$

The next proposition adds two equivalent conditions to those in Theorems 5.37 and 5.44 for a Banach space to be WLD. Some other equivalent conditions will be given in Theorem 5.51.

Proposition 5.45. *Given an M-basis $\{x_\gamma; x_\gamma^*\}_{\gamma \in \Gamma}$ in a Banach space X, then the following are equivalent.*

(i) *X is WLD.*
(ii) *$\operatorname{span}\{x_\gamma^*; \ \gamma \in \Gamma\}$ is sequentially dense in $\left(X^*, w(X^*, \operatorname{span}\{x_\gamma; \gamma \in \Gamma\})\right)$.*
(iii) *$\operatorname{span}\{x_\gamma^*; \ \gamma \in \Gamma\}$ is countably dense in (X^*, w^*), i.e., every element of X^* is in the closure in this topology of a countable set in $\operatorname{span}\{x_\gamma^*; \ \gamma \in \Gamma\}$.*

If an M-basis in X has one of the properties (ii) or (iii) above, then all M-bases in X share the same property.

Proof. (i)⇒(ii) Let $\{x_\gamma; x_\gamma^*\}_{\gamma \in \Gamma}$ be any M-basis in $X \times X^*$. Fix any $x^* \in X^*$. Enumerate the set $\{\gamma \in \Gamma; \ \langle x_\gamma, x^* \rangle \neq 0\}$ as $\{\gamma_1^o, \gamma_2^o, \ldots\}$. Find $x_1^* \in \operatorname{span}\{x_\gamma^*; \ \gamma \in \Gamma\}$ so that $|\langle x_{\gamma_1^o}, x^* - x_1^* \rangle| < 1$. Enumerate $\{\gamma \in \Gamma; \ \langle x_\gamma, x_1^* \rangle \neq 0\}$ by $\{\gamma_1^1, \gamma_2^1, \ldots\}$. Find $x_2^* \in \operatorname{span}\{x_\gamma^*; \ \gamma \in \Gamma\}$ so that $|\langle x_{\gamma_1^o}, x^* - x_2^* \rangle| < \frac{1}{2}$, $|\langle x_{\gamma_2^o}, x^* - x_2^* \rangle| < \frac{1}{2}$, $|\langle x_{\gamma_1^1}, x^* - x_2^* \rangle| < \frac{1}{2}$, and $|\langle x_{\gamma_2^1}, x^* - x_2^* \rangle| < \frac{1}{2}$. Assume that for some $i \in \mathbb{N}$ we found x_j^* with "support" on Γ given by $\{\gamma_1^j, \gamma_2^j \ldots\}$, $j = 1, 2, \ldots, i$. Then find $x_{i+1}^* \in \operatorname{span}\{x_\gamma^*; \ \gamma \in \Gamma\}$ so that $\left| \langle x_{\gamma_l^j}, x^* - x_{i+1}^* \rangle \right| < \frac{1}{i+1}$ for all $j = 0, 1, \ldots, i$ and $l = 1, 2, \ldots, i$. Then we can easily see that $\langle x_\gamma, x^* - x_i^* \rangle \to 0$ as $i \to \infty$ for every $\gamma \in \Gamma$, and (ii) is proved.

(ii)⇒(iii)$_p$ is obvious, where (iii)$_p$ is the statement (iii) for the topology $w(X^*, \operatorname{span}\{x_\gamma; \gamma \in \Gamma\})$.

(iii)$_p$ ⇒(i) If (iii)$_p$ holds, the set $\{x_\gamma; \gamma \in \Gamma\}$ countably supports X^*. It follows from Theorem 5.37 that X is WLD.

(i)⇒(iii) Let Y denote the set of all $x^* \in X^*$ That lie in the weak*-closure of a countable subset of $\operatorname{span}\{x_\gamma^*; \ \gamma \in \Gamma\}$. We want to show that $Y = X^*$. Clearly, Y is linear. Let ξ be any element of the weak*-closure of B_Y. (i) guarantees that (B_{X^*}, w^*) is a Corson compact, and hence ξ can be reached as the weak*-limit of a sequence $(x_i^*)_{i=1}^\infty$ in B_Y. Now, for every $i \in \mathbb{N}$, we can find a suitable at most countable set $C_i \subset \operatorname{span}\{x_\gamma^*; \ \gamma \in \Gamma\}$ so that x_i^* lies in the weak*-closure of C_i. Then ξ lies in the (at most countable) set $\bigcup_{i=1}^\infty C_i$, and so $\xi \in Y$. Now, the Banach-Dieudonné theorem guarantees that Y is weak*-closed. But Y contains $\{x_\gamma^*; \gamma \in \Gamma\}$. Therefore $Y = X^*$.

(iii)⇒(iii)$_p$ is obvious.   □

Definition 5.46. *We say that X is a* Plichko space *if X^* has a 1-norming Σ-subspace.*

Observe that the defining condition for WLD spaces (Definition 3.32) is of isomorphic nature, while for Plichko spaces it is isometric. It is also easy to observe from this definition and Theorem 5.37 that the class of Plichko spaces contains the class of WLD spaces and that Plichko spaces have a dual ball that, endowed with its w^*-topology, is a Valdivia compact. The class WLD contains all Vašák spaces (Theorem 6.25) and henceforth all WCG spaces. The class of Plichko spaces is strictly (check $\ell_1(\Gamma)$ for uncountable Γ) larger than the class of WLD spaces. An M-basis characterization of Plichko spaces is given in Theorem 5.63.

Remark 5.47. Kubiś [Kub] has recently constructed a Banach space of density ω_1, that is not a Plichko space under any renorming (this space even enjoys the SCP property but admits no PRI under any equivalent norm; see Proposition 5.48) and is a subspace of a Plichko space. In particular, the Plichko class is not closed with respect to taking subspaces.

Proposition 5.48. *Let* $\operatorname{dens} X = \omega_1$. *If X admits a PRI, then X is a Plichko space.*

Proof. By the regularity of ω_1, for every $x \in X$ there exists some $\alpha < \omega_1$ such that $x = P_\alpha(x)$. Thus $X = \bigcup_{\alpha < \omega_1} P_\alpha(X)$. Consequently, using the M-basis constructed in Proposition 5.6, we see that $\bigcup_{\alpha < \omega_1} P^*(X^*)$ is a 1-norming Σ-subset of B_{X^*}. □

The next statement follows from the fact that every $(C[0, \alpha], \|\cdot\|_\infty)$, with α an ordinal, has a PRI [DGZ93a, Examp. VI.8.6].

Example 5.49. $(C[0, \omega_1], \|\cdot\|_\infty)$ is a Plichko space that is not WLD and does not contain a copy of ℓ_1.

Example 5.50. Let $X = (C_0[0, \omega_1], \|\cdot\|_\infty)$. Then (B_{X^*}, w^*) is not a Valdivia compact. In particular, X has no PRI.

Proof. We have $\delta_\alpha \xrightarrow{w^*} 0$ as $\alpha \to \omega_1$. Also, $\frac{1}{2}(\delta_{n+1} - \delta_n) \xrightarrow{w^*} 0$ as $n \to \omega$. Since $M = \{\delta_\alpha\}_{\omega < \alpha < \omega_1} \cup \{\frac{1}{2}(\delta_{n+1} - \delta_n)\}_{n \in \omega} \cup \{0\}$ belong to the sequential closure of the G_δ points of (B_{X^*}, w^*) (Lemma 5.29), $M \subset A$ for all Σ-subsets A of (B_{X^*}, w^*). However, 0 is not contained in the sequential closure of $\{\delta_\alpha\}_{\omega < \alpha < \omega_1}$, which is a contradiction with the existence of any Σ-subset A. □

Kalenda proved in [Kal00a] that *there is an equivalent renorming of the space* $X := C[0, \omega_1]$, *so that* (B_{X^*}, w^*) *is a Valdivia compact, but* X *is not a Plichko space.*

Theorem 5.51 (Kalenda [Kal00a], [Kal00b]). *The following conditions are equivalent for a Banach space* $(X, \|\cdot\|)$:

(i) $(X, \|\cdot\|)$ *is WLD.*

(ii) $(X, \|\|\cdot\|\|)$ *has, for every equivalent norm* $\|\|\cdot\|\|$, *a countably 1-norming M-basis.*

(iii) $(B_{(X^*, \|\|\cdot\|\|)}, w^*)$ *is a Valdivia compact for every equivalent norm* $\|\|\cdot\|\|$.

Under the assumption dens $X = \omega_1$, *the conditions above are also equivalent to the following:*

(iv) $(X, \|\|\cdot\|\|)$ *has PRI for every equivalent norm* $\|\|\cdot\|\|$.

Proof. (i)⇒(ii) follows by a simple argument from part (v) in Theorem 5.37. In fact, every weakly Lindelöf M-basis is 1-countably norming.

(ii)⇒(iii) is trivial.

The main result is (iii)⇒(i). We start with a lemma.

Lemma 5.52. *Let* $(X, \|\cdot\|)$ *be a Banach space such that* (B_{X^*}, w^*) *is a Valdivia compact that is not a Corson compact. Then there exists a* w^*-*compact and convex set* $L \subset B_{X^*}$ *that is not a Valdivia compact.*

Proof. By Theorem 5.31, there exists $H := \{f_\alpha\}_{0 \le \alpha \le \omega_1} \subset (B_{X^*}, w^*)$ which is homeomorphic to $[0, \omega_1]$. We may without loss of generality assume that $f_{\omega_1} = 0$. Moreover, fix a Schauder basic sequence $\{e_k\}_{k=1}^\infty \subset X$. Since $e_k(f_\alpha) \to 0$ as $\alpha \to \omega_1$, we have that $e_k(f_\alpha) = 0$ for all sufficiently large $\alpha < \omega_1$. By passing to a subsequence of $\{f_\alpha\}_{0 \le \alpha < \omega_1}$, we have without loss of generality that $e_k(f_\alpha) = 0$ for all $k \in \mathbb{N}, \alpha \in [0, \omega_1]$. Thus $\overline{\mathrm{conv}}^{w^*} H \subset \overline{\mathrm{span}}\{e_k; k \in \mathbb{N}\}^\perp = Z$ is contained in a w^*-closed subspace $Z \hookrightarrow X^*$ of infinite codimension. Choose by a standard argument a sequence $\{g_k\}_{k=1}^\infty \subset X^* \setminus Z$, $\|g_k\| \le \frac{1}{2^k}$, such that $g_k(e_l) = 0$ if and only if $k \ne l$. Then $\{g_k\}_{k=1}^\infty \cup \{0\}$ are extremal points of the (norm) compact and convex set $\overline{\mathrm{conv}}(\{0\} \cup \{g_k\}_{k=1}^\infty)$. Moreover, g_k are strongly exposed by e_k. Let $L = \overline{\mathrm{conv}}^{w^*}(\{g_k\}_{k=1}^\infty \cup \{f_\alpha\}_{0 \le \alpha \le \omega_1})$. Note that if $C \subset (X^*, w^*)$ is a scattered compact, then every element of $\overline{\mathrm{conv}}^{w^*} C = \overline{\mathrm{conv}} C$ is representable by a Radon probability measure on C, which can be expressed as an element of $\ell_1(C)$ with positive coordinates of sum 1. In particular, if C is countable, we have that $\overline{\mathrm{conv}}^{w^*} C$ is norm separable. We apply this observation in the following way. For every $\beta < \omega_1$, $L_\beta = \overline{\mathrm{conv}}^{w^*}(\{g_k\}_{k=1}^\infty \cup \{f_\alpha\}_{\alpha < \beta})$ is norm separable and thus metrizable, and thus there exists $\beta < \gamma_\beta < \omega_1$ such that $\omega_1 > \xi \ge \gamma_\beta$ implies $f_\xi \notin \mathrm{conv} L_\beta$. Indeed, assuming the contrary, an uncountable sequence $\{f_\xi\} \subset \mathrm{conv} L_\beta$ would contain a subsequence converging to f_{ω_1}, which is a contradiction. By the separation theorem, choose $x_\beta \in X$, $0 \le \sup x_\beta(L_\beta) = a_\beta < b_\beta = x_\beta(f_{\gamma_\beta})$. Since $\lim_{\alpha \to \omega_1} x_\beta(f_\alpha) = 0$, we have that $x_\beta(f_\alpha) = 0$ for all α large enough. Passing to a transfinite subsequence of H, based on the above, we may without loss of generality assume that for every nonlimit ordinal α, there exist an $x_\alpha \in X$ and $0 \le a_\alpha < b_\alpha \in \mathbb{R}$ such that $x_\alpha(\mathrm{conv}(\{g_k\}_{k=1}^\infty \cup \{f_\beta\}_{\beta < \alpha})) \le a_\alpha$, $x_\alpha(f_\alpha) = b_\alpha$, $x_\alpha(f_\gamma) = 0$ for all $\gamma > \alpha$. Thus, in particular, all f_α, α nonlimit ordinals, are extremal points of L, strongly exposed by x_α, so they are also norm (and so w^*)-G_δ points. The same is true for all g_k since their strongly exposing functionals $e_k \in X$ satisfy $e_k(\overline{\mathrm{span}}\{f_\alpha; 0 \le \alpha < \omega_1\}) = 0$. We claim that L is not a Valdivia compact. Suppose the contrary, and let $A \subset L$ be a dense Σ-subset. By (2) in Lemma 5.29, we have that $\{g_k\}_{k=1}^\infty \cup \{f_\alpha\}_{0 \le \alpha < \omega_1} \subset A$, and since $g_k \to f_{\omega_1}$, it follows that $f_{\omega_1} \in A$. Put $C = \{f_\alpha\}_{0 \le \alpha < \omega_1} \subset A$. Then $f_{\omega_1} \in \overline{C}$, but no ω-sequence from C converges to f_{ω_1}. This contradicts (2) in Lemma 5.29. $\square$

We continue with the proof of Theorem 5.51. By the same reasoning as above, we may without loss of generality assume that $L \subset \mathrm{Ker}\,(e)$ for some $e \in S_X$. Let $h \in B_{X^*}$ be such that $h(e) = 1$. Consider the set

$$B := \overline{\mathrm{conv}}^{w^*}\left((L + h) \cup (-L - h) \cup \frac{1}{2} B_{X^*}\right)$$
$$= \mathrm{conv}\left((L + h) \cup (-L - h) \cup \frac{1}{2} B_{X^*}\right).$$

Then B is a convex symmetric w^*-compact set such that $\frac{1}{2} B_{X^*} \subset B \subset 2 B_{X^*}$, so there is an equivalent norm $|\cdot|$ on X such that B is its dual unit ball. It remains to show that B is not a Valdivia compact. To this end, it suffices to prove that $\{h+g_k\}_{k=1}^{\infty} \cup \{h+f_\alpha\}_{\alpha<\omega_1}$ are G_δ points of B. Let $\{U_n\}_{n=1}^{\infty}$ be a sequence of w^*-open sets such that $f_\alpha \in U_n$ and $U_n \cap L$ are shrinking to $f_\alpha \in L$. We claim that for a fast enough growing sequence $\{N(n)\}_{n=1}^{\infty}$, the sequence of w^*-open sets $V_n = \{f; e(f) > 1 - \frac{1}{N(n)}\} \cap (h+U_n)$ is a shrinking sequence of open neighborhoods of $h+f_\alpha \in B$. We can without loss of generality assume that $\overline{U_{n+1}}^{w^*} \cap L \subset U_n \cap L$ for all $n \in \mathbb{N}$. In particular, for some $\varepsilon_n \searrow 0$, we have $(U_{n+1} + \varepsilon_n B_{X^*}) \cap L \subset U_n \cap L$. Let $f^n = \alpha_1 f_1^n + \alpha_2 f_2^n + \alpha_3 f_3^n \in V_{n+1} \cap B$, where $\alpha_i \geq 0$, $\sum_{i=1}^{3} \alpha_i = 1$, $f_1^n \in h+L$, $f_2^n \in \frac{1}{2}B_{X^*}$, $f_3^n \in -h-L$. We have $\alpha_1 \to 1$ as $N(n) \to \infty$. Thus, if $N(n+1)$ is large enough, we have, from $f_1^n = f^n + (1-\alpha_1)f_1^n - \alpha_2 f_2^n - \alpha_3 f_3^n$, that $\|f_1^n - f^n\| < \frac{\varepsilon_n}{3}$, and so

$$f_1^n \in (h+L) \cap h + U_{n+1} + \frac{\varepsilon_n}{3} B_{X^*} \subset (h+L) \cap (h+U_n).$$

In particular, $w^*\text{-}\lim_{n\to\infty} f_1^n \to f$, and so $w^*\text{-}\lim_{n\to\infty} f^n \to f$ as claimed. The case of $h+g_k$ is similar. The rest of the proof follows again by Lemma 5.29.

(i)$\Rightarrow$(iv) is trivial.

(iv)$\Rightarrow$(iii) follows from the fact that in this case, and for a PRI $\{P_\alpha; \omega \leq \alpha \leq \omega_1\}$, the subspace $(P_{\alpha+1} - P_\alpha)X$ is separable for all $\alpha < \omega_1$; then (B_{X^*}, w^*) is a Valdivia compact. $\square$

Note that there exist examples [Kal00a] of non-WLD spaces of density $\aleph_2$ with PRI's under every equivalent renorming.

Example 5.53 (Plichko [Plic81a]). There is a Banach space X failing SCP such that (B_{X^*}, w^*) is angelic compact but not Corson compact.

Proof (Sketch). Recall [Fa~01, Exer. 6.49–6.52] that the James tree space JT is a separable Banach space not containing a copy of ℓ_1, with a predual JT_*. For its dual, we have $JT^*/JT_* \cong \ell_2(c)$ and $JT^{**} \cong JT \oplus \ell_2(c)$. We set $X = JT^*$. By [BFT78], (B_{X^*}, w^*) is angelic. If (B_{X^*}, w^*) were Corson, then, by Theorem 5.36 and Theorem 3.42, X would have SCP. Thus there would have to exist a separable and complemented space $Y \hookrightarrow X$, $JT_* \hookrightarrow Y$, such that $X/Y \cong \ell_2(c)$, and so $X \cong Y \oplus \ell_2(c)$. This, however, is a contradiction with the fact that X^* (as a bidual to a separable space) is w^*-separable. $\square$

5.5 $C(K)$ Spaces that Are WLD

In this section, we are going to investigate the structure of $C(K)$ spaces when K is a Corson or Valdivia compact. It turns out that, analogously to the existence of uncountable biorthogonal systems in general Banach spaces, the structure of these spaces also depends on additional set-theoretical axioms.

Definition 5.54. *Let K be a compact set and μ be a nonnegative Radon measure on K. We define the* support *of μ by*

$$\operatorname{supp}(\mu) := \{x \in K; \mu(U) > 0 \text{ for all open sets } U \text{ containing } x\}.$$

For a general Radon measure μ, we define $\operatorname{supp}(\mu) = \operatorname{supp}(\mu^+) \cup \operatorname{supp}(\mu^-)$, where μ^+ and μ^- are, respectively, the positive and negative parts of the measure μ.

It is standard to check that $\operatorname{supp}(\mu)$ is a compact set and that $\int f d\mu = 0$ for every $f \in C(K)$, $\operatorname{supp}(f) \cap \operatorname{supp}(\mu) = \emptyset$. Let μ be a positive Radon measure on K, $L = \operatorname{supp}(\mu)$. It is clear that for every nonempty and open $U \subset L$, $\mu(U) > 0$. Thus L has the CCC property; i.e., every system $\{U_\alpha\}_{\alpha \in A}$ of disjoint nonempty and open subsets of L is at most countable. Indeed, note that $\mu(L) \geq \sum_{\alpha \in A} \mu(U_\alpha)$, and the right-hand summation can be at most countable.

We say that a compact K *admits* (or *supports*) a strictly positive measure if there exists a Radon measure μ on K such that $\operatorname{supp}(\mu) = K$. As remarked, a necessary condition for K to admit a strictly positive measure is the property CCC.

Theorem 5.55 (Kalenda [Kal00a]). *Let K be a Valdivia compact and let A be a dense Σ-subset of K. Then the set*

$$S := \{\mu \in C(K)^*; \operatorname{supp}(\mu) \text{ is a separable subset of } A\}$$

is a 1-norming Σ-subspace of $C(K)^*$. In particular, if K is a Valdivia compact, then $C(K)$ is a Plichko space.

Proof. Let $h : K \to \mathbb{R}^\Gamma$ be a homeomorphic injection with $h(A) = h(K) \cap \Sigma(\Gamma)$. For $\gamma \in \Gamma$, let $f_\gamma = \pi_\gamma \circ h$, where π_γ denotes the projection of $\mathbb{R}^\Gamma$ onto the γ-coordinate. The family $\{f_\gamma; \gamma \in \Gamma\}$ separates the points of K. Let $\tilde{\Gamma}$ be the set of all (possibly empty) finite sequences of elements of Γ. For $\tilde{\gamma} \in \tilde{\Gamma}$, let us define

$$g_{\tilde{\gamma}} := \begin{cases} 1_K & \text{if } \tilde{\gamma} = \emptyset, \\ \prod_{i=1}^n f_{\gamma_i} & \text{if } \tilde{\gamma} = (\gamma_1, \ldots, \gamma_n). \end{cases}$$

By the Stone-Weierstrass theorem, $\overline{\operatorname{span}}\{g_{\tilde{\gamma}}; \tilde{\gamma} \in \tilde{\Gamma}\} = C(K)$, and hence the family $\{g_{\tilde{\gamma}}; \tilde{\gamma} \in \tilde{\Gamma}\}$ separates points of $C(K)^*$. We define a linear w^*-continuous injection $\tilde{h} : C(K)^* \to \mathbb{R}^{\tilde{\Gamma}}$ by the formula $\tilde{h}(\mu)(\tilde{\gamma}) = \langle \mu, g_{\tilde{\gamma}} \rangle$. Put $S = \tilde{h}^{-1}(\Sigma(\tilde{\Gamma}))$. This is clearly a Σ-subspace of $C(K)^*$. Moreover, it contains the Dirac measure δ_x for every $x \in A$. Indeed, if for $\tilde{\gamma} \in \tilde{\Gamma}$ $g_{\tilde{\gamma}}(x) \neq 0$, then either $\tilde{\gamma} = \emptyset$ or $\tilde{\gamma} = (\gamma_1, \ldots, \gamma_n)$, where $f_{\gamma_i}(x) \neq 0$, $i = 1, \ldots, n$. The set of such $\tilde{\gamma}$ is countable. It follows that S is 1-norming.

It remains to prove that S coincides with all Radon measures with separable support on A. Let $\mu \in C(K)^*$ have a separable support $\operatorname{supp}(\mu) \subset A$. Hence $F = \operatorname{supp}(\mu)$ is a separable Corson compact set, and it is metrizable since it is homeomorphic to a subset of $\mathbb{R}^{\mathbb{N}}$. Without loss of generality, we may assume that μ is a probability. The topology w^* on $P(F)$ (the probability measures on F) coincides with the restriction of the topology w^* on $P(K)$ to $P(F)$. By the metrizability of $(S_{C(F)^*}, w^*)$ and the w^*-density of finite linear combinations of Dirac measures there, there exists a sequence μ_n of finite convex combinations of Dirac functionals such that $\mu_n \to \mu$ in $P(K)$. It follows that $\mu \in S$ since $\sum(\tilde{\Gamma})$ is countably closed.

In order to prove the opposite inclusion, put $S' = \overline{\text{span}}\{\delta_x; x \in A\}$. Then $S' \subset S$ and S' is 1-norming. It follows that $S' \cap B_{C(K)^*}$ is w^*-dense in $B_{C(K)^*}$. In particular, every $\mu_0 \in S \cap B_{C(K)^*}$ lies in the w^*-closure of $S' \cap B_{C(K)^*}$. By the Fréchet-Urysohn property of S, there exists a sequence $\mu_n \in S' \cap B_{C(K)^*}$ w^*-convergent to μ_0. In order to establish the support of μ_0, note that trivially the set C of all $x \in K$ such that $\mu_n(x) \neq 0$ for some $n \in \mathbb{N}$ is at most countable. Using the definition of support, μ_0 is supported by $\overline{C}$. Again, $\overline{C} \subset A$ by angelicity. Now $\overline{C}$ is a separable Corson compact set. Thus, it is metrizable. It follows that $\text{supp}(\mu_0) \subset \overline{C}$ is separable as well. □

Definition 5.56. *We say that a Corson compact K has property* (M) *if the support of every Radon probability measure on K is separable.*

Theorem 5.57 (Argyros, Mercourakis and Negrepontis [AMN89]). *The following are equivalent for a compact space K:*

(i) K *is a Corson compact with property* (M).
(ii) $C(K)$ *is WLD.*

Proof. (i)⇒(ii) follows directly from Theorem 5.55 and the Riesz representation theorem.

(ii)⇒(i) $(B_{C(K)^*}, w^*)$ is a Corson compact, so $K \subset (B_{C(K)^*}, w^*)$ is a Corson compact as well. Let S be the 1-norming Σ-subspace defined in the proof of Theorem 5.55. As $(B_{C(K)^*}, w^*)$ is angelic, $(B_{C(K)^*}, w^*)$ is the sequential closure of $S \cap B_{C(K)^*}$, so $S = B_{C(K)^*}$. Hence K has property (M). □

Under CH, there exists a Corson compact K failing (M). The first such example was constructed by Kunen in 1975 and published in [Kun81]. We refer to [Negr84] for related examples. Below we present a simpler construction, based on a space of Erdős, due to Argyros, Mercourakis, and Negrepontis [AMN89]. Let us start by making some simple remarks. Recall that a topological space (T, τ) has *caliber* Γ if, for any family $\{U_\alpha\}_{\alpha < \Gamma}$ of nonempty open sets in T, there exists $A \subset \Gamma$ with card $(A) = \Gamma$ such that $\bigcap_{\alpha \in A} U_\alpha \neq \emptyset$.

Fact 5.58. *The following are equivalent for a Corson compact K:*

(i) K *is separable.*
(ii) K *is metrizable.*
(iii) K *has caliber* ω_1.

Proof. (i)⇔(ii) is in [Fa˜01, p. 428].

(i)$\Rightarrow$(iii) is immediate. Indeed, choosing a $\{x_n\}_{n=1}^{\infty} \subset K$ dense, it is clear that at least one of the sets $A_n = \{\alpha; x_n \in U_\alpha\}$ is uncountable.

(iii)$\Rightarrow$(i) Consider $K \subset \Sigma(\Gamma)$ such that for every $\alpha \in \Gamma$ there exists $x \in K$ for which $x_\alpha \neq 0$. Setting $U_\alpha := \{x \in K; x_\alpha \neq 0\}$, for $\alpha \in \Gamma$, we get that card $\Gamma = \omega$. $\square$

Let $I = [0,1]$ and λ be a Lebesgue measure on I. The Boolean algebra M_λ/N_λ, where M_λ is the Boolean algebra of λ measurable sets and N_λ is the ideal of null sets, has a corresponding Stone space Ω, $i : M_\lambda/N_\lambda \to \Omega$. This is a compact totally disconnected space [Dies84, p. 78], [Wal74, p. 51], that inherits a unique Radon measure $\tilde{\lambda}$ determined by the condition $\tilde{\lambda}(V) = \lambda(U)$, where $V \subset \Omega$ is clopen, $U \subset I$ is measurable, and $V = i(U)$. Note that $\tilde{\lambda}$ is a strictly positive measure on Ω. The space Ω is known as the *Erdős space*. A system of open sets $\{O_\alpha\}$ in a topological space T is called a *pseudobase* if for every open $U \subset T$ there exists $O_\alpha \subset U$.

Lemma 5.59 (Erdős; see [AMN89] (CH)).   *The Erdős space Ω has a pseudobase $\{V_\alpha\}_{\alpha<\omega_1}$, witnessing that Ω fails to have caliber ω_1.*

Proof. Let $\{x_\alpha\}_{\alpha<\omega_1}$ be a well-ordering of I and $\{K_\alpha\}_{\alpha<\omega_1}$ be the set of all compact subspaces of I with $\lambda(K_\alpha) > 0$. For every $\alpha < \omega_1$, we choose a compact $U_\alpha \subset I$ such that

(a) $U_\alpha \subset \{x_\beta; \alpha < \beta < \omega_1\}$.
(b) $U_\alpha \subset K_\alpha$.
(c) $\lambda(U_\alpha) > 0$.

Such a choice is clearly possible using the standard regularity properties of λ. The family of clopen sets $\{V_\alpha\}_{\alpha<\omega_1}$ representing $\{U_\alpha\}_{\alpha<\omega_1}$ in Ω is the pseudobase that witnesses the failure of caliber ω_1. Indeed, let $V = i(U)$ be clopen in Ω, where $U \subset I$. Since $\tilde{\lambda}(V) = \lambda(U) > 0$, there exists $\alpha_0 < \omega_1$ with $K_{\alpha_0} \subset U$, and therefore $U_{\alpha_0} \subset U$. Thus $V_{\alpha_0} \subset V$.

Now if $\bigcap_{\alpha \in A} V_\alpha \neq \emptyset$ for some uncountable set $A \subset \omega_1$, then the family $\{U_\alpha\}_{\alpha \in A}$ has the finite intersection property. But this family consists of compact subsets of I, so $\bigcap_{\alpha \in A} U_\alpha \neq \emptyset$, which is impossible according to (a). $\square$

Theorem 5.60 (Kunen [Kun81] (CH)).   *There exists a nonseparable Corson compact space L with property CCC that fails property (M). The space L supports a strictly positive Radon measure.*

Proof ([AMN89]). Set $\mathcal{A} = \{A \subset \omega_1; \bigcap_{\alpha \in A} V_\alpha \neq \emptyset\}$. Using Lemma 5.59, we see that $\mathcal{A}$ is an adequate family (i.e., closed with respect to taking subsets) of countable subsets of ω_1. We claim that $\mathcal{A} \subset \Sigma([0,\omega_1)) \cap \{0,1\}^{[0,\omega_1)}$ is a closed subset of $\{0,1\}^{[0,\omega_1)}$. Indeed, a compactness argument gives that for $B \in \{0,1\}^{[0,\omega_1)} \setminus \mathcal{A}$ there exists a finite set $\{\beta_1, \dots, \beta_l\} \subset B$ such that $\bigcap_{j=1}^{l} V_{\beta_j} = \emptyset$. Thus $\mathcal{A} \subset \{0,1\}^{[0,\omega_1)}$ represents a Corson compact denoted by K.

We define a continuous mapping $T : \Omega \to K$ by

$$T(x)(\alpha) := \begin{cases} 1, & \text{for } x \in V_\alpha, \\ 0, & \text{for } x \notin V_\alpha, \end{cases}$$

and set $L = T(\Omega) \subset K$. We will prove that L is the desired Corson compact. Clearly, $\{T(V_\alpha)\}_{\alpha < \omega_1}$ is a pseudobase of L, and $T(\tilde{\lambda})$ is a strictly positive measure on L. It remains to show that L is not separable. It suffices to show that $\{T(V_\alpha)\}_{\alpha < \beta}$ is not a pseudobase of L for any $\beta < \omega_1$. The last fact follows from the existence of a Borel set $Z \subset I$, $\lambda(Z) > 0$, and such that $\lambda(U_\alpha \setminus Z) > 0$ for all $\alpha < \beta$. Indeed, we have $U_\gamma \subset Z$ for some $\gamma > \beta$, and so $T(V_\gamma)$ is an open set in L such that $T(V_\alpha) \nsubseteq T(V_\gamma)$ for all $\alpha < \beta$. Lastly, the existence of Z is clear; choose for instance $Z := I \setminus \bigcup_{\alpha < \beta} S_\alpha$, where $S_\alpha \subset U_\alpha$ is a compact set chosen so that $0 < \lambda(S_\alpha) < \varepsilon_\alpha$, where $\sum_{\alpha < \beta} \varepsilon_\alpha < \frac{1}{2}$.   □

Corollary 5.61 ((CH)). *There exists a Corson compact L of weight ω_1 such that $C(L)$ has an equivalent renorming without PRI.*

Proof. Use the space in Theorem 5.60. Since $C(L)$ is not WLD (Theorem 5.57), by Theorem 5.51 there exists a renorming $(C(L), ||| \cdot |||)$ for which the dual is not a Valdivia compact, and so by Proposition 5.48, $(C(L), ||| \cdot |||)$ has no PRI.   □

The next theorem is a consequence of some independent work of Archangelskii, Šapirovskii, and Kunen; we refer to [Frem84].

Theorem 5.62 (Archangelskii, Šapirovskii, and Kunen; see [Frem84] (MA_{ω_1})**).** *Every Corson compact K has property (M). Consequently, $C(K)$ has PRI for every equivalent renorming.*

Proof ([AMN89]). By Fact 5.58, it suffices to show that every Corson compact K with property CCC has a caliber ω_1. Let $\{U_\alpha\}_{\alpha < \omega_1}$ be a system of nonempty open subsets of K. We put $V_\alpha = \overline{\bigcup_{\alpha \le \gamma < \omega_1} U_\gamma}$. Clearly, $V_\beta \subseteq V_\alpha$ whenever $\alpha < \beta$. The CCC condition implies that for some $\alpha < \omega_1$ we have $V_\beta = V_\alpha$ whenever $\beta > \alpha$. Indeed, if $V_\beta \subsetneq V_\alpha$, then there exists an open set $H \subset V_\alpha \setminus V_\beta$ because $(K \setminus V_\beta) \cap \bigcup_{\alpha \le \gamma < \omega_1} U_\gamma \ne \emptyset$ and the latter union is an open set contained in V_α. Due to property CCC, this proper inclusion can happen at most countably many times. Thus $\mathcal{U}_\beta = \bigcup_{\beta \le \gamma < \omega_1} U_\gamma$, where $\beta > \alpha$, is a system of open and dense subsets of the CCC compact V_α. Using the topological version of MA_{ω_1} (Fact 4.43), there exists $x \in \bigcap_{\alpha < \beta < \omega_1} \mathcal{U}_\beta$, and so there exist uncountable $A \subset [\alpha, \omega_1)$ such that $x \in U_\gamma$ for every $\gamma \in A$.   □

5.6 Extending M-bases from Subspaces

The main topic of the present section is that WLD spaces behave nicely with respect to M-basis extensions from their subspaces. We also include the complementation result for $c_0(\Gamma)$ in a WLD overspace when the cardinality of Γ

is less than $\aleph_\omega$. As an application, let us mention that a subspace on which an M-basis can be extended to the whole space is quasicomplemented in the overspace; this topic will be investigated in the next section.

Theorem 5.63 (Valdivia [Vald91]). *The following are equivalent for a Banach space X:*

(i) *X is Plichko.*
(ii) *There is a set $M \subset X$ such that $\overline{\mathrm{span}}(M) = X$ and that $W := \{x^* \in X^*; \mathrm{card}\,\{x; x \in M, \langle x, x^* \rangle \neq 0\} \leq \omega\}$ is a 1-norming subspace of X^*.*
(iii) *X has a countably 1-norming M-basis.*

Moreover, Plichko spaces form a $\mathcal{P}$-class.

Proof. (i)$\Rightarrow$(ii) By assumption, there exists $T : X^* \to \mathbb{R}^\Gamma$, a linear, one-to-one, and w^*-continuous mapping, and a 1-norming linear space $S \hookrightarrow X^*$, such that $T(S) \subset \sum(\Gamma)$. Let $c_\gamma : \mathbb{R}^\Gamma \to \mathbb{R}$ be the γ-th coordinate function, $\gamma \in \Gamma$. Then $x_\gamma := c_\gamma \circ T \in X$ for all $\gamma \in \Gamma$. It suffices to put $M = \{x_\gamma\}_{\gamma \in \Gamma}$. It is clear that $\{\gamma; \langle x_\gamma, x^* \rangle \neq 0\}$ is at most countable for every $x^* \in S$, so $S \subset W$ and W is 1-norming. If $\overline{\mathrm{span}}(M) \neq X$, there would have to exist a nonzero functional $x^* \in B_{X^*}$, $x^* \upharpoonright M \equiv 0$. Then $Tx^* = 0$ and, by the injectivity of T, $x^* = 0$, a contradiction.

(ii)$\Rightarrow$(iii) For every $w \in W$, put $\Phi(w) = \{m \in M; \langle m, w \rangle \neq 0\}$. Then (W, Φ) is a projectional generator. By Theorem 3.42, there exists a PRI $\{P_\alpha; \omega \leq \alpha \leq \mu\}$ on X satisfying $M \subset \bigcup_{\omega \leq \alpha < \mu} X_\alpha$, where $X_\alpha := (P_{\alpha+1} - P_\alpha)X$ for $\omega \leq \alpha < \mu$. Let $M_\alpha := M \cap X_\alpha$. Then $X_\alpha = \overline{\mathrm{span}}(M_\alpha)$ by the density of M in X. Thus $W_\alpha := \{w \upharpoonright X_\alpha; w \in W\}$ is a 1-norming Σ-subspace of X_α^* witnessed by the operator $T_\alpha : X_\alpha^* \to \mathbb{R}^{M_\alpha}$ given by $T_\alpha(x^*) := x^* \upharpoonright M_\alpha$, $x^* \in X_\alpha^*$, and this happens for all $\omega \leq \alpha < \mu$. This implies that all X_α are Plichko spaces, and so Plichko spaces form a $\mathcal{P}$-class. To finish the implication, it suffices to use transfinite induction on the density character of X: M-bases are constructed in each X_α with the property sought; it is enough to glue all the M-bases together.

(iii)$\Rightarrow$(i) is clear. $\qquad\square$

Theorem 5.64 (Valdivia [Vald91], [Vand95]). *Let X be a Plichko space with a WLD subspace $Y \hookrightarrow X$. Any M-basis of Y can be extended to an M-basis of X. If, in addition, dens $X < \aleph_\omega$, then any bounded M-basis of Y can be extended to a bounded M-basis on X.*

Proof. We will show the bounded case, as the sole extensions are obtained by a similar and easier argument. If dens $X = \aleph_0$, we are done by Theorem 1.50, according to which every K-bounded M-basis of a subspace of a separable Banach space can be extended to a $13K$-bounded M-basis of the whole space. We prove by induction that every K-bounded M-basis $\{y_\alpha; g_\alpha\}_{\alpha \in \Lambda}$ of $Y \hookrightarrow X$, dens $X = \aleph_n$, can be extended to a $2^n 13K$-bounded M-basis of X. The inductive step from n to $n + 1$ follows. Since Y is WLD (Theorem 5.37), we have that for every $\phi \in X^*$, card $\{\alpha; \phi(y_\alpha) \neq 0\} \leq \aleph_0$. Let $\tilde{M} \subset X$ be linearly dense and $S \subset X^*$ be a 1-norming subspace for which card $\{x \in \tilde{M}; f(x) \neq 0\} \leq \aleph_0$, $f \in S$. The last property remains valid if we replace $\tilde{M}$ by $M = \tilde{M} \cup \{y_\alpha\}_{\alpha \in \Lambda}$. By Theorem 3.42, there exists a PRI $\{P_\gamma; \omega \leq \gamma \leq \aleph_{n+1}\}$, $X_\gamma = (P_{\gamma+1} - P_\gamma)X$, dens $X_\gamma \leq \aleph_n$ on X for which $M \subset \bigcup_{\gamma < \aleph_{n+1}} X_\gamma$. Let $\{y_\alpha^\gamma\}_{\alpha \in \Lambda_\gamma} = \{y_\alpha\}_{\alpha \in \Lambda} \cap X_\gamma$. By the inductive hypothesis, there exists a $2^n 13K$-bounded extension of $\{y_\alpha^\gamma; g_\alpha^\gamma\}_{\alpha \in \Lambda_\gamma}$ into an M-basis $\{x_\alpha^\gamma; f_\alpha^\gamma\}_{\alpha \in \tilde{\Lambda}_\gamma}$ of X_γ. Finally, $\{x_\alpha^\gamma; (P_{\gamma+1} - P_\gamma)f_\alpha^\gamma\}_{\gamma, \alpha \in \tilde{\Lambda}_\gamma}$ is the $2^{n+1} 13K$-bounded M-basis of X sought. $\square$

Remark 5.65. Note that the PRI defined in the proof of Theorem 5.64 is subordinated to the set $\{x_\alpha^\gamma\}_{\gamma, \alpha \in \tilde{\Lambda}_\gamma}$; see Definition 3.41. In particular, the PRI fixes the subspace Y; i.e., $P_\gamma Y \subset Y$ for every γ.

Corollary 5.66 (Godefroy et al.[GKL00], Argyros et al.[ACGJM02]).
Let $c_0(\Gamma) \hookrightarrow X$, where X is a Plichko space and card $\Gamma < \aleph_\omega$. Then $c_0(\Gamma)$ is complemented in X.

Proof. By the theorem above, the canonical basis $\{e_\gamma\}_{\gamma \in \Gamma}$ of $c_0(\Gamma)$ can be extended to a bounded M-basis $\{e_\alpha; g_\alpha\}_{\alpha \in \Lambda}$ of Y. Without loss of generality, there is $C \in \mathbb{R}$ such that for all α, $\|e_\alpha\| = 1$, $C > \|g_\alpha\|$, and $\Gamma \subset \Lambda$. The formal operator $Ty = (g_\alpha(y)e_\alpha)_{\alpha \in \Gamma}$ is easily seen to be a projection onto $c_0(\Gamma)$. $\square$

We are going to show that the restriction on the cardinality of Γ in the theorem above is necessary. Denote by $\exp \alpha = 2^\alpha$, $\exp^{n+1} \alpha = \exp(\exp^n \alpha)$, where α is a cardinal. For a set S, let $[S]^n = \{X \subset S; \text{card } X = n\}$. We will use the following result, which in the language of partition relations claims that $(\exp^{n-1} \alpha)^+ \to (\alpha^+)_\alpha^n$ ([EHMR84, p. 100]).

Theorem 5.67 (Erdős and Rado; see [EHMR84] (GCH)). *Let α be an infinite cardinal, $n \in \mathbb{N}$, $\kappa = (\exp^{n-1} \alpha)^+$, and $\{G_\gamma\}_{\gamma < \alpha}$ be a partition of $[\kappa]^n$. Then there exist $M \subset \kappa$, card $M = \alpha^+$, and $[M]^n \subset G_\gamma$ for some $\gamma < \alpha$.*

Proof (Sketch). Suppose first that $n = 2$. Let us identify κ with the least ordinal of the same cardinality. By transfinite induction on the levels, we will construct a partially ordered set (a tree) $(T, \prec)$ consisting of pairs (t, N_t), $t \in \kappa$, $t \notin N_t \subset \kappa$ satisfying the following conditions. For every node $n \in T$, the set of its predecessors is order-isomorphic to an ordinal, every nonterminal node n of level $|n| = \lambda$ has a corresponding set of its immediate successors denoted by $\{n'\}$ of level $\lambda + 1$, and $(T, \prec)$ is uniquely determined by the following conditions:

(i) 0 is the root of T, $N_0 = \kappa \setminus \{0\}$.
(ii) (t, N_t), $t \in \kappa$ has at most α successors $\{(t_\gamma', N_{t_\gamma'})\}_{\gamma < \alpha}$, satisfying $t_\gamma' \in N_t$.
(iii) $\{t_\gamma'\} \cup N_{t_\gamma'} = \{s \in N_t; \{t, s\} \in G_\gamma\}$, and $t_\gamma' < s$ for every $s \in N_{t_\gamma'}$.

(iv) For a limit level λ of a tree, we form first the $\tilde{N}_B = \bigcap_{t \in B} N_t$ for every branch B constructed so far, then pick the minimal element $t_B \in \tilde{N}_B$, and set $N_B = \tilde{N}_B \setminus \{t_B\}$.

We will continue the process until we have exhausted the supply of nodes. It is rather straightforward to check that every $t \in \kappa$ eventually becomes the first coordinate of some node $(t, N_t) \in T$ of level $|(t, N_t)| < \kappa^+$. We claim that T has a branch of length at least α^+. If this is not the case, then $\kappa = \bigcup_{\beta < \alpha^+} \{t; |(t, N_t)| = \beta\}$. However, the cardinality of nodes of level β is easily estimated (using the coding of the inductive construction) by $\alpha^\beta \leq \alpha^\alpha \leq 2^\alpha$ ([Je78, p. 42]). Thus $(2^\alpha)^+ = \kappa \leq 2^\alpha \cdot \alpha^+$. By the regularity of κ, as a successor cardinal ([Je78, p. 27]), we have reached a contradiction so the long branch B exists. Now B can be split into disjoint sets B_γ, $\gamma < \alpha$, such that if $(t, N_t) \in B_\gamma$, then $\{t, s\} \in G_\gamma$ for all $s > t, (s, N_s) \in B$. At least one of these sets must satisfy card $B_\gamma \geq \alpha^+$, and it suffices to put $M = B_\gamma$. Let us describe the inductive step from n to $n + 1$. Fix α and $n + 1$, and identify κ with the least ordinal of cardinality $\kappa = (\exp^n \alpha)^+$.

We again construct a tree $(T, \prec)$ consisting of pairs (t, N_t), $t \in \kappa$, $t \notin N_t \subset \kappa$ satisfying the following conditions. This time, we will construct the tree only up to the level $(\exp^{n-1} \alpha)^+$ (noninclusive). This choice allows conditions (ii) and (iii) below to accommodate all possible $\alpha^{\exp^{n-1} \alpha} = \exp^n \alpha$ combinations (indexed by $\xi < \exp^n \alpha$) of values of $\gamma(\xi, t_1, \ldots, t_n)$ for all choices of $t_1 < \cdots < t_n \leq t$.

(i) 0 is the root of T, $N_0 = \kappa \setminus \{0\}$.
(ii) $(t, N_t) \in \kappa$ has at most $\exp^n \alpha$ successors $\{(t'_\xi, N_{t'_\xi})\}_{\xi < \exp^n \alpha}$ satisfying $t'_\xi \in N_t$.
(iii) $\{t'_\xi\} \cup N_{t'_\xi} = \{s \in N_t; (\forall t_1 < \ldots t_n \leq t)\{t_1, \ldots, t_n, s\} \in G_{\gamma(\xi, t_1, \ldots, t_n)}\}$, and $t'_\xi < s$ for every $s \in N_{t'_\xi}$.
(iv) For a limit level λ of a tree, we form first $\tilde{N}_B := \bigcap_{t \in B} N_t$ for every branch B constructed so far, then pick the minimal element $t_B \in \tilde{N}_B$, and set $N_B = \tilde{N}_B \setminus \{t_B\}$.

The cardinality of the set of all nodes of level $\lambda < (\exp^{n-1} \alpha)^+$ (so card $\lambda \leq \exp^{n-1} \alpha$) used in the process is estimated by $(\exp^n \alpha)^\lambda = \exp^n \alpha$. The last equality is the place where GCH is used. It follows from ([Je78, p. 49]) and the fact that $\exp^n \alpha$ is a regular cardinal. (If one is interested in a ZFC result, here is the place to further increase the cardinality of κ in order to make the inductive argument work; see [EHMR84].) Consequently, the cardinality of all nodes of level less than $(\exp^{n-1} \alpha)^+$ is at most $\exp^n \alpha \cdot \exp^{n-1} \alpha^+ = \exp^n \alpha < \kappa$. Therefore, T must have a branch B of length $(\exp^{n-1} \alpha)^+$. Note the crucial property that for every collection of first coordinates $t_1 < \cdots < t_n < s$ of some nodes from B, $\{t_1, \ldots, t_n, s\} \in G_\gamma$ is valid independently of s. Thus we are in a position to apply the inductive assumption to the set $S \subset \kappa$ of the first coordinates of the branch B and the splitting of $[S]^n$ into $\{G_\gamma\}_{\gamma < \alpha}$, which is obtained by simply ignoring the largest element in the $n + 1$ tuples in the original splitting. □

Theorem 5.68 (Argyros et al. [ACGJM02] (GCH)). *There exists an Eberlein compact space K (so $C(K)$ is WCG) such that $c_0(\aleph_\omega) \cong Y \hookrightarrow C(K)$, but Y is not complemented in $C(K)$.*

Proof. Let $K_n = \{\chi_A; A \subset \aleph_n, \text{card } A \leq n\} \subset c_0(\aleph_n)$ be a weakly compact set in $c_0(\aleph_n)$. Clearly, K_n is scattered and its n-th derived set $K_n^{(n+1)} = \emptyset$. Put $A_n = K_n \setminus K_n'$, and denote by X_n the isometric copy of $c_0(A_n) \cong c_0(\aleph_n)$ in $C(K_n)$. Let $P : C(K_n) \to X_n$ be a projection. It is immediate that, for $t \in A_n$, $P^*(\delta_t) = \delta_t + \mu_t$, where supp $\mu_t \subset K_n'$.

Lemma 5.69. *There exists $t = \chi_A \in A_n$ such that $|\mu_t(\{r\})| \geq \frac{1}{2}$ for every $r = \chi_B$, card $B = n - 1, B \subset A$.*

Proof. Using the linear ordering of $\aleph_n$, we may consider every set $A = \{x_1, \ldots, x_n\} \subset \aleph_n$ as ordered $x_1 < \cdots < x_n$. Proceeding by contradiction, we partition $[\aleph_n]^n$ into $G_1, \ldots, G_n$ so that for $A = \{x_1, \ldots, x_n\} \in G_i$ we have that $|\mu_{\chi_A}(\{r\})| < \frac{1}{2}$, where $r = \chi_{\{x_1, \ldots, x_{i-1}, x_{i+1}, \ldots, x_n\}}$. By Theorem 5.67, there exists a set $I \subset \aleph_n$, card $I \geq \aleph_1$, such that $[I]^n \subset G_i$ for some i. Therefore there exist some increasing sequence $\{r_k\}_{k=1}^\infty \subset I$, $r_k < r_{k+1}$, and $s_1 < \cdots < s_{i-1} < r_k < s_{i+1} < \ldots s_n$ in I. Denote $t_k = \chi_{\{s_1, \ldots, r_k, \ldots, s_n\}} \in A_n$ and $t = \chi_{\{s_1, \ldots, s_{i-1}, s_{i+1}, \ldots, s_n\}} \in K_n'$, and let F be a characteristic function of a clopen set in K_n that contains $\{t_k\}_{k=1}^\infty \cup \{t\}$ and whose intersection with K_n' is $\{t\}$. Since $\delta_{t_k} \xrightarrow{w^*} 0$ in X_n^*, we have that $P^*(\delta_{t_k}) = \delta_{t_k} + \mu_{t_k} \xrightarrow{w^*} 0$ in $C(K_n)$. Thus $\langle P^*(\delta_{t_k}), F\rangle \to 0$, and so $1 + \langle \mu_{t_k}, F\rangle = 1 + \mu_{t_k}(\{t\}) \to 0$, which is a contradiction as $|\mu_{t_k}(\{t\})| < \frac{1}{2}$ for all k. □

Having found t, there are n distinct points r satisfying the previous lemma, so it follows immediately that $\|P^*(\delta_t)\| \geq 1 + \frac{n}{2}$. Thus $\|P\| \geq 1 + \frac{n}{2}$. To finish the proof of the theorem, let K be the one-point compactification of the disjoint union of $\{K_n\}_{n=1}^\infty$ and $Y = c_0(\bigcup_{n=1}^\infty A_n)$. □

Theorem 5.70 ([Vand95]). *Suppose $Z \hookrightarrow X$ admits an M-basis $\{z_j; f_j\}_{j\in J}$ and X/Z is separable. Then any M-basis of Z can be extended to an M-basis of X.*

Proof (By transfinite induction on dens X). Let $Y_0 \subset X^*$ be a set of extensions of $\{f_j\}_{j\in J}$ (denoted the same) and $\{x_n\}_{n=1}^\infty \subset X$ be a sequence whose image in X/Z is dense. Our inductive assumption (true for separable spaces by Theorem 1.45) is that there exists an extension of $\{z_j\}_{j\in J}$ into an M-basis of X by adding vectors from span$\{z_j, x_n\}$. In the inductive step, set $Y = \overline{\text{span}}(Y_0 \cup Z^\perp)$. It is easy to check that Y with $S = \{z_j\}_{j\in J} \cup \{x_n\}_{n=1}^\infty$ satisfies the following conditions:

(i) $\overline{\text{span}}S = X$.

(ii) For each $x \notin Z$, there is an $f \in Y \cap Z^{\perp}$ such that $f(x) \neq 0$.

(iii) card $\{n \in \mathbb{N}; f(x_n) \neq 0\} \cup \{j \in J; f(z_j) \neq 0\} \leq \aleph_0$ for each $f \in Y$.

By (ii) it is clear that Y is total on X, so we may define a (not necessarily equivalent) norm $|\cdot|$ on X by $|x| = \sup\{f(x); f \in Y \cap B_{X^*}\}$. Denote by $\tilde{X}$ the completion of $(X, |\cdot|)$. We have that $i : X \to \tilde{X}$ is continuous, dens $\tilde{X} =$ dens X, and Y is 1-norming for $\tilde{X}$. Now Y is also a Σ-subspace of $\tilde{X}^*$, by (iii), so $\tilde{X}$ is Plichko and, moreover, $\Phi : Y \to 2^{i(S)}$ defined as $\Phi(f) = \{i(x_n); n \in \mathbb{N}, f(x_n) \neq 0\} \cup \{i(z_j); j \in J, f(z_j) \neq 0\}$ is a projectional generator. By Theorem 3.42, there exists a PRI $\{P_\alpha; \omega \leq \alpha \leq \Lambda\}$ on $\tilde{X}$ for which $i(S) = \bigcup_{\alpha < \Lambda}(P_{\alpha+1} - P_\alpha)i(S)$. We can now use the inductive assumption and extend the respective subsets of $i(S)$ of the original M-basis in the respective spaces $(P_{\alpha+1} - P_\alpha)\tilde{X}$ of lower density and preserving the linear spans. Putting the partial M-bases together yields an M-basis of $\tilde{X}$, $\{i(z_j)\}_{j \in J} \cup \{\tilde{y}_l\}_{l \in L}$, with its biorthogonal functionals $\{f_j\}_{j \in J} \cup \{g_l\}_{l \in L} \subset X^*$ such that there exist elements $y_l \in i^{-1}\tilde{y}_l$, $y_l \in \operatorname{span}\{z_j, x_n\} \subset X$. Clearly, $\{z_j\}_{j \in J} \cup \{y_l\}_{l \in L}$ together with $\{f_j\}_{j \in J} \cup \{g_l\}_{l \in L} \subset X^*$ form a fundamental biorthogonal system in X. It remains to see that the system is also total. However, this follows from the fact that the dual functionals (which separate elements of Z by assumption) also separate elements of the sequence $\{x_n\}_{n=1}^\infty$ (this follows from the preservation of linear spans).   □

Theorem 5.71. (♣) *There is a Banach space Z with a complemented subspace E where both E and Z have M-bases, but no M-basis of E can be extended to an M-basis of Z.*

Proof (see [Plic86b]). Let X be the $C(L)$ space constructed in Theorem 4.41. Recall that X^* is w^*-separable. Therefore, there exists a Banach space $Z := X \oplus E$, where Z and E have M-bases (see Theorem 2 in [Plic84b] and also the proof of [Fa~01, Thm. 6.45]). Assume that $\{u_\alpha; f_\alpha\}$ is an M-basis of E that can be extended to an M-basis $\{z_\beta; g_\beta\} \cup \{u_\alpha; \hat{f}_\alpha\}$ in $Z \times Z^*$. Write the added elements as $z_\beta = x_\beta + e_\beta$, where $x_\beta \in X$ and $e_\beta \in E$. Then $\{x_\beta\}$ is densely spanning X. Moreover, $g_\beta(u) = 0$ for all $u \in E$ and all β. Therefore, $g_\beta(x_\beta) = 1$ and $g_\beta(x_\alpha) = 0$ for $\alpha \neq \beta$. Then $\{x_\beta; g_\beta\}$ is necessarily an uncountable fundamental biorthogonal system in X, a contradiction.   □

Theorem 5.72 ([Vand95]). *Let X be a separable nonreflexive Banach space and let Z be a 1-codimensional subspace of X. Then X can be renormed so that there is a 1-norming M-basis on Z that cannot be extended to a 1-norming M-basis on X.*

Proof. Let $F \in X^{**} \setminus X$. Then $Y = \operatorname{Ker} F$ is a proper norming subspace (Lemma 2.25). Renorm Z by $\|\cdot\|$ so that $Y \upharpoonright Z$ is 1-norming, and choose a 1-norming M-basis $\{z_n; f_n\}$ of Z, where $f_n \in Y$. By the Bishop-Phelps theorem, choose $\phi \in B_{Z^*}$, which attains its norm at $z_0 \in B_Z$ and dist$(\phi, Y) > \frac{3}{4}$. Fix $x_0 \in X \setminus Z$ and define the norm $\|\|\cdot\|\|$ on X as the Minkowski functional of the ball

$$B = \{z + tx_0; \|z\| \leq 1, |\phi(z)| + |t| \leq 1\}.$$

Observe that $\|\|\cdot\|\|$ extends $\|\cdot\|$. Now $\|\|\frac{z_0}{2} + \frac{x_0}{2}\|\| = 1$. Proceed by proving that for any $\tilde{Y} \subset X^*$, $\tilde{Y} \upharpoonright Z = Y$, we have $f(\frac{z_0}{2} + \frac{x_0}{2}) \leq \frac{7}{8}$ for all $f \in \tilde{Y} \cap (B_{X^*}, \|\|\cdot\|\|)$. To see this, we first show that for $f \in \tilde{Y} \cap (B_{X^*}, \|\|\cdot\|\|)$,

$$f(z) > \frac{1}{4} \text{ for some } z \in \operatorname{Ker}\phi \cap B_Z \text{ if } f(z_0) > \frac{3}{4}.$$

Indeed, otherwise we have $f \in \tilde{Y} \cap (B_{X^*}, \|\|\cdot\|\|)$ with $f(z_0) > \frac{3}{4}$, while $f(z) \leq \frac{1}{4}$ for all $z \in \operatorname{Ker}\phi \cap B_Z$. Thus, for $w \in S_Z$ fixed, we have

$$|(f - \phi)(w)| \leq |(f - \phi)(\phi(w)z_0)| + |(f - \phi)(w - \phi(w)z_0)|$$

$$\leq \frac{1}{4}|\phi(w)| + \frac{1}{4}\|w - \phi(w)z_0\| \leq \frac{3}{4}.$$

Because this holds for all $w \in S_Z$, this contradicts $\|f \upharpoonright Z - \phi\| > \frac{3}{4}$. This is a contradiction, so the claim holds. Now, for $f \in \tilde{Y} \cap (B_{X^*}, \|\|\cdot\|\|)$, if $f(z_0) \leq \frac{3}{4}$, then $f(\frac{z_0}{2} + \frac{x_0}{2}) \leq \frac{3}{4}\frac{1}{2} + \frac{1}{2} \leq \frac{7}{8}$. On the other hand, if $f(z_0) > \frac{3}{4}$, then we can choose $w \in \operatorname{Ker}\phi \cap B_Z$ with $f(w) > \frac{1}{4}$. Now $\|\|w + x_0\|\| = 1$ and so $f(x_0) \leq 1 - f(w) < \frac{3}{4}$. Hence $f(\frac{z_0}{2} + \frac{x_0}{2}) \leq \frac{7}{8}$ in this case as well.   □

5.7 Quasicomplements

In this section, we are going to investigate the notion of a quasicomplemented subspace. We show that every subspace of a WLD space is quasicomplemented. Furthermore, every subspace of ℓ_∞ is also quasicomplemented. Asplund subspaces of $\ell_\infty(\Gamma)$ are quasicomplemented if and only if their dual is w^*-separable. We also give several examples, due to Godun, of unexpected pairs of quasicomplemented subspaces in ℓ_∞.

Recall (Definition 1.52) that a subspace $Y \hookrightarrow X$ is called *quasicomplemented* if there exists a subspace $Z \hookrightarrow X$ (and Z is called a *quasicomplement* of Y) such that $Y \cap Z = \{0\}$ and $\overline{Y + Z} = X$.

Proposition 5.73. *Let $\{x_\gamma; f_\gamma\}_{\gamma \in \Gamma}$ be an M-basis of a Banach space X. Then, for every partition of $\Gamma = A \cup B$, the spaces $Y := \overline{\operatorname{span}}\{x_\gamma; \gamma \in A\}$ and $Z := \overline{\operatorname{span}}\{x_\gamma; \gamma \in B\}$ are quasicomplements in X.*

Proof. Obviously, $\overline{\operatorname{span}}\{x_\gamma; \gamma \in A\} + \overline{\operatorname{span}}\{x_\gamma; \gamma \in B\}$ is dense in X. On the other hand, if $x \in \overline{\operatorname{span}}\{x_\gamma; \gamma \in A\} \cap \overline{\operatorname{span}}\{x_\gamma; \gamma \in B\}$, then $f_\gamma(x) = 0$ for all $\gamma \in \Gamma$, and thus $x = 0$.   □

Applying Theorem 5.64 and Theorem 5.70, we obtain immediately the following corollary.

Corollary 5.74. *Any subspace of a WLD Banach space admits a quasicomplement.*

Proof. It follows from Theorem 5.64 and Proposition 5.73.   □

Corollary 5.75. *Let* $Y \hookrightarrow X$ *be a subspace with an M-basis and* X/Y *separable. Then* Y *is quasicomplemented.*

Proof. It follows from Theorem 5.70.   □

Recall that Banach spaces X and Y are called *totally incomparable* if there is no infinite-dimensional space Z isomorphic to a subspace of both X and Y.

Lemma 5.76. *Let* X, Y *be totally incomparable subspaces of* Z. *Then* $X + Y$ *is norm closed. If, moreover,* $X \cap Y = \{0\}$, *then* $X + Y = X \oplus Y$.

Proof. Clearly, $\dim(X \cap Y) < \infty$, so without loss of generality $X \cap Y = \{0\}$. We will prove that $\inf_{x \in S_X, y \in S_Y} \|x - y\| > \varepsilon$. Once this condition is satisfied, it is clear that the formal operator $P : X + Y \to X$, defined as $P(x + y) = x$, is bounded and $X + Y = X \oplus Y$. By contradiction, suppose that there exist sequences $\{x_n\} \subset S_X$ and $\{y_n\} \subset S_Y$ for which $\|x_n - y_n\| < \frac{1}{2^n}$. Without loss of generality, these sequences form δ-separated sets for some $\delta > 0$. Using the basic construction of Schauder basic sequences, [Fa~01, Prop. 6.13], by passing simultaneously to subsequences, assume that both $\{u_n\} = \{x_{n+1} - x_n\}$ and $\{v_n\} = \{y_{n+1} - y_n\}$ are seminormalized Schauder basic sequences. Since $\sum_{n=1}^{\infty} \|u_n - v_n\| < \infty$, using the basis perturbation [Fa~01, Thm. 6.18], we obtain that $\overline{\operatorname{span}}\{u_n; n \geq N\}$ and $\overline{\operatorname{span}}\{v_n \ n \geq N\}$ are isomorphic Banach spaces, a contradiction.   □

Proposition 5.77 (Rosenthal [Rose69a]).
Let X *be a separable Banach space and* $Y \hookrightarrow X^*$ *be reflexive. Then* Y *is separable. More generally, let* $Y \hookrightarrow X^*$ *be reflexive. Then* $\operatorname{dens} Y \leq \operatorname{dens} X$.

Proof. By Lemma 5.11, Y is w^*-closed, so $(X/Y_\perp)^* \cong Y$. Thus $\operatorname{dens} Y = \operatorname{dens} X/Y_\perp \leq \operatorname{dens} X$.   □

Lemma 5.78. Y *is quasicomplemented in* X *if and only if there exists a* w^*-*closed subspace* $E \hookrightarrow X^*$ *such that* $E \cap Y^\perp = \{0\}, E_\perp \cap Y = \{0\}$. *In fact,* E *has these properties if and only if* $E_\perp$ *is a quasicomplement for* Y.

Proof. Suppose first that E satisfies the conditions. It suffices to show that $(E_\perp + Y)^\perp = \{0\}$. If $f \in (E_\perp + Y)^\perp$, then $f \in (E_\perp)^\perp \cap Y^\perp$. Since E is assumed w^*-closed, $(E_\perp)^\perp = E$ so $f = 0$. Conversely, if Z is a quasicomplement for Y, then setting $E = Z^\perp$, we have that $E \cap Y^\perp = \{0\}$ since $\overline{Y + Z} = X$. Of course, $E_\perp = (Z^\perp)_\perp = Z$.   □

Theorem 5.79 (Lindenstrauss and Rosenthal; see [John73]**).** *Let* X *be a Banach space and* $Y \hookrightarrow X$ *be such that* Y^* *is* w^*-*separable and* X/Y *has an infinite-dimensional separable quotient. Then* Y *is quasicomplemented in* X.

Proof. Since X/Y has a separable quotient, there is a biorthogonal sequence $\{x_n; f_n\}_{n=1}^\infty$ in X with $\{f_n; n \in \mathbb{N}\} \subset Y^\perp$, w^*-basic, and such that $\|x_n\| = 1$ [LiTz77, p. 11]. As Y^* is w^*-separable, Y has a total biorthogonal sequence $\{y_n; \tilde{g}_n\}_{n=1}^\infty$. Extending $\tilde{g}_n$ to $g_n \in X^*$, we obtain a biorthogonal system $\{y_n; g_n\}_{n=1}^\infty$ with $\{g_n\} \subset X^*$, $Y \cap \{g_n; n \in \mathbb{N}\}_\perp = \{0\}$. Assume without loss of generality that $\|g_n\| = 1$ ([LiTz77, p. 43]). Define an operator $T : X \to Y$, $Tx = \sum_{n=1}^\infty 2^{-n-1} g_n(x) x_n$. Then $\|T\| \leq \frac{1}{2}$ and hence $I + T$ is an isomorphism of X. Thus $(I + T)^*$ is an isomorphism of X^*. Thus $\{f_n + T^* f_n\}$ is a w^*-basic sequence equivalent to $\{f_n\}$. We have $T^* f_n = 2^{-n-1} g_n$. We claim that $\{f_n + 2^{-n-1} g_n; n \in \mathbb{N}\}_\perp$ is a quasicomplement of Y in X. Let $x^* \in Y^\perp \cap \overline{\text{span}}^{w^*}\{f_n + 2^{-n-1} g_n\}$. Then $x^* = w^*\text{-}\lim_n (\sum_{i=1}^n \alpha_i f_i + \sum_{i=1}^n 2^{-n-1} \alpha_i g_i)$ for some $\alpha_i \in \mathbb{R}$. As $x^* \in Y^\perp$, $x^*(y_n) = 2^{-n-1} \alpha_n = 0$ for every n, and thus $x^* = 0$. Now let $y \in Y \cap \{f_n + 2^{-n-1} g_n; n \in \mathbb{N}\}_\perp$. Then $f_n(y) = 0$ and hence $g_n(y) = 0$, and this shows that $y \in \{g_n; n \in \mathbb{N}\}_\perp \cap Y = \{0\}$. Lemma 5.78 concludes the proof. □

Corollary 5.80. *A Banach space X has a separable and infinite-dimensional quotient if and only if it has a separable and infinite-dimensional quasicomplemented subspace.*

Proof. If Y, Z are quasicomplemented in X, Y separable, then X/Z is infinite-dimensional and separable. If X/Z is separable for some Z, then choose a separable subspace $Y \hookrightarrow Z$, and apply Theorem 5.79 to show that Y is quasicomplemented in X. □

The problem of the existence of a separable and infinite-dimensional quotient for every Banach space is still open. This problem is equivalent to whether in every Banach space X there is an increasing sequence (E_n) of distinct subspaces such that $\overline{\bigcup_n E_n} = X$ (see, e.g., [Muj97]).

Corollary 5.81 (Rosenthal [Rose69a]). *Let Y be a subspace of X such that Y^* is w^*-separable. If $Y^\perp$ contains a reflexive subspace, then Y is quasicomplemented.*

Proof. Since $(X/Y)^* = Y^\perp$ contains a reflexive subspace, X/Y has a quotient that is reflexive. However, all reflexive spaces have a separable quotient. The result follows by Theorem 5.79. □

Theorem 5.82 (Rosenthal [Rose69a]). *Let X be a Banach space such that X^* contains a nonseparable reflexive subspace. Then every separable $Y \hookrightarrow X$ is quasicomplemented. Similarly, if X^* contains a reflexive subspace of density larger than c, then every subspace $Y \hookrightarrow X$, with w^*-separable dual Y^* is quasicomplemented in X.*

Proof. Let $Z \hookrightarrow X^*$ be reflexive and nonseparable (resp. dens $Z > c$), $Y \hookrightarrow X$. In order to apply Corollary 5.81, we only need to show that $Y^\perp$ contains a reflexive subspace. If $Y^\perp$ contains a reflexive subspace, there is nothing to do.

If not, $Y^\perp$ and Z are totally incomparable, and thus, by Lemma 5.76, without loss of generality, $Y^\perp + Z = Y^\perp \oplus Z$. Next, $Y^* \cong X^*/Y^\perp$, and $Z \hookrightarrow Y^*$. In the case where Y is separable, we are done by Proposition 5.77. In the case when Y^* is w^*-separable, it suffices to observe that dens $Y \le c$, which is a contradiction. $\qquad\square$

Theorem 5.83 (Rosenthal [Rose68b]). *Let X be a subspace of $\ell_\infty(\Gamma)$. Then $X^\perp$ contains a reflexive subspace. Consequently, if X has a w^*-separable dual, then X is quasicomplemented. In particular, every subspace of ℓ_∞ is quasicomplemented.*

Proof. We have $(\ell_\infty(\Gamma)/X)^* \cong X^\perp$. If $\ell_\infty(\Gamma)/X$ is reflexive, so is $X^\perp$. If it is not, then the quotient mapping is not weakly compact, and so by [LiTz77, 2.f.4] $\ell_\infty(\Gamma)/X$ contains a copy of ℓ_∞. This copy is complemented by the injectivity of ℓ_∞. Now ℓ_∞ (and thus also $\ell_\infty(\Gamma)/X$) has a further quotient isomorphic to $\ell_2(c)$ (see Theorem 4.22). Thus $X^\perp$ contains a reflexive subspace. The rest follows by Corollary 5.81. $\qquad\square$

The following example sheds light on the fundamental difference between the notions of complemented and quasicomplemented subspaces.

Example 5.84 (Godun [Godu84]). There exist X and Y, quasicomplements in ℓ_∞, such that $\ell_\infty/X \cong \ell_\infty/Y \cong \ell_2$.

Proof. By Theorem 4.22, let $\ell_2 \cong H \hookrightarrow \ell_\infty^*$, with a canonical basis $\{h_i\}_{i=1}^\infty$. Denote by $\{f_i\}_{i=1}^\infty$ the coordinate functionals of ℓ_∞ (which form a canonical basis of $\ell_1 \hookrightarrow \ell_\infty^*$). Since ℓ_1 and ℓ_2 are totally incomparable, we may without loss of generality assume that $H \cap \overline{\operatorname{span}}\{f_i; i \in \mathbb{N}\} = \{0\}$. By the basis perturbation theorem [Fa~01, Thm. 6.18], for some $\varepsilon_i \searrow 0$, the sequence $\tilde{h}_i = h_i + \varepsilon_i f_i$ is equivalent to h_i, and $\tilde{H} = \overline{\operatorname{span}}\{\tilde{h}_i; i \in \mathbb{N}\} \cong \ell_2$. It is clear that $H \cap \tilde{H} = \{0\}$. By Lemma 5.11, both H and $\tilde{H}$ are w^*-closed, so using Lemma 5.78, we see that $X = H_\perp \hookrightarrow \ell_\infty$ and $Y = \tilde{H}_\perp \hookrightarrow \ell_\infty$ are quasicomplements. Lastly, $\ell_\infty/X \cong X^\perp = H \cong \ell_2$ and also $\ell_\infty/Y \cong \tilde{H} \cong \ell_2$. $\qquad\square$

Going in the opposite direction, we have the following theorem.

Theorem 5.85. *Let Y be a quasicomplemented subspace of a Grothendieck space X. Suppose that the unit ball of Y^* is w^*-sequentially compact. Then either Y is reflexive and complemented or $Y^\perp$ contains a reflexive subspace.*

Proof. Suppose that Z is a quasicomplement of Y. Let $Q : X \to X/Z$ be a quotient mapping. Denote $S := Q \upharpoonright Y$. As $Q(Y) = S(Y)$ is dense in the range, $S^* : Z^\perp \to Y^*$ is injective. It follows that $B_{Z^\perp}$ is w^*-sequentially compact. By the Grothendieck property of X, $B_{Z^\perp}$ is also weakly sequentially compact, and hence $Z^\perp$ is reflexive. Now, if $Y^\perp$ contains no reflexive subspace, then $Y^\perp + Z^\perp$ is norm closed, so $S^* : Z^\perp \to X^*/Y^\perp$ has a closed range. Therefore S also has a closed range (e.g., [Fa~01, Exer. 2.39]). Using the open mapping theorem, it is easy to see that $Q(Y)$ is closed in the range if and only if $Y + Z$ is a closed space. Hence $Y + Z = X$, Y is complemented, and Y^* is isomorphic to a reflexive space $Z^\perp$. Hence Y is reflexive.   □

Recall that a set $D \subset X$ is called X^*-*limited* (see Definition 3.9) if any w^*-convergent sequence from X^* converges uniformly on D. A space X is called a *Gelfand-Phillips* space if relatively compact sets are the only X^*-limited sets in the space X.

Lemma 5.86. *Suppose that B_{X^*} is w^*-sequentially compact. Then X is a Gelfand-Phillips space.*

Proof. Suppose D is a bounded subset of X that is not relatively compact. Then we can choose a sequence (x_n) in D so that $\liminf d(x_n, E_{n-1}) > 0$, where $E_{n-1} = \text{span}\{x_1, x_2, \ldots, x_{n-1}\}$. Let $0 < \varepsilon < \liminf_n d(x_n, E_{n-1})$. Now choose $f_n \in S_{X^*}$ so that $f(x_n) > \epsilon$, while $f_n(x_k) = 0$ for $k < n$. Because B_{X^*} is w^*-sequentially compact, there exist $f \in B_{X^*}$ and a subsequence (f_{n_k}) that weak*-converges to f. However, the convergence is not uniform on D because $f_{n_k}(x_k) > \epsilon$, while $f(x_j) = \lim_{j \to \infty} f_{n_j}(x_k) = 0$.   □

Lemma 5.87 (Josefson [Jos02]). *A subset $D \subset \ell_\infty(\Gamma)$ is $\ell_\infty^*(\Gamma)$-limited if and only if it is bounded and does not contain a sequence equivalent to the unit basis of ℓ_1.*

Proof. First we sketch the proof that in any Banach space X, an X^*-limited set $D \subset X$ cannot contain a sequence equivalent to the canonical basis of ℓ_1. To this end, it is sufficient to extend the canonical injection $i : \ell_1 \to c_0$ to a bounded operator $I : X \to c_0$, $I(x) = (f_n(x))_{n \in \mathbb{N}}$. Indeed, $f_n \in \lambda B_{X^*}$ is a w^*-null sequence that is not uniformly convergent on the ℓ_1 basis inside D. The extension consists of using the operator $R : \ell_1 \to L_\infty[0,1]$, defined by $Re_n = r_n$, where r_n is the n-th Rademacher function on $[0,1]$. Since $L_\infty[0,1] \cong \ell_\infty$ is an injective space, there exists an extension $R : X \to L_\infty[0,1]$. Consider $L : L_\infty[0,1] \to c_0$ defined by $Lf = \int_0^1 f(t)r_n(t)dt$. It is well known that L is bounded, $Lr_n = e_n$. Thus it suffices to extend i by $I = L \circ R$.

To prove the opposite implication, suppose that D does not contain a sequence equivalent to the unit basis of ℓ_1. Let $f_n \xrightarrow{w^*} 0$ and assume by contradiction that for some $\varepsilon > 0$ and a sequence $x_n \in D$, $f_n(x_n) > \varepsilon$. Without loss of generality, $|f_k(x_n)| < \frac{\varepsilon}{2}$ for $k > n$. Also, by Rosenthal's ℓ_1 theorem, we may without loss of generality assume that (x_n) is weakly Cauchy, so $x_n - x_{n+1} \xrightarrow{w} 0$ as $n \to \infty$. By the Grothendieck property of $\ell_\infty(\Gamma)$, $f_n \xrightarrow{w} 0$. We have $f_n(x_n - x_{n+1}) > \frac{\varepsilon}{2}$ for all $n \in \mathbb{N}$. This is, however, a contradiction with the Dunford-Pettis property of $\ell_\infty(\Gamma)$ (see Definition 3.12). Indeed, $\ell_\infty(\Gamma)$ is isomorphic to some $C(K)$ space [Fa~01, 11.34, 11.36].   □

Theorem 5.88 (Josefson [Jos02]). *Let $Y \hookrightarrow \ell_\infty(\Gamma)$ be a Banach space whose dual ball B_{Y^*} is w^*-sequentially compact, and, moreover, $\ell_1 \not\hookrightarrow Y$. If Y is quasicomplemented, then Y^* is w^*-separable.*

Proof. Assume that Y is quasicomplemented by Z and Y^* is not w^*-separable. Let $T : \ell_\infty(\Gamma) \to \ell_\infty(\Gamma)/Z$ be the quotient map. Then $S = T \upharpoonright Y$ is injective and has a dense range in $\ell_\infty(\Gamma)/Z$. Thus $S^* : Z^\perp \to Y^*$ is injective and has a w^*-dense range in Y^*. Thus $\ell_\infty(\Gamma)/Z$ has a w^*-sequentially compact dual ball, and so it is a Gelfand-Phillips space. By Lemma 5.87, B_Y is a $\ell_\infty^*(\Gamma)$-limited set in $\ell_\infty(\Gamma)$. Since Y^* is not w^*-separable, S^* and also S are noncompact operators. Therefore $T(B_Y)$ is not $(\ell_\infty(\Gamma)/Z)^*$-limited in $\ell_\infty(\Gamma)/Z$. Pick a sequence $f_n \overset{w^*}{\to} 0$ from $Z^\perp = (\ell_\infty(\Gamma)/Z)^*$ that is not uniformly convergent on $T(B_Y)$. Then $f_n \circ T \overset{w^*}{\to} 0$ on $\ell_\infty(\Gamma)$, but it is not uniformly convergent on B_Y, a contradiction with B_Y being a $\ell_\infty^*(\Gamma)$-limited set in $\ell_\infty(\Gamma)$.   □

Corollary 5.89 (Josefson [Jos02], Lindenstrauss [Lind68]). *An Asplund space $X \hookrightarrow \ell_\infty(\Gamma)$ is quasicomplemented if and only if X^* is w^*-separable. In particular, $c_0(\Gamma), \ell_p(\Gamma), 1 < p < \infty$, Γ uncountable are not quasicomplemented in $\ell_\infty(\Gamma)$.*

Proof. Asplund spaces have a w^*-sequentially compact dual ball [Fab97, p. 38] and do not contain a copy of ℓ_1.   □

Theorem 5.90 (James [Jam72c], Johnson [John73], Plichko [Plic75]). *Let Z, Y be quasicomplemented but not complemented subspaces of X. Then there exists $Y \hookrightarrow Y_1$, $\dim Y_1/Y = \infty$, such that Y_1 and Z are quasicomplemented.*

Proof. We start the proof with the following lemma.

Lemma 5.91. *Let Z, Y be quasicomplemented but not complemented subspaces of X. Then there exists a countable-dimensional linear space $E \hookrightarrow X$ such that, for every $0 \neq x \in E$, there exist no bounded sequences $\{z_n\}_{n=1}^\infty \subset Z$, $\{y_n\}_{n=1}^\infty \subset Y$, for which $\lim_{n \to \infty} \|y_n + z_n - x\| = 0$.*

Proof. Denote

$$X_n := \{x \in X; \text{there exist sequences } (y_n) \text{ in } Y, \text{ and } (z_n) \text{ in } Z,$$

$$\|y_n\|, \|z_n\| \leq n\|x\|, \text{ such that } \lim_{n \to \infty} \|y_n + z_n - x\| = 0\}.$$

It is clear that $X_n \hookrightarrow X$ is a closed subspace. We claim that X_n has an empty interior. Indeed, if $B(x, \rho) \subset X_n$, then a standard argument gives that $X = X_m$ for some $m \in \mathbb{N}$. Thus, for every $0 \neq x \in X$, we can construct by induction sequences $(x_n)_{n=0}^\infty \subset X, (y_n)_{n=0}^\infty \subset Y, (z_n)_{n=0}^\infty \subset Z$ such that

(i) $x_0 = 0$, $\lim_{n \to \infty} x_n = x$, $\|x_{n+1} - x_n\| < \frac{1}{2^n}$ for $n \geq 1$,
(ii) $x_{n+1} - x_n = y_n + z_n$ for $n \geq 0$ and $\|y_n\|, \|z_n\| \leq \frac{m}{2^n}$ for $n \geq 1$.

We immediately obtain that $x = y + z$, where $y = \sum_{n=0}^\infty y_n \in Y$, $z = \sum_{n=0}^\infty z_n \in Z$. This implies that $Y + Z = X$ so Y and Z are complemented, which is a contradiction. Clearly, $F = \bigcup_{n=1}^\infty X_n$ is a dense linear subspace of X

of first category (by Baire's theorem). Thus there exists $e_1 \in X \setminus F$. Repeating the argument for the nested sequence of subspaces $\tilde{X}_n = \overline{\mathrm{span}}\{X_n \cup \{e_1\}\}$, we obtain that $\bigcup_{n=1}^{\infty} \tilde{X}_n$ is a dense linear subspace of X of first category, and so there exists $e_2 \in X \setminus \bigcup_{n=1}^{\infty} \tilde{X}_n$. Proceeding inductively, we obtain an infinite sequence $\{e_i\}_{i=1}^{\infty}$ of vectors linearly independent on F. To finish, put $E := \mathrm{span}\{e_i\}_{i \in \mathbb{N}}$. □

We continue the proof of Theorem 5.90. We have that $\overline{\mathrm{span}}(E + Y)/Y$ is separable, and so is $V = Z \cap \overline{\mathrm{span}}(E + Y)$. Since $Y \cap V = \{0\}$, there exists a sequence $\{f_i\}_{i=1}^{\infty} \subset Y^{\perp}$ for which $f_{2i+1} = -f_{2i}$ and $Y \subset \overline{\mathrm{span}}\{f_i; i \in \mathbb{N}\}_{\perp}$ and $\overline{\mathrm{span}}\{f_i; i \in \mathbb{N}\}_{\perp} \cap V = \{0\}$. Let $W_i^j = \{x \in V; \|x\| \le j, f_i(x) \ge \frac{1}{i}\}$ be convex and closed sets for $i, j \in \mathbb{N}$. For convenience, let us reindex the family $\{W_i^j\}_{i,j \in \mathbb{N}}$ as $\{V_n\}_{n \in \mathbb{N}}$. By the choice of f_i, we have $\bigcup_{n=1}^{\infty} V_n = V \setminus \{0\}$. Next, we construct, by induction in n, sequences $\{u_i\}_{i=1}^{\infty} \subset E$ and $\{H_i\}_{i=1}^{\infty}$, where H_i are closed hyperplanes in X, that satisfy the following conditions for $i \le n$:

(i) u_i are linearly independent.
(ii) $V_i \cap H_i = \emptyset$.
(iii) $Y + \mathrm{span}\{u_j\}_{j=1}^{n} \subset H_i$.

For $n = 1$, $\mathrm{dist}(V_1, Y) > 0$, so the separation theorem [Fa~01, Thm. 2.13] gives a closed hyperplane $H_1 \supset Y$, $H_1 \cap V_1 = \emptyset$. Since E is infinite-dimensional, there exists $0 \ne u_1 \in H_1 \cap E$. Let us assume now that $\{u_i\}_{i=1}^{n} \subset E$ and $\{H_i\}_{i=1}^{n}$ satisfying (i)–(iii) have been constructed. In order to proceed with the inductive step, we claim that $\mathrm{dist}(V_{n+1}, Y + \mathrm{span}\{u_j\}_{j=1}^{n}) > 0$. Assuming the contrary, there exist $a_l \in V_{n+1}, b_l \in Y, c_l \in \mathrm{span}\{u_j\}_{j=1}^{n}$ such that $\lim_{l \to \infty} \|a_l - b_l - c_l\| = 0$. Now $\{a_l\}_{l=1}^{\infty}$ is bounded and so must be $\{b_l + c_l\}_{l=1}^{\infty}$. Since $Y + \mathrm{span}\{u_j\}_{j=1}^{n}$ forms a topological sum, both $\{b_l\}_{l=1}^{\infty}$ and $\{c_l\}_{l=1}^{\infty}$ are bounded. By compactness, without loss of generality, $c_l = c \in E$. Thus $\lim_{l \to \infty} \|a_l - b_l - c\| = 0$, which is a contradiction with the properties of E, and the claim is established. Next we apply the separation theorem to obtain a hyperplane $H_{n+1} \supset Y + \mathrm{span}\{u_j\}_{j=1}^{n}$, $H_{n+1} \cap V_{n+1} = \emptyset$. Choose $z_{n+1} \in E \cap \bigcap_{i=1}^{n+1} H_i \setminus \mathrm{span}\{z_i; i \le n\}$. This finishes the inductive step. To end the proof, we set $Y_1 = \overline{\mathrm{span}}(Y \cup \{u_i\}_{i=1}^{\infty})$. It is left to observe that $Y_1 \cap Z = \{0\}$. Since $Y_1 \cap Z \subset V$, $Y_1 \subset \bigcap_{i=1}^{\infty} H_i$, and every element of V belongs to some V_n, an appeal to condition (ii) finishes the proof. □

5.8 Exercises

5.1. Prove that if P is a projection on the dual of a WCG space X generated by a weakly compact (absolutely convex) set K, and $\|P\| = \|P\|_K = 1$, where $\|\cdot\|_K := \sup\{|\langle k, \cdot \rangle|; k \in K\}$, then P is a dual projection.

Hint. P^* preserves K.

5.2. Let X be WLD of density ω_1 and $\{P_\alpha; \omega \leq \alpha \leq \omega_1\}$ be a PRI in X. Show that $\bigcup P_\alpha^*(X^*) = X^*$. Note that this is not true in $X = \ell_1(\omega_1)$.

5.3. Prove that, for uncountable Γ, the space $\ell_1(\Gamma)$ is not WLD.

Hint. Check the nonangelicity of $(B_{\ell_\infty(\Gamma)}, w^*)$ by looking at $B_{c_0(\Gamma)}$.

5.4. Let the norm $|\cdot|$ on X be Gâteaux differentiable, and suppose B_{X^*} in its w^*-topology is a Valdivia compact. Then prove that X is WLD.

Hint. Let B_{X^*} mean the dual unit ball corresponding to $|\cdot|$. Let $\varphi : (B_{X^*}, w^*) \to \mathbb{R}^\Gamma$ be a continuous injection such that $\varphi(B_{X^*}) \cap \Sigma(\Gamma)$ is pointwise dense in $\varphi(B_{X^*})$. Take any $0 \neq x \in X$ and denote $\xi = |\cdot|'(x)$. Then ξ is a G_δ point in (B_{X^*}, w^*). More concretely, $\{\xi\} = \bigcap_{n=1}^\infty \{x^* \in B_{X^*} : \langle x^*, x \rangle > 1 - \frac{1}{n}\}$. From this we can easily deduce that $\varphi(\xi)$ can be written as a limit of a sequence of elements from $\Sigma(\Gamma)$ and hence $\varphi(\xi) \in \Sigma(\Gamma)$. Now, the Bishop-Phelps theorem guarantees that $\varphi(\xi) \in \Sigma(\Gamma)$ for every $\xi \in S_{X^*}$, where S_{X^*} is the dual unit sphere with respect to the norm $|\cdot|$. According to the Josefson-Nissenzweig theorem (see Theorem 3.27), S_{X^*} is weak*-sequentially dense in B_{X^*}. Therefore $\varphi(B_{X^*}) \subset \Sigma(\Gamma)$. This means that (B_{X^*}, w^*) is a Corson compact and so X is weakly Lindelöf determined.

5.5. Show that there exists an LUR renormable Asplund space without an M-basis.

Hint. Consider the $C(K)$ space of Ciesielski-Pol [DGZ93a, Examp. VI.8.8]. As $K^{(3)} = \emptyset$, $C(K)$ has an LUR renorming. However, this space admits no bounded injection into $c_0(\Gamma)$.

5.6 (Godun [Godu84]). The space ℓ_∞ has quasicomplementary subspaces X, Y, both admitting a fundamental biorthogonal system, such that for no fundamental systems $\{x_\gamma\}$ and $\{y_\gamma\}$ of X and Y is the system $\{x_\gamma\} \cup \{y_\gamma\}$ minimal.

Hint. Consider the spaces X, Z from Example 5.84. They have a fundamental biorthogonal system by Corollary 4.19. Assume by contradiction that $(\{x_\gamma\} \cup \{y_\gamma\}; \{f_\gamma\} \cup \{g_\gamma\})$ is a biorthogonal system in ℓ_∞ obtained from the union of fundamental systems for X and Y. Since $f_\gamma(y_\lambda) = 0$, we have that $f_\gamma \in Y^\perp$ for all γ. This implies that a separable space $Y^\perp \cong \ell_2$ has an uncountable biorthogonal system, a contradiction.

5.7 (Rosenthal [Rose69a]). Let $\phi : K \to L$ be a surjective continuous mapping between compact spaces, $Y = \{h \in C(K); h = g \circ \phi, g \in C(L)\}$. If there exists $p \in L$ such that $M = \phi^{-1}(p)$ is an infinite perfect set, then $\ell_2 \hookrightarrow Y^\perp$.

Hint. As an infinite perfect set, $M = \phi^{-1}(p)$ contains a homeomorphic copy of the Cantor discontinuum $i : \mathbb{D} \to M$. The set $\mathbb{D}$ supports a Haar measure

μ, that maps via i onto a measure λ supported by $i(\mathcal{C})$. By the Khintchine inequality, the space $L_1(d\mu)$ contains a sequence $\{r_n\}$ of functions $\int r_n d\mu = 0$ equivalent to the basis of ℓ_2. Thus the Radon measures $\{\mu_n\}$ supported by $i(\mathbb{D})$ and represented by $\mu_n(A) = \int_{i^{-1}(A)} r_n d\mu$ for $A \subset i(\mathcal{C})$ are easily verified to form a sequence from $Y^\perp$ equivalent to the unit basis of ℓ_2.

5.8 (Lacey [La72]; see Mujica [Muj97]). Let K be any compact space. Then $C(K)$ has a quotient isomorphic to either c_0 or ℓ_2.

Hint. If K contains a nonconstant convergent sequence $p_n \to p$, then $T(f) := (f(p_1), f(p_2), \dots)$, $f \in C(K)$, is a quotient operator from $C(K)$ onto c. Otherwise, use Exercise 5.7.

5.9. Let $Y \hookrightarrow X$, $Y \cong c_0$ and either (B_{X^*}, w^*) be sequentially compact or $\ell_1 \not\hookrightarrow X$. Then Y contains a copy of c_0 complemented in X.

Hint. Let $\{e_n\}_{n=1}^\infty$ be the canonical basis of $Y \cong c_0$, and denote by $\{f_n\}_{n=1}^\infty$ the dual functionals (equivalent to ℓ_1 basis) extended to the whole X. In the sequentially compact case, we may without loss of generality assume that $\{f_n\}_{n=1}^\infty$ is w^*-convergent, so $v_n = f_{2n+1} - f_{2n}$ is w^*-null. The desired projection is $P(x) = \sum v_n(x)e_{2n}$. The ℓ_1 case rests on the following result of Hagler and Johnson (see [Dies84, p. 219]): *Suppose that X^* contains a copy of ℓ_1, without a w^*-null normalized block sequence. Then $\ell_1 \hookrightarrow X$.* In our situation, this implies that there does exist some w^*-null sequence of normalized blocks of $\{f_n\}_{n=1}^\infty$, and we are again done.

5.10. A Banach space X satisfying any of the following conditions has a separable quotient:

(i) $c_0 \hookrightarrow X^*$.
(ii) $\ell_1 \hookrightarrow X^*$.
(iii) $Z \hookrightarrow X^*$ for a reflexive space Z.

Hint. (i) By a result of Bessaga and Pełczyński (see, e.g., [Fa~01, Thm. 6.39]), ℓ_1 is a complemented subspace of X.

(ii) Let $\{f_n\}_{n=1}^\infty$ be the basis of ℓ_1. If $\{f_n\}_{n=1}^\infty$ is w^*-null, then by a result of Johnson and Rosenthal ([LiTz77, Thm. 2.e.9]), c_0 is a quotient of X. Otherwise, by a result of Hagler and Johnson, $\ell_1 \hookrightarrow X$, and so by Pełczyński's result ℓ_2 is a quotient of X.

(iii) Let $i : Z \to X^*$ be the embedding. Using the reflexivity of Z, show that $i^*(Z) = X$.

5.11 (Plichko). Let B_{X^*} be w^*-angelic. Then, for every $Y \hookrightarrow X$, there exists a 1-complemented $Z \hookrightarrow X$, $Y \hookrightarrow Z$, dens $Z \leq m := \max\{\,\text{dens}\,Y, c\}$.

Hint. The construction follows the same pattern as getting a PRI from a projective generator (see Theorem 3.42). Transfinite sequences $Z_\alpha \hookrightarrow X$, dens $Z_\alpha \leq m$ and $F_\alpha \hookrightarrow X^*$, $1 \leq \alpha < \omega_1$ have to be constructed such that:

(1) $Z_1 = Y$, $Z_\alpha \subset Z_\beta, F_\alpha \subset F_\beta$ for $\alpha < \beta$.
(2) F_α 1-norms Z_α and $Z_{\alpha+1}$ 1-norms F_α.
(3) Z_α is closed and $\overline{B_{F_\alpha}}^{w^*} \subset F_{\alpha+1}$.

The angelicity assumption is used in order to keep the density of F_α not more than m. Indeed, $m^\omega = m$, so card $B_{F_\alpha} = $ dens B_{F^α}. Next, every element of $\overline{B_{F_{\alpha+1}}}^{w^*}$ can be coded using a convergent sequence from B_{F_α}. It follows that card $\overline{B_{F_{\alpha+1}}}^{w^*} \leq m$.

5.12. Recall that $B_{(\ell_\infty, \|\cdot\|_\infty)^*}$ is weak*-separable by Goldstine's theorem. On the other hand, let Y be the subspace of ℓ_∞ constructed in [JoLi74, Example 2]. Let $\|\|\cdot\|\|$ be an equivalent norm on ℓ_∞ such that its restriction to Y is $\|\cdot\|$. Then $B_{(\ell_\infty, \|\|\cdot\|\|)^*}$ is not weak*-separable.

Hint. This subspace is not separable and admits an equivalent norm $\|\cdot\|$ such that the corresponding dual norm is locally uniformly rotund. Assume that the dual unit ball $B_{(\ell_\infty, \|\|\cdot\|\|)^*}$ is weak*-separable. Then the dual unit ball $B_{(Y, \|\cdot\|)^*}$ would be weak*-separable. Since the dual norm to $\|\cdot\|$ on Y^* is locally uniformly rotund, the corresponding unit sphere in Y^* would then be norm separable. Hence Y would be separable, a contradiction.

6

Weak Compact Generating

The delicate gradation of the different subclasses of the class of weakly Lindelöf determined (WLD) spaces is shown in this chapter through the optic of M-bases. It is known that WCG spaces can be characterized by the existence of a weakly compact M-basis, although not every M-basis in such a space is necessarily weakly compact (that it is so when using some more precise orthogonal structures will be investigated in Chapter 7). The subtle distinction between WCG spaces and subspaces of WCG spaces was brought to light when, in [Rose74], Rosenthal produced an example of the latter that was not WCG. Examples separating different important classes (WCG, their subspaces, Vašák, WLD) appear in the work of Rosenthal, Talagrand, Argyros, Mercourakis, and others. Here we reflect on these distinctions in terms of the existing M-bases using countable splittings of the systems.

More precisely, in this chapter we show that the class of WLD spaces and some of its subclasses, such as weakly compactly generated spaces and their subspaces, Vašák (i.e., WCD) spaces, and Hilbert generated spaces and their subspaces, can be characterized by the existence of M-bases with special covering properties. Moreover, in many cases, these properties of M-bases are then shared by all M-bases in the space. As applications of this, we present short proofs of the (uniform) Eberlein property for continuous images of (uniform) Eberlein spaces. Results of this nature were initially proved by using infinite combinatorial methods. We conclude this chapter with some results on spaces that are strongly generated by reflexive (resp. superreflexive) spaces. Using these results, we provide some short proofs of results on weakly compact sets in $L_1(\mu)$ spaces where μ is a finite measure.

6.1 Reflexive and WCG Asplund Spaces

In the first part of this section, reflexive spaces are characterized in terms of the existence of a particular class of M-basis in the spirit of James' characterization of reflexivity by means of Schauder bases. From this, a renorming characterization is derived. In the second part, Asplund spaces that are weakly compactly generated are described in terms of the existence of a shrinking M-basis.

Theorem 6.1. *Let X be a Banach space. Then the following are equivalent*

(i) X *is reflexive.*
(ii) *There is an M-basis $\{x_\gamma; f_\gamma\}_{\gamma \in \Gamma}$ for X that is both shrinking and boundedly complete. Equivalently, every M-basis for X is both shrinking and boundedly complete.*
(iii) [HaJo04] X *admits an equivalent w-2R norm (see Definition 3.47).*
(iv) [OdSc98] *If X is separable, then the above are equivalent to the existence of an equivalent 2R norm (see Definition 3.47).*

Proof. (i)$\Rightarrow$(ii) Assume that X is reflexive. Let $\{x_\gamma; f_\gamma\}_{\gamma \in \Gamma}$ be an M-basis for X; its existence follows from Theorem 5.1. Since $\overline{\operatorname{span}}^{w^*}\{f_\gamma\}_{\gamma \in \Gamma} = X^*$ and X is reflexive, we have $\overline{\operatorname{span}}\{f_\gamma\}_{\gamma \in \Gamma} = X^*$, so the M-basis is shrinking.

Let (y_i) be a bounded sequence in X. Assume $\lim_i f_\gamma(y_i) = a_\gamma$ for every $\gamma \in \Gamma$. By the weak compactness of B_X, we can extract a subsequence (y_{i_j}) of (y_i) that w-converges to an element $y \in X$. Then $f_\gamma(y) = \lim_i f_\gamma(y_i) = a_\gamma$ for every $\gamma \in \Gamma$. Therefore the M-basis $\{x_\gamma; f_\gamma\}_{\gamma \in \Gamma}$ is boundedly complete.

(ii)$\Rightarrow$(i) Assume that the M-basis $\{x_\gamma; f_\gamma\}_{\gamma \in \Gamma}$ in $X \times X^*$ is both shrinking and boundedly complete. Let (y_n) be a bounded sequence in X. Given $x \in X$, there is a countable set $\Gamma_{\{x\}} \subset \Gamma$ such that $f_\gamma(x) = 0$ if $\gamma \notin \Gamma_{\{x\}}$; this follows from the definition of an M-basis. By the Cantor diagonal procedure, we can extract a subsequence (n_i) of the natural numbers such that $a_\gamma := \lim_i f_\gamma(y_{n_i})$ exists for each $\gamma \in \bigcup_n \Gamma_{\{y_n\}}$. For $\gamma \notin \bigcup_n \Gamma_{\{y_n\}}$, the sequence $(f_\gamma(y_{n_i}))_i$ is identically zero. Put $a_\gamma := 0$ in this case. By the bounded completeness, there is $y \in X$ such that $f_\gamma(y) = a_\gamma$ for each $\gamma \in \Gamma$. Then we have $\lim_i f_\gamma(y_{n_i}) = a_\gamma = f_\gamma(y)$ for every $\gamma \in \Gamma$. Since (y_{n_i}) is bounded and $\{x_\gamma; f_\gamma\}_{\gamma \in \Gamma}$ is shrinking, we have that $\lim_i f(y_{n_i}) = f(y)$ for every $f \in X^*$. Therefore B_X is weakly compact by the Eberlein-Šmulyan theorem, and X is thus reflexive.

(i)$\Rightarrow$(iii) Let X be a reflexive space and let $\{x_\gamma; x_\gamma^*\}_{\gamma \in \Gamma}$ be an M-basis in $X \times X^*$. Let T be the one-to-one bounded linear operator from X into $c_0(\Gamma)$ defined by $T(x) := (\langle x, x_\gamma^* \rangle)_{\gamma \in \Gamma}$. Let $\| \cdot \|_D$ be the Day norm on $c_0(\Gamma)$ from Lemma 3.54. Finally, let $\| \cdot \|$ be the norm on X defined by $\|x\|^2 = \|x\|_0^2 + \|Tx\|_D^2$ for $x \in X$, where $\| \cdot \|_0$ is the original norm of X. We will show that the norm $\| \cdot \|$ has the desired property. Indeed, let $\lim_{m,n \to \infty} 2\|x_n\|^2 + 2\|x_m\|^2 - \|x_n + x_m\|^2 = 0$ for some bounded sequence (x_n). Then a similar fact holds for (Tx_n). Moreover, as B_X is weakly compact, we have that (Tx_n) has a weak cluster point Tx in $c_0(\Gamma)$. Thus, from Lemma 3.54, we get that $\lim Tx_n = Tx$ in the supremum norm of $c_0(\Gamma)$ and thus $f(x_n - x) \to 0$ uniformly on $f \in \{T^* e_\gamma\}_{\gamma \in \Gamma}$, where e_γ are the unit vectors in $l_1(\Gamma)$. Thus, in particular, $x_n \to x$ weakly in X as the linear hull of $\{T^* e_\gamma\}$ is $\| \cdot \|$-dense in X^* due to the reflexivity of X.

(iii)$\Rightarrow$(i) Assume that X has a norm $\|\cdot\|$ as in (iii). Let $f \in S_{X^*}$ be given. Let $x_n \in S_X$ be such that $f(x_n) \to 1$. Then $2 \geq \|x_n + x_m\| \geq f(x_n + x_m) \to 2$. Thus x_n is weakly convergent to some $x \in B_X$ and $f(x) = 1$. Therefore each $f \in S_{X^*}$ attains its norm, and X is reflexive by the James theorem.

(i)$\Rightarrow$(iv) We follow [JoLi01h, Chap. 18]. If $\|\cdot\|$ denotes some equivalent norm on a separable reflexive Banach space X and $x \in X$, we define the *symmetrized type norm* $\|\cdot\|_x$ on X by

$$\|y\|_x := \left\| x\|y\| + y \right\| + \left\| x\|y\| - y \right\|, \quad y \in X.$$

For every $x \in X$, $\|\cdot\|_x$ is an equivalent norm on X such that $2\|y\| \leq \|y\|_x \leq (2 + 2\|x\|)\|y\|$ for all y. To check the triangle inequality, one uses the fact that for fixed u and v in X, the function $s(r) = \|ru + v\| + \|ru - v\|$ is convex and even on $\mathbb{R}$ and thus is increasing on $\mathbb{R}^+$.

We now fix a countable dense $\mathbb{Q}$-linear subspace $Z \subset X$, and we choose a sequence $(p_z)_{z \in Z}$ of positive real numbers such that $\sum_{z \in Z} p_z(1 + \|z\|) < \infty$. We define a map Δ from the set of equivalent norms on X into itself as follows:

$$\Delta(\|\cdot\|)(x) := \sum_{z \in Z} p_z \|x\|_z, \quad \|\cdot\| \text{ any equivalent norm on } X, \ x \in X.$$

Let $\|\cdot\|$ be a strictly convex equivalent norm. It turns out that $\|\cdot\|_M =: \Delta(\Delta(\|\cdot\|))$ satisfies (iv). This relies on the following crucial fact.

Fact 6.2. *Let us denote $\|\cdot\|_1 = \Delta(\|\cdot\|)$. Let $(x_n) \subset X$ be a sequence such that $\|x_n\| = 1$ for all n and*

$$\lim_m \lim_n \|x_m + x_n\|_1 = 2 \lim_n \|x_n\|_1.$$

Then there is a subsequence (x'_n) of (x_n) such that, for all $y \in X$ and $\gamma, \beta \geq 0$, we have

$$\lim_m \lim_n \|y + \gamma x'_m + \beta x'_n\| = \lim_m \|y + (\gamma + \beta)x'_m\|.$$

To show this fact, we extract a subsequence (x'_n) such that, for every $z \in Z$, $y \in Z$, and $\gamma, \beta \in \mathbb{Q}^+$, the limits

$$\lim_m \lim_n \|y + \gamma x'_m + \beta x'_n\|_z$$

exist, which is easy through a diagonal argument. The assumptions of the fact and a classical convexity argument imply that, for all $z \in Z$,

$$\lim_m \lim_n \|x'_m + x'_n\|_z = 2 \lim_n \|x'_n\|_z.$$

If we let $z = 0$, we obtain, since (x_n) is normalized, that

$$\lim_m \lim_n \|x'_m + x'_n\| = 2,$$

and thus for $\gamma, \beta \geq 0$ one has

$$\lim_m \lim_n \|\gamma x'_m + \beta x'_n\| = \gamma + \beta. \tag{6.1}$$

Similarly, we have, for all $z \in Z$, that

$$\lim_m \lim_n \|\gamma x'_m + \beta x'_n\|_z = (\gamma + \beta) \lim_m \|x'_m\|_z. \tag{6.2}$$

Let $y \in Z$ and $\gamma, \beta \in \mathbb{Q}^+$. We apply (6.2) to $z := (\gamma + \beta)^{-1} y$. Using (6.1), we obtain that

$$\lim_m \lim_n \left(\|y + \gamma x'_m + \beta x'_n\| + \|y - \gamma x'_m - \beta x'_n\| \right)$$
$$= \lim_m \left(\|y + (\gamma + \beta) x'_m\| + \|y - (\gamma + \beta) x'_m\| \right). \tag{6.3}$$

By the triangle inequality, we have for $\eta \in \{-1,\ 1\}$ that

$$\lim_m \lim_n \|y + \eta(\gamma x'_m + \beta x'_n)\|$$
$$\leq \lim_m \left\| y \frac{\gamma}{\gamma + \beta} + \eta \gamma x'_m \right\| + \lim_n \left\| y \frac{\beta}{\gamma + \beta} + \eta \beta x'_n \right\|$$
$$\leq \lim_m \|y + \eta(\gamma + \beta) x'_m\|,$$

and now it follows from (6.3) that

$$\lim_m \lim_n \|y + \gamma x'_m + \beta x'_n\| = \lim_m \|y + (\gamma + \beta) x'_m\|.$$

We proved the fact for $y \in Z$ and $\gamma, \beta \in \mathbb{Q}^+$. An obvious density argument concludes the proof of the fact in general.

In the notation of the fact and under the same assumptions, one obtains through an inductive procedure the following statement: given $\epsilon > 0$, there is a subsequence (x''_n) of (x_n) such that for all $k \in \mathbb{N}$ and all positive real numbers $\gamma_1,\ \gamma_2, \ldots, \gamma_k$, one has

$$\left\| \sum_{i=1}^k \gamma_i x''_i \right\| \geq (1 - \epsilon) \sum_{i=1}^k \gamma_i. \tag{6.4}$$

In particular, it follows from (6.4) and Mazur's theorem that

$$(x''_n) \text{ has no weakly null subsequence.} \tag{6.5}$$

We now conclude the proof of (i) implies (iv). Starting from a strictly convex norm $\|\cdot\|$, we let $\|\cdot\|_1 = \Delta(\|\cdot\|)$ and we denote $\|\cdot\|_M = \Delta(\|\cdot\|_1)$. We consider a sequence (x_n) such that

$$\lim_m \lim_n \|x_m + x_n\|_M = 2 \lim_n \|x_n\|_M.$$

Since the norm $\|\cdot\|_M$ is strictly convex, it suffices to show that any such sequence has a norm-convergent subsequence. Since X is reflexive, passing to a subsequence, we may assume that $x_n = x + y_n$, where y_n is weakly null and $\lim \|y_n\| = A$ exists. If $A = 0$, we are done. Assume it is not so. Then we may also assume that $\|y_n\|_1 = 1$ for all n.

We now apply the fact to $\|\cdot\|_1$ and $\|\cdot\|_M = \Delta(\|\cdot\|_1)$ to find a further subsequence, which we still denote (x_n), such that, for all $y \in X$, one has

$$\lim_m \lim_n \|y + x_m + x_n\|_1 = 2 \lim_m \left\|\frac{y}{2} + x_m\right\|_1.$$

Letting $y = -2x$, it follows with the notation above that

$$\lim_m \lim_n \|y_m + y_n\|_1 = 2 \lim_m \|y_m\|_1 = 2. \tag{6.6}$$

Choose y'_n such that $\|y'_n\| = 1$ and $\lim \|Ay'_n - y_n\| = 0$. It follows from (6.6) that

$$\lim_m \lim_n \|y'_m + y'_n\|_1 = 2 \lim_m \|y'_m\|_1.$$

We may now again apply the fact, this time with $\|\cdot\|$ and $\|\cdot\|_1 = \Delta(\|\cdot\|)$, and its consequence (6.5) to conclude that (y'_n) has no weakly null subsequence. But this is a contradiction since (y_n) is weakly null.

(iv)$\Rightarrow$(iii) This is obvious. $\square$

It is an open problem if every reflexive Banach space has an equivalent 2R norm.

Theorem 6.3. *Let X be a Banach space. Then the following are equivalent:*

(1) X *admits a shrinking M-basis.*
(2) X *is WCG and Asplund.*
(3) X *is WLD and Asplund.*
(4) X *is WLD and has an equivalent norm whose dual norm is LUR.*
(5) X *is WLD and has an equivalent Fréchet differentiable norm.*

Proof. (1)$\Rightarrow$(2) follows from the fact that for a shrinking M-basis $\{x_\gamma; x_\gamma^*\}_{\gamma \in \Gamma}$ in $X \times X^*$, the set $\{x_\gamma; \gamma \in \Gamma\} \cap \{0\}$ is weakly compact and from the Remark after Definition 6.2.3 in [Fab97]. (2)$\Rightarrow$(3) is trivial. (3)$\Rightarrow$(1) is in [Fab97, Thm. 8.3.3]. (2)$\Rightarrow$(4) is in [DGZ93a, Cor. VII.1.13]. (4)$\Rightarrow$(5) is trivial. (5)$\Rightarrow$(3) follows from the fact that every Banach space with a Fréchet norm is Asplund (see, e.g., [DGZ93a, Thm. II.5.3]). $\square$

It is not known if in Theorem 6.3 WLD can be replaced by the existence of a norming M-basis for X.

Since the notions of Asplund property and WLD property are both hereditary, we have the following.

Corollary 6.4 ([JoZi77]). *The property of having a shrinking M-basis is hereditary. In particular, for every Γ, any subspace of $c_0(\Gamma)$ is WCG.*

Argyros and Mercourakis proved, in [ArMe05a], that *there is a WCG space X with unconditional basis and a subspace Y of X with unconditional basis such that Y is not WCG.*

6.2 Reflexive Generated and Vašák Spaces

In this section, WCG spaces are characterized in terms of the existence of a weakly compact M-basis. Then we use various versions of the notion of σ-shrinkable M-bases to provide characterizations of subspaces of weakly compactly generated spaces and Vašák spaces. As an application, we give a short proof of the Eberlein property of continuous images of Eberlein compacta. The technique of projectional resolutions of the identity developed in Chapter 3 plays a substantial role here.

Definition 6.5. *We will say that a Banach space X is* reflexive generated *(resp.* Hilbert generated*) if there is a reflexive (resp. Hilbert) Banach space Y and a bounded linear operator $T : Y \to X$ with a dense range.*

Definition 6.6. (i) *Let M be a bounded linearly dense set in X. We will say that the norm $\|\cdot\|$ on X is* dually M-2-rotund *(dually M-2R, for short) if a sequence (f_n) converges to some $f \in B_{X^*}$ uniformly on M whenever $f_n \in S_{X^*}$ are such that $\lim_{n,m\to\infty} \|f_n + f_m\| = 2$.*

(ii) *If M is a bounded linearly dense set in X, we will say that the norm $\|\cdot\|$ on X is* Fréchet M-smooth *if, at each point of S_X, the norm is differentiable uniformly in the directions of M, i.e.,*

$$\lim_{t\to 0+} \sup_{h\in M} \frac{1}{t}(\|x + th\| + \|x - th\| - 2) = 0$$

for each $x \in S_X$.

Note that any dually M-2R norm is Fréchet M-smooth by Šmulyan's lemma [DGZ93a, Thm. I.1.4]. In particular, if a norm is dually B_X-2R (i.e., dually 2R), it is Fréchet smooth. An instance of a norm dually 2R appeared already in Theorem 6.1 (iv).

Definition 6.7. *An M-basis $\{x_\gamma; x_\gamma^*\}_{\gamma\in\Gamma}$ for a Banach space X is called* weakly compact *if $\{x_\gamma; \gamma \in \Gamma\} \cup \{0\}$ is a weakly compact set.*

The M-basis is called σ-weakly compact if $\{x_\gamma; \gamma \in \Gamma\} \cup \{0\}$ is a σ-weakly compact set; i.e., a countable union of weakly compact sets.

Remark 6.8. Every σ-weakly compact M-basis is easily transformed into a weakly compact M-basis using suitable scalar multiplications, but not every M-basis in a WCG Banach space is necessarily σ-weakly compact. Indeed, suppose that X is WCG and $Y \hookrightarrow X$ is a non-WCG subspace (see, e.g.,

[Rose74]). Using the basis extension in Theorem 5.64, there exists an M-basis $\{x_\gamma; x_\gamma^*\}_{\gamma \in \Gamma}$ in $X \times X^*$ that extends an M-basis $\{x_\gamma; x_\gamma^*\}_{\gamma \in \Gamma_0}$ of Y. If $\{x_\gamma; x_\gamma^*\}_{\gamma \in \Gamma}$ were σ-weakly compact, $\{x_\gamma; x_\gamma^*\}_{\gamma \in \Gamma_0}$ would be also, which is a contradiction. On the other hand, Johnson (see Theorem 7.40) showed that every unconditional basis of a WCG space is necessarily σ-weakly compact.

The following result, partly coming from [FMZ05], [FMZ04a], [FHMZ05] and [FGMZ04], collects equivalent conditions for weakly compact generating of Banach spaces.

Theorem 6.9. *Let X be a Banach space. Then the following are equivalent:*

(i) *X is WCG.*
(ii) *X admits a (σ-) weakly compact M-basis.*
(iii) *X admits an M-basis $\{x_\gamma; x_\gamma^*\}_{\gamma \in \Gamma}$ such that $X^* = \overline{\operatorname{span}}^{\,\mathcal{T}}\{x_\gamma^*; \gamma \in \Gamma\}$, where $\mathcal{T}$ is the topology in X^* of uniform convergence on the set $\{x_\gamma; \gamma \in \Gamma\}$.*
(iv) *X admits an equivalent dually M-2R norm for some bounded linearly dense set M in X.*
(v) *X is WLD and admits an equivalent Fréchet M-smooth norm for some bounded linearly dense M in X.*
(vi) *X is generated by a reflexive space (with an unconditional basis). Indeed, for every weakly compact M-basis $\{x_\gamma; x_\gamma^*\}_{\gamma \in \Gamma}$, there is a reflexive space R with bounded unconditional basis $\{z_\gamma\}_{\gamma \in \Gamma}$ and a bounded linear operator $T : R \to X$ such that $Tz_\gamma = x_\gamma$ for all $\gamma \in \Gamma$.*

In the proof, we shall need the following lemma.

Lemma 6.10. *Let K be a bounded separable set in a Banach space $(X, \|\cdot\|)$. Then K is weakly relatively compact if and only if X admits an equivalent norm $\|\|\cdot\|\|$ such that (f_n) is a sequence that is uniformly Cauchy on K whenever $f_n \in S_{X^*}$ are such that $\lim_{m,n\to\infty} \|\|f_m + f_n\|\| = 2$.*

Proof. Assume that K is weakly relatively compact. Let Z be a separable reflexive Banach space and T be a bounded linear operator from Z into X such that $TB_Z \supset K$ ([DFJP74]). Let $\|\|\cdot\|\|$ be a norm on Z such that its dual norm is 2R (see Theorem 6.1 (iv)). Then a simple convexity argument proves that the norm on X^* defined by $\|\|f\|\|^2 := \|f\|^2 + \|T^*f\|^2$ is the dual of the required norm on X^*.

We will now prove sufficiency (even without the separability assumption on K) by assuming the existence of a norm $\|\|\cdot\|\|$ on X with the required properties and referring the rest of the argument to this norm. Let $S \subset K$ be a countable subset of K, and assume that $s^{**} \in \overline{S}^{w^*} \subset X^{**}$ does not belong to X. Let $F \in S_{X^{***}}$ be such that $F \in X^\perp$ and $F(s^{**}) = \operatorname{dist}(s^{**}, X) > 0$. Let $\{y_i; i \in \mathbb{N}\} \subset B_{X^{**}}$ such that $\sup_i F(y_i) = 1$. From a "metrizable version" of the Goldstine theorem (see, e.g., [Fa~01, Thm. 3.27]), we can find $f_n \in S_{X^*}$ such that $\lim_n (f_n - F)(x) = 0$ for all $x \in S \cup \{y_i; i \in \mathbb{N}\} \cup \{s^{**}\}$. Then

$\lim_{m,n} \||f_n + f_m\|| = 2$ and thus, by the rotundity assumed, $\lim_n (f_n - F)(x) = 0$ uniformly on $S \cup \{s^{**}\}$. As all f_n are continuous on $(S \cup \{s^{**}\}, \hat{w}^*)$, where $\hat{w}^*$ denotes the restriction to $S \cup \{s^{**}\}$ of the w^*-topology, so is their uniform limit on this set, which is not the case, as F is zero on S and $F(s^{**}) > 0$. Therefore $s^{**} \in X$ and thus K is weakly relatively countably compact and thus weakly relatively compact by the Eberlein-Šmulyan theorem.   □

Proof of Theorem 6.9. (i)⇒(ii) In fact, if K is an absolutely convex weakly compact linearly dense subset of X, the M-basis $\{x_\gamma; x_\gamma^*\}_{\gamma \in \Gamma}$ can be chosen such that $\{x_\gamma; \gamma \in \Gamma\} \subset K$. This follows from Proposition 3.43, Theorem 3.44, and by a standard induction process, a modification of the proof of Theorem 5.1. Indeed, given a separable Banach space X, Lemma 1.21 gives (after homogenizing) an M-basis $\{x_n\}_{n=1}^\infty \subset K$. If the result holds for every WCG Banach space of density less than a certain uncountable cardinal $\aleph$, and if X is a WCG of density $\aleph$, a PRI $\{P_\alpha; \omega \leq \alpha \leq \mu\}$ on X exists such that $P_\alpha(K) \subset K$ for all α (see Theorem 3.44). For $\omega \leq \alpha < \mu$, an M-basis in $(P_{\alpha+1} - P_\alpha)X$ can be found, by the induction hypothesis, in $(P_{\alpha+1} - P_\alpha)(K) \subset K$. Finally, put together all those bases in one as in the proof of Theorem 5.1.

(ii)⇒(i) This is trivial in the case of weakly compact M-bases; if $\{x_\gamma; x_\gamma^*\}_{\gamma \in \Gamma}$ is a normalized σ-weakly compact M-basis, apply what has been said in Remark 6.8.

(i)⇔(iii) If $\{x_\gamma; \gamma \in \Gamma\}$ is bounded, an M-basis $\{x_\gamma; x_\gamma^*\}_{\gamma \in \Gamma}$ in a Banach space X is weakly compact if and only if $X^* = \overline{\text{span}}^{\mathcal{T}}\{x_\gamma^*; \gamma \in \Gamma\}$, where $\mathcal{T}$ is the topology of the uniform convergence on the set $\{x_\gamma; \gamma \in \Gamma\}$. Indeed, if the M-basis is weakly compact, we can use the Mackey-Arens theorem (Theorem 3.2) to show the statement. On the other hand, if the condition holds, given $f \in X^*$ and given $\varepsilon > 0$, find $g \in \text{span}\{x_\gamma^*; \gamma \in \Gamma\}$ such that $\sup_{\gamma \in \Gamma} |\langle x_\gamma, f - g\rangle| < \varepsilon$. Let $(\gamma_n)_{n=1}^\infty$ be a sequence of distinct points in Γ. There exists $n_0 \in \mathbb{N}$ such that, for every $n \geq n_0$, $\langle x_{\gamma_n}, g\rangle = 0$, due to the orthogonality of the system, so $\limsup_{n\to\infty} |\langle x_{\gamma_n}, f\rangle| \leq \varepsilon$ and this implies that $x_{\gamma_n} \xrightarrow{\omega} 0$. Thus the M-basis is weakly compact by the Eberlein-Šmulyan theorem.

(iv)⇒(i) See Lemma 6.10 (separability is not needed in this direction).

(ii)⇒(iv) Assume that X has a weakly compact M-basis $\{x_\gamma; x_\gamma^*\}_{\gamma \in \Gamma}$. Then the operator $Tf := (\langle x_\gamma, f\rangle)_{\gamma \in \Gamma}$, $f \in X^*$, is a bounded linear weak*-weak continuous map from X^* into $c_0(\Gamma)$ (see Theorem 5.3). Let $\|\cdot\|_D$ be the Day norm on $c_0(\Gamma)$. By [HaJo04] (see Lemma 3.54), the norm $\||\cdot\||$ defined on X^* by $\||f\||^2 := \|f\|_1^2 + \|Tf\|_D^2$ is dually M-2R for $M := \{T^*(e_\gamma); \gamma \in \Gamma\}$, where $\{e_\gamma; \gamma \in \Gamma\}$ is the set of unit vectors in $\ell_1(\Gamma)$. By Lemma 6.10, the set M is weakly relatively compact in X, and the closed linear hull of it equals X.

(iv)⇒(v) It has already been proved that (iv) implies that X is WCG and hence X is WLD. Moreover, every dually M-2-rotund norm is M-Fréchet smooth by the Šmulyan's lemma [DGZ93a, Thm. I.1.4].

(v)⇒(i) We will show that (v) implies the existence in X of a linearly dense set $G \subset B_X$ such that for every $\varepsilon > 0$ and for every $x^* \in B_{X^*}$, card $\{x \in$

$G; |\langle x, x^* \rangle| > \varepsilon\} < \omega$. This will imply (i) as $G \cup \{0\}$ is then weakly compact by the Eberlein-Šmulyan theorem.

In order to construct the set G, we assume that the set M in (v) is convex symmetric and closed. We shall find the set G satisfying the assertion (ii) by a transfinite induction. If X is separable, then we can take $G = \{\frac{1}{n}x_n; \ n \in \mathbb{N}\}$, where $\{x_n; \ n \in \mathbb{N}\}$ is any dense countable set in M. Let $\aleph$ be an uncountable cardinal and assume that we already found a set $G \subset M$ as in the assertion (ii) whenever the density of X was less than $\aleph$. Now assume that X has density $\aleph$.

There is a PRI $\{P_\alpha; \omega \le \alpha \le \mu\}$ on $(X, \|\cdot\|)$ such that $P_\alpha M \subset M$ for every $\alpha \in [\omega, \mu)$ (see Theorem 3.44). For $\alpha \in [\omega, \mu)$, denote $Q_\alpha = P_{\alpha+1} - P_\alpha$; observe that then $Q_\alpha X$ has density less than $\aleph$ and the norm $\|\cdot\|$ restricted to this subspace is $Q_\alpha M$-smooth. For every $\alpha \in [\omega, \mu)$, find, by the induction assumption, a linearly dense set $G_\alpha \subset \frac{1}{2} Q_\alpha M \ (\subset M)$ satisfying (ii).

Put $G = \bigcup_{\alpha < \mu} G_\alpha$. It remains to verify the stated property for this set. As the set $\bigcup_{\alpha < \mu} Q_\alpha X$ is linearly dense in X, so is the set G. Fix any $\varepsilon > 0$ and any $x^* \in B_{X^*}$. We have to show that the set $\{x \in G; \ |\langle x, x^* \rangle| > \varepsilon\}$ is finite. In order to do so, we shall be proving the statement

$$\text{card}\{\alpha \in [\omega, \mu); \ \langle x, P_\beta^* x^* \rangle > \varepsilon \ \text{for some} \ x \in G_\alpha\} < \aleph_0 \qquad (\beta)$$

for all $\beta \in [\omega, \mu]$. Clearly, (ω) is valid. Also, since $P_{\beta+1} \circ Q_{\beta+1} = 0$, we have that (β) implies $(\beta+1)$ for every $\beta < \mu$. Now let $\lambda \le \mu$ be any limit ordinal and assume that we verified (β) for every $\beta < \lambda$. Find $\beta < \lambda$ so that $\sup\langle M, P_\lambda^* x^* - P_\beta^* x^* \rangle < \varepsilon$. This follows from the M-smoothness of $\|\cdot\|$, the Šmulyan lemma, and the Bishop-Phelps theorem. We observe that if $|\langle x, P_\lambda^* x^* \rangle| > \varepsilon$ for some $x \in G_\alpha$, where $\alpha \in [\omega, \mu)$, then, as $x \in M$,

$$|\langle x, P_\beta^* x^* \rangle| \ge |\langle x, P_\lambda^* x^* \rangle| - |\langle x, P_\lambda^* x^* - P_\beta^* x^* \rangle| > \varepsilon - \varepsilon = 0,$$

so we must have $\alpha < \beta$. Thus

$$\text{card}\{\alpha \in [\omega, \mu); \ |\langle x, P_\lambda^* x^* \rangle| > \varepsilon \ \text{for some} \ x \in G_\alpha\}$$

$$= \text{card}\{\alpha \in [\omega, \mu); \ |\langle x, P_\beta^* x^* \rangle| > \varepsilon \ \text{for some} \ x \in G_\alpha\} < \aleph_0$$

and hence (λ) holds. We thus proved (β) for every $\beta \le \mu$. In particular, (μ) holds; that is, given any $\varepsilon > 0$, the set

$$F = \{\alpha \in [\omega, \mu); \ |\langle x, x^* \rangle| > \varepsilon \ \text{for some} \ x \in G_\alpha\}$$

is finite and so

$$\text{card}\{x \in G; \ |\langle x, x^* \rangle| > \varepsilon\} = \sum_{\alpha \in F} \text{card}\{x \in G_\alpha; \ |\langle x, x^* \rangle| > \varepsilon\} < \aleph_0$$

by the induction assumption.

(vi)⇒(i) is trivial.

(i)⇒(vi) This is a result of Davis, Figiel, Johnson and Pełczyński; see e.g., [Fa⁻01, Thm. 11.17]. For the unconditional basis amendment, we refer to [DFJP74].

This finishes the proof of Theorem 6.9.   □

Definition 6.11. *A nonempty subset G of a Banach space X will be called σ-shrinkable if $G = \bigcup_{n=1}^{\infty} G_n$ so that for every neighborhood U of the origin in $(X^{**}, \|\cdot\|)$ and for every $x \in G$ there is $n \in \mathbb{N}$ such that $x \in G_n$ and $G'_n \subset U$, where A' denotes the set of all accumulation points of a set $A \subset X$ in (X^{**}, w^*). An M-basis $\{x_\gamma; x_\gamma^*\}_{\gamma \in \Gamma}$ in $X \times X^*$ is called σ-shrinkable if the set $\{x_\gamma; \gamma \in \Gamma\}$ is σ-shrinkable.*

Remark 6.12. By the Hahn-Banach theorem, a nonempty subset G of a Banach space X is σ-shrinkable if and only if, for every $\varepsilon > 0$, $G = \bigcup_{n=1}^{\infty} G_n^\varepsilon$, so that $\operatorname{card}\{x \in G_n^\varepsilon; |\langle x, x^*\rangle| \geq \varepsilon\} < \aleph_0$ for all $x^* \in B_{X^*}$ and for all $n \in \mathbb{N}$.

The following result, partly coming from [FMZ05] and [FMZ04a], collects several equivalent conditions for being a subspace of a weakly compactly generated Banach space.

Theorem 6.13. *Let X be a Banach space. Then the following are equivalent:*

(i) *X is a subspace of a WCG Banach space.*
(ii) *X admits a σ-shrinkable M-basis.*
(iii) *There exists a σ-shrinkable linearly dense subset G of X.*
(iv) *(B_{X^*}, w^*) is an Eberlein compact.*
(v) *For every $\varepsilon > 0$, $B_X = \bigcup_n B_n^\varepsilon$, so that $\overline{B_n^\varepsilon}^{w^*} \subset X + \varepsilon B_{X^{**}}$ for every n.*

Moreover, if this is the case, then every linearly dense subset G of X countably supporting X^* (in particular, every M-basis of X) is σ-shrinkable.

In order to prove the theorem, we need to develop some material. The following result, based on Theorems 3.44 and 5.64, will frequently be used.

Lemma 6.14. *Let Z be a WCG Banach space generated by a weakly compact absolutely convex set K and X be a subspace of Z. Then any M-basis $\{x_\gamma; x_\gamma^*\}_{\gamma \in \Gamma_1}$ in $X \times X^*$ can be extended to an M-basis $\{x_\gamma; x_\gamma^*\}_{\gamma \in \Gamma}$ in $Z \times Z^*$, and a PRI $\{P_\alpha; \omega \leq \alpha \leq \mu\}$ subordinated to $\{x_\gamma; \gamma \in \Gamma\}$ can be constructed on Z such that $P_\alpha(K) \subset K$ for all $\omega \leq \alpha \leq \mu$. In particular, $P_\alpha X \subset X$ for all $\omega \leq \alpha \leq \mu$.*

Definition 6.15. *We will say that a PRI $\{P_\alpha; \omega \leq \alpha \leq \mu\}$ on a Banach space X is σ-shrinkable if there is a countable collection $\{B_n\}_{n=1}^{\infty}$ of subsets of B_X such that for every $x_0 \in B_X$ and for every $\varepsilon > 0$, there is $n_0 \in \mathbb{N}$ such that $x_0 \in B_{n_0}$ and $\limsup_{\alpha \uparrow \beta} \sup |\langle B_{n_0}, (P_\alpha^* - P_\beta^*)f\rangle| \leq \varepsilon$, for all $f \in B_{X^*}$ and all limit ordinals $\beta \in (\omega, \mu]$.*

Proposition 6.16. *Let X be a Banach space. Let G be a linearly dense subset of X that countably supports X^*. Let $\{P_\alpha; \omega \leq \alpha \leq \mu\}$ be a PRI on X subordinated to G. Then G is σ-shrinkable if and only if $(P_\alpha)_{\omega \leq \alpha \leq \mu}$ is σ-shrinkable.*

Proof. Assume first that G is σ-shrinkable. We may and do assume that $G \subset B_X$. Let $(G_n)_{n=1}^\infty$ be the covering of G given by the definition of σ-shrinkable. Given $\varepsilon > 0$, let $n \in \mathbb{N}$ be such that $G'_n \subset \varepsilon B_{X^{**}}$. Suppose that, for some limit ordinal $\omega < \beta \leq \mu$ and some $x^* \in B_{X^*}$,

$$\limsup_{\alpha \uparrow \beta} \ \sup_{x \in G_n} |\langle x, (P_\beta^* - P_\alpha^*)(x^*)\rangle| > \varepsilon.$$

Then we can find an increasing net $(\alpha_i)_{i \in I}$ in $[\omega, \beta)$ such that $\alpha_i \to \beta$ and elements $x_i \in G_n$ such that

$$|\langle x_i, (P_\beta^* - P_{\alpha_i}^*)x^*\rangle| = |\langle (P_\beta - P_{\alpha_i})x_i, x^*\rangle| > \varepsilon \text{ for all } i \in I.$$

If $P_\beta(x_i) = 0$, then $P_\alpha(x_i) = 0$ for all $\alpha \leq \beta$, so $P_\beta(x_i) = x_i$ and $P_{\alpha_i}(x_i) = 0$ for all $i \in I$. It follows that $|\langle x_i, x^*\rangle| > \varepsilon$ for all $i \in I$. Let x^{**} be an accumulation point of $\{x_i; \ i \in I\}$ in (X^{**}, w^*). Then $|\langle x^{**}, x^*\rangle| \geq \varepsilon$, a contradiction. It follows that

$$\limsup_{\alpha \uparrow \beta} \ \sup_{x \in G_n} |\langle x, (P_\beta^* - P_\alpha^*)x^*\rangle| \leq \varepsilon \quad \text{for all } x^* \in B_{X^*}.$$

Now, a simple argument involving sets of the form

$$\overline{\text{span}}\{a_1 G_1 + a_2 G_2 + \ldots + a_m G_m + \varepsilon B_X\} \cap B_X,$$

where $a_i \in \mathbb{Q}$, $\sum_{j=1}^m |a_j| \leq K$, $\varepsilon > 0$, $m \in \mathbb{N}$, $K > 0$, proves that $(P_\alpha)_{\omega \leq \alpha \leq \mu}$ is σ-shrinkable.

Assume now that $\{P_\alpha; \omega \leq \alpha \leq \mu\}$ is a σ-shrinkable long sequence of projections on X that satisfies all properties of a PRI but not necessarily the requirement that μ be the first ordinal of cardinality dens X (let us call it, from now on, a PRI′ on X), and let G be a linearly dense subset of X countably supporting X^* and subordinated to the PRI′. We shall prove that G is σ-shrinkable. This will be done by transfinite induction on the density of X. If X is separable, then G is countable, and the result is obvious. Assume that the result has been proved for every Banach space of density less than $\aleph$, a certain uncountable cardinal, having a σ-shrinkable PRI′. Let X be a Banach space of density $\aleph$ with a σ-shrinkable PRI′, and let G be a linearly dense subset of X countably supporting X^* and subordinated to the PRI′. We may and do assume that $G \subset B_X$.

Given $x \in G$, let $b(x)$ be the first ordinal α in $(\omega, \mu]$ such that $P_\alpha(x) = x$. Then $b(x)$ has a predecessor $a(x)$; it follows that, for all $x \in G$, $x \in (P_{a(x)+1} - P_{a(x)})(X)$. Define a well-order in each of the sets $\{x \in G; \ a(x) = \alpha\}$, $\alpha \in [\omega, \mu)$. This induces a lexicographic well-order $\prec$ in Γ, and the mapping

$a : \Gamma \to [\omega, \mu)$ is obviously increasing for this order. Given $\varepsilon > 0$, we can write $B_X = \bigcup_{n \in \mathbb{N}} B_n^\varepsilon$ and

$$\limsup_{\alpha \uparrow \beta} \sup_{b \in B_n^\varepsilon} |\langle b, (P_\beta^* - P_\alpha^*)x^* \rangle| \le \varepsilon$$

for all limit ordinals $\beta \in (\omega, \mu]$ and $x^* \in B_{X^*}$.

Define $G_n^\varepsilon := G \cap B_n^\varepsilon$, $n \in \mathbb{N}$. It follows that $G = \bigcup_{n \in \mathbb{N}} G_n^\varepsilon$. We fix $\varepsilon > 0$ and $n \in \mathbb{N}$. Let $x^{**} \in (G_n^\varepsilon)'$ (the set of all accumulation points of G_n^ε in (X^{**}, w^*)). Let $\mathcal{W}$ be the family of neighborhoods of x^{**} in (X^{**}, w^*) partially ordered by inclusion; i.e., $W_1 \le W_2$ if and only if $W_2 \subset W_1$. Given $W \in \mathcal{W}$, let $g(W)$ be the first element (in the order $\prec$) in $G_n^\varepsilon \cap W$. The net $\{g(W); W \in (\mathcal{W}, \le)\}$ is w^*-convergent to x^{**}, and the mapping $g : \mathcal{W} \to G_n^\varepsilon$ is increasing. It follows that the mapping $a \circ g : \mathcal{W} \to [\omega, \mu)$ is also increasing. Let $\beta := \lim_{W \in \mathcal{W}} (a \circ g(W) + 1)$. If β is not a limit ordinal, then consider the Banach space $P_\beta(X)$ (whose density is less than $\aleph$) and the long sequence $\{P_\alpha; \omega \le \alpha \le \beta\}$ of projections on it (a σ-shrinkable PRI' on $P_\beta(X)$ for the sets $B_n^\varepsilon \cap P_\beta(X)$), and carry on the construction in this setting to get, by the induction hypothesis, $\|x^{**}\| \le \varepsilon$. If β is a limit ordinal, given $x^* \in B_{X^*}$, we get

$$\langle g(W), x^* \rangle = \langle (P_\beta - P_{a \circ g(W)})g(W), x^* \rangle = \langle g(W), (P_\beta^* - P_{a \circ g(W)}^*)x^* \rangle,$$

and

$$\langle g(W), x^* \rangle \to \langle x^{**}, x^* \rangle.$$

Since $g(W) \in B_n^\varepsilon$, we get $|\langle x^{**}, x^* \rangle| \le \varepsilon$ for all $x^* \in B_{X^*}$, so $\|x^{**}\| \le \varepsilon$.   □

We will use the following statement.

Lemma 6.17. *Let X be a Banach space, W be an absolutely convex and weakly compact subset of X, and $\{P_\alpha; \omega \le \alpha \le \mu\}$ be a PRI on X such that $P_\alpha(W) \subset W$ for all α. Then, given $x^* \in X^*$ and a limit ordinal $\beta \in (\omega, \mu]$, $P_\alpha^* x^* \to P_\beta^* x^*$ uniformly on W when $\alpha \uparrow \beta$.*

Proof. Obviously, $P_\alpha^* x^* \overset{w^*}{\to} P_\beta^* x^*$ when $\alpha \uparrow \beta$, so

$$P_\beta^* x^* \in \overline{\bigcup_{\alpha < \beta} P_\alpha^* X^*}^{w^*} = \overline{\bigcup_{\alpha < \beta} P_\alpha^* X^*}^{\tau(X^*, X)},$$

where $\tau(X^*, X)$ is the Mackey topology on X^* associated to the dual pair $\langle X^*, X \rangle$ (see Definition 3.1).

Given $\varepsilon > 0$, find $y^* \in X^*$ and $\alpha_0 < \beta$ such that $\sup |\langle W, P_\beta^* x^* - P_{\alpha_0}^* y^* \rangle| < \varepsilon$. Let $\alpha_0 \le \alpha < \beta$. Then $\sup |\langle P_\alpha(W), P_\beta^* x^* - P_{\alpha_0}^* y^* \rangle| < \varepsilon$, as $P_\alpha(W) \subset W$. This implies $\sup |\langle W, P_\alpha^* x^* - P_{\alpha_0}^* y^* \rangle| < \varepsilon$. Then

$$\sup |\langle W, P_\beta^* x^* - P_\alpha^* x^* \rangle|$$

$$\le \sup |\langle W, P_\beta^* x^* - P_{\alpha_0}^* y^* \rangle| + \sup |\langle W, P_\alpha^* x^* - P_{\alpha_0}^* y^* \rangle| < 2\varepsilon.$$ □

Lemma 6.18. *Let X be a WCG Banach space. Let $W \subset X$ be an absolutely convex weakly compact and linearly dense set in X. Let $\{P_\alpha; \omega \leq \alpha \leq \mu\}$ be a PRI on X such that $P_\alpha(W) \subset W$ for all α. Then $(P_\alpha)_{\omega \leq \alpha \leq \mu}$ is σ-shrinkable. If X is a subspace of a WCG Banach space, then X has a σ-shrinkable PRI.*

Remark 6.19. By Proposition 3.43 and Theorem 3.44, a Banach space X generated by W as in Lemma 6.18 has a PRI $\{P_\alpha; \omega \leq \alpha \leq \mu\}$ such that $P_\alpha(W) \subset W$ for all α.

Proof of Lemma 6.18. Given $\varepsilon > 0$, let $B_n^\varepsilon := (nW + \varepsilon B_X) \cap B_X$, $n \in \mathbb{N}$. Given $x \in B_X$, we can find $y \in \mathrm{span}(W)$ such that $\|x - y\| < \varepsilon$. Now, $y \in nW$ for some $n \in \mathbb{N}$, so $x \in B_n^\varepsilon$. By Lemma 6.17, we get

$$\sup |\langle nW, P_\beta^* x^* - P_\alpha^* x^* \rangle| \to 0, \text{ when } \alpha \uparrow \beta \text{ for all } \omega < \beta \leq \mu.$$

Then there exists $\alpha_0 < \beta$ such that

$$\sup |\langle nW, P_\beta^* x^* - P_\alpha^* x^* \rangle| < \varepsilon \text{ for all } \alpha \text{ such that } \alpha_0 \leq \alpha < \beta,$$

so

$$\sup |\langle B_n^\varepsilon, P_\beta^* x^* - P_\alpha^* x^* \rangle| < 2\varepsilon, \text{ for all } \alpha \text{ such that } \alpha_0 \leq \alpha < \beta,$$

and this proves the first part.

In order to prove the second part, observe first that if a Banach space X is a subspace of a WCG Banach space E, then it is also a subspace of a WCG Banach space Z of density $\mathrm{dens}\, X$. Indeed, this follows from the existence of a PG Φ in E (see Proposition 3.43) and the construction of a first norm-1 projection on E using $\Delta := X$ and $\nabla := \{0\}$ in Lemmas 3.33 and 3.34, where $\Psi(e) \subset S_{E^*}$, $\mathrm{card}\, \Psi(e) \leq \aleph_0$, and $\|e\| = \sup |\langle e, \Psi(e) \rangle|$ for all $e \in E$. Let μ be the first ordinal of cardinality $\mathrm{dens}\, X$. By Lemma 6.14, we can find a PRI $\{P_\alpha; \omega \leq \alpha \leq \mu\}$ on Z such that $P_\alpha(K) \subset K$ and $P_\alpha(X) \subset X$ for all α and thus σ-shrinkable by the first part of the proof. It follows that $\{P_\alpha \upharpoonright X; \omega \leq \alpha \leq \mu\}$ is a σ-shrinkable PRI on X.   □

Corollary 6.20. *Let X be a subspace of a WCG Banach space. Then every linearly dense subset of G that countably supports X^* (in particular, every M-basis in $X \times X^*$) is σ-shrinkable.*

Proof. It is enough to put together Lemma 6.14, Proposition 6.16, and Lemma 6.18. The particular case of an M-basis follows from the fact that it countably supports X^* (see Lemma 5.35).   □

We will now give an elementary proof to the following lemma.

Lemma 6.21. *Let X be a Banach space with a σ-shrinkable linearly dense set $G \subset X$ (in particular, assume that there is a σ-shrinkable M-basis in $X \times X^*$). Then (B_{X^*}, w^*) is an Eberlein compact.*

Proof. Let G be a σ-shrinkable linearly dense subset of X. We will construct a homeomorphism of (B_{X^*}, w^*) onto a subset of $c_0(\Delta)$ in its weak topology for some set Δ.

Given $n \in \mathbb{N}$, let $\{G_n^{1/m}\}_{n=1}^{\infty}$ be the sets that cover G for $U := (1/m)B_{X^{**}}$ (see Definition 6.11). For $i \in \mathbb{N}$, let the real-valued function τ_i be defined on the real numbers by

$$\tau_i(t) := \begin{cases} t + (1/i) & \text{if } t \le -(1/i), \\ 0 & \text{if } t \in [-(1/i), (1/i)], \\ t - (1/i) & \text{if } t \ge (1/i). \end{cases} \tag{6.7}$$

The set Δ will be an infinite matrix whose first row is a display of G_1^1, followed by a disjoint display of G_2^1, then G_3^1, etc. The second row is the display of $G_1^{1/2}$ followed by a disjoint display of $G_2^{1/2}$, etc.

If $f \in B_{X^*}$ and $\delta \in \Delta$ is in the i-th row, in the display $G_k^{1/i}$, we put $\Phi f(\delta) := 2^{-(i+k)} \tau_i(f(\delta))$. Then it is easy to see that Φ maps B_{X^*} into $c_0(\Delta)$. Indeed, due to the "weights" 2^{-i}, it suffices to note that, on each row, the values are in c_0. This holds due to the properties of $G_n^{1/m}$ and due to the weights 2^{-k}. The map Φ is weak*-to-pointwise continuous and thus weak*-to-weak continuous. The one-to-one property follows from the observation that if t_1 and t_2 are two different real numbers, then for sufficiently large i, $\tau_i(t_1) \ne \tau_i(t_2)$. Hence B_{X^*} in its weak*-topology is homeomorphic to a weakly compact set in $c_0(\Delta)$. □

We can proceed now with the proof of Theorem 6.13.

Proof. (i)$\Rightarrow$(ii) Let X be a subspace of the WCG Banach space Z. Then X admits an M-basis (see [Reif74] or, more generally, Theorem 5.1). Take any M-basis in X. This basis can be extended to an M-basis of Z (see Lemma 6.14). By Corollary 6.20, this extended M-basis is σ-shrinkable, so the original M-basis on X is σ-shrinkable, too.

(ii)$\Rightarrow$(iii) is obvious.

(iii)$\Rightarrow$(iv) This is Lemma 6.21.

(iv)$\Rightarrow$(i) If K is an Eberlein compact, then $C(K)$ is WCG (see, e.g., [Fa~01, p. 392]) and $X \subset C(B_{X^*})$.

(i)$\Rightarrow$(v) Assume that X is a subspace of a WCG Banach space $(Z, \|\cdot\|)$. Let K be a linearly dense and weakly compact subset of Z. By Krein's theorem (see, e.g., [Fa~01, Theorem 3.58]), we may and do assume that K is convex and symmetric. For $n, p \in \mathbb{N}$, put

$$M_{n,p} = \left(nK + \frac{1}{4p} B_Z\right) \cap B_X.$$

Since the sets $nK + \frac{1}{4p} B_Z$ are obviously $\frac{1}{4p}$-weakly compact in Z, Proposition 3.62 guarantees that the sets $M_{n,p}$ are $\frac{1}{p}$-weakly compact in X. That they satisfy the remaining properties in (iv) can be easily checked.

(v)⇒(iii) [FMZ04a] Instead of this implication, we shall prove the following
*Claim: There exists a symmetric and linearly dense set $G \subset B_X$ such that for
every $\varepsilon > 0$ there are sets $G_m^\varepsilon \subset G$, $m \in \mathbb{N}$, with $\bigcup_{m=1}^\infty G_m^\varepsilon = G$, such that*

$$\forall x^* \in B_{X^*}, \quad \forall m \in \mathbb{N}, \quad \text{card}\left\{x \in G_m^\varepsilon; \; |\langle x, x^* \rangle| > \varepsilon\right\} < \aleph_0.$$

This claim obviously implies (iii).

We shall prove the claim above by induction over the density of X. If X is
separable, we can obviously take for G any countable symmetric dense subset
of B_X. Furthermore, let $\aleph$ be any uncountable cardinal, and assume that we
have verified the claim for all Banach spaces whose density is less than $\aleph$.
Now, let X be a Banach space of density $\aleph$, and assume that we have at hand
the sets $M_{n,p}$ with the properties stated in (v). A simple argument produces
new $M_{n,p}$'s that have the additional properties that each of them is closed
and has a nonempty interior and that $\bigcup_{n=1}^\infty M_{n,p} = B_X$ for every $p \in \mathbb{N}$.

We observe that for every $x \in X$ and every $x^{**} \in X^{**} \setminus X$ there exist
$m, n, p \in \mathbb{N}$ such that $x \in m\overline{M_{n,p}}^{w^*}$ and $x^{**} \notin m\overline{M_{n,p}}^{w^*}$. Because each set
$m\overline{M_{n,p}}^{w^*}$ is weak*-compact, the space X is Vašák.

By Theorem 3.44, we can find a "long sequence" $\{P_\alpha; \omega \leq \alpha \leq \mu\}$ of
projections on X such that, for every $\omega \leq \alpha \leq \mu$ and for every $n, p \in \mathbb{N}$, we
have $P_\alpha(M_{n,p}) \subset M_{n,p}$.

Fix any $\omega \leq \alpha < \mu$ and put $Q_\alpha := P_{\alpha+1} - P_\alpha$. Define

$$M_{n,p}^\alpha = Q_\alpha(X) \cap M_{n,4p}, \quad n, p \in \mathbb{N}.$$

This family, in the subspace $Q_\alpha(X)$, satisfies all the properties from (v) (see
Proposition 3.62 for the last one). Now, since the subspace $Q_\alpha(X)$ has density
less than $\aleph$, there exists, by the induction assumption, a set $G_\alpha \subset B_{Q_\alpha(X)}$
that is symmetric, linearly dense in $Q_\alpha(X)$, and has the property that for
every $\varepsilon > 0$ there are sets $G_m^{\alpha,\varepsilon} \subset G_\alpha$, $m \in \mathbb{N}$, with $\bigcup_{m=1}^\infty G_m^{\alpha,\varepsilon} = G_\alpha$, such
that

$$\forall y^* \in B_{(Q_\alpha X)^*} \quad \forall m \in \mathbb{N} \quad \text{card}\left\{x \in G_m^{\alpha,\varepsilon}; \; |\langle x, y^* \rangle| > \varepsilon\right\} < \aleph_0. \qquad (6.8)$$

Put $G = \bigcup_{\alpha < \mu} G_\alpha$. Since $\{P_\alpha; \omega \leq \alpha \leq \mu\}$ is a projectional resolution of the
identity on X, G is a linearly dense subset of X. We shall show that this set
fits our needs.

So fix an arbitrary $\varepsilon > 0$. Find $p \in \mathbb{N}$ so large that $p > \frac{6}{\varepsilon}$. Then put

$$G_{m,n}^\varepsilon = \bigcup_{\alpha < \mu} G_m^{\alpha,\varepsilon} \cap M_{n,p}, \quad m, n \in \mathbb{N};$$

this is a countable family. Let us check that its union is all of G. Indeed, fix
any $x \in G$. Then $x \in G_\alpha$ for a suitable $\alpha < \mu$. Thus $x \in G_m^{\alpha,\varepsilon}$ for a suitable
$m \in \mathbb{N}$. Also, $x \in B_X = \bigcup_{n=1}^\infty M_{n,p}$. Thus $x \in M_{n,p}$ for a suitable $n \in \mathbb{N}$.
Therefore $x \in G_{m,n}^\varepsilon$, and the equality $\bigcup_{m,n=1}^\infty G_{m,n}^\varepsilon = G$ is verified.

Fix any $m, n \in \mathbb{N}$. It remains to prove that the set $\{x \in G^\varepsilon_{m,n}; \ |\langle x, x^*\rangle| > \varepsilon\}$ is finite for every $x^* \in B_{X^*}$. So fix an arbitrary $x^* \in B_{X^*}$.

We subclaim that *for every* $\alpha \le \mu$

$$\operatorname{card}\{x \in G^\varepsilon_{m,n}; \ |\langle x, P^*_\alpha x^*\rangle| > \varepsilon\} < \aleph_0. \tag{6.9}$$

Having this proved and recalling that $P^*_\mu x^* = x^*$, our claim will also be verified.

The subclaim will be proved by induction over the ordinal α. Expression (6.9) is true for $\alpha = 0$ since $P^*_0 = 0$. Consider any ordinal $\beta \le \mu$, and assume that (6.9) was verified for every $\alpha < \beta$. Denote

$$S := \{\alpha < \mu; \ |\langle x, P^*_\beta x^*\rangle| > \varepsilon \ \text{ for some } \ x \in G^{\alpha,\varepsilon}_m \cap M_{n,p}\}.$$

Assume that S is infinite. Then there exist ordinals $\alpha_1 < \alpha_2 < \cdots < \mu$ and $x_i \in G^{\alpha_i,\varepsilon}_m \cap M_{n,p}$ such that $\left|\langle x_i, P^*_\beta x^*\rangle\right| > \varepsilon$ for every $i \in \mathbb{N}$. We observe that $\beta > \alpha_i$ for every $i \in \mathbb{N}$. Indeed, $\beta \le \alpha_i$ would imply that $\langle x_i, P^*_\beta x^*\rangle = \langle P_\beta x_i, x^*\rangle = 0$ as $x_i \in Q_{\alpha_i}(X)$ and $P_\beta \circ Q_{\alpha_i} = 0$. Put $\lambda = \lim_{i \to \infty} \alpha_i$; thus $\lambda \le \beta$. If $\lambda < \beta$, then we would have

$$\varepsilon < \left|\langle x_i, P^*_\beta x^*\rangle\right| = \left|\langle P_\beta x_i, x^*\rangle\right| = \left|\langle P_\lambda x_i, x^*\rangle\right| = \left|\langle x_i, P^*_\lambda x^*\rangle\right|$$

for every $i \in \mathbb{N}$, which contradicts (6.9), valid for $\alpha := \lambda$ (indeed, $i \ne j$ implies $\alpha_i \ne \alpha_j$ and hence $x_i \ne x_j$; thus the set $\{x_1, x_2, \ldots\}$ is infinite). Therefore $\lambda = \beta$.

Now, we shall show that there exists $j \in \mathbb{N}$ such that

$$\sup\left|\langle M_{n,p}, P^*_\beta x^* - P^*_{\alpha_j} x^*\rangle\right| < \frac{6}{p}. \tag{6.10}$$

Once this has been proved, then for all $i \in \mathbb{N}$ with $i > j$, we have

$$\varepsilon < \left|\langle x_i, P^*_\beta x^*\rangle\right| = \left|\langle x_i, P^*_\beta x^* - P^*_{\alpha_j} x^*\rangle\right|$$
$$\le \sup\left|\langle M_{n,p}, P^*_\beta x^* - P^*_{\alpha_j} x^*\rangle\right| < \tfrac{6}{p} < \varepsilon,$$

a contradiction.

Let $Y = \bigcup_{\alpha < \lambda} P^*_\alpha X^*$; its closure is a subspace of X^*. As $M_{n,p}$, the closed unit ball of the norm $\|\cdot\|_{n,p}$, is $\frac{1}{p}$-weakly compact, and $P^*_\lambda x^* \in \overline{B_Y}^{w^*}$, Lemma 3.63 reveals that the $\|\cdot\|_{n,p}$-distance from $P^*_\lambda x^*$ to Y is at most $\frac{2}{p}$. Hence, there exists $y^* \in Y$ such that $\|P^*_\lambda x^* - y^*\|_{n,p} < \frac{3}{p}$. Find $\beta < \lambda$ such that $y^* \in P^*_\beta X^*$. Then, for all $\beta \le \alpha < \lambda$, we have $y^* \in P^*_\alpha X^*$, and so

$$\sup\left|\langle M_{n,p}, P^*_\lambda x^* - P^*_\alpha x^*\rangle\right|$$
$$= \|P^*_\lambda x^* - P^*_\alpha x^*\|_{n,p} \le \|P^*_\lambda x^* - y^*\|_{n,p} + \|y^* - P^*_\alpha x^*\|_{n,p}$$
$$= \|P^*_\lambda x^* - y^*\|_{n,p} + \|P^*_\alpha(y^* - P^*_\lambda x^*)\|_{n,p}$$
$$\le 2\|y^* - P^*_\lambda x^*\|_{n,p} < \tfrac{6}{p};$$

here we used that $P_\alpha(M_{n,p}) \subset M_{n,p}$.

Therefore the set S above must be finite. But for every $\alpha \in S$ the inequality (6.8) says that the set $\left\{x \in G_m^{\alpha,\varepsilon}; \; \left|\langle x, P_\beta^* x^* \rangle\right| > \varepsilon\right\}$ is finite. This proves the claim. $\qquad\square$

As a consequence of Theorem 6.13, we get the following well-known fact. For earlier proofs of it, see [BRW77], [Gul77], [MiRu77], [NeTs81].

Corollary 6.22. *A continuous image of an Eberlein compact is Eberlein.*

Proof. Let φ be a continuous mapping from an Eberlein compact K onto a (compact) L. Then $C(L)$ is isometric to a subspace of $C(K)$ via the mapping $f \mapsto f \circ \varphi$, $f \in C(L)$, and so $C(K)$ is weakly compactly generated (see, e.g., [Fa~01, Thm. 12.12]). Then, by (iv) in Theorem 6.13, the dual unit ball in $C(L)^*$ with the weak*-topology is an Eberlein compact. Thus, L, homeomorphic to a subspace of the latter, is also an Eberlein compact. $\qquad\square$

Definition 6.23. *A nonempty subset G of a Banach space X will be called* weakly σ-shrinkable *if $G = \bigcup_{n=1}^\infty G_n$ so that, for every neighborhood U of the origin in (X^{**}, w^*) and for every $x \in G$, there is $n \in \mathbb{N}$ such that $x \in G_n$ and $G_n' \subset U$, where A' is the set of all accumulation points in (X^{**}, w^*) of a set $A \subset X$. In particular, an M-basis $\{x_\gamma; x_\gamma^*\}_{\gamma \in \Gamma}$ in $X \times X^*$ will be called* weakly σ-shrinkable *if the set $\{x_\gamma; \gamma \in \Gamma\}$ is weakly σ-shrinkable.*

Remark 6.24. By the Hahn-Banach theorem, a set $G \subset X$ is weakly σ-shrinkable if and only if $G = \bigcup_{n \in \mathbb{N}} G_n$ such that, for each $\varepsilon > 0$, for each $x_0 \in G$, and for each $x^* \in B_{X^*}$ there is $n \in \mathbb{N}$ such that $x_0 \in G_n$ and $\{x \in G_n; |\langle x, x^* \rangle| \geq \varepsilon\}$ is finite. As a consequence, note that *if G is a weakly σ-shrinkable subset of X, G countably supports X^*.* In particular, if G is linearly dense, (B_{X^*}, w^*) is a Corson compact and thus X is WLD (see Theorem 5.37). This happens, for example, if X has a weakly σ-shrinkable M-basis.

Theorem 6.25 ([FGMZ04], [FHMZ]). *Let X be a Banach space. Then the following are equivalent:*

(i) *X is a Vašák space.*
(ii) *X contains a linearly dense and weakly σ-shrinkable subset.*
(iii) *X admits a weakly σ-shrinkable M-basis.*

Moreover, if this is the case, then every linearly dense subset of X that countably supports X^* (in particular, every M-basis in $X \times X^*$) is weakly σ-shrinkable. Henceforth, every Vašák space is WLD.

Proof. (i)$\Rightarrow$(ii) Let $K_m \subset B_{X^{**}}$, $m \in \mathbb{N}$, be weak*-closed sets as in the definition of Vašák space; i.e., for every $x \in B_X$, there is $N \subset \mathbb{N}$ so that $x \in \bigcap_{m \in N} K_m \subset X$. We may and do assume that, for all $m, n \in \mathbb{N}$, if $K_m \cap K_n \neq \emptyset$, then there is $l \in \mathbb{N}$ such that $K_m \cap K_n = K_l$. Let $\{P_\alpha; \omega \leq \alpha \leq \mu\}$ be a *separable* PRI on X (see Theorem 3.46). We recall that one of the features of such a PRI is that the range of the projection $Q_\alpha := P_{\alpha+1} - P_\alpha$ is separable for every $\alpha \in [\omega, \mu)$. For each such α, we find a dense subset $\{v_n^\alpha; \ n \in \mathbb{N}\}$ in $B_{Q_\alpha X}$. Then put $G = \bigcup_{n,m=1}^\infty G_{m,n}$, where

$$G_{m,n} := \{v_n^\alpha; \ \alpha \in [\omega, \mu)\} \cap K_m, \quad m, n \in \mathbb{N}.$$

Clearly, G is total in X.

Now, fix any $\varepsilon > 0$, any $x^* \in X^*$, and any $x \in G$. Find a set $N \subset \mathbb{N}$ such that $x \in \bigcap_{m \in N} K_m \subset X$. We can then choose a sequence $(m_i)_{i \in \mathbb{N}}$ in N (not necessarily injective) such that $K_{m_1} \supset K_{m_2} \supset \cdots$ and $\bigcap_{i=1}^\infty K_{m_i} \subset X$. Find $n \in \mathbb{N}$ and a (unique) $\alpha \in [\omega, \mu)$ such that $x = v_n^\alpha$. We claim that there is $j \in \mathbb{N}$ such that

$$\text{card} \left\{x' \in G_{m_j,n}; \ |\langle x', x^* \rangle| > \varepsilon\right\} < \aleph_0.$$

Once we have this, (ii) will be proved since clearly $x \in G_{m_i,n}$ for every $i \in \mathbb{N}$.

Assume that the claim is false. Then subsequently pick $x_1 \in G_{m_1,n}$ with $|\langle x_1, x^* \rangle| > \varepsilon$, $x_2 \in G_{m_2,n} \backslash \{x_1\}$ with $|\langle x_2, x^* \rangle| > \varepsilon, \ldots$, and $x_{i+1} \in G_{m_{i+1},n} \backslash \{x_k; 1 \leq k \leq i\}$ with $|\langle x_{i+1}, x^* \rangle| > \varepsilon, \ldots$. For every $i \in \mathbb{N}$, find a (unique) $\alpha_i < \mu$ such that $x_i = v_n^{\alpha_i}$. Let x^{**} be a weak*-cluster point of the sequence $(x_i)_{i \in \mathbb{N}}$. Then, necessarily, $x^{**} \in \bigcap_{i=1}^\infty K_{m_i} \subset X$. Fix for a while any $\beta < \mu$. We recall that the sequence $(x_i)_{i \in \mathbb{N}}$ is injective. Hence so is the sequence $(\alpha_i)_{i \in \mathbb{N}}$. Then we have $Q_\beta \circ Q_{\alpha_i} = 0$ for all large $i \in \mathbb{N}$. Hence $Q_\beta x^{**} = 0$. This holds for every $\beta \in [\omega, \mu)$. Therefore $x^{**} = 0$. However, $|\langle x_i, x^* \rangle| > \varepsilon$ for every $i \in \mathbb{N}$, and so $(0 =) \ |\langle x^{**}, x^* \rangle| \geq \varepsilon > 0$, a contradiction.

(ii)$\Rightarrow$(iii) Assume that (ii) holds. In order to obtain a weakly σ-shrinkable M-basis and to prove the "moreover" part in the statement, observe that any M-basis $\{x_\gamma; x_\gamma^*\}_{\gamma \in \Gamma}$ in $X \times X^*$ countably supports X^* (see Theorem 5.37). There is a separable PRI $\{P_\alpha; \omega \leq \alpha \leq \mu\}$ on X as above, with the additional property that $P_\alpha(x_\gamma) \in \{x_\gamma, 0\}$ for every $\alpha \in [\omega, \mu)$ and every $\gamma \in \Gamma$ (see Theorem 3.46). For every $\alpha \in [\omega, \mu)$, the set $\{x_\gamma; \gamma \in \Gamma\} \cap Q_\alpha X$ is countable; this can be seen as follows: $Q_\alpha X$ is separable, and hence there exists a w^*-dense subset $\{x_n^* : \ n \in \mathbb{N}\}$ of $(Q_\alpha(X))^*$. Let $S_n := \{\gamma \in \Gamma; x_\gamma \in Q_\alpha X : \langle x_\gamma, x_n^* \rangle \neq 0\}$. Then S_n is countable for all $n \in \mathbb{N}$. If $x_\gamma \in Q_\alpha X \backslash \bigcup_{n=1}^\infty \{x_\gamma; \gamma \in S_n\}$, we have $\langle x_\gamma, x_n^* \rangle = 0$ for all $n \in \mathbb{N}$, and hence $x_\gamma = 0$. Enumerate $\{x_\gamma; \gamma \in \Gamma\} \cap Q_\alpha X$ as $\{v_n^\alpha; \ n \in \mathbb{N}\}$ and proceed as above to prove that the M-basis (in fact, any M-basis) is weakly σ-shrinkable.

(iii)$\Rightarrow$(i) Assume that the space X contains a weakly σ-shrinkable M-basis $\{x_\gamma, x_\gamma^*\}_{\gamma \in \Gamma}$, and let $\Gamma = \bigcup_{n=1}^\infty \Gamma_n$ from the definition. We proceed similarly to the proof of Lemma 6.21. For $i \in \mathbb{N}$, let $\tau_i(t)$ be a function on the real line such that $\tau_i = 0$ on $[-\frac{1}{i}, +\frac{1}{i}]$ and $\tau_i(t) = t - \frac{1}{i}$ on $[\frac{1}{i}, \infty)$ and $\tau_i(t) = t + \frac{1}{i}$ on $(-\infty, \frac{-1}{i}]$. Let Δ be the infinite matrix whose first row consists of countably many disjoint copies of Γ_1 (call them Γ_1^1, Γ_1^2, etc.) whose second row consists of countably many disjoint copies of Γ_2 (call them Γ_2^1, Γ_2^2, etc.) and so on. Define the map φ from B_{X^*} into $\ell_\infty(\Delta)$ by $\varphi(x^*)(\gamma_n^i) = \tau_i(\langle x_{\gamma_n^i}, x^* \rangle)$, where γ_n^i is an element of Γ_n^i. Then it can be checked that φ is a one-to-one continuous map from the weak*-topology of B_{X^*} into the pointwise topology of $\ell_\infty(\Delta)$. Thus X is a Vašák space by [Fab97, Thm. 7.2.5 (vi)]. For the last assertion, we refer to Remark 6.24. □

Theorem 6.26 (Mercourakis;, see, e.g., VII.1.17 in [DGZ93a]). *Every Vašák space admits a norm the dual of which has the following property: $f_n - f \to 0$ in the weak*-topology of X^* whenever $f_n, f \in S_{X^*}$ are such that $\|f_n + f\| \to 2$.*

Proof. This follows from the fact that every Sokolov subspace of $\ell_\infty(\Gamma)$ for any Γ admits a pointwise LUR norm (see Theorem 3.51). Indeed, if $\{x_\gamma, f_\gamma\}_{\gamma \in \Gamma}$ is a weakly σ-shrinkable basis for X, then the operator $Tf = \{f(x_\gamma)\}$ clearly maps X^* onto a Sokolov subspace of $\ell_\infty(\Gamma)$. □

Theorem 6.27. *There are Sokolov subspaces of some $\ell_\infty(\Gamma)$ that do not inject into any $c_0(\Gamma')$.*

Proof. Let R be the non-WCG subspace of the WCG space $L_1(\mu)$ for a probability μ [Rose74]. As R has an unconditional basis, R^* does not inject into any $c_0(\Gamma)$ by Theorem 7.40.

However, as R is a subspace of a WCG space, R^* injects weak*-pointwise into a Sokolov subspace S by Theorem 6.13. Therefore this Sokolov subspace S cannot inject into any $c_0(\Gamma)$. □

It was proved in [Haj94] that *there exists a nonseparable reflexive Banach space with a symmetric basis such that no nonseparable subspace of which can be mapped by a one-to-one bounded linear operator into some superreflexive space.*

6.3 Hilbert Generated Spaces

In this section, we give an M-basis characterization of Hilbert generated spaces and a characterization of subspaces of such spaces, this time also in terms of the existence of a uniformly Gâteaux (UG) differentiable equivalent renorming. As an application, we present a short proof to the uniform Eberlein property of all continuous images of uniform Eberlein compacta and to the impossibility of renorming some reflexive spaces with a UG norm.

Recall that a Banach space X is said to be *Hilbert generated* if there is a Hilbert space H and a bounded linear operator $T : H \to X$ with dense range (see Definition 6.5).

Theorem 6.28 ([FGHZ03], [FHMZ]). *Let X be a Banach space with dens $X \leq \omega_1$. Then the following are equivalent:*

(i) *X is Hilbert generated.*

(ii) X *admits an M-basis* $\{x_\gamma; x_\gamma^*\}_{\gamma \in \Gamma}$ *and a bounded linear operator* $T :$ $\ell_2(\Gamma) \to X$ *such that* $x_\gamma = Te_\gamma$ *for every* $\gamma \in \Gamma$, *where* e_γ *are the unit vectors in* $\ell_2(\Gamma)$.

Proof. (ii)$\Rightarrow$(i) This is trivial.

(i)$\Rightarrow$ (ii) Assume for simplicity that $\ell_2(\Gamma)$ is a dense subset of X and that $\|f\| \leq \|f\|_{\ell_2}$ for every $f \in \ell_2(\Gamma)$. Fix any $x^* \in X^*$. Then the restriction $x^* \upharpoonright \ell_2(\Gamma)$ lies in $\ell_2(\Gamma)^*$ ($\equiv \ell_2(\Gamma)$). Thus the set $\{\gamma \in \Gamma;\ \langle e_\gamma, x^* \rangle \neq 0\}$ is at most countable, which means that the set $\{e_\gamma;\ \gamma \in \Gamma\}$ countably supports all elements of X^*. There is a separable PRI $\{P_\alpha; \omega \leq \alpha \leq \mu\}$ on X subordinated to the set $G := \{e_\gamma;\ \gamma \in \Gamma\}$ (see Theorem 3.46). Fix any $\alpha \in [\omega, \mu)$. Put $G_\alpha := (P_{\alpha+1} - P_\alpha)G$. Note that $G_\alpha \subset G \cup \{0\}$ and that G_α is linearly dense in the (separable) subspace $(P_{\alpha+1} - P_\alpha)X$. We find an M-basis $\{x_{\alpha,n}; x_{\alpha,n}^*\}_{n \in \mathbb{N}}$ in the subspace $(P_{\alpha+1} - P_\alpha)X$ such that $x_{\alpha,n} \in \operatorname{span} G_\alpha$ ($\subset \ell_2(\Gamma)$) and $\|x_{\alpha,n}\|_{\ell_2} = 1$ for every $n \in \mathbb{N}$. Define $Q_\alpha : X \to (P_{\alpha+1} - P_\alpha)X$ by $Q_\alpha x = (P_{\alpha+1} - P_\alpha)x$, $x \in X$. Performing this for every $\omega \leq \alpha < \mu$, we get the system $\{\frac{1}{n}x_{\alpha,n}; nQ_\alpha^* x_{\alpha,n}^*\}_{n \in \mathbb{N},\ \omega \leq \alpha < \mu}$, which is an M-basis in X.

For every element $(a_{\alpha,m};\ \omega \leq \alpha < \mu,\ m \in \mathbb{N})$ of $\ell_2([\omega,\mu) \times \mathbb{N})$ with finite support, we define

$$T(a_{\alpha,m}) := \sum_{m=1}^{\infty} \sum_{\omega \leq \alpha < \mu} a_{\alpha,m} \frac{1}{m} x_{\alpha,m}.$$

This is a linear mapping from a dense subset of $\ell_2([\omega,\mu) \times \mathbb{N})$ into X. Now, using the Hölder inequality and a disjoint support argument in the last of the following inequalities, we can estimate

$$\left\| T(a_{\alpha,m}) \right\| \leq \sum_{m=1}^{\infty} \frac{1}{m} \left\| \sum_{\omega \leq \alpha < \mu} a_{\alpha,m} x_{\alpha,m} \right\|$$

$$\leq \left(\sum_{m=1}^{\infty} \frac{1}{m^q} \right)^{\frac{1}{2}} \left(\sum_{m=1}^{\infty} \left\| \sum_{\omega \leq \alpha < \mu} a_{\alpha,m} x_{\alpha,m} \right\|^p \right)^{\frac{1}{2}}$$

$$\leq C \left(\sum_{m=1}^{\infty} \left\| \sum_{\omega \leq \alpha < \mu} a_{\alpha,m} x_{\alpha,m} \right\|_{\ell_2}^2 \right)^{\frac{1}{2}}$$

$$\leq C \left(\sum_{m=1}^{\infty} \sum_{\omega \leq \alpha < \mu} |a_{\alpha,m}|^2 \right)^{\frac{1}{2}} = C \left\| (a_{\alpha,m}) \right\|_{\ell_2},$$

where $\left(\sum_{m=1}^{\infty} \frac{1}{m^2} \right)^{\frac{1}{2}} = C$. Therefore, the mapping T can be extended to the whole space $\ell_2([\omega,\mu) \times \mathbb{N})$. Now, every canonical basic vector from this space is mapped by T to $\frac{1}{m} x_{\alpha,m}$ with a suitable $m \in \mathbb{N}$ and $\omega \leq \alpha < \mu$. Therefore, the range of T is dense in X and the proof is finished. $\qquad \square$

Definition 6.29. *A compact space K is a* uniform Eberlein compact *if K is homeomorphic to a weakly compact subset of a Hilbert space $\ell_2(\Gamma)$ taken in its weak topology.*

Theorem 6.30 ([FHZ97], [FGZ01], [FGHZ03], [FHMZ]). *The following conditions are equivalent for a Banach space X:*

(i) *X admits a UG norm.*
(ii) *There exists an M-basis $\{x_\gamma; x_\gamma^*\}_{\gamma \in \Gamma}$ in $X \times X^*$ such that $\{x_\gamma; \gamma \in \Gamma\} \subset B_X$ and, for every $\varepsilon > 0$, we have $\Gamma = \bigcup_{n=1}^{\infty} \Gamma_n^\varepsilon$ satisfying*

$$\mathrm{card}\left\{\gamma \in \Gamma_n^\varepsilon; \ |\langle x_\gamma, x^*\rangle| > \varepsilon\right\} < n, \ \forall n \in \mathbb{N} \ \ \forall x^* \in B_{X^*}.$$

(iii) *There exists a linearly dense set $G \subset B_X$ such that, for every $\varepsilon > 0$, we have $G = \bigcup_{n=1}^{\infty} G_n^\varepsilon$ satisfying*

$$\mathrm{card}\left\{x \in G_n^\varepsilon; \ |\langle x, x^*\rangle| > \varepsilon\right\} < n, \ \forall n \in \mathbb{N} \ \ \forall x^* \in B_{X^*}.$$

(iv) *(B_{X^*}, w^*) is a uniform Eberlein compact.*
(v) *X is a subspace of a Hilbert generated space.*

Moreover, if one of the above holds, then every M-basis $\{x_\gamma; x_\gamma^*\}_{\gamma \in \Gamma}$ in $X \times X^*$ (resp. every set $G \subset B_X$ linearly dense and countably supporting X^*) satisfies the assertion in (ii) *(resp. in* (iii)*).*

For the proof of Theorem 6.30, we need the following lemma.

Lemma 6.31. *Let $\|\cdot\|$ be a uniformly Gâteaux smooth norm on a Banach space X. Then, for every $\epsilon > 0$, there are sets $S_i^\epsilon \subset S_X$, $i \in \mathbb{N}$, such that $\bigcup_{i=1}^{\infty} S_i^\epsilon = S_X$ and*

$$\|x_1 + \ldots + x_i\| < \epsilon i$$

whenever $x_1, \ldots, x_i \in S_i^\epsilon$ and $x_{j+1} \perp \mathrm{sp}\{x_1, \ldots, x_j\}$, $j = 1, \ldots, i - 1$.

Proof. Fix any $\varepsilon > 0$ and any $i \in \mathbb{N}$. If $\varepsilon i \leq 2$, put $S_i^\epsilon = \emptyset$. Otherwise, let S_i^ϵ be the set of all $x \in S_X$ such that, for every $y \in S_X$ with $x \perp y$ and every $0 \neq \tau \in (-\frac{2}{\varepsilon i - 2}, \frac{2}{\varepsilon i - 2})$,

$$\frac{1}{\tau}\left(\|y + \tau x\| - 1\right) < \frac{\varepsilon}{2}.$$

The uniform Gâteaux smoothness and the orthogonality guarantee that $S_X = \bigcup_{i=1}^{\infty} S_i^\epsilon$.

Take $\varepsilon > 0$ and $i \in \mathbb{N}$ such that $\varepsilon i > 2$, and choose $x_1, \ldots, x_i \in S_i^\epsilon$ as in the lemma. Put $v_j = x_1 + \ldots + x_j$, $j = 1, \ldots, i$. We shall show by induction that

$$\|v_j\| < \frac{\varepsilon}{2}(i + j), \quad j = 1, \ldots, i.$$

Trivially, this is true for $j = 1$. Let it hold for some $j < i$. If $\|v_j\| > \frac{\varepsilon i}{2} - 1$, then $\|v_j\|^{-1} < \frac{2}{\varepsilon i - 2}$ and so

$$\|v_{j+1}\| = \|v_j + x_{j+1}\| = \|v_j\| \left\| \frac{v_j}{\|v_j\|} + \frac{1}{\|v_j\|} x_{j+1} \right\|$$

$$< \|v_j\| \left(1 + \frac{\varepsilon}{2} \frac{1}{\|v_j\|} \right) < \frac{\varepsilon}{2} (i+j) + \frac{\varepsilon}{2} = \frac{\varepsilon}{2}(i+j+1).$$

On the other hand, if $\|v_j\| \le \frac{\varepsilon i}{2} - 1$, then

$$\|v_{j+1}\| = \|v_j + x_{j+1}\| \le \frac{\varepsilon i}{2} - 1 + 1 = \frac{\varepsilon i}{2} < \frac{\varepsilon}{2}(i+j+1).$$

In particular, for $j = i$, we have $\|v_i\| < \frac{\varepsilon}{2}(i+i) = \varepsilon i$.   □

Proof of Theorem 6.30. (i)⇒(ii) First we will show that (i) implies that X is a Vašák space. By using the Šmulyan duality lemma (see, e.g., [DGZ93a, Chap. II]), we have that $f_n - g_n \to 0$ in the weak*-topology whenever $f_n, g_n \in S_{X^*}$ are such that $\|f_n + g_n\| \to 2$.

For $\varepsilon > 0$ and $n \in \mathbb{N}$, put

$$B_n^\varepsilon := \left\{ x \in B_X \,;\, |(f - g)(x)| < \varepsilon \text{ if } f, g \in B_{X^*} \text{ satisfy } \|f + g\| > 2 - \frac{1}{n} \right\}.$$

We have, for every $\varepsilon > 0$, that $\bigcup_n B_n^\varepsilon = B_X$.

We claim that for each $\varepsilon > 0$ and each n,

$$\overline{B_n^\varepsilon}^{w^*} \subset X + 4\varepsilon B_{X^{**}}.$$

Indeed, if not, take $x_0 \in \overline{B_n^\varepsilon}^{w^*} \subset X^{**}$ with the distance greater than 2ε from X. Then take $F \in S_{X^{***}}$ such that F equals 0 on X and $F(x_0) = 2\varepsilon$. Let $f_\alpha \in S_{X^*}$ be such that $f_\alpha \to F$ in the weak*-topology of X^{***}. Then $\|f_\alpha + f_\beta\| \to 2$ and thus $|(f_\alpha - f_\beta)(x)| < \varepsilon$ for all $x \in B_n^\varepsilon$ for large α, β. As f_α weak*-converges to F, we have $|(f_\alpha - F)(x)| \le \varepsilon$ for all $x \in B_n^\varepsilon$ for large α. Since $F = 0$ on X, in particular on B_n^ε, we have $|f_\alpha(x)| \le \varepsilon$ for every $x \in B_n^\varepsilon$, and thus $|f_\alpha(x_0)| \le \varepsilon$ for large α from the continuity of f_α in the weak*-topology of X^{**}. Since $f_\alpha \to F$ in the weak*-topology of X^{***}, we get $|F(x_0)| \le \varepsilon$, which is a contradiction. Therefore X is Vašák and thus it admits a PRI in any equivalent norm (see Theorem 3.44). We shall prove the existence of an M-basis satisfying the assertion in (ii) by transfinite induction. First, if X is separable, then clearly every M-basis in $X \times X^*$ satisfies (ii). Let $\aleph$ be an uncountable cardinal and assume that the implication has already been verified for every space of density less than $\aleph$ having an equivalent UG norm. Now assume that a Banach space X, of density $\aleph$, has an equivalent UG norm, say $\|\cdot\|$. Let $\{P_\alpha; \omega \le \alpha \le \mu\}$ be a PRI on X given by the fact that X is Vašák. Put $Q_\alpha := P_{\alpha+1} - P_\alpha$ for $\omega \le \alpha < \mu$. By the induction assumption, find an M-basis $\{x_{\alpha,\gamma}; x_{\alpha,\gamma}^*\}_{\gamma \in \Gamma_\alpha}$ in $Q_\alpha X \times Q_\alpha^* X^*$ for $\omega \le \alpha < \mu$. We can assume that the index sets $\{\Gamma_\alpha\}_{\omega \le \alpha < \mu}$ are pairwise disjoint. Put $\{x_{\alpha,\gamma}; Q_\alpha^* x_{\alpha,\gamma}^*\}_{\gamma \in \Gamma_\alpha, \omega \le \alpha < \mu}$; this is an M-basis, as it is standard to check. Call this M-basis $\{x_\gamma; x_\gamma^*\}_{\gamma \in \Gamma}$, where α, γ, for $\gamma \in \Gamma_\alpha$, has been written just as γ. We shall show that this M-basis satisfies the assertions in (ii).

So, fix any $0 < \varepsilon < 1$. For every $\alpha \in [\omega, \mu)$ and every $n \in \mathbb{N}$, find the set $\Gamma_{\alpha,n}^\varepsilon \subset \Gamma_\alpha$ as it is stated in the assertion (ii). For $n, m \in \mathbb{N}$, put

$$\Gamma_{n,m}^\varepsilon := \bigcup_{\alpha \in [\omega, \mu)} \Gamma_{\alpha,n}^\varepsilon \cap \{\gamma \in \Gamma; x_\gamma \in B_m^{\varepsilon/2}\} \backslash (\Gamma_{n,m-1}^\varepsilon \cup \cdots \cup \Gamma_{n,1}^\varepsilon \cup \{0\});$$

this is a countable family of mutually disjoint sets since $\Gamma_\alpha \cap \Gamma_\beta = \emptyset$ if $\alpha \neq \beta$. Also, we can easily verify that $\bigcup_{n,m=1}^\infty \Gamma_{n,m}^\varepsilon = \Gamma$.

Fix any $n, m \in \mathbb{N}$ and any $x^* \in B_{X^*}$. We shall show that

$$\text{card}\left\{\gamma \in \Gamma_{n,m}^\varepsilon; |\langle x_\gamma, x^* \rangle| > \varepsilon\right\} < \frac{4mn}{\varepsilon^2},$$

and thus the assertion (ii) will be almost proved. Define

$$F := \left\{\alpha \in [\omega, \mu); |\langle x_\gamma, x^* \rangle| > \varepsilon \text{ for some } \gamma \in \Gamma_{n,m}^\varepsilon \cap \Gamma_\alpha\right\}.$$

We claim that $\text{card } F < \frac{4m}{\varepsilon^2}$; then we easily get, by the induction assumption,

$$\text{card}\left\{\gamma \in \Gamma_{n,m}^\varepsilon; |\langle x_\gamma, x^* \rangle| > \varepsilon\right\}$$
$$\leq \sum_{\alpha \in F} \text{card}\left\{\gamma \in \Gamma_{\alpha,n}^\varepsilon; |\langle x_\gamma, x^* \upharpoonright Q_\alpha X \rangle| > \varepsilon\right\} < \text{card } F \cdot n < \frac{4mn}{\varepsilon^2}.$$

Let us prove the claim. If the set F is infinite, let N be any fixed positive integer. Otherwise, denote $N = \text{card } F$. Find $\alpha_1 < \alpha_2 < \cdots < \alpha_N < \mu$ such that $F \supset \{\alpha_1, \alpha_2, \ldots, \alpha_N\}$. For $j = 1, \ldots, N$, find $\gamma_j \in \Gamma_{n,m}^\varepsilon \cap \Gamma_{\alpha_j}$, with $|\langle x_{\gamma_j}, x^* \rangle| > \varepsilon$, and write $v_j = x_{\gamma_1} + \cdots + x_{\gamma_j}$. Find $i \in \mathbb{N}$ so that $\frac{m}{\varepsilon} \leq i \leq \frac{2m}{\varepsilon}$. If $i \geq N$, then $N \leq \frac{2m}{\varepsilon} < \frac{4m}{\varepsilon^2}$. Further assume that $i < N$. Since $\|v_i\| \geq |\langle v_i, x^* \rangle| > i\varepsilon \geq m$, $P_{\alpha_{i+1}} \circ Q_{\alpha_{i+1}} = 0$, and $x_{\gamma_{i+1}} \in B_m^{\varepsilon/2}$, the convexity of $\|\cdot\|$ yields

$$\|v_{i+1}\| = \|v_i\| \left(\left\|\frac{v_i}{\|v_i\|} + \frac{x_{\gamma_{i+1}}}{\|v_i\|}\right\| - 1\right) + \|v_i\|$$

$$\leq m\left(\left\|\frac{v_i}{\|v_i\|} + \frac{x_{\gamma_{i+1}}}{m}\right\| - 1\right) + \|v_i\|$$

$$\leq m\left(\left\|\frac{v_i}{\|v_i\|} + \frac{x_{\gamma_{i+1}}}{m}\right\| + \left\|\frac{v_i}{\|v_i\|} - \frac{x_{\gamma_{i+1}}}{m}\right\| - 2\right) + \|v_i\|$$

$$< \frac{\varepsilon}{2} + \|v_i\|.$$

Similarly, we get

$$\|v_{i+2}\| < \frac{\varepsilon}{2} + \|v_{i+1}\| < 2\frac{\varepsilon}{2} + \|v_i\|, \quad \ldots, \quad \|v_N\| < (N-i)\frac{\varepsilon}{2} + \|v_i\| < N\frac{\varepsilon}{2} + i.$$

Thus

$$N\varepsilon < |\langle v_N, x^* \rangle| \leq \|v_N\| < N\frac{\varepsilon}{2} + i,$$

and so $N < \frac{2}{\varepsilon}i \leq \frac{4m}{\varepsilon^2}$. This also shows that the set F cannot be infinite. Hence card $F = N < \frac{4m}{\varepsilon^2}$ and the claim is proved. Now, it remains to enumerate the (countable) family $\Gamma_{n,m}^{\varepsilon}$, $n, m \in \mathbb{N}$, by one index running throughout $\mathbb{N}$ and to insert eventually "a few" empty sets. This will yield (ii).

To prove the "moreover" part of the statement, assume that we have already given an M-basis $\{x_\gamma; x_\gamma^*\}_{\gamma \in \Gamma}$ in $X \times X^*$ such that $\{x_\gamma : \gamma \in \Gamma\} \subset B_X$. We note that any M-basis countably supports X^* (see Theorem 5.37). We get a PRI $\{P_\alpha; \omega \leq \alpha \leq \mu\}$ on X as above, with the additional property that it is subordinated to the set $\{x_\gamma; \gamma \in \Gamma\}$. For every $\alpha \in [\omega, \mu)$, put $\Gamma_\alpha = \Gamma \cap \{\gamma \in \Gamma; x_\gamma \in Q_\alpha X\}$; the set $\{x_\gamma; \gamma \in \Gamma_\alpha\}$ is linearly dense in $Q_\alpha X$, which countably supports $(Q_\alpha X)^*$. The rest of the proof is as above.

(ii)⇒(iii) is trivial.

(iii)⇒(iv) Note that the set G given in (iii) countably supports X^*. Let $\tau_i : \mathbb{R} \to \mathbb{R}$, $i \in \mathbb{N}$, be the functions defined in the proof of Lemma 6.21. Define $\Phi : B_{X^*} \to \mathbb{R}^{G \times \mathbb{N}}$ by

$$\Phi(x^*)(x, i) = \frac{1}{2^n 2^i \sqrt{n}} \tau_i(\langle x, x^* \rangle) \quad \text{if} \quad x \in G_n^{1/i}, \quad n \in \mathbb{N}, \quad \text{and} \quad i \in \mathbb{N}.$$

Clearly, Φ is weak*-to-pointwise continuous. The injectivity of Φ can be checked exactly as in Lemma 6.21.

It remains to prove that $\Phi(B_{X^*}) \subset \ell_2(G \times \mathbb{N})$. Fix an arbitrary $x^* \in B_{X^*}$. We observe that for every $n, i \in \mathbb{N}$

$$\text{card} \left\{ x \in G_n^{1/i}; \ \Phi(x^*)(x, i) \neq 0 \right\} \leq \text{card} \left\{ x \in G_n^{1/i}; \ |\langle x, x^* \rangle| > \frac{1}{i} \right\} < 2n.$$

Therefore

$$\sum \left\{ \left(\Phi(x^*)(x, i) \right)^2; \ (x, i) \in G \times \mathbb{N} \right\} = \sum_{i,n=1}^{\infty} \sum \left\{ \left(\Phi(x^*)(x, i) \right)^2; \ x \in G_n^{1/i} \right\}$$

$$\leq \sum_{i,n=1}^{\infty} \frac{1}{4^n 4^i n} \cdot \text{card} \left\{ x \in G_n^{1/i}; \ \Phi(x^*)(x, i) \neq 0 \right\} < \sum_{i,n=1}^{\infty} \frac{2}{4^n 4^i} = \frac{2}{9} < +\infty,$$

and hence $\Phi(x^*) \in \ell_2(G \times \mathbb{N})$.

(iv)⇒(v) Assume that (B_{X^*}, w^*) is a uniform Eberlein compact. A result of Benyamini, Rudin, and Wage [BRW77] says that *the space of continuous functions on this compact, endowed with the supremum norm, is Hilbert generated*; see also [Fa~01, Thm. 12.17]. But X is isomorphic to a subspace of this space. Thus we get (v).

(v)⇒(i) Assume that a Banach space $(X, \|\cdot\|)$ is a subspace of a Hilbert generated space $(Z, \|\cdot\|)$. Find a Hilbert space H and a bounded linear mapping $T : H \to Z$ with dense range. Define $\|\|\cdot\|\|^*$ on Z^* by

$$|||z^*|||^{*2} = \|z^*\|^{*2} + \|T^*z^*\|^{*2}, \quad z^* \in Z^*;$$

this is an equivalent dual norm on Z^*. A convexity argument guarantees that this norm is uniformly $T(B_H)$-rotund. Hence, by a Šmulyan duality argument, the predual norm $||| \cdot |||$ on Z is uniformly $T(B_H)$-smooth;, that is,

$$\sup \left\{ |||z + tTh||| + |||z - tTh||| - 2; \; z \in Z, \; |||z||| = 1, \; h \in B_H \right\} = o(t) \quad \text{as} \quad t \downarrow 0.$$

Now, since $T(H)$ is dense in Z and the norm $||| \cdot |||$ (like any norm) is Lipschitzian, we get that this norm is uniformly Gâteaux smooth. Then the restriction of $||| \cdot |||$ to the subspace X gives (i). □

The following result completes Theorem 6.30. For the proof, see, e.g., [Fa~01, Thm. 12.18].

Theorem 6.32 ([FGZ01]). *Let K be a compact space. $C(K)$ admits an equivalent UG-smooth norm if and only if K is a uniform Eberlein compact.*

Recall that a compact space K is a *Gul'ko compact* if $C(K)$ is a Vašák space.

Theorem 6.33. *Let Γ be an uncountable set and $K \subset \Sigma(\Gamma) \cap [-1,1]^\Gamma$ be a compact set.*

(i) ([Farm87]) *K is a (uniform) Eberlein compact if and only if, for every $\varepsilon > 0$, we have $\Gamma = \bigcup_{n=1}^\infty \Gamma_n^\varepsilon$ such that*

$$\forall n \in \mathbb{N}, \quad \forall k \in K, \quad \operatorname{card} \{\gamma \in \Gamma_n^\varepsilon; \; |k(\gamma)| > \varepsilon\} < \aleph_0 \quad (< n).$$

(ii) ([FMZ04b]) *K is a Gul'ko compact if and only if there are sets $\Gamma_n \subset \Gamma$, $n \in \mathbb{N}$, such that*

$$\forall \varepsilon > 0, \quad \forall k \in K, \quad \forall \gamma \in \Gamma, \quad \exists n \in \mathbb{N} \quad \text{such that}$$

$$\gamma \in \Gamma_n \quad \text{and} \quad \operatorname{card} \{\gamma' \in \Gamma_n; \; |k(\gamma')| > \varepsilon\} < \aleph_0.$$

Proof. Necessity. Denote $\Gamma_0 = \{\gamma \in \Gamma; \; k(\gamma) = 0 \text{ for every } k \in K\}$. For $\gamma, \gamma' \in \Gamma$ we write $\gamma \sim \gamma'$ if $k(\gamma) = k(\gamma')$ for every $k \in K$; this is a relation of equivalence. For $\gamma \in \Gamma \backslash \Gamma_0$, let $[\gamma] = \{\gamma' \in \Gamma \backslash \Gamma_0; \; \gamma' \sim \gamma\}$. Denote $\Lambda = \{[\gamma]; \; \gamma \in \Gamma \backslash \Gamma_0\}$. Since $K \subset \Sigma(\Gamma)$, we get that every $\lambda \in \Lambda$ consists of at most countably many elements; let us enumerate it as $\lambda = \{\gamma_1^\lambda, \gamma_2^\lambda, \dots\}$ (the enumeration may not be injective). For $i \in \mathbb{N}$, then put $\Gamma_i = \{\gamma_i^\lambda; \; \lambda \in \Lambda\}$. Clearly $\Gamma = \Gamma_0 \cup \Gamma_1 \cup \Gamma_2 \cup \cdots$. For $\gamma \in \Gamma$, we define $\pi_\gamma(k) = k(\gamma)$, $k \in K$; then, clearly, $\pi_\gamma \in C(K)$. It may happen that the correspondence $\gamma \mapsto \pi_\gamma$ is not injective; however, its restriction to each Γ_i is one-to-one for each $i \in \mathbb{N}$.

Fix for a while any $i \in \mathbb{N}$. Put $\widetilde{\Gamma}_i = \{\pi_\gamma; \gamma \in \Gamma_i\}$ and let X_i denote the closed subspace of $C(K)$ generated by $\widetilde{\Gamma}_i$. Fix any $x^* \in B_{X_i^*}$. We claim that

$$\operatorname{card} \{\tilde{\gamma} \in \widetilde{\Gamma}_i; \; \langle \tilde{\gamma}, x^* \rangle \neq 0\} \leq \aleph_0. \tag{6.11}$$

Indeed, find $y^* \in B_{C(K)^*}$ such that $y^*|_X = x^*$. If $y^* = \delta_k$, the point mass at some $k \in K$, then (6.11) holds trivially. Also, (6.11) holds if y^* is equal to a finite linear combination of point masses. Note that $(B_{C(K)^*}, w^*)$ is a Corson compact. Hence every element of $B_{C(K)^*}$ lies in the weak*-closure of a *countable* subset of the linear span of $\{\delta_k; \ k \in K\}$ (see Proposition 5.27). Therefore (6.11) holds for any $x^* \in X_i^*$. We have thus proved that the set $\widetilde{\Gamma}_i$, which is linearly dense in X_i, countably supports X_i^* and the claim is proved.

Consider first the case of (uniform) Eberlein compacta. Fix any $\varepsilon > 0$. Let $\widetilde{\Gamma}_i = \bigcup_{n=1}^\infty \widetilde{\Gamma}_{i,n}^\varepsilon$ be provided by (iii) in Theorem 6.13 (in Theorem 6.30). Then put

$$\Gamma_{i,n}^\varepsilon := \{\gamma \in \Gamma_i; \ \pi_\gamma \in \widetilde{\Gamma}_{i,n}^\varepsilon\}, \quad \varepsilon > 0, \quad n \in \mathbb{N}.$$

Clearly, $\Gamma_i = \bigcup_{n=1}^\infty \Gamma_{i,n}^\varepsilon$. Now fix any $n \in \mathbb{N}$ and any $k \in K$. Then, using the injectivity of the mapping $\gamma \mapsto \pi_\gamma$ between Γ_i and $\widetilde{\Gamma}_i$, we have

$$\text{card}\,\{\gamma \in \Gamma_{i,n}^\varepsilon; \ |k(\gamma)| > \varepsilon\}$$
$$= \text{card}\,\{\gamma \in \Gamma_{i,n}^\varepsilon; \ \langle\pi_\gamma, \delta_k\rangle > \varepsilon\} + \text{card}\,\{\gamma \in \Gamma_{i,n}^\varepsilon; \ \langle\pi_\gamma, -\delta_k\rangle > \varepsilon\}$$
$$= \text{card}\,\{\tilde{\gamma} \in \widetilde{\Gamma}_{i,n}^\varepsilon; \ \langle\tilde{\gamma}, \delta_k\rangle > \varepsilon\} + \text{card}\,\{\tilde{\gamma} \in \widetilde{\Gamma}_{i,n}^\varepsilon; \ \langle\tilde{\gamma}, -\delta_k\rangle > \varepsilon\}$$
$$< \aleph_0 \quad (< 2n).$$

This holds for every $i \in \mathbb{N}$. We note that $\Gamma = \bigcup_{i=0}^\infty \Gamma_i = \bigcup_{i,n=1}^\infty \Gamma_{i,n}^\varepsilon \cup \Gamma_0$. It remains to enumerate the family Γ_0, $\Gamma_{i,n}^\varepsilon$, $i, n \in \mathbb{N}$, by elements of $\mathbb{N}$, to "make" it pairwise disjoint and in the uniform case to insert "a few" empty sets. This proves the necessity in (i).

In the case of Gul'ko compacta, we find for every $i \in \mathbb{N}$ sets $\widetilde{\Gamma}_{i,n} \subset \widetilde{\Gamma}$, $n \in \mathbb{N}$, as stated in Theorem 6.25 (ii). Then putting $\Gamma_{i,n} := \{\gamma \in \Gamma_i; \ \pi_\gamma \in \widetilde{\Gamma}_{i,n}\}$, $i, n \in \mathbb{N}$, we get a countable family that, together with the set Γ_0, obviously satisfies the necessary condition in (ii).

Sufficiency. Let Γ_n^ε, $\varepsilon > 0$, $n \in \mathbb{N}$, be as in (i). Let τ_i, $i \in \mathbb{N}$, be the functions defined in the proof of Lemma 6.21. Then define $\Phi : K \to \mathbb{R}^{\Gamma \times \mathbb{N}}$ by

$$\Phi(k)(\gamma, i) = \frac{1}{2^n 2^i \sqrt{n}} \tau_i(k(\gamma)) \quad \text{if} \quad \gamma \in \Gamma_n^{1/i}, \quad n \in \mathbb{N}, \quad \text{and} \quad i \in \mathbb{N}$$

for $k \in K$. Clearly, Φ is continuous. It is also injective. And $\Phi(K)$ is a subset of $c_0(\Gamma \times \mathbb{N})$ (of $\ell_2(\Gamma \times \mathbb{N})$). Therefore K is a (uniform) Eberlein compact.

Now let the condition in (ii) be satisfied. Let X be the subspace of $C(K)$ generated by the set $\widetilde{\Gamma} := \{\pi_\gamma; \ \gamma \in \Gamma\}$. Then we can easily check that this set, equipped with the weak topology of X, is $\mathcal{K}$-countably determined or $\mathcal{K}$-analytic in $(B_{X^{**}}, w^*)$, see [Fab97, Def. 7.1.2] and the proof of Theorem 6.25. And since the set $\widetilde{\Gamma}$ separates the points of K, [Tala79, Thm. 3.4(iii)] guarantees that the whole $C(K)$ is a Vašák space. $\qquad\square$

Theorem 6.34 (Benyamini, Rudin and Wage [BRW77]). *A continuous image of a uniform Eberlein compact is uniform Eberlein.*

Proof. $C(\varphi(K))$ is a subspace of $C(K)$ so, by Theorem 6.30, $C(\varphi(K))$ admits a UG norm. Thus $B_{C(\varphi(K))^*}$ is uniform Eberlein, and thus it is K. □

Theorem 6.35 (Kutzarova and Troyanski [KuTr82]). *There is a reflexive Banach space with an unconditional basis that does not admit a UG norm.*

Proof. Let K be an Eberlein compact that is not a uniform Eberlein compact (see, e.g., [DGZ93a, Chap. IV], [Fa~01, p. 419]). Then $C(K)$ is WCG ([AmLi68]) and does not admit a uniformly Gâteaux differentiable norm (Theorem 6.30). By a standard method (see, e.g., [DGZ93a, Chap. II]), neither does a reflexive space that factorizes through $C(K)$ (see [DFJP74]). This factorization result says that for every WCG space X there is a reflexive space Z with unconditional basis and a bounded linear operator from Z onto a dense set in X. □

Argyros and Mercourakis proved in [ArMe05b] that *there is a Banach space X such that X^* is UG, although there is no bounded linear injection of X into any $c_0(\Gamma)$.* Thus there is a Banach space X with X^* a subspace of WCG such that there is no bounded linear injection of X into $c_0(\Gamma)$.

6.4 Strongly Reflexive and Superreflexive Generated Spaces

In this section, we study Banach spaces that are strongly generated by reflexive or superreflexive spaces. The results developed here are then used to give short proofs of some results on weak compact sets in $L_1(\mu)$ spaces for finite measures μ due to Rosenthal, Argyros, and Farmaki.

Definition 6.36. *We will say that a Banach space X is* strongly generated by *an absolutely convex weakly compact set $K \subset X$ (or that K strongly generates X) if for every weakly compact set $W \subset X$ and every $\varepsilon > 0$ there is $m \in \mathbb{N}$ such that $W \subset mK + \varepsilon B_X$. We will say that a Banach space X is* strongly generated by *a Banach space Z (or that Z strongly generates X) if there exists a bounded linear operator $T : Z \to X$ such that $T(B_Z)$ strongly generates X. If this is the case and Z is a reflexive (resp. superreflexive) space, we will say that X is* strongly reflexive *(resp.* strongly superreflexive*) generated.*

Theorem 6.37 (Schlüchtermann and Wheeler [ScWh88]). *The following are equivalent for a Banach space X:*

(i) *X is strongly reflexive generated.*
(ii) *There is a weakly compact (absolutely convex) set $K \subset X$ that strongly generates X.*
(iii) *$(B_{X^*}, \tau(X^*, X))$ is (completely) metrizable, where $\tau(X^*, X)$ is the Mackey topology on X^* associated to the dual pair $\langle X^*, X \rangle$.*

(iv) *There is a weakly compact (absolutely convex) set K in X such that for every weakly null sequence (x_n) in X and for every $\varepsilon > 0$ there is $m \in \mathbb{N}$ so that $\{x_n; n \in \mathbb{N}\} \subset mK + \varepsilon B_X$.*

In statements (ii) *and* (iv), *the existence of K can be replaced by the existence of a sequence $(K_n)_{n=1}^\infty$ of weakly compact (absolutely convex) sets in X such that for every weakly compact set (resp. weakly null sequence) L in X and every $\varepsilon > 0$ there is $m \in \mathbb{N}$ such that $L \subset K_m + \varepsilon B_X$.*

Proof. It is simple to prove the validity of the statement in the last sentence. We shall proceed with the proof of the equivalences.

(i)$\Rightarrow$(ii) If X is strongly generated by a reflexive space Z by an operator T, we put $K := TB_Z$ in (ii).

(ii)$\Rightarrow$(i) Assuming (ii), there is by [DFJP74] a reflexive space Z and a bounded operator T from Z into X such that $K \subset TB_Z$.

(ii)$\Rightarrow$(iii) We shall prove that the topology $\mathcal{T}_K$ of the uniform convergence on K coincides on B_{X^*} with the topology $\tau(X^*, X)$. In view of Lemma 3.6, it is enough to prove that the restriction of both topologies to B_{X^*} coincide at the element 0. Obviously, $\mathcal{T}_K$ is coarser than $\tau(X^*, X)$. Let L be an absolutely convex w-compact subset of X. From (ii) we can find $m \in \mathbb{N}$ such that $L \subset mK + (1/2)B_X$. We shall check that $(2mK)^\circ \cap B_{X^*} \subset L^\circ$. To that end, take $x^* \in (2mK)^\circ \cap B_{X^*}$. For $x \in L$, put $x = mk + b/2$, where $k \in K$ and $b \in B_X$. Then

$$|\langle x, x^* \rangle| = |\langle mk + b/2, x^* \rangle|$$
$$\leq |\langle 2mk, x^*/2 \rangle| + |\langle b/2, x^* \rangle| \leq \frac{1}{2} + \frac{1}{2} = 1.$$

This proves the assertion.

(iii)$\Rightarrow$(ii) If $(B(X^*), \tau(X^*, X))$ is metrizable, there exists a countable basis of neighborhood of zero; i.e., a family $\{K_n^\circ \cap B_{X^*}\}_{n=1}^\infty$, where $\{K_n\}_{n=1}^\infty$ is a certain family of absolutely convex and weakly compact subsets of X. Put $p_n := \sup\{\|x\| : x \in K_n\}$, $n \in \mathbb{N}$. We may assume that $p_n \neq 0$ for all $n \in \mathbb{N}$. The set $K := \sum_{n=1}^\infty (n)^{-2} p_n^{-1} K_n$ is closed and absolutely convex. It is weakly compact by Lemma 3.20. Given $m \in \mathbb{N}$ and an absolutely convex weakly compact set $L \subset X$, there exists $n \in \mathbb{N}$ such that $K_n^\circ \cap mB_{X^*} \subset L^\circ \cap mB_{X^*}$. By taking polars, $(L \subset) \overline{\mathrm{conv}}\,(L \cup \frac{1}{m} B_X) \subset \overline{\mathrm{conv}}\,(K_n \cup \frac{1}{m} B_X)(\subset K_n + \frac{1}{m} B_X)$. This is (ii).

(ii)$\Rightarrow$(iv) is trivial.

(iv)$\Rightarrow$(ii) Assume that (iv) holds and let K be the absolutely convex w-compact subset of X given by (iv). This K works also for (ii). Indeed, assume the contrary; then there exists $\varepsilon > 0$ and a w-compact subset L such that $L \not\subset mK + \varepsilon B_X$ for all $m \in \mathbb{N}$. By the Eberlein-Šmulyan theorem, we obtain a w-convergent sequence (x_n) in L such that $x_n \notin m_n K + \varepsilon B_X$ for some increasing sequence (m_n), a contradiction. $\qquad\square$

Theorem 6.38 (Schlüchtermann and Wheeler [ScWh88]). *Any strongly reflexive generated space is weakly sequentially complete.*

Proof. Let (x_n) be a Cauchy sequence in X. For $n \in \mathbb{N}$, put $D_n := \overline{\Gamma}\{x_p - x_q; p, q \geq n\}$, where $\Gamma(S)$ denotes the absolutely convex hull of a set $S \subset X$. Obviously, $X^* = \bigcup_{n \in \mathbb{N}} D_n^\circ$. In particular, $mB_{X^*} = \bigcup_{n \in \mathbb{N}}(D_n^\circ \cap mB_{X^*})$ for every $m \in \mathbb{N}$. From (iii) in Theorem 6.37, $(B_{X^*}, \mu(X^*, X))$ is a complete metrizable space. Fix $m \in \mathbb{N}$. The sets $(D_n^\circ \cap mB_{X^*})$ are $\mu(X^*, X)$-closed; hence, by the Baire category theorem, there exists $n(m) \in \mathbb{N}$ and an absolutely convex weakly compact subset K_m of X such that

$$(K_m^\circ \cap mB_{X^*}) \subset (D_{n(m)}^\circ \cap mB_{X^*}).$$

By taking polars in X, we get

$$\overline{\operatorname{conv}}\left(D_{n(m)} \cup \frac{1}{m}B_X\right) \subset \overline{\operatorname{conv}}\left(K_m \cup \frac{1}{m}B_X\right) \ (\subset K_m + \frac{1}{m}B_X).$$

In particular, $x_p - x_q \in K_m + \frac{1}{m}B_X$ for every $p, q \geq n(m)$. Let x^{**} be the w^*-limit of (x_n) in X^{**}. Then $x^{**} - x_q \in K_m + \frac{1}{m}B_{X^{**}}$ for every $q \geq n(m)$, and we obtain $x^{**} \in X + \frac{1}{m}B_{X^{**}}$. This happens for every $m \in \mathbb{N}$, so $x^{**} \in X$. □

Corollary 6.39. *Let X be a strongly reflexive generated Banach space that does not contain a copy of ℓ_1. Then X is reflexive.*

Proof. Let (x_n) be a bounded sequence in X. By Rosenthal's ℓ_1 theorem, (x_n) contains a weakly Cauchy subsequence. Since X is weakly sequentially complete (Theorem 6.38), (x_n) contains a weakly convergent subsequence and thus X is reflexive by the Eberlein-Šmulyan theorem. □

Corollary 6.40. *Every separable Banach space with the Schur property is strongly reflexive generated.*

Proof. (B_{X^*}, w^*) is metrizable. A w^*-convergent sequence in X^* converges uniformly on X^*-limited subsets of X. Since X has the Schur property, w-compact subsets of X are $\|\cdot\|$-compact and hence X^*-limited (see Remark 3.10). It follows that the topology $\tau(X^*, X)$ coincides in B_{X^*} with the w^*-topology. To finish the proof, we use Theorem 6.37 (iii). □

Proposition 6.41. *The space $L_1(\mu)$, where μ is a finite measure, is strongly generated by a Hilbert space.*

Proof. We will use [JoLi01h, Chap. 1, p. 17]. Assume without loss of generality that μ is a probability measure on a set Ω. By using the identity operator, $B_{L_\infty(\mu)} \subset B_{L_2(\mu)} \subset B_{L_1(\mu)}$. Let K be a weakly compact set in the unit ball of $L_1(\mu)$. Then K is uniformly integrable (see Theorem 3.24).

Put, for $k \in \mathbb{N}$ and for $x \in K$, $M_k(x) := \{t \in \Omega; |x(t)| \geq k\}$. Then $x = x_1 + x_2$, where $x_1 := x.\chi(\Omega \setminus M_k(x))$ and $x_2 := x.\chi(M_k(x))$ (here $\chi(S)$ denotes the characteristic function of the set $S \subset \Omega$). Let $a_k(x) := \|x_2\|_1$, $a_k(K) := \sup\{a_k(x); x \in K\}$. Then

$$K \subset kB_{L_\infty(\mu)} + a_k(K)B_{L_1(\mu)} \subset kB_{L_2(\mu)} + a_k(K)B_{L_1(\mu)}.$$

We have $k\mu(M_k(x)) \leq a_k(x) \leq 1$, and hence $\mu(M_k(x)) \leq 1/k$ for all $x \in K$. From the uniform integrability of K, we get that $a_k(K) \to 0$ when $k \to \infty$. This finishes the proof. $\square$

Mercourakis and Stamati proved in [MeSt] that *there is a subspace of $L_1[0,1]$ that is not strongly reflexive generated*. On the other hand, we have the following proposition.

Proposition 6.42. *Assume that a strongly superreflexive generated space X does not contain a copy of ℓ_1. Then X is superreflexive.*

Proof. The space X is reflexive by Corollary 6.39. A reflexive Banach space X is a quotient of a Banach space Z if Z strongly generates X. This follows from the fact that if a set M is an ε-net for S_X, then the absolutely closed convex hull of M contains 0 as an interior point if $\varepsilon > 0$ is small; see, e.g., [Fa~01, Exer. 8.77]. To finish the proof, note that a quotient of a superreflexive space is superreflexive [DGZ93a, Cor. IV.4.6]. $\square$

Theorem 6.43. *Assume that X is strongly superreflexive generated. Then X has an equivalent norm $\|\cdot\|$ whose dual norm satisfies the following property: $f_n - g_n \to 0$ uniformly on any weakly compact set in X whenever $f_n, g_n \in S_{X^*}$ are such that $\|f_n + g_n\| \to 2$.*

Proof. Assume that Z is a superreflexive space that strongly generates X. Without loss of generality, we may assume that the norm of Z is uniformly Fréchet differentiable; see, e.g., [DGZ93a, Cor. IV.4.6]. Put $W := T(B_Z) \subset X$.

Then, by a standard argument (see, e.g., [DGZ93a, Chap. II]), the norm defined on X^* by $\|f\|^2 := \|f\|_1^2 + \|T^*(f)\|_2^2$ for all $f \in X^*$, where $\|\cdot\|_1$ is the norm in X^* dual to the original norm in X and $\|\cdot\|_2$ is the dual norm of Z^*, has the property that $\sup_{x \in W} |(f_n - g_n)x| \to 0$ whenever f_n, g_n are uniformly bounded in X^* and $f_n, g_n \in X^*$ satisfy $2\|f_n\|^2 + 2\|g_n\|^2 - \|f_n + g_n\|^2 \to 0$.

We will show that the predual norm to $\|\cdot\|$ satisfies our property. Indeed, we need to show that given two sequences (f_n) and (g_n) in S_{X^*} such that

$$2\|f_n\|^2 + 2\|g_n\|^2 - \|f_n + g_n\|^2 \to 0, \tag{6.12}$$

then $\sup_{x \in K} |(f_n - g_n)x| \to 0$ for each weakly compact set K in X.

For showing this, let a weakly compact K and $\varepsilon > 0$ be given. From the definition of strong generating, given $\varepsilon > 0$, find m_0 such that $K \subset m_0 W + \varepsilon B_X$.

Then from (6.12), we find n_0 such that

$$\sup_{x \in W} |(f_n - g_n)x| \leq \varepsilon/m_0$$

for each $n > n_0$.

Then, for each $n > n_0$,

$$\sup_{x \in K} |(f_n - g_n)x| \leq \sup_{x \in m_0 W} |(f_n - g_n)x| + \sup_{x \in \varepsilon B_X} |(f_n - g_n)x|$$
$$\leq m_0 \varepsilon / m_0 + 2\varepsilon = 3\varepsilon.$$

$\square$

The following result was motivated by [GiSci96] and [BoFi93].

Corollary 6.44. *Let X be a strongly superreflexive generated space. Then there is an equivalent norm on X the restriction of which to any subspace Y of X that does not contain a copy of ℓ_1 is uniformly Fréchet differentiable. In particular, any such subspace Y is superreflexive.*

Proof. The space X is weakly sequentially complete (Theorem 6.38). Thus, by Rosenthal's ℓ_1 theorem, Y is reflexive, so B_Y is weakly compact and the restriction of the norm from Theorem 6.43 to Y is, by Šmulyan's lemma [DGZ93a, Thm. I.1.4], uniformly Fréchet differentiable. Thus Y is superreflexive [DGZ93a, Cor. IV.4.6V]. $\square$

Theorem 6.45. *Let X be a strongly superreflexive generated space. Then any weakly compact subset K of X is a uniform Eberlein compact in its weak topology.*

Proof. Let $W := T B_Z$. Let $\{e_\gamma; f_\gamma\}_{\gamma \in \Gamma}$ be an M-basis for X with $\{f_\gamma; \gamma \in \Gamma\} \subset B_{X^*}$. Given $\varepsilon > 0$, find m so that $K \subset mW + (\varepsilon/4)B_X$. The set mW is a uniform Eberlein compact (the unit ball of a superreflexive space is a uniform Eberlein compact (Theorem 6.30)) and a continuous image of a uniform Eberlein compact is uniform Eberlein (Theorem 6.34). The map $x \to (f_\gamma(x))_{\gamma \in \Gamma}$ maps X into $c_0(\Gamma)$ (see Theorem 5.3).

Therefore, by using Theorem 6.33, for every $\varepsilon > 0$, we can write $\Gamma = \bigcup_{n=1}^\infty \Gamma_n^\varepsilon$ such that

$$\forall n \in \mathbb{N}, \quad \forall w \in mW, \quad \operatorname{card}\{\gamma \in \Gamma_n^\varepsilon; \ |f_\gamma(w)| > \varepsilon/4\} < n.$$

Now, if $x \in K$, then $x = w + y$, where $w \in mW$ and $y \in (\varepsilon/4)B_X$, and if $|f_\gamma(x)| > \varepsilon$, then easily $|f_\gamma(w)| \geq \frac{3}{4}\varepsilon$. There are only less than n members $\gamma \in \Gamma_n^\varepsilon$ with this property. Thus, for $\varepsilon > 0$, the sets Γ_n^ε can be used in Theorem 6.33 for the set K. This shows that K is a uniform Eberlein compact. $\square$

Corollary 6.46 (Rosenthal [Rose73]). *Let X be a subspace of $L_1(\mu)$ for a finite measure μ. If X does not contain ℓ_1, then X is superreflexive.*

Proof. This follows from Proposition 6.41 and Corollary 6.44. $\square$

Corollary 6.47 (Argyros and Farmaki [ArFa85]). *Every weakly compact set in the space $L_1(\mu)$, for a finite measure μ, is a uniform Eberlein compact.*

Proof. This follows from Theorem 6.45 and Corollary 6.44.   □

Definition 6.48 (Kalton [Kalt74]). *Let X be a separable Banach space. Let $\{e_n; f_n\}$ be a Schauder basis for X. We will say that $\{e_n; f_n\}$ is* almost shrinking *if, for each $f \in X^*$, $P_n^* f \to f$ in the sense of the Mackey topology $\tau(X^*, X)$ of X^*, where (P_n) is the sequence of projections on X associated to the Schauder basis.*

Theorem 6.49 (Kalton [Kalt74]). *Every unconditional basis in a separable Banach space is almost shrinking.*

Proof. If $\{e_i; f_i\}$ is an unconditional basis for a Banach space X, then for every $f \in X^*$, $\sum f(e_i) f_i = f$ in the w^*-sense. Since the basis is unconditional, this series converges subseries; i.e., for any increasing sequence $\{k_n\}$ of integers, $\sum f(e_{k_n}) f_{k_n}$ is w^*-convergent. Then, by McArthur's version of the Orlicz-Pettis theorem ([Arth67]), $\sum f(e_{k_n}) f_{k_n}$ converges in the Mackey topology $\tau(X^*, X)$. Therefore $f = \sum f(e_i) f_i$ in the sense of the topology $\tau(X^*, X)$.   □

Theorem 6.50 (Kalton [Kalt74]). *Let $(X, \|\cdot\|)$ be the space $(C[0, 1], \|\cdot\|_\infty)$. Then:*

(i) *$(X^*, \tau(X^*, X))$ is a complete separable locally convex space.*
(ii) *$(X^*, \tau(X^*, X))$ has no Schauder basis in the locally convex setting.*
(iii) *$(X^*, \tau(X^*, X))$ is not sequentially separable.*
(iv) *$(B_{X^*}, \tau(X^*, X))$ is not metrizable.*

Proof. (i) The completeness follows from the general fact mentioned in Subsection 3.1. The separability follows from the fact that $(X, \|\cdot\|)$ is separable, so (X^*, w^*) is separable and thus $\tau(X^*, X)$-separable.

(ii) Assume that $\{f_n; f_n'\}$ is a Schauder basis for $(X^*, \tau(X^*, X))$. Then, for any $f \in X^*$, we have $f = \sum f_n'(f) f_n$ in the $\tau(X^*, X)$ sense and thus in the sense of the weak topology of X^* by Theorem 3.26. Thus X^* is w-separable and thus $\|\cdot\|$-separable, a contradiction.

(iii) $X^* := C(K)^*$ has the same $\tau(X^*, X)$ and weak convergent sequences by Theorem 3.26. Thus the space X^* would be w-separable, and hence $\|\cdot\|$-separable, a contradiction.

(iv) We use Theorem 6.38 and the fact that $C[0, 1]$ is not weakly sequentially complete.   □

Corollary 6.51. *The space $C[0, 1]$ does not have any unconditional basis and is not strongly reflexive generated.*

Proof. We use Theorem 6.49, Theorem 3.26, and Theorem 6.38.   □

Remark 6.52. An example of a WLD space that is not a Vašák space can be found, e.g., in [Fab97, Thm. 7.3.2 and 7.3.4]. An example of a Vašák space that is not a subspace of a WCG space can be found, e.g., in [Fab97, Thm. 8.4.6]. An example of a subspace of a WCG space that is not WCG was first

given by Rosenthal [Rose74] (see Exercise 7.13). An example of a scattered
Eberlein compact that is not a uniform Eberlein compact can be found, e.g.,
in [Fa~01, Exer. 12.11]. An example of a reflexive generated Banach space
that is not a subspace of a Hilbert generated space is, e.g., in Theorem 6.35.
An example of a subspace of a Hilbert generated space that is not Hilbert
generated is Rosenthal's example in [Rose74].

Every reflexive space that is not superreflexive (for example, $(\sum \ell_\infty^n)_2$) gives
an example of a space that is reflexive generated and not superreflexive generated (Proposition 6.42).

For more counterexamples in this area, we refer to [ArMe93] and [FGHZ03].

6.5 Exercises

6.1. Prove the following result: a norming M-basis $\{x_\gamma; x_\gamma^*\}_{\gamma\in\Gamma}$ of a Banach
space X is σ-shrinkable if and only if, given $\varepsilon > 0$, $\Gamma = \bigcup_{n\in\mathbb{N}} \Gamma_n^\varepsilon$, so that for
each $n \in \mathbb{N}$, $\{x_\gamma; \gamma \in \Gamma_n^\varepsilon\}' \subset X + \varepsilon B_{X^{**}}$.

Hint. In order to prove one implication, let $\{x_\gamma; x_\gamma^*\}_{\gamma\in\Gamma}$ be a λ-norming M-
basis and $\Gamma_0 \subset \Gamma$ a set such that $\{x_\gamma; \gamma \in \Gamma_0\}' \subset X + \varepsilon B_{X^{**}}$. Show that

$$\{x_\gamma; \gamma \in \Gamma_0\}' \subset \varepsilon(1+1/\lambda)B_{X^*}.$$

Let $x^{**} \in \{x_\gamma; \gamma \in \Gamma_0\}'$. Then $x^{**} = x + u^{**}$, where $x \in X$ and $u^{**} \in \varepsilon B_{X^{**}}$.
Choose $x^* \in \operatorname{span}\{x_\gamma; \gamma \in \Gamma\} \cap B_{X^*}$. Then

$$0 = \langle x^{**}, x^* \rangle = \langle x, x^* \rangle + \langle u^{**}, x^* \rangle,$$

so $|\langle x, x^* \rangle| < \varepsilon$. As the basis is norming, we get $\|x\| < \varepsilon/\lambda$, so $\|x^{**}\| < \varepsilon(1 + 1/\lambda)$. The reverse implication is obvious.

6.2. Show that every bounded shrinking M-basis is a weakly compact M-basis.

6.3. Does there exist a bounded operator from $c_0(\omega_1)$ onto a dense set in ℓ_∞?

Hint. No, ℓ_∞ is not WCG.

6.4. Does there exist a bounded operator from $C[0, \omega_1]$ onto a dense set in
$\ell_\infty(\mathbb{N})$?

Hint. No; use the nonweak*-separability of $C[0, \omega_1]^*$.

6.5. Does there exist an operator from JL_0 onto a dense set in ℓ_∞?

Hint. No; ℓ_∞ does not have a Gâteaux smooth equivalent norm, while JL_0
does.

6.6. Let X be separable and let X^* contain a nonseparable subspace Y with
an M-basis. Show that then $\ell_1 \subset X$.

Hint. Otherwise, $(B_{X^{**}}, w^*)$ is angelic, so it is (B_{Y^*}, w^*), and then the M-basis in Y countably supports Y^* and so Y is WLD. (X^{**}, w^*) is separable, so it is (Y^*, w^*) and then Y is separable, which is not the case [BFT78].

6.7. Assume that K is either a scattered compact or a Corson compact. Show that $C(K)$ is isomorphic to its hyperplanes.

Hint. In both cases, c_0 is isomorphic to a complemented subspace of $C(K)$.

6.8. Suppose that a reflexive Z strongly generates X and that X does not contain an isomorphic copy of ℓ_1. Show that X is a quotient of Z.

Hint. X is weakly sequentially complete by Theorem 6.38. Moreover, it does not contain ℓ_1. So, by Rosenthal's ℓ_1 theorem, X is reflexive and we can use the proof of Proposition 6.42.

6.9. Show that any WCG Banach space with the Schur property is separable.

Hint. X has a weakly compact M-basis $\{e_\alpha\}_{\alpha \in \Gamma}$. For an uncountable subset $\Gamma_1 \subset \Gamma$, $\|e_\alpha\| \geq \varepsilon > 0$ for any $\gamma \in \Gamma_1$ and some $\varepsilon > 0$. The point 0 is in the weak closure of $\{e_\alpha\}_{\alpha \in \Gamma_1}$; therefore there is a sequence e_{γ_i}, $\gamma_i \in \Gamma_1$ with $e_{\gamma_i} \to 0$ weakly and thus in norm, as the weak compact sets are angelic, a contradiction.

6.10. Let X be a Banach space generated by a Banach space Z. Show that if Z is (a) a subspace of a WCG space, (b) a Vašák space, (c) WLD, then so is X.

6.11. Let X have an M-basis. Let T be a one-to-one operator from X onto a dense subset of a Banach space Y. Does Y necessarily have an M-basis?

Hint. No. Let (f_i) be a w^*-dense sequence in $B_{\ell_1^*(c)}$. Consider an operator $T : \ell_1(c) \to \ell_\infty \oplus \ell_2$ defined by $T(x) := (q(x), \frac{1}{2^i} f_i(x))_{i=1}^\infty$, where $q : \ell_1(c) \to \ell_\infty$ is a quotient map. Use Theorem 5.10.

6.12. Use Corollary 6.20 and Theorem 5.64 to give an alternative proof that a continuous image of an Eberlein compact is an Eberlein compact.

Transfinite Sequence Spaces

In this chapter, we discuss another biorthogonalization-like principle, namely Rosenthal's principle of disjointization of measures. We then use this to prove the Pełczyński and Rosenthal results on nonweakly compact operators on $C(K)$ spaces. We give Rosenthal's characterization of $C(K)$ spaces that contain a copy of $c_0(\Gamma)$ for uncountable Γ. We then present results of Pełczyński, Talagrand, and others on characterizations of spaces containing a copy of $\ell_1(c)$. In the latter part of this chapter, we present characterizations of spaces with long unconditional bases that are weakly compactly generated (Johnson), weakly Lindelöf determined (Argyros, Mercourakis), or that admit uniformly Gâteaux differentiable norms (Troyanski). We also include some renorming results on spaces with long symmetric bases due to Troyanski.

7.1 Disjointization of Measures and Applications

In this section, we study Rosenthal's principle of disjointization of measures and its application to nonweakly compact operators on $C(K)$ spaces. We also present Pełczyński's and Rosenthal's results on operators fixing c_0 subspaces and on the containment of $c_0(\Gamma)$ and ℓ_∞ spaces in dual spaces. As an application, we show Grothendieck's theorem on the Grothendieck property of ℓ_∞ and the Dieudonné-Phillips lemmas on measures. We also include Rosenthal's characterization of $C(K)$ spaces containing a copy of nonseparable $c_0(\Gamma)$. We also list some recent results on scattered Eberlein compacta due to Marciszewski and others.

Theorem 7.1 (Rosenthal [Rose70b]). *Let Γ be an infinite set and let $\{\mu_\gamma; \gamma \in \Gamma\}$ be a family of nonnegative finitely additive measures defined on all subsets of Γ such that*

$$\sup_{\gamma \in \Gamma} \mu_\gamma(\Gamma) < \infty.$$

Then, for all $\varepsilon > 0$, there exists a set $\Delta \subset \Gamma$ such that card $\Delta =$ card Γ and such that

$$\mu_\gamma(\Delta \setminus \{\gamma\}) < \varepsilon \text{ for all } \gamma \in \Delta.$$

Proof (Kupka [Ku74]). Assume by contradiction that for some $\varepsilon > 0$ no such set Δ exists. As Γ is infinite, we have card $\Gamma = $ card $\Gamma \times \Gamma$. Hence $\Gamma = \bigcup\{\Delta_\gamma; \gamma \in \Gamma\}$, where Δ_γ are pairwise disjoint, and card $\Delta_\gamma = $ card Γ for all $\gamma \in \Gamma$. We claim that there is a $\gamma_0 \in \Gamma$ such that $\mu_\gamma(\Gamma \setminus \Delta_{\gamma_0}) \geq \varepsilon$ for all $\gamma \in \Delta_{\gamma_0}$. Indeed, otherwise, we could find, for all $\gamma \in \Gamma$, an $\alpha_\gamma \in \Delta_\gamma$ such that $\mu_{\alpha_\gamma}(\Gamma \setminus \Delta_\gamma) < \varepsilon$. The set $\Delta = \{\alpha_\gamma; \gamma \in \Gamma\}$ then satisfies the conclusion of Theorem 7.1, contrary to our assumption.

Repeat this procedure with Δ_{γ_0} in place of Γ. Iterate this process. After finitely many steps, we violate the uniform boundedness of all μ_γ. $\square$

Corollary 7.2 (Rosenthal [Rose70b]). *Let Λ be a discrete set and $\{\mu_\alpha; \alpha \in \Gamma\}$ be an infinite family of finitely additive positive measures on Λ such that*

$$\sup_{\alpha \in \Gamma} \mu_\alpha(\Lambda) < \infty.$$

and let $\{E_\alpha, \alpha \in \Gamma\}$ be a family of disjoint subsets of Λ. Then, for all $\varepsilon > 0$, there exists a $\Gamma' \subset \Gamma$ with card $\Gamma' = $ card Γ such that

$$\mu_\alpha\left(\bigcup\{E_\beta; \beta \in \Gamma', \beta \neq \alpha\}\right) < \varepsilon \text{ for all } \alpha \in \Gamma'.$$

Corollary 7.3 (Rosenthal [Rose70b]). *Let (μ_n) be a bounded sequence in $C(K)^*$ and let (E_n) be a sequence of pairwise disjoint Borel subsets of K. Then, for every $\varepsilon > 0$, there exists an increasing sequence of integers (n_i) such that*

$$\sum_{i \neq j} |\mu_{n_j}|(E_{n_i}) < \varepsilon \quad \text{for all } j.$$

Proof. For each $n \in \mathbb{N}$, define the set function ν_n on the discrete set $\mathbb{N}$ by

$$\nu_n(F) := \mu_n(\overline{\bigcup_{m \in F} E_m})$$

for all $F \subset \mathbb{N}$.

By Corollary 7.2 applied to the family $\{\nu_n; n \in \mathbb{N}\}$ and the family $\{\{n\}; n \in \mathbb{N}\}$ (the family of singletons), for every $\varepsilon > 0$, there is an infinite subset $\mathbb{N}' \subset \mathbb{N}$ such that $\nu_n(\mathbb{N}' \setminus \{n\}) < \varepsilon$ for all $n \in \mathbb{N}'$. This finishes the proof. $\square$

Theorem 7.4. *Let X be a Banach space with the Dunford-Pettis property and let T be a weakly compact operator from X into a Banach space Y. Then T is a Dunford-Pettis operator.*

Proof. By contradiction, assume that T is not Dunford-Pettis. This means that for some $\delta > 0$ and some sequence (x_n) such that $x_n \to 0$ weakly, we have $\|Tx_n\| \geq \delta$ for all n. Let $x_n^* \in S_{X^*}$ be such that $x_n^*(Tx_n) = \|Tx_n\|$ for all n. Since T^* is weakly compact by the Gantmacher theorem, we may assume that, for some $x^* \in X^*$, $T^*(x_n^*) \to x^*$ in the weak topology of X^*. Since X has the Dunford-Pettis property and $x_n \to 0$ weakly in X and $T^*x_n^* \to x^*$ weakly in X^*, we have

$$0 = \lim(T^*x_n^* - x^*)(x_n) = \lim(x_n^*(Tx_n) - x^*(x_n)) = \lim \|Tx_n\|.$$

This contradicts that $\|Tx_n\| \geq \delta$ for all n, proving that T is a Dunford-Pettis operator. $\qquad\square$

Corollary 7.5. *Let K be a compact space and T be a weakly compact operator from $C(K)$ into a Banach space X. Then T is a Dunford-Pettis operator.*

Proof. The space $C(K)$ has the Dunford-Pettis property by [Fa~01, p. 376]. $\qquad\square$

Theorem 7.6 (Pełczyński [Pelc65]). *Let K be a compact space and let $T : C(K) \to X$ be a nonweakly compact operator. Then $C(K)$ contains a subspace isomorphic to c_0 on which T acts as an isomorphism. In particular, X contains a subspace isomorphic to c_0. If K is scattered, it suffices to assume that T is a noncompact operator.*

Proof. Put $W = T^*(B_{X^*})$. Since T^* is not weakly compact, by Theorem 3.26 we can choose $\eta > 0$, a sequence $O_1, O_2, \ldots$ of disjoint open sets in K, and a sequence $\mu_1, \mu_2, \ldots$ in W such that

$$|\mu_j|(O_j) > \eta \quad \text{for all } j.$$

Let $0 < \varepsilon < \eta$. By Corollary 7.3, by passing to a subsequence, we may assume that

$$\sum_{i \neq j} |\mu_j|(O_i) < \varepsilon \quad \text{for all } j.$$

For each j, choose $f_j \in C(K)$ of norm 1 with $0 \leq f_j \leq 1$ and f_j supported in O_j such that

$$\left| \int f_j d\mu_j \right| > \eta.$$

Then $Z = \overline{\text{span}}\{f_j\}$ is equivalent to the standard basis of c_0. Thus, given n and scalars $c_1, c_2, \ldots c_n$, we have

$$\left\| T\left(\sum_{j=1}^n c_j f_j \right) \right\| \leq \|T\| \max_j |c_j|.$$

Moreover, for each j,

$$\left\| T\left(\sum_{i=1}^n c_i f_i \right) \right\| \geq \sup_{x^* \in B_{X^*}} \left| (T^*x^*) \left(\sum_{i=1}^n c_i f_i \right) \right|$$

$$\geq \left| \int \left(\sum_{i=1}^{n} c_i f_i \right) d\mu_j \right| \geq |c_j| \left| \int f_j d\mu_j \right| - \sum_{i \neq j} |c_i| \int |f_i| d\mu_j|$$

$$\geq |c_j| \eta - \max_i |c_i| \sum_{i \neq j} |\mu_j|(O_i) \geq |c_j| \eta - \max_i |c_i| \varepsilon.$$

By taking the maximum over all j, we get

$$\left\| T(\sum_{i=1}^{n} c_i f_i) \right\| \geq (\eta - \varepsilon) \max_i |c_i|.$$

Thus T is an isomorphism on Z.

If K is scattered, then every weakly compact operator from $C(K)$ into X is norm compact. Indeed, its dual operator is weakly compact by Gantmacher's theorem and thus norm compact by the Schur property of $(C(K))^*$. It follows that the operator itself is norm compact by Schauder's theorem. □

In particular, we have the following corollaries.

Corollary 7.7 (Pełczyński [Pelc65]). *Assume that $T : c_0 \to X$ is a non-compact operator. Then there is a subspace Z of c_0 that is isomorphic to c_0 such that T is an isomorphism on Z.*

Corollary 7.8 (Pełczyński [Pelc65]). *Every infinite-dimensional complemented subspace of $C(K)$ contains a subspace isomorphic to c_0.*

Proof. Let P be a projection of $C(K)$ onto an infinite-dimensional subspace $X \subset C(K)$. Then P is not weakly compact. Indeed, since $C(K)$ has the Dunford-Pettis property, a weakly compact projection $P^2 = P$ would be compact and then X would be finite-dimensional. To finish the proof, apply Theorem 7.6. □

Theorem 7.9 (Pełczyński [Pelc65]). *Let K be a compact space, X be a Banach space and $T : C(K) \to X$ be a bounded operator. Then T is weakly compact if and only if T is strictly singular, i.e., T is an isomorphism on no infinite-dimensional subspace of $C(K)$.*

Proof. Assume that T is weakly compact. The space $C(K)$ has the Dunford-Pettis property and hence, by Theorem 7.4, T is a Dunford-Pettis operator. Assume that for some subspace Y of $C(K)$, $T \upharpoonright Y$ is an isomorphism. Now $T(B_Y)$ is w-compact and hence so is B_Y. Let (y_n) be a sequence in B_Y. By the Eberlein-Šmulyan theorem, it has a w-convergent subsequence (y_{n_k}). Then (Ty_{n_k}) is $\| \cdot \|$-convergent and so is (y_{n_k}). Thus Y is finite-dimensional.

The sufficient condition follows from Theorem 7.6. □

Theorem 7.10 (Rosenthal [Rose70b]). *Let X be a Banach space. Assume $T : \ell_\infty(\Gamma) \to X$ is such that $\inf_{\gamma \in \Gamma} \|T(e_\gamma)\| > 0$ (where e_γ is the unit vector in $\ell_\infty(\Gamma)$). Then there is a set $\Gamma' \subset \Gamma$ with $\mathrm{card}\, \Gamma' = \mathrm{card}\, \Gamma$ such that $T \upharpoonright \ell_\infty(\Gamma')$ is an isomorphism. This holds, in particular, if $T \upharpoonright c_0(\Gamma)$ is an isomorphism.*

Proof. Let $\inf_{\gamma \in \Gamma} \|Te_\gamma\| \geq \frac{1}{K} > 0$. Fix $\gamma \in \Gamma$. By the Hahn-Banach theorem, choose $f_\gamma \in X^*$ with $\|f_\gamma\| \leq K$ and $f_\gamma(Te_\gamma) = 1$. Define the set function μ_γ by

$$\mu_\gamma(E) = T^* f_\gamma(\chi_E)$$

for all $E \subset \Gamma$, where χ_E is the characteristic function of E in Γ.

It is well known that μ_γ is a finitely additive measure with $\|\mu\| = \|T^* f_\gamma\|$ and $\sup_{\gamma \in \Gamma} \|\mu_\gamma\| \leq \|T\| K$.

Letting $E_\gamma = \{\gamma\}$ in Theorem 7.1 for each $\gamma \in \Gamma$, we have that there is a set $\Gamma' \subset \Gamma$ with card $\Gamma' =$ card Γ such that $|\mu_\alpha|(\Gamma' \setminus \{\alpha\}) < \frac{1}{2}$ for each $\alpha \in \Gamma'$.

If $\varphi \in \ell_\infty(\Gamma')$ and if $\alpha \in \Gamma'$, then

$$\left| \int \varphi \, d\mu_\gamma \right| = \left| \varphi(\alpha) + \int_{\Gamma' \setminus \{\alpha\}} \varphi \, d\mu_\gamma \right| \geq |\varphi(\alpha)| - \frac{1}{2} \|\varphi\|_\infty$$

since $\mu_\alpha(\alpha) = T^* f_\alpha(e_\alpha) = 1$.

Thus

$$\|T\varphi\| \geq \frac{1}{K} \sup_{\gamma \in \Gamma} |f_\gamma(T\varphi)| \geq \frac{1}{2K} \|\varphi\|_\infty.$$

Therefore $T \restriction \ell_\infty(\Gamma')$ is an isomorphism. $\square$

Similarly, we obtain the following theorem.

Theorem 7.11 (Rosenthal [Rose70b]). *Let $T : c_0(\Gamma) \to X$ be such that for some $\varepsilon > 0$, $\|T(e_\gamma)\| > \varepsilon$. Then there exists $\Gamma' \subset \Gamma$ of the same cardinality as Γ such that T restricted to $c_0(\Gamma')$ is an isomorphism.*

Theorem 7.11 has recently been improved in [ACGJM02] in the following way. *Let $T : c_0(I) \to X$ be a bounded linear operator such that, for some $\delta > 0$, $\|T(e_i)\| \geq \delta$ for all $i \in I$, where e_i are the unit vectors in $c_0(I)$. Then there is a finite partition $\{I_1, \ldots, I_n\}$ of I such that the operator $T \restriction c_0(I_k)$ is an isomorphism for $k = 1, 2, \ldots, n$.*

Theorem 7.12 (Rosenthal [Rose70b]). *Let X and E be Banach spaces. Let Γ be an infinite set and let $T : X^* \to E$ be an operator such that there is a subspace Z of X^* isomorphic to $c_0(\Gamma)$ and the restriction of T to Z is an isomorphism. Then there exists a subspace Y of X^* isomorphic to $\ell_\infty(\Gamma)$ such that the restriction of T to Y is an isomorphism.*

Proof. We first observe that there is an operator $S : \ell_\infty(\Gamma) \to X^*$ such that $S \restriction c_0(\Gamma)$ is an isomorphism onto Z. Indeed, choose an isomorphism $i : c_0(\Gamma) \to Z$. Let P be a projection from X^{***} onto X^*. Then $S = Pi^{**}$ is the desired operator. Then TS is an operator from $\ell_\infty(\Gamma)$ into E such that $TS \restriction c_0(\Gamma)$ is an isomorphism. By Theorem 7.10, there is $\Gamma' \subset \Gamma$ with card $\Gamma' =$ card Γ such that $TS \restriction \ell^\infty(\Gamma')$ is an isomorphism. Thus $S \restriction \ell^\infty(\Gamma')$ and $T \restriction S(\ell^\infty(\Gamma'))$ are both isomorphisms. Thus, putting $Y = S(\ell_\infty(\Gamma'))$, the result follows. $\square$

Corollary 7.13 (Rosenthal [Rose70b]). *Let X be a Banach space. If Γ is an infinite set and X^* contains an isomorphic copy of $c_0(\Gamma)$, then X^* contains an isomorphic copy of $\ell_\infty(\Gamma)$.*

Corollary 7.14 (Rosenthal [Rose70b]). *Let T be a nonweakly compact operator from ℓ_∞ into a Banach space X. Then there is a subspace Y of X isomorphic to ℓ_∞ such that $T \upharpoonright Y$ is an isomorphism. In particular, X contains a subspace isomorphic to ℓ_∞.*

Proof. By Theorem 7.6, there is a subspace A of ℓ_∞ such that A is isomorphic to c_0 and $T \upharpoonright A$ is an isomorphism. Therefore the result follows from Theorem 7.12. □

Corollary 7.15 (Phillips). *The space c_0 is not complemented in ℓ_∞.*

Proof. Let P be a projection of ℓ_∞ onto c_0. Then P is not weakly compact, as c_0 is not reflexive. Thus, by Corollary 7.14, c_0 contains a subspace isomorphic to ℓ_∞, and this is impossible. □

Rosenthal, in [Rose72], proved the following result: *Let K be a compact space, X be a Banach space, and $T : C(K) \to X$ be an operator such that $T^*(X^*)$ is nonseparable. Then there is a subspace $Z \subset C(K)$ isomorphic to $C[0, 1]$ such that $T \upharpoonright Z$ is an isomorphism.*

Johnson and Zippin proved, in [JoZi89], the following result: *Let Γ be any set, T be a bounded linear operator from a subspace $Z \in c_0(\Gamma)$ into some $C(K)$ space, and $\varepsilon > 0$. Then T can be extended to a linear operator $\tilde{T}$ from $c_0(\Gamma)$ into $C(K)$ so that $\|\tilde{T}\| \leq (1 + \varepsilon)\|T\|$.*

We will now discuss some applications.

Theorem 7.16. *Let μ_n be a sequence of finitely additive set functions defined on the discrete set Λ. Then:*

(i) (Dieudonné). *If $\sup_n |\mu_n(E)| < \infty$ for all $E \subset \Lambda$, then*

$$\sup_n \|\mu_n\| < \infty.$$

(ii) (Phillips). *If $\lim_n \mu_n(E) = 0$ for all $E \subset \Lambda$, then*

$$\lim_n \sum_{j \in \Lambda} |\mu_n(j)| = 0.$$

Proof. (i) (Rosenthal). Assume $\sup_n \|\mu_n\| = \infty$. For each $E \subset \Lambda$, put $\lambda(E) := \sup_n |\mu_n(E)|$ and choose a subsequence ν_n of μ_n and $E_i \subset \Lambda$ so that, for all $n > 1$,

$$|\nu_n(E_n)| \geq \|\nu_n\|/5 \geq n + 2\Sigma_{j=1}^{n-1}\lambda(E_j).$$

Put $F_1 = E_1$ and $F_n = E_n \setminus \bigcup_{j=1}^{n-1} E_j$ for $n > 1$. Then, for all n, $|\nu_n(F_n)| \geq \|nu_n\|/10$ and $F_n \cap F_m = \emptyset$ for $n \neq m$.

By Corollary 7.3, choose $n_1 < n_2 < \cdots$ such that, for all i,

$$\left|\nu_{n_i}\right|\left(\bigcup_{j\neq i} F_{n_j}\right) < \frac{1}{2}\left|\nu_{n_i}(F_{n_i})\right|.$$

Put

$$F = \bigcup_{j=1}^{\infty} F_{n_j}.$$

Then

$$\left|\nu_{n_i}(F)\right| \geq \|\nu_{n_i}\|/10, \quad \text{and thus} \quad \left|\nu_{n_i}(F)\right| \to \infty,$$

a contradiction.

(ii) (Rosenthal). Due to (i), we can use Corollary 7.3. Assume the conclusion is false. By a sliding hump argument again, we could choose a $\delta > 0$, a subsequence (ν_n) of (μ_n), and a sequence (E_n) of disjoint finite subsets of Λ such that, for all n, $|\mu_n(E_n)| \geq \delta$. By Corollary 7.3, there would exist an increasing sequence of indices (n_i) such that, for all i,

$$\left|\nu_{n_i}\right|\left(\bigcup_{j\neq i} E_{n_j}\right) < \delta/2.$$

Put $E = \bigcup_{j=1}^{\infty} E_{n_j}$. Then, for all i,

$$\left|\nu_{n_i}(E)\right| > \delta/2,$$

which contradicts the assumptions of (ii). $\square$

Call a series $\sum x_n$ *subseries convergent* (resp. *weakly subseries convergent*) if for each increasing sequence k_n of integers, the series $\sum x_{k_n}$ is convergent (resp. weakly convergent).

Theorem 7.17 (Orlicz, Pettis). *Every weakly subseries convergent series in a Banach space is subseries convergent.*

Proof. Without loss of generality, assume that X is separable. Assume, by contradiction, that there is a series $\sum x_n$ that is weakly subseries convergent but not subseries convergent. Therefore there is an increasing sequence (k_n) of integers for which $\sum x_{k_n}$ is not Cauchy, so there is an $\varepsilon > 0$ and a sequence (F_n) of finite subsets of $\{k_n; n \in \mathbb{N}\}$ such that $\max F_n < \min F_{n+1}$ and such that $\|\sum_{i \in F_n} x_{k_i}\| \geq \varepsilon$. Let $y_n := \sum_{i \in F_n} x_{k_i}$. Thus $\sum y_n$ is weakly subseries convergent; in particular, $y_n \to 0$ weakly. On the other hand, we have $\|y_n\| \geq \varepsilon$.

For each n, choose $y_n^* \in B_{X^*}$ so that $y_n^*(y_n) = \|y_n\|$. Without loss of generality, assume that $y_n^* \xrightarrow{w^*} y_0^*$. For each $\Delta \subset \mathbb{N}$, the series $\sum_{n \in \Delta} y_n$ converges weakly to some $\sigma_\Delta \in X$.

Define $\mu_n \in \ell_\infty^*$ at $\Delta \subset \mathbb{N}$ by

$$\mu_n(\Delta) = (y_n^* - y_0^*)(\sigma_\Delta).$$

As $y_n^* \to y_0^*$, we get

$$\lim_n \mu_n(\Delta) = 0 \quad \text{for each} \quad \Delta.$$

By Theorem 7.16, we get

$$\lim_n \sum_k |\mu_n(\{k\})| = 0.$$

On the other hand, as (y_n) is weakly null and $\|y_n\| \geq \varepsilon$ for each n, we have for n large enough, $|\mu_n(n)| = |(y_n^* - y_0^*)(y_n)| \geq \varepsilon/2$. Thus, for n large enough,

$$\sum_k |\mu_n(\{k\})| \geq |\mu_n(\{n\})| \geq \varepsilon/2$$

This contradiction proves Theorem 7.17. □

Theorem 7.18 (Grothendieck). *The space ℓ_∞ has the Grothendieck property; i.e., in ℓ_∞^*, the weak*-convergent sequences are weak convergent.*

Proof (Sketch). Identify ℓ_∞ with the space $C(K)$, where $K = \beta\mathbb{N}$. Let (μ_n) be a sequence that is weak*-null in $C(K)^*$. It suffices to show that (μ_n) is weakly relatively compact. By contradiction, using Theorem 3.26, there is $\varepsilon > 0$ and a sequence of open disjoint subsets of K and a subsequence (ν_n) of (μ_n) such that

$$|\nu_n(O_n)| \geq \varepsilon.$$

Assume that O_n are clopen (K is extremely disconnected; i.e., closures of open sets are open and ν_n are regular).

We can define set functions $\tilde{\nu}_n$ on the collection of subsets of integers by

$$\tilde{\nu}_n(\Delta) := \nu_n\Big(\sup_{k \in \Delta} O_k\Big)$$

for any $\Delta \subset \mathbb{N}$.

Since (ν_n) is weak*-null, and $\sup_{k \in \Delta} O_k$ is a clopen set in K for any $\Delta \subset \mathbb{N}$,

$$\lim_n \tilde{\nu}_n(\Delta) = \lim_n \nu_n\Big(\sup_{k \in \Delta} O_k\Big)$$

for any Δ. From Theorem 7.16, we get

$$\lim_n \sum_k |\nu_n(O_k)| = \lim_n \sum_k |\tilde{\nu}_n(\{k\})| = 0,$$

which is a contradiction with the fact that $|\nu_n(O_n)| \geq \varepsilon$ for each n. □

Theorem 7.19 (see [Gr98]). *A subspace Y of $c_0(\Gamma)$ is complemented in $c_0(\Gamma)$ if and only if it is isomorphic to $c_0(\Gamma')$ for some $\Gamma' \subset \Gamma$.*

Proof. Assume without loss of generality that dens $Y = $ card I. Let $\{y_j ; j \in J\}$ be a maximal family in the unit sphere of Y with disjoint supports. Then card $J = $ card I. Assume, by contradiction, that card $J < $ card I. Then, denoting $I_0 := \bigcup\{\mathrm{supp}\, y_j ; j \in J\}$, we have that card $I_0 = $ card J. Therefore there is $y \in S_Y$ such that supp $y \cap I_0 = \emptyset$. Indeed, otherwise, if each element of S_Y has a nonzero coordinate in I_0, the coordinates in I_0 would provide a weak*-dense set in $c_0(J)$ of smaller cardinality, which is impossible.

For each $j \in J$, pick $y_j^* \in S_{c_0(I)^*}$ such that $y_j^*(y_j) = 1$ and supp $y_j^* \subset$ supp y_j. Let Z be the closed linear span of $\{y_j ; j \in J\}$. Then Z is isometric to $c_0(J)$, and the operator $P : Y \to Z$ defined by $P(y) = \sum_{j \in J} y_j^*(y) y_j$ is a projection. Thus Z is complemented in Y and, by the assumption, Y is complemented in $c_0(I)$. Therefore, by Pełczyński's decomposition method, Y is isomorphic to $c_0(I)$.

If Y is isomorphic to $c_0(J)$, then card $J \leq$ card Γ. By the result of Johnson and Zippin quoted before Theorem 7.16, the identity operator from Y into itself can be extended into an operator T from $c_0(I)$ into Y. This means that Y is complemented in $c_0(I)$. $\square$

From the proof of Theorem 7.19, we get the following theorem.

Theorem 7.20. *Every subspace Z of $c_0(\Gamma)$ contains a subspace that is complemented in $c_0(\Gamma)$ and has density* dens Z.

We will need the following lemma.

Lemma 7.21 (Rosenthal [Rose70a]). *Assume that a compact space K satisfies the CCC property and that $\mathcal{F}$ is an uncountable family of open subsets of K. Then there is an infinite sequence (F_i) of distinct members of $\mathcal{F}$ with $\bigcap_{i=1}^{\infty} F_i \neq \emptyset$.*

Proof. If $\mathcal{A}$ is a family of subsets of K and n is a positive integer, let $\mathcal{A}_n$ be the family of all sets of the form $F_1 \cap F_2 \cap \cdots \cap F_n$, where $F_1, \ldots, F_n$ are n distinct members of $\mathcal{A}$. Put $\mathcal{A}^* = \bigcup_n \mathcal{A}_n$. In other words, $\mathcal{A}^*$ is the family of all finite intersections of members of $\mathcal{A}$. We have card $\mathcal{A}^* = $ card $\mathcal{A}$.

We claim that for all n, $(\mathcal{A}_n)_2 \subset \mathcal{A}_{n+1}^*$. Indeed, let A and B be distinct members of $\mathcal{A}$ with $A = \bigcap_{i=1}^{n} F_i$ and $B = \bigcap_{i=1}^{n} G_i$. Since $A \neq B$, there must exist indices i with $1 \leq i \leq n$ such that $G_i \neq F_j$ for any j with $1 \leq j \leq n$. Let $i_1 < i_2 < \ldots i_k$ be an enumeration of this set of such indices. Then, for each r with $1 \leq r \leq k$, $F_1 \cap \cdots \cap F_n \cap G_{i_k}$ is a member of $\mathcal{A}_{n+1}$ and $A \cap B = \bigcap_{r=1}^{k}(F_1 \cap \cdots \cap F_n \cap G_{i_r})$, and thus $A \cap B \in \mathcal{A}_{n+1}^*$.

Claim. (*) Either some nonempty member of $\mathcal{A}_2$ is contained in uncountably many members of $\mathcal{A}$ or $\mathcal{A}_2$ is uncountable.

In order to prove the claim, let $\mathcal{H}$ denote the class of all sets F in $\mathcal{A}$ such that there exists a G in $\mathcal{A}$ with $G \neq F$ and $G \cap F \neq \emptyset$. Then $\mathcal{H}$ is uncountable.

Indeed, $\mathcal{A} \setminus \mathcal{H}$ is a disjoint family of open sets, hence at most countable. Now, for each $A \in \mathcal{A}_2$, let $\mathcal{A}_A$ denote the class of all sets $F \in \mathcal{A}$ with $F \subset A$. Then we have that $\mathcal{H} = \cup\{\mathcal{A}_A; A \in \mathcal{A}_2, A \neq \emptyset\}$. Thus, if $\mathcal{A}_2$ is countable, $\mathcal{A}_A$ must be uncountable for some nonempty $A \in \mathcal{A}_2$, and the claim follows.

From $(^*)$, we deduce by induction that:

$(^{**})$ If $\mathcal{B}$ is an uncountable family of open subsets of K and n is a positive integer, then there are uncountably many distinct n-tuples $(B_1, \ldots, B_n)$ in $\mathcal{B}$ (i.e., $B_i \neq B_j$ if $i \neq j$) with $\bigcap_{i=1}^n B_i \neq \emptyset$.

To see this, let us assume that no nonempty member of $\mathcal{B}^*$ is contained in uncountably many members of $\mathcal{B}$ (since otherwise $(^{**})$ holds automatically). We will then show that $\mathcal{B}_n$ is uncountable for all n, from which $(^{**})$ follows immediately.

Observe that $\mathcal{B}_1$ is trivially uncountable. Suppose we have proved that $\mathcal{B}_n$ is uncountable. Then, if $\mathcal{B}_{n+1}$ were countable, $\mathcal{B}_{n+1}^*$ and consequently $(\mathcal{B}_n)_2$ would also be countable by our preliminary observations. Thus, by $(^*)$, there would exist A and B in $\mathcal{B}_n$ with $A \cap B$ nonempty and contained in uncountably many members of $\mathcal{B}_n$. But if $E \in \mathcal{B}_n$ and $A \cap B \subset E$, then E is a finite intersection of members of $\mathcal{B}$ each of which contains $A \cap B$. Hence $A \cap B$ would be contained in uncountably many members of $\mathcal{B}$, and of course $A \cap B \in \mathcal{B}^*$, so our assumption on $\mathcal{B}$ would be contradicted. Thus $(^{**})$ has been established by finite induction.

To finish the proof, let $\mathcal{F}$ be as in the statement of Lemma 7.21. For every positive integer n, let G_n be the set of all points in K that are contained in at most n distinct members of $\mathcal{F}$. Put G_n^0 equal to the interior of G_n, and let $\mathcal{G}_n := \{F \in \mathcal{F}; F \cap G_n^0 \neq \emptyset\}$. Fixing n, we claim that $\mathcal{G}_n$ is at most countable. Indeed, denoting $\mathcal{G}_n \cap G_n^0 := \{F \cap G_n^0; F \in \mathcal{G}_n\}$, we have that no $n+1$ distinct elements of $\mathcal{G}_n \cap G_n^0$ have a point in common. Thus, by $(^{**})$, $\mathcal{G}_n \cap G_n^0$ is at most countable. But each member of $\mathcal{G}_n \cap G_n^0$ is contained in at most n members of $\mathcal{G}_n$ and $\mathcal{G}_n = \{F \in \mathcal{F}; \exists A \in \mathcal{G}_n \cap G_n^0 \text{ with } F \supset A\}$. Thus, since $\mathcal{G}_n$ is countable for all n, $\bigcup_{n=1}^\infty \mathcal{G}_n$ is countable. Thus there exists a nonempty $F \in \mathcal{F}$ with $F \notin \bigcup_{n=1}^\infty \mathcal{G}_n$. It is easily seen that G_n is closed for all n and hence there exists an $s \in F$ with $s \notin \bigcup_{n=1}^\infty G_n \setminus G_n^0$ by the Baire category theorem (here we use that K is compact). Then $s \notin \bigcup_{n=1}^\infty G_n$ by the definition of F, so s belongs to infinitely many members of $\mathcal{F}$. □

Theorem 7.22 (Rosenthal [Rose70a]). *Let K be a compact space. Then the following are equivalent:*

(i) *$C(K)$ contains a nonseparable WCG space.*
(ii) *$C(K)$ contains a nonseparable $c_0(\Gamma)$.*
(iii) *K does not satisfy the CCC property.*

Proof. (i)⇒(iii) Assume that $C(K)$ contains a subspace with a weakly compact M-basis $\{e_\gamma; f_\gamma\}_{\gamma \in \Gamma_1}$, where Γ_1 is uncountable. As $\Gamma_1 = \bigcup_{n=1}^\infty \{\gamma \in \Gamma_1; \|e_\gamma\| \geq \frac{1}{n}\}$, we can assume that for an uncountable $\Gamma \subset \Gamma_1$, and some $\delta > 0$, we have $\|e_\gamma\| \geq \delta$ for all $\gamma \in \Gamma$. For each $\gamma \in \Gamma$, put $U_\gamma = \{x \in K; |e_\gamma(x)| > \frac{\delta}{2}\}$. Then there is an infinite sequence (γ_i) of distinct elements of Γ such that $\bigcap_{i=1}^\infty U_{\gamma_i} \neq \emptyset$. This follows from Lemma 7.21. Since $e_{\gamma_i} \to 0$ weakly, we have $e_{\gamma_i}(x) \to 0$ for all $x \in K$, which is impossible for $x \in \bigcap_{i=1}^\infty U_{\gamma_i}$.

(iii)$\Rightarrow$(ii) Now assume that the compact set K fails the CCC property. Then there is an uncountable family $\{U_\gamma; \gamma \in \Gamma\}$ of pairwise disjoint nonempty sets of K with $U_\gamma \neq U_{\gamma'}$ if $\gamma \neq \gamma'$. For each $\gamma \in \Gamma$, choose $e_\gamma \in C(K)$ with $\|e_\gamma\| = 1$, and e_γ is 0 on the complement of U_γ. Then the closed linear hull of $\{e_\gamma\}$ is isometric to $c_0(\Gamma)$.

(ii)$\Rightarrow$(i) This is trivial because $c_0(\Gamma)$ is a nonseparable WCG space. $\square$

We will now briefly discuss scattered Eberlein compacta.

Theorem 7.23 (Godefroy, Kalton, and Lancien [GKL00]). *Let K be an Eberlein compact of weight $< \aleph_\omega$ and finite height. Then $C(K)$ is isomorphic to $c_0(\Gamma)$ for some Γ.*

Proof. We prove the result for the weight ω_1 and height n by induction on n. If $n = 1$, then K is finite and the statement is obvious.

Assume the statement holds when $L^{(n)} = \emptyset$ and pick K such that $K^{(n+1)} = \emptyset$. Put $L = K'$ and $X := \{f \in C(K); f \upharpoonright L = 0\}$. The space X is isometric to $c_0(K \setminus L)$ and, by Tietze's theorem, $C(K)/X$ is isometric to $C(L)$, which is, by the induction hypothesis, isomorphic to a space $c_0(\Gamma)$.

By Corollary 5.66, X is complemented in $C(K)$. Thus we have that $C(K)$ is isomorphic to $X \oplus C(L)$, which is in turn isomorphic to $c_0(K \setminus L) \oplus c_0(\Gamma)$. $\square$

Concerning Eberlein compacta, Marciszewski proved in [Mar03] that, given a compact space K, the following three statements are equivalent: *(i) $C(K)$ is isomorphic to $c_0(\Gamma)$ for some Γ, (ii) $C(K)$ is isomorphic to a subspace of $c_0(\Gamma)$ for some Γ, and (iii) K can be embedded into the space $[X]^{\leq n}$ for some set X and some $n \in \mathbb{N}$, where $[X]^{\leq n}$ denote the subspace of the product 2^X consisting of all characteristic functions of sets of cardinality $\leq n$.*

Argyros and Godefroy (see [BeMa]) proved that *every Eberlein compact of weight $< \aleph_\omega$ and finite height can be embedded into $[X]^{\leq n}$ for some set X and some integer n.*

Bell and Marciszewski [BeMa] proved that *there is an Eberlein compact K of weight $\aleph_\omega$ and of height 3 that cannot be embedded into any $[X]^{\leq n}$.*

Benyamini and Starbird [BeSt76] proved that *there are Eberlein compacta of weight $\omega + 2$ that are not uniform Eberlein compacta.*

Bell and Marciszewski [BeMa] proved that *any Eberlein compact of height at most $\omega + 1$ is a uniform Eberlein compact.*

We finish this section by mentioning some properties of $\ell_p(\Gamma)$ spaces for uncountable Γ. We refer to Köthe [Ko66], Rosenthal [Rose70b], and Rodríguez-Salinas [Rod94] for the following statements: *Any complemented subspace of $\ell_p(\Gamma)$ is isomorphic to some $\ell_p(\Gamma')$ and any $\ell_p(\Gamma)$, $p \geq 1$ contains a subspace of the same density that is complemented in $\ell_p(\Gamma)$.*

It is well known that separable L_p spaces admit unconditional bases if $1 < p < \infty$ (Paley). This is no longer true if $L_p(\mu)$ is nonseparable and $p \neq 2$ ([EnRo73], [FGK]).

7.2 Banach Spaces Containing $\ell_1(\Gamma)$

This section contains results of Talagrand, Pełczyński, and others showing that X contains a copy of $\ell_1(c)$ if and only if ℓ_∞ is a quotient of X if and only if the dual ball of X^* in its weak*-topology contains a copy of the Čech-Stone compactification of the integers.

We use the symbol $\mathcal{T}_p$ for the pointwise topology.

Definition 7.24. *Let S be a set and $\{(A_\alpha, B_\alpha)\}_{\alpha \in \Gamma}$ be a system of disjoint pairs of subsets of S. We say that this system is* independent *if, for every finite set of distinct indices $\{\alpha_i\}_{i=1}^n \cup \{\beta_j\}_{j=1}^m \subset \Gamma$, we have $\bigcap_{i=1}^n A_{\alpha_i} \cap \bigcap_{j=1}^m B_{\beta_j} \neq \emptyset$.*

The following basic criterion shows the importance of this concept.

Proposition 7.25 (Rosenthal [Rose77]). *Let S be a set, $\{f_\alpha\}_{\alpha \in \Gamma} \subset B_{\ell_\infty(S)}$. Assume that there exist numbers $a < b$ such that the system of sets $\{(A_\alpha, B_\alpha)\}_{\alpha \in \Gamma}$, where $A_\alpha = f_\alpha^{-1}[b, 1]$, $B_\alpha = f_\alpha^{-1}[-1, a]$, is independent. Then $\{f_\alpha\}_{\alpha \in \Gamma}$ is equivalent to the canonical basis of $\ell_1(\Gamma)$.*

Proof. For every finite set of distinct indices $\{\alpha_i\}_{i=1}^n \cup \{\beta_j\}_{j=1}^m \subset \Gamma$, and real numbers $\{a_i\}_{i=1}^n \cup \{b_j\}_{j=1}^m$, $a_i > 0$, $b_j < 0$, assuming without loss of generality that $\sum a_i \geq -\sum b_j$, choose $s \in \bigcap_{i=1}^n A_{\alpha_i} \cap \bigcap_{j=1}^m B_{\beta_j}$. Then

$$\sum_{i=1}^n a_i f_{\alpha_i}(s) + \sum_{j=1}^m b_j f_{\beta_j}(s) \geq \sum_{i=1}^n a_i(b-a) \geq \frac{(b-a)}{2} \left(\sum_{i=1}^n |a_i| + \sum_{j=1}^m |b_j| \right). \quad \square$$

The following consequence of Lemma 4.21 will be very useful.

Fact 7.26. *Let τ be an infinite cardinal. Then*

(1) $w^*\text{-}\mathrm{dens}\, B_{\ell_\infty(2^\tau)} = \tau$.
(2) $\ell_1(2^\tau) \hookrightarrow \ell_\infty(\tau)$.

Proof. (1) The claim is equivalent to the existence of a dense subset of cardinality τ, for the set $([0,1]^{2^\tau}, \mathcal{T}_p)$. We prove that $(\{0,1\}^{2^\tau}, \mathcal{T}_p)$ contains a dense subset of cardinality τ. The full statement then follows using standard arguments. Consider the system $\mathcal{C}$ of cardinality 2^τ, of uniformly independent subsets of τ from Lemma 4.21, and the set

$$\mathcal{S} := \{f_t; f_t(A) = 1 \text{ if and only if } t \in A, t \in \tau, A \in \mathcal{C}\} \subset \{g; g : \mathcal{C} \to \{0,1\}\}.$$

It is now easy to verify, using the uniform independence of the system $\mathcal{C}$, that $\mathcal{S}$ is dense in $\{g; g : \mathcal{C} \to \{0, 1\}\} \cong \{0, 1\}^{2^\tau}$.

(2) We rely again on the uniformly independent family $\mathcal{C} \subset 2^\tau$ of cardinality 2^τ to define a subspace $Z := \overline{\text{span}}\{\chi_X - \chi_{\tau \setminus X}; X \in \mathcal{C}\} \hookrightarrow \ell_\infty(\tau)$. Rosenthal's criterion gives that $\{\chi_X - \chi_{\tau \setminus X}\}_{X \in \mathcal{C}}$ is equivalent to the canonical basis of $\ell_1(\mathcal{C})$, so $Z \cong \ell_1(2^\tau)$. $\qquad\square$

Proposition 7.27. *Let X be a Banach space, and let τ be an infinite cardinal. The following are equivalent:*

(1) $\ell_\infty(\tau)$ *is a quotient of X.*
(2) $\ell_1(2^\tau) \hookrightarrow X$.

Proof. (1)$\Rightarrow$(2) By Fact 7.26, $\ell_\infty(\tau)$ contains a copy of $\ell_1(2^\tau)$, which can be lifted to X.

(2)$\Rightarrow$(1) Clearly, dens $\ell_\infty(\tau) = 2^\tau$, so there exists a quotient $Q : \ell_1(2^\tau) \to \ell_\infty(\tau)$. As $\ell_\infty(\tau)$ is an injective space, Q can be extended to X. $\qquad\square$

Proposition 7.28 (Pełczyński [Pelc68]). *Let τ be an infinite cardinal. Then $i : \ell_1(\tau) \hookrightarrow X$ implies $\ell_1(2^\tau) \hookrightarrow X^*$.*

Proof. We have $i^*(X^*) = \ell_\infty(\tau)$. By Fact 7.26, there exists $\ell_1(2^\tau) \cong Z \hookrightarrow i^*(X^*)$. By the lifting property of $\ell_1(\Gamma)$ (for all Γ), we obtain immediately the desired conclusion. $\qquad\square$

The example $X = c_0(\Gamma)$ shows that the statement above cannot be reversed in general. However, for separable X it can be, as the following result shows.

Theorem 7.29 (Pełczyński [Pelc68], Hagler [Hag73]). *The following are equivalent for a Banach space X:*

(1) $\ell_1 \hookrightarrow X$.
(2) $L_1[0, 1] \hookrightarrow X^*$.

If X is separable, the above are equivalent to

(iii) $\ell_1(c) \hookrightarrow X^*$.

Proof. (2)$\Rightarrow$(1) Denote $i : L_1[0, 1] \hookrightarrow X^*$. Assume, by contradiction, that $\ell_1 \not\hookrightarrow X$. By Rosenthal's theorem, every sequence from B_X contains a weakly Cauchy subsequence, and so also must $i^*(B_X) \subset L_\infty[0, 1]$. By one of the equivalent formulations of the Dunford-Pettis property, which is shared by $L_1[0, 1]$, we have that $f_n \to 0$ uniformly on $i^*(B_X)$ for all $\{f_n\}_{n=1}^\infty \subset L_1[0, 1]$, $f_n \xrightarrow{w} 0$. Note that the set $i^*(B_X)$ is norming for $L_1[0, 1]$, and also there exists a weakly null sequence for which $\|f_n\| = 1$ (e.g., the sequence of Rademacher functions from $L_1[0, 1]$). This contradiction finishes the proof of one implication.

(1)$\Rightarrow$(2) is shown below in greater generality.

If X is separable, we have by the Odell-Rosenthal theorem that $\ell_1 \not\hookrightarrow X$ implies that $(B_{X^{**}}, w^*)$ is angelic, so in particular card $X^{**} = c$. However, condition (3) implies that $\ell_\infty(c)$ is a quotient of X^{**}. The rest of the proof follows from Proposition 7.28. $\qquad\square$

Theorem 7.30 (Hagler and Stegall [HagSt73]). *Let τ be an infinite cardinal. Suppose that $i : \ell_1(\tau) \hookrightarrow X$. Then $L_1(\{0,1\}^\tau) \hookrightarrow X^*$.*

Proof. The map $i^* : X^* \to \ell_\infty(\tau)$ is a quotient map. Checking the densities, it is easy to see that there exists a quotient map $Q : \ell_1(\tau) \to C\{0,1\}^\tau$, so in particular $Q^* : M\{0,1\}^\tau \hookrightarrow \ell_\infty(\tau)$ is an isometry. Since $L_1\{0,1\}^\tau \hookrightarrow M\{0,1\}^\tau$ is a subspace, we have that there exists an isometry $j : L_1\{0,1\}^\tau \hookrightarrow \ell_\infty(\tau)$. Consider $r > 1$ and a net of finite-dimensional subspaces of $Z = j(L_1\{0,1\}^\tau)$ (ordered by inclusion), $\{Z_\alpha\}_\alpha$, such that the Banach-Mazur distance $d(Z_\alpha, \ell_1^{d(\alpha)}) \le r$ for some integer $d(\alpha)$, and $Z = \bigcup_\alpha Z_\alpha$. By the lifting property of ℓ_1, there exist the corresponding system of injections $I_\alpha : Z_\alpha \to X^*$, which are bounded below by $\frac{1}{r}$ and above by r, and lifting i^*. We extend I_α to the whole Z, preserving the notation, by putting $I_\alpha(z) = 0$ for all $z \notin Z_\alpha$. In this way, we consider $\{I_\alpha\}_\alpha$ as a net in the compact space of all functions $\mathcal{K} = \{f; f : (B_Z, \mathcal{T}_p) \to (rB_{X^*}, w^*)\}$. By a standard compactness argument, there exists an operator $I : Z \to X^*$ that is a cluster point of the system $\{I_\alpha\}_\alpha$. It is now standard to check that $I : Z \cong L_1(\{0,1\}^\tau) \hookrightarrow X^*$ is the embedding sought. □

Let us now state without proof the following reverse statement (Haydon's earlier result needed stronger assumptions on τ).

Theorem 7.31 (Argyros [Ar82], Haydon [Ha77]). *Let $\tau > \omega_1$ be a cardinal. Suppose that $L_1(\{0,1\}^\tau) \hookrightarrow X^*$. Then $\ell_1(\tau) \hookrightarrow X$.*

Moreover, under MA_{ω_1}, Argyros [Ar82] has proved that the previous theorem remains true also for $\tau = \omega_1$. Combining this with Theorem 7.29, we obtain that, under MA_{ω_1}, the theorem of Hagler and Stegall is in fact a characterization of $\ell_1(\tau) \hookrightarrow X$ for all infinite τ. On the other hand, under the continuum hypothesis, Haydon [Ha77] has constructed a Banach space X such that $L_1\{-1,1\}^{\omega_1} \hookrightarrow X^*$ but $\ell_1(\omega_1) \not\hookrightarrow X$.

Proposition 7.32. *Let K be a compact and τ an infinite cardinal. The following are equivalent:*

(1) *K contains a subset homeomorphic to $\beta\tau$.*
(2) *There exists a continuous and surjective mapping $Q : K \to [0,1]^{2^\tau}$.*

Proof. (2)$\Rightarrow$(1) First note that $\beta\tau$ is a closed subset of $[0,1]^{2^\tau}$. It suffices to identify every element of $\mathcal{F} \in \beta\tau$ (an ultrafilter on τ), $\mathcal{F} = \{A; A \subset \tau\}$ with the element $f : 2^\tau \to \{0,1\}$, $f(A) = 1$, if and only if $A \in \mathcal{F}$. It is clear that, with this identification, the restriction of the pointwise topology $\mathcal{T}_p$ to $\{f; f \text{ corresponds to an ultrafilter } \mathcal{F}\}$ coincides with the topology generated by all subsets $\{\mathcal{F}; A \in \mathcal{F}, A \subset \tau\}$, which verifies the claim. It remains to observe that $\beta\tau$ can always be lifted from the quotients. Indeed, pick a function $\phi : \tau \to K$ so that $Q \circ \phi = Id \restriction \tau$. There is a unique continuous extension $\phi : \beta\tau \to K$ that necessarily satisfies $Q \circ \phi = Id \restriction \beta\tau$. Thus $\phi : \beta\tau \to K$ is a homeomorphism.

(1)$\Rightarrow$(2) To prove the opposite implication, note that Fact 7.26 implies that $([0,1]^{2^\tau}, \mathcal{T}_p)$ contains a dense subset of cardinality τ. So there exists a continuous surjection $Q : \beta\tau \to [0,1]^{2^\tau}$ that which can be extended to the whole K since the target space is a universal retract. $\qquad\square$

If $\ell_1(\Gamma) \hookrightarrow X$, then $\ell_\infty(\Gamma)$ is a quotient of X^*, so in particular there exists a continuous surjective mapping $Q : (B_{X^*}, w^*) \to [0,1]^\Gamma$. The reverse implication (Theorem 7.35) turns out to be valid for infinite cardinals with cof $\tau > \omega$.

In the proof of this, we use the following combinatorial result on the existence of a free set. For the proof, see, e.g., [Will77, p. 64]. Given a set S and a cardinal κ, denote $\mathcal{P}_\kappa(S) = \{A; A \subset S, \text{card}(A) < \kappa\}$.

Theorem 7.33 (Hajnal;, see [Will77]). *Let α be an infinite cardinal, $\kappa < \alpha$, and $f : \alpha \to \mathcal{P}_\kappa(\alpha)$ be a function such that $\xi \notin f(\xi)$ for all $\xi < \alpha$. Then there exists $A \subset \alpha$, card $A = \alpha$, such that $\xi \notin f(\zeta)$ for all $\xi, \zeta \in A$. A set with this property is called a* free set *for f.*

Lemma 7.34 (Talagrand [Tala81]). *Let α be a cardinal with cof $\alpha > \omega$, S be a set, $\{(A_i, B_i); i < \alpha\}$ be an independent family on S, $n \in \mathbb{N}$, and $A_{i,m} \subset A_i$, $B_{i,m} \subset B_i$ for $i < \alpha$, $1 \leq m \leq n$, be such that*

$$A_i \times B_i = \bigcup_{1 \leq m \leq n} (A_{i,m} \times B_{i,m}) \text{ for } i < \alpha.$$

Then there is $1 \leq m \leq n$, and $I \subset \alpha$, with $|I| = \alpha$, such that the family $\{(A_{i,m}, B_{i,m}); i \in I\}$ is independent on S.

Proof (Argyros). (By induction in n). Suppose the theorem holds for n (the case $n = 1$ is trivial), and

$$A_i \times B_i = \bigcup_{1 \leq m \leq n+1} (A_{i,m} \times B_{i,m}) \text{ for } i < \alpha.$$

We define $T_i : S \to \{1, 0, -1\}$ by $T_i \upharpoonright A_i = 1$, $T_i \upharpoonright B_i = -1$, and $T_i \upharpoonright S \setminus (A_i \cup B_i) = 0$. We define

$$T : \beta S \to \{-1, 0, 1\}^\alpha$$

by $T := \prod_{i<\alpha} \tilde{T}_i$, where $\tilde{T}_i : \beta S \to \{-1, 0, 1\}$ is the unique continuous extension of T_i. It is clear, since the family $\{(A_i, B_i); i < \alpha\}$ is independent, that $T(\beta S) \supset \{1, -1\}^\alpha$. We set

$$Z := T^{-1}(\{1, -1\}^\alpha),$$

which is a closed subset of βS. By a well-known and easy fact (based on Zorn's lemma), there exists a closed set $Y \subset Z$ such that $T(Y) = \{1, -1\}^\alpha$, and $T \upharpoonright Y$ is irreducible; i.e., $T(V) \neq \{1, -1\}^\alpha$ whenever $V \subsetneq Y$ is a closed set. Note that every subset $A \subset S$ corresponds canonically to an open set in βS consisting of all ultrafilters containing A. We set

$$\tilde{A}_i := \text{cl}_{\beta S} A_i \bigcap Y, \quad \tilde{A}_{i,m} := \text{cl}_{\beta S} A_{i,m} \bigcap Y,$$

$$\tilde{B}_i := \text{cl}_{\beta S} B_i \bigcap Y, \quad \tilde{B}_{i,m} := \text{cl}_{\beta S} B_{i,m} \bigcap Y,$$

for $i < \alpha$, and $1 \le m \le n$, and we note that

$$\tilde{A}_i \times \tilde{B}_i = \bigcup_{i \le m \le n+1} \tilde{A}_{i,m} \times \tilde{B}_{i,m} \text{ for } i < \alpha.$$

Claim 1. For every $i < \alpha$, either $\tilde{B}_{i,m} = \tilde{B}_i$ for all $1 \le m \le n$ or there is a clopen subset C_i of $\tilde{B}_i$ and m_0, $1 \le m_0 \le n$, such that

$$\tilde{A}_i \times C_i = \bigcup_{m=1, m \ne m_0}^{n} \tilde{A}_{i,m} \times \tilde{B}_{i,m}.$$

Indeed, if the first possibility fails, then there is m_0 with $\tilde{B}_{i,m_0} \ne \tilde{B}_i$. Set $C_i = \tilde{B}_i \setminus \tilde{B}_{i,m_0}$.

Claim 2. Let $\{(D_i, C_i); i < \alpha\}$ be a family of ordered pairs of clopen subsets of Y such that $D_i \bigcap C_i = \emptyset$ for $i < \alpha$ and either $D_i = \tilde{A}_i$ or $C_i = \tilde{B}_i$ for $i < \alpha$. Then there is $I \subset \alpha$, with $|I| = \alpha$, such that $\{(D_i, C_i); i \in I\}$ is independent.

Without loss of generality, assume that $D_i = \tilde{A}_i$ for $i < \alpha$. Since $T \upharpoonright Y$ is irreducible, there is a clopen subset W_i of $\{1, -1\}^\alpha$ with $(T \upharpoonright Y)^{-1}(W_i) \subset C_i$. Let F_i be the finite subset of α on which W_i depends. By the fact that the space $\{1, -1\}^\alpha$ has caliber α (for all cardinals α with cof $\alpha > \omega$; see, [CoNe82, Thm. 3.18(a)]) and Hajnal's theorem above, it follows that there is $I \subset \alpha$, with $|I| = \alpha$, such that $\{W_i; i \in I\}$ has the finite intersection property, and $i \notin F_j$ for $i, j \in I$, $i \ne j$. It is now easy to prove that if $i_1, \ldots, i_p, j_1, \ldots, j_q$ are distinct elements of I, then

$$\bigcap_{k=1}^{p} \pi_{i_k}^{-1}(\{1\}) \cap \bigcap_{l=1}^{q} W_{j_l} \ne \emptyset.$$

It follows that

$$\emptyset \ne (T \upharpoonright Y)^{-1} \left(\bigcap_{k=1}^{p} \pi_{i_k}^{-1}(\{1\}) \cap \bigcap_{l=1}^{q} W_{j_l} \right) \subset \left(\bigcap_{k=1}^{p} D_{i_k} \cap \bigcap_{l=1}^{q} C_{j_l} \right).$$

The lemma follows by induction using Claims 1 and 2. □

Theorem 7.35 (Talagrand [Tala81]). *Let X be a Banach space, and let α be a cardinal with cof $\alpha > \omega$. Suppose there is a quotient $\phi : B_{X^*} \to [0, 1]^\tau$. Then $\ell_1(\tau) \hookrightarrow X$.*

Proof. Note the basic fact that $X \hookrightarrow C(B_{X^*}, w^*)$ separates the points of $K = (B_{X^*}, w^*)$. For $i \in \alpha$, we denote $\pi_i : [0,1]^\alpha \to [0,1]$ the natural projection on the i-th coordinate and define an independent family $X_i = (\pi_i \circ \phi)^{-1}(\{0\})$, $Y_i = (\pi_i \circ \phi)^{-1}(\{1\})$. Since $X_i \times Y_i$ is compact and X is separating, there exists a finite set $\{f_l\}_{l=1}^N \subset X$ and rationals $p_l < q_l$ such that

$$ X_i \times Y_i \subset \bigcup_{l=1}^N \left(f_l^{-1}(-\infty, p_l) \times f_l^{-1}(q_l, \infty) \right). $$

Using the assumption cof $\alpha > \omega$, we may without loss of generality assume that N, p_l, and q_l are independent of i. The rest of the proof follows using Lemma 7.34 together with Proposition 7.25. □

The next theorem summarizes the previous results in the special case where $\tau = 2^\omega$.

Theorem 7.36. *The following are equivalent for a Banach space X.*

(i) $\ell_1(c) \hookrightarrow X$.
(ii) $\beta\mathbb{N} \subset (B_{X^*}, w^*)$.
(iii) ℓ_∞ *is a quotient of X.*
(iv) $[0,1]^c$ *is a continuous quotient of (B_{X^*}, w^*).*
(v) *There is a bounded linear operator T from X onto a dense set in ℓ_∞.*

In the rest of the section, we describe some related results.

Proposition 7.37 (Pełczyński [Pelc68]). *Let Γ be uncountable. Assume that X generates Y and that Y contains a subspace isomorphic to $\ell_1(\Gamma)$. Then X contains a subspace isomorphic to $\ell_1(\Gamma_1)$ for some uncountable Γ_1. If the cardinality of Γ is regular, then Γ_1 can be chosen to be such that* card $\Gamma_1 =$ card Γ.

Proof. Let $K > 0$ and $\{y_\gamma\} \subset Y$ be such that

$$ K^{-1} \sum_{\gamma \in \Gamma} |t(\gamma)| \leq \left\| \sum_{\gamma \in \Gamma} t(\gamma) y_\gamma \right\| \leq K \sum_{\gamma \in \Gamma} |t(\gamma)| \quad \text{for} \quad \{t(\gamma)\} \in \ell_1(\Gamma). $$

For each $\gamma \in \Gamma$, choose $x_\gamma \in X$ so that $\|y_\gamma - Tx_\gamma\| < (2K)^{-1}$, where T is a bounded linear operator from X onto a dense set in Y.

Put $\Gamma_n = \{\gamma \in \Gamma; \|x_\gamma\| \leq n\}$ for $n = 1, 2, \ldots$. Since $\bigcup_n \Gamma_n = \Gamma$ and Γ is uncountable, there is n_0 so that Γ_{n_0} is uncountable.

Now let $\Gamma_1 = \Gamma_{n_0}$. For $\{t(\gamma)\} \in \ell_1(\Gamma_1)$, we have

$$n_0 \sum_{\gamma \in \Gamma_1} |t(\gamma)| \geq \left\| \sum_{\gamma \in \Gamma_1} t(\gamma) x_\gamma \right\| \geq \|T\|^{-1} \left\| \sum_{\gamma \in \Gamma_1} t(\gamma) T x_\gamma \right\|$$

$$\geq \|T\|^{-1} \left(\left\| \sigma_{\gamma \in \Gamma_1} t(\gamma) y_\gamma \right\| - \sum_{\gamma \in \Gamma_1} |t(\gamma)| \, \|T x_\gamma - y_\gamma\| \right)$$

$$\geq \left((K\|T\|)^{-1} - (2K\|T\|)^{-1} \right) \sum_{\gamma \in \Gamma_1} |t(\gamma)|$$

$$= \left(2K\|T\| \right)^{-1} \sum_{\gamma \in \Gamma_1} |t(\gamma)|.$$

Therefore the operator $S : \ell_1(\Gamma_1) \to X$ defined by

$$S\left(\{t(\gamma)\}_{\gamma \in \Gamma_1} \right) = \sum_{\gamma \in \Gamma_1} t(\gamma) x_\gamma \quad \text{for} \quad \{t(\gamma)\}_{\gamma \in \Gamma_1} \in \ell_1(\Gamma_1)$$

is the required isomorphic embedding. The rest follows from the definition of a regular cardinal. $\square$

Proposition 7.38 (Rosenthal [Rose70b]). *Let X be a Banach space and Γ be an infinite set. Then the following are equivalent:*

(i) *There is a bounded linear operator T from X onto a dense set in $\ell_\infty(\Gamma)$.*
(ii) *There is a bounded linear operator Q from X onto $\ell_\infty(\Gamma)$.*

Proof. Assume (i) is true and let card $\Gamma = \aleph$. It is well known that $\ell_\infty(\Gamma_1)$ is isometric to a subspace of $\ell_\infty(\Gamma)$ for some Γ_1 with card $\Gamma_1 = 2^\aleph$. By Proposition 7.37, $\ell_1(\Gamma_1)$ is isomorphic to a subspace Z of X (note that $2^\aleph$ is a regular cardinal). It is well known that there is a bounded linear operator Q from Z onto $\ell_\infty(\Gamma)$. Since $\ell_\infty(\Gamma)$ is an injective space, the operator Q can be extended to a bounded linear operator from X onto $\ell_\infty(\Gamma)$. $\square$

We finish this section by mentioning the following three results. Pełczyński proved in [Pelc68] that *if a separable Banach space contains an isomorphic copy of ℓ_1, then $C[0,1]$ is isomorphic to a quotient of X*. This result follows from another of Pełczyński's results, in [Pelc68b], saying that *a separable Banach space X that contains an isomorphic copy Y of $C[0,1]$ contains a subspace $Z \hookrightarrow Y$ that is isomorphic to $C[0,1]$ and complemented in X*. Enflo and Rosenthal proved in [EnRo73] the following result: *Let $1 \leq p, r < \infty$ and Γ be an uncountable set. Then there is a probability measure μ so that $\ell_p(\Gamma)$ is isomorphic to a subspace of $L_r(\mu)$ if and only if $r < p < 2$ or $p = 2$ and r is arbitrary.* Let us just mention that $\ell_1(\omega_1)$ is not isomorphic to a subspace of $L_1(\mu)$ for a probability measure μ since $\ell_1(\omega_1)$ does not have an equivalent Gâteaux differentiable norm [DGZ93a, p. 59] and $L_1(\mu)$ does, as it is WCG.

7.3 Long Unconditional Bases

This section contains necessary and sufficient conditions for a Banach space X with a long unconditional basis to be weakly compactly generated, respectively weakly Lindelöf determined, respectively admit a uniformly Gâteaux differentiable norm. The results are due to Johnson, Argyros, and Mercourakis and Troyanski, respectively.

Let X be a Banach space. Recall that $\{e_\gamma\}_{\gamma \in \Gamma}$ is called an *unconditional Schauder basis* of X if for every $x \in X$ there is a unique family of real numbers $\{a_\gamma\}_{\gamma \in \Gamma}$ such that $x = \sum a_\gamma e_\gamma$ in the sense that for every $\varepsilon > 0$ there is a finite set $F \subset \Gamma$ such that $\left\| x - \sum_{\gamma \in F'} a_\gamma e_\gamma \right\| \leq \varepsilon$ for every $F' \supset F$. Note that, for every $x \in X$, only countably many coordinates a_γ are nonzero. Indeed, given $n \in \mathbb{N}$, there is a finite set $F \subset \Gamma$ such that $\left\| \sum_{\gamma \in F'} a_\gamma e_\gamma \right\| \leq \frac{1}{n}$ for every finite F' disjoint from F. Applying this to $F' = \{\gamma\}$ and assuming $\|e_\gamma\| = 1$, we get $\{\gamma;\ |a_\gamma| \geq \frac{1}{n}\} \subset F$.

Clearly, an unconditional Schauder basis is—under every reordering—a long Schauder basis.

Theorem 7.39. *Assume that X is a Banach space with an unconditional basis $\{e_\gamma; f_\gamma\}_{\gamma \in \Gamma}$. Then the following are equivalent:*

(i) *X is Asplund.*
(ii) *The basis $\{e_\gamma; f_\gamma\}_{\gamma \in \Gamma}$ is shrinking.*
(iii) *X does not contain an isomorphic copy of ℓ_1.*
(iv) *X admits a Fréchet differentiable norm.*

Proof. (i)$\Rightarrow$(iii) is trivial, as ℓ_1 is not Asplund and to be Asplund is a hereditary property.

(iii)$\Rightarrow$(ii) If $\{e_\gamma; f_\gamma\}_{\gamma \in \Gamma}$ is not shrinking, then there is $f \in B_{X^*}$ and $\varepsilon > 0$ such that for every finite set F of the coordinates, $\sup\{f(x); x \in \text{span } \{e_\gamma; \gamma \notin F\} \cap B_X\} > \varepsilon$. From this we construct, by a sliding hump argument, disjoint finite blocks $\{u_j\}_{j \in \mathbb{N}}$ in B_X with $f(u_j) > \frac{1}{2}\varepsilon$ for all $j \in \mathbb{N}$ (see, e.g., [Fa~01, Thm. 6.35]). Then $\{u_j\}$ is equivalent to the unit vector basis of ℓ_1.

(ii)$\Rightarrow$(iv) follows from Troyanski's renorming theorem (Theorem 3.48).

(iv)$\Rightarrow$(i) is in [DGZ93a, Thm. II.5.3]. $\qquad\square$

Theorem 7.40 (Johnson; see [Rose74]). *Let X be a Banach space with an unconditional basis $\{e_\gamma; f_\gamma\}_{\gamma \in \Gamma}$. Then X is WCG if and only if there is a bounded linear one-to-one operator T from X^* into $c_0(\Delta)$ for some set Δ. If this happens, then $\{e_\gamma\}_{\gamma \in \Gamma} \cup \{0\}$ is σ-weakly compact; i.e., a countable union of weakly compact sets.*

Proof. If such an operator exists, we will prove that $\{e_\gamma\}_{\gamma \in \Gamma} \cup \{0\}$ is σ-weakly compact. Let $\Gamma_j := \{\gamma \in \Gamma; \|Tf_\gamma\| \geq \frac{1}{j}\}$, $j \in \mathbb{N}$. As T is one-to-one, $\bigcup \Gamma_j = \Gamma$. We will show that $\{e_\gamma; \gamma \in \Gamma_j\} \cup \{0\}$ is weakly compact. By the Eberlein-Šmulyan theorem, it is enough to show that, for fixed j, it cannot happen that for some $f \in X^*$ and some $\delta > 0$, for a sequence of distinct elements $\{\gamma_i\}$ of Γ_j, that $|f(e_{\gamma_i})| > \delta$ for all $i \in \mathbb{N}$. Proceed by contradiction. We have that $\{e_{\gamma_i}\}$ is equivalent to the canonical basis of ℓ_1 and thus $\{f_{\gamma_i}\}$ is equivalent to the canonical basis of c_0, in particular, $f_{\gamma_i} \to 0$ weakly in X^*. If T were weakly compact, then the restriction of T to $\overline{\text{span}}\{f_{\gamma_i}\}$ would be a completely continuous operator since c_0 has the Dunford-Pettis property. Since $(f_{\gamma_i}) \to 0$ weakly, then $Tf_{\gamma_i} \to 0$ in norm, a contradiction with the choice of Γ_j. Therefore, T is not weakly compact. The restriction to $\overline{\text{span}}\{e_{\gamma_i}; i \in \mathbb{N}\}$ of the quotient mapping $q : X \to X/\{f_{\gamma_i}; i \in \mathbb{N}\}_\perp$ is one-to-one and onto and thus an isomorphism. Hence, if Z denotes the weak*-closure of $\overline{\text{span}}\{f_{\gamma_i}\}$ in X^*, then Z is isomorphic to ℓ_∞. Therefore, T carries some subspace W of Z that is isomorphic to ℓ_∞ isomorphically onto a subspace of $c_0(\Gamma)$. This is impossible, as $c_0(\Gamma)$ is Asplund and ℓ_∞ is not. Therefore we again reached a contradiction, meaning that $\Gamma_j \cup \{0\}$ is weakly compact.

To prove the reverse implication, assume that X is WCG. Then there is a bounded one-to-one weak*-weak continuous operator from X^* into some $c_0(\Gamma)$ (see Theorem 6.9). □

Theorem 7.41 (Mercourakis and Stamati [MeSt]). *Let X be a Banach space strongly generated by an absolutely convex weakly compact set K, with a normalized unconditional basis $\{x_\gamma; x_\gamma^*\}_{\gamma \in \Gamma}$. Then, given a set $A \subset \Gamma$, the set $\{x_\gamma; \gamma \in A\}$ is weakly relatively compact if and only if*

$$\inf_{\gamma \in A} \sup_{x \in K} \{|\langle x, x_\gamma^* \rangle|\} > 0.$$

Proof. Let $T : X^* \to C(K)$ be the operator of the restriction of X^* to K, i.e., $T(x^*) = x^* \upharpoonright K, x^* \in X^*$. From Grothendieck's theorem on the coincidence of weak and pointwise compactness in $C(K)$ spaces (see, e.g., [Fa~01, Thm. 12.1],) it follows that $T(B_{X^*})$ is a weakly compact set and T is thus a weakly compact operator. For $n \in \mathbb{N}$, put $\Gamma_n := \{\gamma \in \Gamma; \|T(x_\gamma^*)\|_\infty \geq \frac{1}{n}\}$, where $\|\cdot\|_\infty$ is the supremum norm in $C(K)$.

From the method of the proof of Theorem 7.40, it follows that every $\{x_\gamma; \gamma \in \Gamma_n\} \cup \{0\}$ is weakly compact and $\Gamma = \bigcup_n \Gamma_n$.

Assume that, for some $A \subset \Gamma$, the set $\{x_\gamma; \gamma \in A\} \cup \{0\}$ is weakly compact. Assume, by contradiction, that the infimum in question is 0. Then there is an infinite sequence $\{\gamma_n\}$ of distinct points in A such that $\sup_{x \in K} |\langle x, x_{\gamma_n}^* \rangle| \to 0$ as $n \to \infty$. From the proof of Theorem 6.37, it follows that the Mackey topology $\tau(X^*, X)$ on B_{X^*} coincides with the metric given by the norm on X^* defined by $\|x^*\|_K := \sup_{x \in K} |\langle x, x^* \rangle|$. Thus $x_{\gamma_n}^* \to 0$ in the topology $\tau(X^*, X)$ and, since $\{x_\gamma; \gamma \in A\}$ is weakly relatively compact, $\sup_{\gamma \in A} |\langle x_\gamma, x_{\gamma_n}^* \rangle| \to 0$. This contradicts the fact that $\sup_{\gamma \in A} |\langle x_\gamma, x_{\gamma_n}^* \rangle| \geq |\langle x_{\gamma_n}, x_{\gamma_n}^* \rangle| = 1$ for every $n \in \mathbb{N}$. Thus, $\inf_{\gamma \in A} \sup_{x \in K} \{|\langle x, x_\gamma^* \rangle|\} > \frac{1}{n}$ for some $n \in \mathbb{N}$, then $A \subset \Gamma_n$, and $\{x_\gamma; \gamma \in A\}$ is weakly relatively compact. □

From the proof of Theorem 7.41 and we have the following corollary.

Corollary 7.42 (Mercourakis and Stamati [MeSt]). *Let X be a strongly reflexive generated space with a normalized unconditional basis $\{x_\gamma; x_\gamma^*\}_{\gamma \in \Gamma}$. Then $\Gamma = \bigcup_n \Gamma_n$ in such a way that*

(i) *$\{x_\gamma; \gamma \in \Gamma_n\} \cup \{0\}$ is weakly compact for each n; and*
(ii) *if $A \subset \Gamma$ is such that $\{x_\gamma; \gamma \in A\}$ is weakly relatively compact, then there is $n \in \mathbb{N}$ such that $A \subset \Gamma_n$.*

Argyros and Mercourakis [ArMe93] proved that *if X is a WLD space, then there is a WLD space Z with an unconditional basis and a bounded linear operator T from Z onto a dense set in X.*

Theorem 7.43 (Argyros and Mercourakis [ArMe93]). *Let X be a Banach space with an unconditional basis $\{e_\gamma; f_\gamma\}_{\gamma \in \Gamma}$. Then the following are equivalent:*

(i) *X is WLD.*
(ii) *X does not contain an isomorphic copy of $\ell_1(\omega_1)$.*
(iii) *For every uncountable subset $\Delta \subset \Gamma$, $\{e_\gamma; \gamma \in \Delta\}$ is not equivalent to the unit vector basis of $\ell_1(\Delta)$.*
(iv) *There is a bounded linear one-to-one operator T from X^* into $\ell_\infty^c(\Delta)$ for some set Δ.*

Proof. (i)⇒(ii) Every subspace of a WLD Banach space is itself WLD (Corollary 5.43) and $\ell_1(\omega_1)$ is not WLD.

(ii)⇒(iii) is trivial.

(iii)⇒(i) If X is not WLD, then for some $f \in B_X$ and some $\varepsilon > 0$, there is an uncountable set $\Delta \subset \Gamma$ such that $|f(e_\gamma)| > \varepsilon$ for every $\gamma \in \Delta$. Take a finite set $F \subset \Delta$ and real numbers $a_\gamma, \gamma \in F$. For each $\gamma \in \Delta$, find $\epsilon_\gamma \in \{-1, 1\}$ such that $|a_\gamma||f(e_\gamma)| = \epsilon_\gamma a_\gamma f(e_\gamma)$. Then

$$\varepsilon \sum_{\gamma \in F} |a_\gamma| \leq \sum_{\gamma \in F} |a_\gamma||f(e_\gamma)| = f\left(\sum_{\gamma \in F} \epsilon_\gamma a_\gamma e_\gamma\right)$$

$$\leq \left\|\sum_{\gamma \in F} \epsilon_\gamma a_\gamma e_\gamma\right\| \leq C \left\|\sum_{\gamma \in F} a_\gamma e_\gamma\right\| \leq C \sum_{\gamma \in F} |a_\gamma|,$$

where C is an unconditional basis constant. It follows that $\{e_\gamma; \gamma \in \Delta\}$ is equivalent to the unit vector basis in $\ell_1(\Delta)$.

(i)⇒(iv) is clear.

(iv)⇒(ii) We will use the following result.

Proposition 7.44 (Argyros and Mercourakis [ArMe93]). *Assume that Γ is uncountable. Then there is no one-to-one operator from $\ell_\infty(\Gamma)$ into $\ell_\infty^c(\Delta)$ for any set Δ.*

Proof. We will use the following result from [DaLi73]: *Let Z be a Banach space such that $c_0(\Gamma) \subset Z \subset \ell_\infty(\Gamma)$ for some uncountable Γ. Let $T : Z \to \ell_\infty(\Delta)$ (for some Δ) be an operator such that $T \upharpoonright c_0(\Gamma)$ is one-to-one. Then there exists an uncountable set $\Gamma_2 \subset \Gamma$, a one-to-one mapping $\delta : \Gamma_2 \to \Delta$, and $\varepsilon > 0$ such that, if $B \subset \Gamma_2$ with $\chi_B \in Z$, then $\delta(B) \subset \sigma_{\varepsilon/2}(T(\chi_B))$, where for $g \in \ell_\infty(\Delta)$, $\sigma_\varepsilon(g) := \{\delta \in \Delta; |g(\delta)| \geq \varepsilon\}$.*

Assume Γ is an ordinal. Let $\gamma_1 = 0$, and choose any $\delta(\gamma_1) \in \operatorname{supp} T\chi_{\{\gamma_1\}}$. Assume that, for some ordinal β, we have already chosen $\{\gamma_\alpha; \alpha < \beta\}$. We have two possibilities:

(a) For all $\gamma \in \Gamma \setminus \{\gamma_\alpha; \alpha < \beta\}$, $\operatorname{supp} T\chi_{\{\gamma\}} \subset \{\gamma_\alpha; \alpha < \beta\}$. In this case, put $\Gamma_0 := \{\gamma_\alpha; \alpha < \beta\}$.

(b) There exists an element $\gamma_\beta \in \Gamma \setminus \{\gamma_\alpha; \alpha < \beta\}$ such that $\operatorname{supp} T\chi_{\{\gamma_\beta\}} \not\subset \{\gamma_\alpha; \alpha < \beta\}$. If this is the case, choose an element $\delta(\gamma_\beta) \in \operatorname{supp} T\chi_{\{\gamma_\beta\}} \setminus \{\gamma_\alpha; \alpha < \beta\}$. This ensures that $\delta(\gamma_\beta) \neq \delta(\gamma_\alpha)$ for all $\alpha < \beta$ and, at the same time, $T\chi_{\{\gamma_\beta\}}(\delta(\gamma_\beta)) \neq 0$. Continue the process.

The process terminates by finding a set $\Gamma_0 \subset \Gamma$ and a one-to-one mapping $\delta : \Gamma_0 \to \Delta$ with the following two properties:

(1) $T\chi_{\{\gamma\}}(\delta(\gamma)) \neq 0$ for all $\gamma \in \Gamma_0$.

(2) $\operatorname{supp} T\chi_{\{\gamma\}} \subset \delta(\Gamma_0)$ for all $\gamma \in \Gamma \setminus \Gamma_0$.

We claim that Γ_0 is uncountable. Assume, by contradiction, that Γ_0 is countable. The set $\{\phi_\gamma; \gamma \in \Gamma_0\} \subset (c_0(\Gamma \setminus \Gamma_0))^*$, where $\langle f, \phi_\gamma \rangle := Tf(\delta(\gamma))$ for all $f \in c_0(\Gamma \setminus \Gamma_0)$, is total. In order to see this, let $f \in c_0(\Gamma \setminus \Gamma_0)$ such that $0 = \langle f, \phi_\gamma \rangle (= Tf(\delta(\gamma)))$ for all $\gamma \in \Gamma_0$. If $\delta \notin \delta(\Gamma_0)$ and $\gamma \notin \Gamma_0$ then $T\chi_{\{\gamma\}}(\delta) = 0$, so $T(f)(\delta) = 0$. It follows that $Tf = 0$ and, by the injectivity of $T \upharpoonright c_0(\Gamma)$, $f = 0$. As a consequence, $(c_0(\Gamma \setminus \Gamma_0))^*$ is w^*-separable, which implies that $\Gamma \setminus \Gamma_0$ is countable. The set Γ is then countable, a contradiction.

Thus, there exists $\varepsilon > 0$ and an uncountable set $\Gamma_1 \subset \Gamma_0$ such that, for all $\gamma \in \Gamma_1$, $T\chi_{\{\gamma\}}(\delta(\gamma)) > \varepsilon$. Let $\alpha_1 \in \Gamma_1$. Denote again by $\delta(\alpha_1)$ the element in $\ell_\infty^*(\Delta)$ given by the evaluation at $\delta(\alpha_1)$. Then $T^*(\delta(\alpha_1)) \in Z^*$. Let μ_{α_1} be a Hahn-Banach extension of $T^*(\delta(\alpha_1))$ to an element in $\ell_\infty(\Gamma)$. We have $\|\mu_{\alpha_1}\| \leq \|T\|$ for all $\alpha_1 \in \Gamma_1$. Regarding the μ_{α_1} as a finitely additive measure on the subsets of Γ, we find, by Rosenthal's theorem (Theorem 7.1), an uncountable subset $\Gamma_2 \subset \Gamma_1$ such that $|\mu_\gamma|(\Gamma \setminus \{\gamma\}) < \frac{\varepsilon}{2}$ for all $\gamma \in \Gamma_2$.

If $B \subset \Gamma_2$ is any subset with $\chi_B \in Z$, then, for all $\gamma \in B$,

$$\begin{aligned}
|\langle T\chi_B, \delta(\gamma)\rangle| &= |\langle \chi_B, T^*\delta(\gamma_2)\rangle| = |\mu_\gamma(B)| \\
&\geq |\mu_\gamma\{\gamma\}| - |\mu|(B \setminus \{\gamma\}) \geq |\mu_\gamma\{\gamma\}| - |\mu|(\Gamma \setminus \{\gamma\}) \\
&= \langle T\chi_{\{\gamma\}}, \delta(\gamma)\rangle - |\mu|(\Gamma_2 \setminus \{\gamma\}) > \varepsilon - \varepsilon/2 = \varepsilon.
\end{aligned}$$

Thus $\delta(B) \subset \sigma_{\frac{\varepsilon}{2}}(T\chi_B)$. This finishes the proof of the result. To prove the proposition, use $Z = \ell_\infty(\Gamma)$. $\square$

We continue with the proof of the implication (iv)$\Rightarrow$(ii). Assume that T is a bounded linear one-to-one operator from X^* into some $\ell_\infty^c(\Delta)$ for some set Δ and that X contains a copy of $\ell_1(\Gamma)$ for some uncountable Γ. Assume without loss of generality that the copy of $\ell_1(\Gamma)$ in X is complemented. Then $\ell_\infty(\Gamma)$ is isomorphic to a complemented subspace of X^*. Thus there is a one-to-one operator from $\ell_\infty(\Gamma)$ into some $\ell_\infty^c(\Delta)$, a contradiction with the lemma above. $\qquad\square$

Corollary 7.45. *Suppose X has an unconditional basis and admits an equivalent Gâteaux differentiable norm. Then X is WLD.*

Proof. The space $\ell_1(\omega_1)$ does not admit a Gâteaux smooth norm [DGZ93a, Examp. I.1.6.c]. $\qquad\square$

Argyros and Mercourakis proved in [ArMe93] that *there is a WLD space with an unconditional basis that does not admit a Gâteaux differentiable norm.*

Definition 7.46. *The norm $\|\cdot\|$ is* uniformly rotund in every direction *(URED) if $\|x_n - y_n\| \to 0$ whenever $x_n, y_n \in S_X$ are such that $x_n - y_n = \lambda_n z$ for some $z \in X$, for some real numbers λ_n, and for $\|x_n + y_n\| \to 2$.*

The following result of Troyanski motivated much of the results in the classifications of subclasses of WLD spaces discussed in Chapter 6.

Theorem 7.47 (Troyanski [Troy77]). *Let X be a Banach space with an unconditional basis $\{e_\gamma\}_{\gamma \in \Gamma}$. Then*

(i) *X admits an equivalent URED norm if and only if for each $\varepsilon > 0$ we can write $\Gamma = \bigcup_i \Gamma_i^\varepsilon$ in such a way that, for each finite set of distinct indices $\{\gamma_j\}_{j=1}^i \subset \Gamma_i^\varepsilon$, we have*

$$\left\|\sum_{j=1}^i e_{\gamma_j}\right\| > \varepsilon^{-1}.$$

(ii) *X admits an equivalent UG norm if and only if for each $\varepsilon > 0$, we can write $\Gamma = \bigcup_i \Gamma_i^\varepsilon$ in such a way that, for each finite set of distinct indices $\{\gamma_j\}_{j=1}^i \subset \Gamma_i^\varepsilon$, we have*

$$\left\|\sum_{j=1}^i e_{\gamma_j}\right\| < \varepsilon i.$$

In the proof we will use the following two lemmas.

Lemma 7.48 (Troyanski [Troy77]). *Let the norm $\|\cdot\|$ of X be UG. Then for every $\varepsilon > 0$, the unit sphere S_X can be written as $S_X = \bigcup_i S_i^\varepsilon$, in such a way that, if $\{x_j\}_{j=1}^i \subset S_i^\varepsilon$ is a finite set of distinct elements, then*

$$\min_{\alpha_j = \pm 1} \left\|\sum_{j=1}^i \alpha_j x_j\right\| < \varepsilon i.$$

Lemma 7.49 (Troyanski [Troy77]). *Let the norm $\| \cdot \|$ of X be URED. Then for every $\varepsilon > 0$, the unit sphere S_X can be written as $S_X = \bigcup_i S_i^\varepsilon$, in such a way that, if $\{x_j\}_{j=1}^i \subset S_i^\varepsilon$ is a finite set of distinct elements, then*

$$\max_{\alpha_j = \pm 1} \left\| \sum_{j=1}^i \alpha_j x_j \right\| > \varepsilon^{-1}.$$

Proof. (Lemma 7.49). Let $\varepsilon < 1$ and $i > 1 - \ln \varepsilon$. Put

$$S_i^\varepsilon := \left\{ x \in S; \inf_{y \in S, |\lambda| \geq \varepsilon} \left(\max_{\alpha = \pm 1} \|y + \alpha \lambda x\| \right) > \frac{i-1}{i-1+\ln\varepsilon} \right\}$$

It follows that $S = \bigcup_i S_i^\varepsilon$. Let $\{x_j\}_{j=1}^i \subset S_i^\varepsilon$. Put $\sigma_1 = x_1$. If $\sigma_1, \dots, \sigma_j$, for $j < i$, have been chosen, put $\sigma_{j+1} = \sigma_j + \alpha_j x_{j+1}$, where $\alpha_j = \pm 1$ is so chosen that

$$\|\sigma_j + \alpha_j x_{j+1}\| \geq \|\sigma_j - \alpha_j x_{j+1}\|$$

Let $\|\sigma_j\| > \varepsilon^{-1}$ for $j < i$. As $\|\sigma_j\|^{-1} > \varepsilon$, then, from the definition of the sets S_i^ε's, it follows that $\|\sigma_{j+1}\|/\|\sigma_j\| > \frac{i-1}{i-1+\ln\varepsilon}$. Thus $\|\sigma_i\| > \frac{i-1}{i-1+\ln\varepsilon} > \varepsilon^{-1}$. $\quad\square$

Proof. (Lemma 7.48). Assume that $\varepsilon i > 2$. Put

$$S_i^\varepsilon := \left\{ x \in S; \sup \left\{ \frac{2}{\tau} \left(\min_{\alpha = \pm 1} \|y + \alpha \tau x\| - 1 \right); y \in S, 0 < \tau < \frac{2}{\varepsilon i - 2} \right\} < \varepsilon \right\}.$$

It follows that $S = \bigcup S_i^\varepsilon$. Let $\{x_j\}_{j=1}^i \subset S_i^\varepsilon$. Put $\sigma_1 = x_1$. If $\sigma_1, \sigma_2, \dots, \sigma_j$, $j < i$ are defined, and put $\sigma_{j+1} = \sigma_j + \alpha_j x_{j+1}$, where $\alpha_j = \pm 1$ is chosen so that $\|\sigma_j + \alpha_j x_{j+1}\| \leq \|\sigma_j - \alpha_j x_{j+1}\|$. We will inductively show that

$$\|\sigma_j\| < \varepsilon(i+j)/2, \ j = 1, 2, \dots, i.$$

It is clear that $\|\sigma_1\| < \varepsilon(i+1)/2$. Assume that this holds for $j < i$. If $\|\sigma_j\| > (\varepsilon i - 2)/2$, then $\|\sigma_j\|^{-1} < 2(\varepsilon i - 2)$. Then, by the definition of the sets S_i^ε's, $\|\sigma_{j+1}\|/\|\sigma_j\| < 1 + \frac{\varepsilon}{2}\|\sigma_j\|$. From this, $\|\sigma_{j+1}\| < \varepsilon(i+j+1)/2$. If $\|\sigma_j\| \leq (\varepsilon i - 2)$, then $\|\sigma_{j1}\| \leq \|\sigma_j\| + \|x_{j+1}\| \leq \varepsilon i/2 < \varepsilon(i+1+1)/2$. $\quad\square$

Proof of Theorem 7.47. (ii) Define the operator $T : X \to \ell^\infty(\Gamma)$ by

$$Tx(\gamma) = e_\gamma^*(x),$$

where $\{e_\gamma^*\}_{\gamma \in \Gamma}$ are the dual coefficients of the basis $\{e_\gamma\}_{\gamma \in \Gamma}$. From Lemma 7.49, the proof is finished by a standard method (see the proof of Theorem 3.51).

(i) Define an operator T from X^* into $\ell^\infty(\Gamma)$ by $Tx^*(\gamma) = x^*(\gamma)$. By Proposition 3.57, X^* admits a W*UR norm. Recall that the norm of X is UG if and only if its dual norm is W*UR; see, e.g., [DGZ93a, Thm. II.6.7]. Therefore X admits a UG norm. $\quad\square$

Theorem 7.50 (Rychtář [Rych00]). *Assume that X has an unconditional basis and that X^* admits a URED norm (not necessarily a dual one). Then X admits a UG norm.*

Proof. Assume that $\{x_\gamma; f_\gamma\}_{\gamma\in\Gamma}$ is a normalized unconditional basis for X that is unconditionally monotone. Assume that $\|\cdot\|_1$ is a URED norm on X^* such that

$$k\|x^*\|_1 \geq \|x^*\| \geq \|x^*\|_1 \text{ for all } x^* \in X^*.$$

Let $\tilde{f}_\gamma = \frac{f_\gamma}{\|f_\gamma\|}$ for all γ. For $\varepsilon > 0$ and $i \in \mathbb{N}$, put

$$\Gamma_i^\varepsilon := \{\gamma \in \Gamma; \tilde{f}_\gamma \in S_i^{\frac{\varepsilon}{k^2}}\},$$

where $S_{(X^*,\|\cdot\|_1)} = \bigcup_{i=1}^\infty S_i^{\frac{\varepsilon}{k^2}}$ by the URED property of the norm $\|\cdot\|_1$. This means that for all distinct $\{x_j^*\}_{j=1}^i \subset S_i^{\frac{\varepsilon}{k^2}}$, we have

$$\max_{\alpha_j=\pm 1}\left\|\sum_{j=1}^i \alpha_j x_j^*\right\|_1 > \frac{k^2}{\varepsilon}.$$

Put

$$\Gamma_{x^*,i}^\varepsilon := \left\{\gamma \in \Gamma_i^\varepsilon; |x^*(\gamma)| > \frac{1}{i}\right\} \tag{7.1}$$

and assume that, for some $\varepsilon > 0$, $x^* \in B_{(X^*,\|\cdot\|)}$, and $i \in \mathbb{N}$, we have card $\Gamma_{x^*,i}^\varepsilon \geq i$. Let $A \subset \Gamma_{x^*,i}^\varepsilon$ be such that card $A = i$. Then

$$1 \geq \|x^*\| \geq \|P_A^*(x^*)\| = \left\|\sum_{\gamma\in A} x^*(x_\gamma)f_\gamma\right\|$$

$$\geq \min_{\gamma\in A}|x^*(x_\gamma)|\left\|\sum_{\gamma\in A}f_\gamma\right\| > \varepsilon\left\|\sum_{\gamma\in A}f_\gamma\right\| = \varepsilon\max_{\alpha_\gamma=\pm 1}\left\|\sum_{\gamma\in A}\alpha_\gamma f_\gamma\right\|$$

$$\geq \varepsilon k^{-1}\max_{\alpha=\pm 1}\left\|\sum_{\gamma\in A}\alpha_\gamma\tilde{f}_\gamma\right\| \geq \varepsilon k^{-2}\max_{\alpha=\pm 1}\left\|\sum_{\alpha\in A}\alpha_\gamma\tilde{f}_\gamma\right\| > 1,$$

which is a contradiction. Then X has a UG norm since (B_{X^*}, w^*) is a uniform Eberlein compact (see Theorem 6.30). $\qquad\square$

Remark 7.51. 1. The space $C[0, \omega_1]$ admits no equivalent UG norm, as it is not a subspace of WCG, yet the dual space, being isomorphic to $\ell_1(\Gamma)$, admits a URED norm. This is due to the fact that $C[0, \omega_1]$ admits no unconditional basis.

2. There is a space with unconditional basis whose dual admits a strictly convex dual norm and yet the space does not admit any UG norm ([ArMe93]).

3. The dual to a James tree space admits no UG norm, yet its dual admits a dual URED norm ([Haj96]).

7.4 Long Symmetric Bases

This section contains, typically, Troyanski's classification of spaces with long symmetric bases that admit uniformly Gâteaux differentiable norms as those spaces that are not isomorphic to $\ell_1(\Gamma)$ spaces. We also present an application of this result to separable Banach spaces.

Recall that an unconditional basis $\{e_\gamma; \gamma \in \Gamma\}$ of a Banach space X is said to be *symmetric* if, for any permutation π of Γ, the basis $\{e_{\pi(\gamma)}\}$ is equivalent to $\{e_\gamma\}$ (i.e., $\sum x_\gamma e_\gamma$ converges if and only if $\sum x_\gamma e_{\pi(\gamma)}$ converges). If π is a permutation of Γ and $x = \sum x_\gamma e_\gamma$, we will denote

$$x_\pi := \sum x_\gamma e_{\pi(\gamma)}.$$

We will call a norm $\|\cdot\|$ on X a *symmetric norm* if, for every $x \in X$ and every permutation π of Γ,

$$\|x\| = \|x_\pi\|.$$

Lemma 7.52 (Troyanski [Troy75]). *Let $\{e_\gamma\}_{\gamma \in \Gamma}$ be a symmetric basis in a Banach space X. Then either for every $\varepsilon > 0$ there exists an integer k such that for all $f \in X^*$ the sets $\{\gamma \in \Gamma; |f(e_\gamma)| > \varepsilon\|f\|\}$ contain at most k elements or the basis $\{e_\gamma\}_{\gamma \in \Gamma}$ is equivalent to the canonical basis of $\ell_1(\Gamma)$.*

Proof. First note that the unconditionality of $\{e_\gamma\}_{\gamma \in \Gamma}$ implies that there is a positive constant c such that for every finite system $\{\gamma_i\}_{i=1}^m \subset \Gamma$ and any finite system $\{a_i\}_{i=1}^m$ of real numbers we have

$$\left\|\sum_{i=1}^m a_i e_{\gamma_i}\right\| \geq c \max_{|\varepsilon_i| \leq 1} \left\|\sum_{i=1}^m \varepsilon_i a_i e_{\gamma_i}\right\|.$$

Also, from the symmetry of the basis, we have that there exists a positive constant d such that, for every finite system $\{\alpha_i\}_{i=1}^m, \{\beta_i\}_{i=1}^m \subset \Gamma$ and every finite system $\{a_i\}_{i=1}^m$ of real numbers, we have

$$\left\|\sum_{i=1}^m a_i e_{\alpha_i}\right\| \geq d \left\|\sum_{i=1}^m a_i e_{\beta_i}\right\|.$$

Suppose that for some $\varepsilon > 0$ there are sequences $\{f_n\} \subset X^*$ and $\{\gamma_i\} \subset \Gamma$ such that, for every $n \in \mathbb{N}$,

$$\|f_n\| = 1, \ |f_n(e_{\gamma_i})| > \varepsilon, \ i = i_n + 1, i_n + 2, \ldots, i_{n+1}, \ i_{n+1} - i_n = n.$$

Then, for any finite system $\{\beta_i\}_{i=1}^n \subset \Gamma$ and any finite system $\{a_i\}_{i=1}^n$ of real numbers, we have

$$\left\|\sum_{i=1}^n a_i e_{\beta_i}\right\| \geq d \left\|\sum_{i=i_n}^{i_{n+1}} a_{i+1-i_n} e_{\gamma_{i+1}}\right\| \geq \varepsilon c d \sum_{i=1}^n |a_i|.$$

Therefore the basis $\{e_\gamma\}_{\gamma \in \Gamma}$ is equivalent to the canonical basis of $\ell_1(\Gamma)$. $\square$

Lemma 7.53 (Troyanski [Troy75]). *Let $\{e_\gamma; e_\gamma^*\}_{\gamma \in \Gamma}$ be a symmetric basis in a Banach space X. Then either for every $\varepsilon > 0$ there exists an integer k such that for all $x \in X$ the sets $\{\gamma; |e_\gamma^*(x)| > \varepsilon \|x\|\}$ contain at most k elements or the basis $\{e_\gamma\}_{\gamma \in \Gamma}$ is equivalent to the canonical basis of $c_0(\Gamma)$.*

Proof. First note that $\{e_\gamma\}_{\gamma \in \Gamma}$ is a symmetric basis of its norm-closed linear hull. By Lemma 7.52, either there is an integer k with the desired property or there is a positive constant b such that, for any finite system $\{\gamma_i\}_{i=1}^n \subset \Gamma$ and any finite system $\{a_i\}_{i=1}^n$ of real numbers, we have

$$\left\| \sum_{i=1}^n a_i e_{\gamma_i}^* \right\| \geq b \sum_{i=1}^n |a_i|.$$

Now take an arbitrary finite subset $B \subset \Gamma$. We can find a finite subset $A \subset \Gamma$ and real numbers $\{a_\alpha\}_{\alpha \in A}$ (by eventually adding zeros we may assume that $A \supset B$) such that

$$\left\| \sum_{\alpha \in A} a_\alpha e_\alpha^* \right\| \leq 1, \text{ and } \left\| \sum_{\beta \in B} e_\beta \right\| \leq \sum_\alpha e_\alpha^* \left(\sum_{\beta \in B} e_\beta \right) + \frac{1}{b}.$$

Thus, from the preceding inequality in this proof, we get

$$\left\| \sum_{\beta \in B} e_\beta \right\| \leq \frac{2}{b}.$$

Therefore $\{e_\gamma\}_{\gamma \in \Gamma}$ is equivalent to the canonical basis of $c_0(\Gamma)$. $\square$

Theorem 7.54 (Troyanski [Troy75]). *Suppose that a nonseparable Banach space X has a symmetric basis. Then:*

(1) *X admits an equivalent symmetric UG norm if and only if X is not isomorphic to $\ell_1(\Gamma)$ for any set Γ.*

(2) *X admits an equivalent symmetric URED norm if and only if X is not isomorphic to $c_0(\Gamma)$ for any Γ.*

Proof. (1) Let $\{e_\gamma\}_{\gamma \in \Gamma}$ be a symmetric basis for X. If X admits a UG norm, then the basis cannot be equivalent to the unit vector basis for $\ell_1(\Gamma)$, as $\ell_1(\Gamma)$ does not admit a Gâteaux differentiable norm (see [DGZ93a, p. 59]).

Suppose the basis is not equivalent to the unit vector basis for $\ell_1(\Gamma)$. Define an operator $T : X^* \to \ell_\infty(\Gamma)$ by $Tf := \{f(e_\gamma)\}_{\gamma \in \Gamma}$. According to Lemma 7.52, $T(X^*) \subset c_0(\Gamma)$. For $f \in X^*$, put

$$\||f\||^2 := \|f\|^2 + \|Tf\|_{\mathcal{D}}^2,$$

where $\| \cdot \|_{\mathcal{D}}$ is Day's norm on $c_0(\Gamma)$ and $\| \cdot \|$ is the original norm on X^*. Obviously, $\|| \cdot \||$ is a dual norm. It follows from Proposition 3.57 and the estimates given in Lemma 7.52 that $\|\|\cdot\|\|$ is W*UR, so the predual norm on X is UG.

(2) If X admits an equivalent URED norm, then X cannot have an unconditional basis equivalent to the unit vector basis in $c_0(\Gamma)$, as $c_0(\Gamma)$ does not admit any URED norm (see [DGZ93a, Prop. II.7.9]).

Assume that X has a symmetric basis $\{e_\gamma\}_{\gamma \in \Gamma}$ that is not equivalent to the unit vector basis of $c_0(\Gamma)$. Define the operator $T : X \to c_0(\Gamma)$ by $Tx := \{e_\gamma^*(x)\}_{\gamma \in \Gamma}$. Define the norm on X by $\|\|x\|\|^2 := \|x\|^2 + \|Tx\|_{\mathcal{D}}^2$, where $\|\cdot\|_{\mathcal{D}}$ is Day's norm on $c_0(\Gamma)$ and $\|\cdot\|$ is the original norm of X. By Proposition 3.57 and the estimates given in Lemma 7.53, the norm $\|\|\cdot\|\|$ is URED.   □

Corollary 7.55 (Troyanski [Troy75]). *Let X be a Banach space with a symmetric basis $\{e_\gamma\}_{\gamma \in \Gamma}$.*

(i) *If X contains a space isomorphic to $c_0(\Delta)$ for some uncountable set Δ, then $\{e_\gamma\}_{\gamma \in \Gamma}$ is equivalent to the unit vector basis of $c_0(\Gamma)$.*
(ii) *If X contains a space isomorphic to $\ell_1(\Delta)$ for some uncountable set Δ, then $\{e_\gamma\}_{\gamma \in \Gamma}$ is equivalent to the unit vector basis of $\ell_1(\Gamma)$.*

Corollary 7.56 (Troyanski [Troy75]). *The Banach space $c_0(\omega_1) \times \ell_1(\omega_1)$ is not isomorphic to a subspace of a space with a symmetric basis.*

This is in contrast with Lindenstrauss' result in separable spaces (see [LiTz77, Thm. 3.b.1]) saying that *every separable Banach space with an unconditional basis is isomorphic to a complemented subspace of a separable space with a symmetric basis.*

We will now extend symmetric norms from some separable Banach spaces to their "canonical" nonseparable "extensions". In this way, we will be able to apply the nonseparable Troyanski results above to the setting of separable spaces. Let $(X, \|\cdot\|)$ be a separable Banach space with a symmetric basis $\{e_i\}$.

If θ is an injection of $\mathbb{N}$ into $\mathbb{N}$, then we denote

$$x_\theta := \sum_{i=1}^{\infty} x_i e_{\theta(i)}.$$

Clearly, $\sum x_i e_{\theta(i)}$ is convergent. We note that if π is a permutation of $\mathbb{N}$ and θ is an injection of $\mathbb{N}$ into $\mathbb{N}$ and $x \in X$, then

$$\|x_\pi\| = \|x_\theta\| = \|x\|.$$

Let Γ be an infinite set. We introduce the space $X(\Gamma)$, with a symmetric basis $\{e_\gamma; \gamma \in \Gamma\}$, as a completion of $c_{00}(\Gamma)$ under the norm

$$\left\|\sum a_{\gamma_i} e_{\gamma_i}\right\|_{X(\Gamma)} := \left\|\sum a_{\gamma_i} e_i\right\|.$$

It is standard to check that $(X(\Gamma), \|\cdot\|_{X(\Gamma)})$ is a well-defined "nonseparable version" of X.

Theorem 7.57 ([HaZi95]). *Let X be a separable Banach space with a symmetric basis. Then X admits an equivalent symmetric Gâteaux differentiable norm if and only if X admits an equivalent symmetric uniformly Gâteaux differentiable norm, if and only if X is not isomorphic to ℓ_1.*

Proof. Let $\{e_i\}$ be a symmetric basis of X. If X is not isomorphic to ℓ_1, then the basis $\{e_i\}$ is not equivalent to the standard unit vector basis of ℓ_1. Then, from Theorem 7.54, it follows that X^* admits an equivalent dual norm that is weak*-uniformly rotund and symmetric on the norm-closed linear hull of the biorthogonal functionals to the basis. Therefore its predual norm is Gâteaux differentiable and symmetric.

On the other hand, assume that $\|\cdot\|$ is an equivalent symmetric and Gâteaux differentiable norm on X and T be an isomorphism of X onto ℓ_1 and u_i be the standard unit vectors in ℓ_1. Define an equivalent norm $\|\|\cdot\|\|$ on ℓ_1, for $y = \sum \lambda_i u_i$ by $\|\|y\|\| := \|\sum \lambda_i e_i\|$. To see that this is a good definition, we observe that the basis $\{T(e_i)\}$ is equivalent to the standard unit vector basis of ℓ_1 because it is a normalized unconditional basis (see, [LiTz77, Prop. 2.b.9]). The norm $\|\|\cdot\|\|$ is symmetric on ℓ_1 with respect to $\{u_i\}$. Therefore a nonseparable extension $(\ell_1[0,\omega_1], \|\|\cdot\|\|)$, which is an equivalent renorming of the usual $\ell_1[0,\omega_1]$, is an equivalent Gâteaux differentiable norm, which is impossible by Day's result (see, e.g., [DGZ93a, p. 59] or [Fa~01, Exer. 10.5]). □

Theorem 7.58 ([HaZi95]). *Let X be a separable Banach space with a symmetric basis. Then X admits an equivalent symmetric Fréchet differentiable norm if and only if X^* is separable.*

Proof. Assume that $\{e_i\}$ is a symmetric basis for X and that X^* is separable. As $\{e_i\}$ is unconditional, we have that $\{e_i\}$ is shrinking (see Theorem 7.39). Then Troyanski's classical construction produces an equivalent norm on X^* that is locally uniformly rotund, and we can check that it is symmetric (see the proof of Theorem 3.48). Then its predual is Fréchet differentiable and symmetric on X. If X^* is not separable, X cannot admit any Fréchet differentiable equivalent norm (see Theorem 7.39). □

It is shown in [HaZi95] that *the space ℓ_∞ admits no equivalent symmetric rotund norm* and that *the space c_0 admits an equivalent symmetric C^∞ norm and admits also an equivalent symmetric rotund norm but admits no equivalent symmetric norm that is at the same time C^2 and rotund.*

James showed (see, e.g., [Fa~01, Cor. 6.36]) that *any nonreflexive separable Banach space with an unconditional basis contains either c_0 or ℓ_1.*

Troyanski proved in [Troy75] that *there exists a nonseparable Banach space X with a symmetric basis that does not contain any subspace isomorphic to $c_0(\Gamma)$ for uncountable Γ, while every infinite-dimensional subspace of X contains a subspace isomorphic to c_0.* This space is thus nonseparable and nonreflexive with a symmetric basis and does not contain an isomorphic copy of $c_0(\Gamma)$ or $\ell_1(\Gamma)$ for uncountable Γ.

7.5 Exercises

7.1. Does there exist an operator from ℓ_∞ onto a dense subset of $L_1(\mu)$ for finite measure μ?

Hint. Yes. By Rosenthal's theorem (Theorem 4.22), there is an operator from ℓ_∞ onto $L_2(\mu)$. Compose it with the canonical embedding of $L_2(\mu)$ into $L_1(\mu)$.

7.2. Does there exist an operator from $L_1(\mu)$, for finite measure μ, onto a dense set in ℓ_∞?

Hint. No. ℓ_∞ is not WCG.

7.3. Does there exist a one-to-one operator from a nonseparable $L_1(\mu)$, for finite measure μ, into ℓ_∞?

Hint. No. Otherwise $(L_1(\mu))^*$ would be weak* separable as is ℓ_∞^*.

7.4. Does there exist an operator from ℓ_∞ onto a dense set of $c_0(c)$?

Hint. Yes, by Rosenthal's theorem (Theorem 4.22).

7.5. Does there exist a one-to-one operator from ℓ_∞ onto a dense set in $c_0(c)$?

Hint. Yes. Consider a disjoint union of copies of c and $\mathbb{N}$ and an operator $Tx = (T_1x, (\frac{1}{i}x_i)_{i=1}^\infty)$, where T_1 is the map from Exercise 7.4.

7.6. Is it true that every WCG space is generated by a uniform Eberlein compact?

Hint. Yes. Let $T : X^* \to c_0(\Gamma)$ be a weak*-weak injection. Then X is generated by $T^*(B_{\ell_1(\Gamma)})$, which is uniform Eberlein. Indeed, $B_{\ell_1(\Gamma)}$ is a uniform Eberlein compact by the formal identity map from $\ell_1(\Gamma)$ into $\ell_2(\Gamma)$. A continuous image of a uniform Eberlein compact is uniform Eberlein compact.

7.7. Prove that if X is ℓ_∞-generated and X does not contain an isomorphic copy of ℓ_∞, then X is WCG.

Hint. (i) If T maps ℓ_∞ onto a dense set in X and X is not WCG, then T is not weakly compact and we use Corollary 7.14.

7.8. Show that every unconditional basis is a norming M-basis. Is the same true for long Schauder bases?

Hint. No. $C[0, \omega_1]$.

7.9. It is well known that being a subspace of c_0 is a three-space property. Is this still true for nonseparable $c_0(\Gamma)$?

Hint. No. Ciesielski-Pol space (see [DGZ93a, Thm. VI.8.8.3]).

7.10. Prove that if a nonseparable Banach space X is generated by $c_0(\omega_1)$, then X contains an isomorphic copy of $c_0(\omega_1)$.

Hint. Let Γ denote the set of all indexes in ω_1 such that the operator $T(e_\gamma) \neq 0$, where T is an operator witnessing the fact that X is $c_0(\omega_1)$ generated. We claim that Γ is uncountable. Assume the contrary. Denote by P the projection in $c_0(\omega_1)$ onto the closed linear span of e_γ for $\gamma \in \Gamma$. Then $T(c(\omega_1)) = TP(c_0(\omega_1)) = Tc_0(\Gamma)$, which must be separable, a contradiction. Therefore, for some $\varepsilon > 0$, there is an uncountable set Γ_1 such that $\|Te_\gamma\| \geq \varepsilon$ for all $\gamma \in \Gamma_1$. By the preceding theorem, there is an uncountable Γ_3 such that the restriction of T to $c_0(\Gamma_3)$ is an isomorphism.

7.11. Does there exist a one-to-one operator from $c_0(\omega_1)$ into ℓ_∞?

Hint. No because of weak*-separability.

7.12. Does there exist an operator from $c_0(\omega_1)$ onto a dense set in ℓ_∞?

Hint. ℓ_∞ is not WCG.

7.13. Follow steps (1) to (6) to show the following example, due to Rosenthal [Rose74], that the strongly Hilbert generated Banach space $L_1(\mu)$ (for some probability space $(\Omega, \mathcal{M}, \mu)$) has a non-WCG subspace. For details, we refer to [Rose74], [MeSt], and [Rose70c].

Let $\mathcal{R} := \{r : [0,1] \to \mathbb{R}, \ r \in L_1[0,1], \ \int_0^1 r\,dx = 0, \ \text{and} \ \int_0^1 |r|\,dx = 1\}$.

(1) Show that the set $\mathcal{R}$ is not a σ-weakly relatively compact subset of $L_1[0,1]$. (Note that $\{f \in L_1[0,1]; \int_0^1 f = 0\}$ is a hyperplane).

(2) Let μ denote the product Lebesgue measure on the compact space $\Omega := [0,1]^{\mathcal{R}}$. To each function $r \in \mathcal{R}$ we associate the μ-integrable function $f_r : [0,1]^{\mathcal{R}} \to \mathbb{R}$ defined by $f_r := r \circ \pi_r$, where π_r is the projection at coordinate r. Show that $\int_\Omega f_r d\mu = 0$ and $\int_\Omega |f_r| d\mu = 1$.

(3) For $(X, \mathcal{M}, \mu)$ a probability measure space, define, for every $f \in L_1(\mu)$,

$$\omega(f, \delta) := \sup \left\{ \int_E |f| d\mu; \ E \in \mathcal{M}, \ \mu(E) \leq \delta \right\}, \ \delta \in [0,1].$$

Show that, for $r \in \mathcal{R}$, $\omega(r, \delta) = \omega(f_r, \delta)$.

(4) Use Theorem 3.24 and (1) to show that the family $\tilde{\mathcal{R}} := \{f_r; r \in \mathcal{R}\}$ is not σ-weakly relatively compact in $L_1(\mu)$.

(5) Show that $\tilde{\mathcal{R}}$ is a family of independent random variables and thus, by basic probability theory, forms an unconditional basis in $Y := \overline{\text{span}}\{\tilde{\mathcal{R}}\}$ in $L_1(\mu)$.

(6) Use Theorem 7.40 to show that Y is not WCG.

8
_____

More Applications

In this chapter, we discuss some further applications of techniques of orthogonalization in the geometry of Banach spaces. In particular, this chapter begins by showing the connection between a form of one-sided biorthogonal systems and the existence of support sets in nonseparable Banach spaces. Set theory once again plays an important role because, building on some previous work of Rolewicz, Borwein, Kutzarova, Lazar, Bell, Ginsburg, Todorčević, and others, recent results of Todorčević and Koszmider demonstrate that the existence of support sets in every nonseparable Banach space is undecidable in ZFC. The second section highlights some work of Granero, Jiménez-Sevilla, Moreno, Montesinos, Plichko, and others on the study of nonseparable Banach spaces that do not admit uncountable biorthogonal systems (the existence of such spaces relies on the use of additional axioms such as ♣); these results include characterizations of spaces in which every dual ball is weak*-separable, as well as an improvement of Sersouri's result showing that such spaces must only contain countable ω-independent families.

The latter part of the chapter presents several applications of various types of biorthogonal systems. In particular, it is shown that fundamental biorthogonal systems (and even weaker systems) have applications in the study of norm-attaining operators as originated by Lindenstrauss. The attention then shifts to the Mazur intersection property, where, among other things, an application of biorthogonal systems to renorming Banach spaces with the Mazur intersection property is presented. The chapter concludes by showing that every Banach space can be renormed to have only trivial isometries; the proof of this relies on the fact that every Banach space has a total biorthogonal system.

8.1 Biorthogonal Systems and Support Sets

This section begins with the definition of support sets. It then proceeds to characterize spaces that admit such sets in terms of a certain type of one-sided biorthogonal system and closes with some remarks on $C(K)$ spaces and on the undecidability, within ZFC, of the existence of support sets in nonseparable spaces.

Definition 8.1. *A closed convex set C in a Banach space is called a* support set *if for every point $x_0 \in C$ there exists $x_0^* \in X^*$ such that $\langle x_0, x_0^* \rangle = \inf_{x \in C} \langle x, x_0^* \rangle < \sup_{x \in C} \langle x, x_0^* \rangle$. A point x_0 in C with such a property is called a* proper support point.

Theorem 8.2 (Rolewicz [Role78]). *If X is a separable Banach space, then it contains no support set.*

Proof. Suppose C is a closed convex subset of X. Let $\{x_n\}_{n=1}^\infty$ be dense in C, and choose $\{b_n\}$ such that $b_n > 0$, $\sum b_n = 1$, and $\sum b_n x_n$ converges. Now let $\bar{x} = \sum_{n=1}^\infty b_n x_n$. Then $\bar{x} \in C$ and suppose that $\phi \in X^*$ attains its minimum on C at $\bar{x}$. Now $\phi(x_n) = \phi(\bar{x})$ for each n, and so $\phi(x) = \phi(\bar{x})$ for each $x \in C$. Thus $\bar{x}$ is not a proper support point of C and thus C is not a support set. □

Fact 8.3. *If X has a support set, then $X \oplus \mathbb{R}$ has a support cone (i.e., a cone that is a support set).*

Proof. Let C be a support set in X. Define $K \subset X \oplus \mathbb{R}$ by $K = \{t(x,1); x \in C, t \geq 0\}$. Consider $t_0(x_0, 1) \in K$. If $t_0 = 0$, then $(0,1) \in X^* \oplus \mathbb{R}$ properly supports K at $(0,0)$. Otherwise, when $t_0 > 0$, choose $\phi_0 \in X^*$ such that ϕ_0 properly supports C at x_0. It is easy to check that $(\phi_0, -\phi_0(x_0)) \in X \oplus \mathbb{R}$ properly supports K at $t_0(x_0, 1)$. □

Fact 8.4. *If X admits a support set, then every subspace of finite codimension in X admits a support set.*

Proof. Suppose C is a support set in X, and $\Lambda \in X^* \setminus \{0\}$. If $C \subset \Lambda^{-1}(\alpha)$ for some α, then, by translation, $\Lambda^{-1}(0)$ has a support set. Otherwise, fix α such that $\inf_C \Lambda < \alpha < \sup_C \Lambda$, and let $C_\alpha = C \cap \Lambda^{-1}(\alpha)$. Let $x_0 \in C_\alpha$ be arbitrary. We wish to show that x_0 is properly supported in C_α. Choose ϕ properly supporting x_0 in C. Now choose $\bar{x} \in C$ such that $\phi(\bar{x}) > \phi(x_0)$. If $\bar{x} \in C_\alpha$, there is nothing further to do. So we suppose $\Lambda(\bar{x}) \neq \alpha$. If $\Lambda(\bar{x}) > \alpha$, choose $y \in C$ such that $\Lambda(y) < \alpha$. Now there is a convex combination $\bar{x}_\alpha = t\bar{x} + (1-t)y$ with $0 < t < 1$ and $\bar{x}_\alpha \in C_\alpha$; clearly $\phi(\bar{x}_\alpha) > \phi(x_0)$. If $\Lambda(\bar{x}) < \alpha$, one chooses $y \in C$ with $\Lambda(y) > \alpha$ and proceeds as above to complete the proof that C_α is a support set. Translating C_α to $\Lambda^{-1}(0)$ proves that $\Lambda^{-1}(0)$ has a support set. The proof then follows by induction. □

Definition 8.5. *Let X be a Banach space. A system $\{x_\alpha; f_\alpha\}_{1 \leq \alpha < \omega_1}$ in $X \times X^*$ will be called an* ω_1-semibiorthogonal system *if $f_\mu(x_\alpha) = 0$ for all $\alpha < \mu$, $f_\alpha(x_\alpha) = 1$, and $f_\mu(x_\alpha) \geq 0$ for all α.*

Theorem 8.6 ([BoVa96]). *For a Banach space X, the following are equivalent:*

(a) *There is an ω_1-semibiorthogonal system $\{x_\alpha, f_\alpha\}_{1\le\alpha<\omega_1}$ in $X \times X^*$.*
(b) *X has a support cone.*
(c) *X has a (bounded) support set.*

Proof. (a)$\Rightarrow$(c) Normalize the system so that $\|x_\alpha\| = 1$ for all α, and let $C = \overline{\text{conv}}(\{x_\alpha\}_{1\le\alpha<\omega_1})$. If $x \in C$, then $x \in \overline{\text{span}}(\{x_\alpha\}_{1\le\alpha\le\mu})$ for some countable ordinal μ. Thus $f_{\mu+1}(x) = 0$, while $f_{\mu+1}(x_{\mu+1}) = 1$, and so C is properly supported by $f_{\mu+1}$ at x.

(c)$\Rightarrow$(b) Write $X = Y \oplus \mathbb{R}$. According to Fact 8.4, Y has a support set. Then Fact 8.3 ensures that X has a support cone.

(b)$\Rightarrow$(a) Let K be a support cone. Fix $x_0 \in K$ and choose $f_1 \in X^*$ such that $f_1(x_0) = \inf_K f_1 < \sup_K f_1$. Note that $f_1(x_0)$ must be 0 since $tK \subset K$ for $t \ge 0$. By scaling f_1 if necessary, we choose $x_1 \in S_X \cap K$ such that $f_1(x_1) = 1$. Suppose μ is a countable ordinal and (x_α, f_α) in $(S_X \cap K) \times X^*$ have been chosen for $\alpha < \mu$ so that $f_\alpha(x_\alpha) = 1$, $f_\beta(x_\alpha) = 0$ for $\alpha < \beta < \mu$, and $f_\beta(x_\alpha) \ge 0$ for all $\alpha, \beta < \mu$. Let $\bar{x} = \sum_{\alpha<\mu} c_\alpha x_\alpha$, where $\sum_{\alpha<\mu} c_\alpha = 1$ and $c_\alpha > 0$ for each α. Choose $f_\mu \in X^*$ such that $f_\mu(\bar{x}) = \inf_K f_\mu < \sup_K f_\mu$. Now $f_\mu(k) \ge 0$ for all $k \in K$, and since each $c_\alpha > 0$ we have $f_\mu(x_\alpha) = 0$ for each $\alpha < \mu$. To complete the inductive step, scaling f_μ if necessary, we choose $x_\mu \in K \cap S_X$ such that $f_\mu(x_\mu) = 1$. □

The additional structure of $C(K)$ spaces allows simple construction of support sets in some cases [Laza81].

Proposition 8.7 (Lazar [Laza81]). *If K is a compact space and F is a closed non-G_δ subset of K, then $C = \{f \in C(K); f(F) = \{0\}, f \ge 0\}$ is a support cone and $C \cap \{f; \|f\|_\infty \le r\}$ is a support set in $C(K)$ for any $r > 0$.*

Proof. Let $f \in C$. Because F is not a G_δ set, F is a proper subset of $f^{-1}(0)$. Thus we choose $p \notin F$ with $f(p) = 0$. Now consider δ_p (the point mass measure at p). Then $\delta_p(f) = f(p) = 0 = \inf_C \delta_p$, while $\sup_C \delta_p = \infty$ (by Tietze's theorem); in the second case, nothing changes except the sup of δ_p, which is now $r > 0$. □

Theorem 8.8 (Granero et al. [GJM98]). *If K is a scattered compact space and $C(K)$ is nonseparable, then $C(K)$ has a support set. In particular, $C(K)$ for the compact space K in Theorem 4.41 has a support set.*

Proof. If $K^{(\omega_1)} \ne \emptyset$, then $\{K \setminus K^{(\alpha)}\}_{\alpha<\omega_1}$ is an open covering of $K \setminus K^{(\omega_1)}$ without a countable subcover. In the other case, there is an $\alpha_0 < \omega_1$ such that $K^{(\alpha_0)} = \emptyset$, and so for some $\alpha < \alpha_0$, $K^{(\alpha)} \setminus K^{(\alpha+1)}$ is uncountable. In either case, K is not hereditarily Lindelöf, and so K has a closed non-G_δ set (see Exercise 8.1). Applying Proposition 8.7 completes the proof. □

Corollary 8.9. (♣) *The nonseparable $C(K)$ space from Theorem 4.41 has a support set but no uncountable biorthogonal system.*

Proposition 8.10 (Lazar [Laza81]). *If K is a compact space with a nonseparable subset, then $C(K)$ has a support set.*

Proof. This follows from Theorem 4.31 and Theorem 8.6.   □

We close this section by mentioning some recent advances in this area.

Theorem 8.11 (Todorčević [Todo06]). *Every function space $C(K)$ of density $> \aleph_1$ contains an uncountable semibiorthogonal system and hence a support set.*

The reader is referred to [Todo06, Thm. 9] for a proof of this result. Moreover, this leads naturally to a question posed in [Todo06, Problem 5] *whether every Banach space X of density $> \aleph_1$ contains a support set.*

However, under additional axioms, one has the following result.

Theorem 8.12 (Todorčević [Todo06]). (MM) *A Banach space is separable if and only if X does not contain a support set.*

Proof. If X is separable, this follows from Theorem 8.2. If X is nonseparable, it contains an uncountable biorthogonal system by Theorem 4.48. According to Theorem 8.6, X has a support set.   □

In contrast to this, under other axioms consistent with ZFC, Todorčević [Todo] and Kozsmider [Kosz04a] have independently shown that there are nonseparable $C(K)$ spaces without support sets; see also Theorem 8.24 below.

8.2 Kunen-Shelah Properties in Banach Spaces

In this section, we study the relationship of conditions that are weaker than the existence of uncountable biorthogonal systems with properties such as the weak*-separability of convex sets in dual spaces and with representations of convex closed sets as zero sets of C^∞-smooth convex nonnegative functions on spaces. We show that in certain nonseparable spaces all ω-independent families are countable (Sersouri).

Lemma 8.13 (Azagra and Ferrera [AzFe02]). *Suppose that a closed convex set C in a Banach space X can be represented as a countable intersection of half-spaces. Then there is a C^∞-smooth convex function $f_C : X \to [0, \infty)$ such that $C = f_C^{-1}(0)$.*

An outline of the basic construction of f_C is as follows. First represent $C = \bigcap_{n=1}^\infty \phi_n^{-1}(-\infty, \alpha_n]$, where $\|\phi_n\| = 1$ for each n; then choose $\theta : \mathbb{R} \to [0, \infty)$ to be an appropriate C^∞-smooth convex function such that $\theta(t) = 0$ for all

$t \leq 0$ and $\theta(t) = t+b$ for all $t > 1$, where $-1 < b < 0$. The C^∞-smooth convex function f_C is defined by

$$f_C(x) := \sum_{n=1}^{\infty} \frac{\theta(\phi_n(x) - \alpha_n)}{(1 + |\alpha_n|)2^n}. \tag{8.1}$$

Lemma 8.14. *Let C be a closed convex subset of X containing the origin. Then C can be written as a countable intersection of half-spaces if and only if its polar $C^o := \{\phi \in X^*; \phi(x) \leq 1 \text{ for all } x \in C\}$ is w^*-separable.*

Proof. $\Rightarrow$: Since $0 \in C$, we can write the countable intersection as $C = \bigcap_{n=1}^{\infty} \phi_n^{-1}(\infty, 1]$. Now let $W = \overline{\text{conv}}^{w^*}(\{\phi_n\} \cup \{0\})$. Because $\phi_n(x) \leq 1$ for all $x \in C$, it follows that $\phi(x) \leq 1$ for all $x \in C$ and all $\phi \in W$; thus $W \subset C^o$. If $W \neq C^o$, then there exist $\phi \in C^o \setminus W$ and $x_0 \in X$ such that $\phi(x_0) > 1 > \sup_W x_0$ (we know this since $0 \in W$). Thus $\phi_n(x_0) < 1$ for all n and so $x_0 \in C$. This with $\phi(x_0) > 1$ contradicts that $\phi \in C^o$. Consequently, $W = C^o$, and so C^o is w^*-separable.

$\Leftarrow$: Let C be a closed convex set containing the origin, and suppose that C^o is w^*-separable. Choose a countable w^*-dense collection $\{\phi_n\}_{n=1}^{\infty} \subset C^o$. Clearly, $C \subset \bigcap_{n=1}^{\infty} \phi_n^{-1}(-\infty, 1]$. Moreover, if $x_0 \notin C$, then there is a $\phi \in X^*$ such that $\phi(x_0) > 1 > \sup_C \phi$. Then $\phi \in C^o$. The w^*-density of $\{\phi_n\}_{n=1}^{\infty}$ in C^o implies there is a ϕ_n such that $\phi_n(x_0) > 1$. Thus $C = \bigcap_{n=1}^{\infty} \phi_n^{-1}(-\infty, 1]$ as desired. $\square$

Theorem 8.15. *Assume that X^* is hereditarily w^*-separable. Let C be a closed convex subset of X. Then there is a C^∞-smooth convex function $f_C : X \to [0, \infty)$ such that $C = f_C^{-1}(0)$.*

Proof. The proof follows from Lemmas 8.13 and 8.14. $\square$

A striking contrast to the previous theorem is provided in [Haj98], where it is shown that if Γ is uncountable, then $c_0(\Gamma)$ admits no C^2-smooth function that would attain its minimum at exactly one point. This answered a question of J.A. Jaramillo. The next result also contrasts with the previous theorem.

Theorem 8.16. *Every Banach space X with a Fréchet differentiable norm such that (X^*, w^*) is hereditarily separable is itself separable.*

Proof. Let B_{X^*} denote the unit ball of a norm that is dual to a Fréchet differentiable norm on X. Let $D = \{f_n\}_{n=1}^{\infty}$ be a countable set that is w^*-dense in B_{X^*}. Suppose $f \in S_{X^*}$ attains its norm at $x \in S_X$. Find a net from D that w^*-converges to f, say $f_{n_\alpha} \to_{w^*} f$. Then $f_{n_\alpha}(x) \to 1$, and so $\|f_{n_\alpha} - f\| \to 0$ by Šmulyan's theorem ([Fa˜01, Lemma 8.4]). Therefore f is in the norm closure of D, and it then follows from the Bishop-Phelps theorem that X^* is separable. $\square$

Note that therefore, for the compact space K in Theorem 4.41, the space $C(K)$ is an Asplund space that admits no equivalent Fréchet differentiable norm, its dual is hereditarily weak*-separable, and for every closed convex subset C there exists a nonnegative C^∞-smooth convex function f on X such that $C = \operatorname{Ker} f$. It is not known if this $C(K)$ admits a C^∞-smooth function with nonempty bounded support. In general, it is an open problem if every Asplund Banach space admits a C^1-smooth function with nonempty bounded support.

Now we give a simple criterion that we will use to build closed convex sets that cannot be expressed as countable intersections of half-spaces.

Lemma 8.17. *Let C be a closed convex subset of a Banach space X. Suppose there is an uncountable sequence $\{x_i\}_{i \in I}$ such that $x_i \notin C$ for all $i \in I$ but $\frac{x_i + x_j}{2} \in C$ for all $i \neq j$. Then C cannot be expressed as a countable intersection of half-spaces.*

Proof. By translation, we may assume $0 \in C$ and thus we may suppose $C = \bigcap_\alpha f_\alpha^{-1}(-\infty, 1]$. We will show that there are uncountably many f_α's in this representation. Indeed, for each i, we find α_i such that $f_{\alpha_i}(x_i) > 1$. Now $\frac{x_i + x_j}{2} \in C$ and so $f_{\alpha_i}(x_j) < 1$ for $i \neq j$. Therefore, if $i \neq j$, one has $f_{\alpha_i}(x_i) > 1$ and $f_{\alpha_j}(x_i) < 1$. This shows $f_{\alpha_i} \neq f_{\alpha_j}$ for $i \neq j$ and so there are necessarily uncountably many f_α's in this representation.   □

All of the equivalent properties given in the next theorem are often referred to as *Kunen-Shelah properties* in reference to the famous examples of Kunen and Shelah showing that there are nonseparable Banach spaces that fail to possess those properties.

Let us single out the central such property in the following definition.

Definition 8.18. *Let X be a Banach space. An uncountable family $\{x_\alpha\}_{\alpha < \omega_1} \subset X$ is called a ω_1-polyhedron if $x_\alpha \notin \overline{\operatorname{conv}}(x_\beta; \beta \neq \alpha)$.*

Theorem 8.19 (Granero et al. [GJMMP03]). *Let X be a Banach space. Then the following are equivalent:*

(a) *X contains a ω_1-polyhedron.*
(b) *There is a bounded closed convex subset of X that cannot be represented as a countable intersection of half-spaces.*
(c) *There is a closed convex subset in X that cannot be represented as a countable intersection of half-spaces.*
(d) *There is a w^*-closed convex subset of X^* that is not w^*-separable.*
(e) *There is a ball of an equivalent dual norm in X^* that is not w^*-separable.*
(f) *There is an equivalent norm on X whose unit ball cannot be represented as a countable intersection of half-spaces.*
(g) *There is a bounded uncountable system $\{x_\alpha; \phi_\alpha\} \subset X \times X^*$ such that $\phi_\alpha(x_\alpha) = 1$ and $|\phi_\alpha(x_\beta)| \leq a$ for some $a < 1$ and all $\alpha \neq \beta$.*

Proof. (a)$\Rightarrow$(b) Suppose (a) holds. Then, for some $N > 0$ there are uncountably many $\{x_\alpha\}$ such that $\|x_\alpha\| < N$, so we may and do assume $\|x_\alpha\| < N$ for all α. By the separation theorem, for each α, we find $f_\alpha \in X^*$ and $\delta_\alpha > 0$ such that $f_\alpha(x_\alpha) > f_\alpha(x_\beta) + \delta_\alpha$ for all $\alpha \neq \beta$. Now let $a_\alpha = f_\alpha(x_\alpha)$ and let

$$C = \{x; f_\alpha(x) \leq a_\alpha - \delta_\alpha/2 \text{ for all } \alpha\} \cap NB_{X^*}.$$

Then $x_\alpha \notin C$ for all α; however, for $\alpha \neq \beta$, we have $f_\mu(\frac{x_\alpha + x_\beta}{2}) \leq a_\mu - \delta_\mu/2$ and so $\frac{x_\alpha + x_\beta}{2} \in C$ for all $\alpha \neq \beta$. Therefore, C cannot be written as a countable intersection of half-spaces by Lemma 8.17.

(b)$\Rightarrow$(c) is trivial, and (c) $\Rightarrow$ (d) follows from Proposition 8.14.

(d)$\Rightarrow$(e) Let W be a w^*-closed convex subset of X^* that is not w^*-separable. Then $W \cap NB_{X^*}$ is not w^*-separable for some $N > 0$ (otherwise W would be a countable union of w^*-separable sets). Thus, we assume without loss of generality that W is bounded. Also, if $W + \epsilon B_{X^*}$ were w^*-separable for each $\epsilon > 0$, then it would follow that W is w^*-separable. (Indeed, if, for each $n \in \mathbb{N}$, $\{w_{n,k} + b_{n,k}\}_{k=1}^\infty$ is w^*-dense in $W + \frac{1}{n}B_{X^*}$, where $w_{n,k} \in W$ and $b_{n,k} \in \frac{1}{n}B_{X^*}$ for all $n, k \in \mathbb{N}$, then $\{w_{n,k}\}_{k,n \in \mathbb{N}}$ is w^*-dense in W.) Thus $W + \epsilon B_{X^*}$ is not w^*-separable for some $\epsilon > 0$, so we may assume without loss of generality that W is w^*-compact convex and has a nonempty norm interior.

Now we construct a dual ball in X^* that is not w^*-separable. Fix $x_0 \in S_X$. Then, for some (rational) number $a \neq 0$ with $\inf_W x_0 < a < \sup_W x_0$, we have that $x_0^{-1}(a) \cap W$ is not w^*-separable; otherwise, W would be w^*-separable. Now let $K := x_0^{-1}(a) \cap W$. Then K is a w^*-compact convex set, and so the symmetric convex set $B := \text{conv}(K \cup (-K))$ is also w^*-compact. Moreover, B has a nonempty norm interior because $x_0^{-1}(a) \cap W$ has a nonempty norm interior relative to $x_0^{-1}(a)$. Finally, if B were w^*-separable, we could find a countable collection $\{\lambda_n x_n^* - (1 - \lambda_n)y_n^*\}_{n=1}^\infty$, where $x_n^*, y_n^* \in K$ and $0 \leq \lambda_n \leq 1$ that is w^*-dense in B. Any net from this collection converging to $k \in K$ has $\lambda_{n_\alpha} \to 1$, and so it follows that $\{x_n^*\}_{n=1}^\infty$ is w^*-dense in K. This contradiction shows that B is not w^*-separable, as desired.

(e)$\Leftrightarrow$(f) This follows from Proposition 8.14(a).

(e)$\Rightarrow$(g) Suppose B_{X^*} is not w^*-separable. For $Y \subset X^*$, let

$$|x|_Y := \sup\{\phi(x); \phi \in Y \cap B_{X^*}\} \quad \text{for } x \in X,$$

and define

$$\lambda := \sup\{\alpha; \alpha\|\cdot\| \leq |\cdot|_Y \text{ where } Y \subset X^* \text{ is a separable subspace}\}. \quad (8.2)$$

Then $\lambda < 1$, or else for some separable subspace Y we would have $|x|_Y \geq \|x\|$ and then $Y \cap B_{X^*}$ would be w^*-dense in B_{X^*}—contradicting that B_{X^*} is not w^*-separable. Now, choose $l > 0$ so that $1 - l > \lambda$. For $Y \subset X^*$ a separable subspace, we define $F_Y := \{x \in S_X; |\phi(x)| \leq 1 - l \text{ for all } \phi \in Y \cap B_{X^*}\}$. Now let

$$\delta := \inf\{e(F_Y, Z); Y \subset X^* \text{ and } Z \subset X \text{ are separable}\}. \quad (8.3)$$

We prove that $\delta > 0$; otherwise, choose Y_n and Z_n such that $e(F_{Y_n}, Z_n) \to 0$ as $n \to \infty$. Letting $Y = \overline{\text{span}}(\bigcup_{n \in \mathbb{N}} Y_n)$ and $Z = \overline{\text{span}}(\bigcup_{n \in \mathbb{N}} Z_n)$, we find that $e(F_Y, Z) = 0$. Because Z is separable, there is a countable set in $S \subset B_{X^*}$ such that $\sup\{\phi(z); \phi \in S\} = \|z\|$ for all $z \in Z$. Let $\widetilde{Y} := \overline{\text{span}}(Y \cup S)$ and let $x \in S_X$. If $x \in F_Y$, then $x \in Z$, and so there exists $\phi \in S \cap B_{X^*} \subset \widetilde{Y} \cap B_{X^*}$ such that $\phi(x) > 1 - l$. If $x \notin F_Y$, we have $\phi \in Y \cap B_{X^*} \subset \widetilde{Y} \cap B_{X^*}$ such that $\phi(x) > 1 - l$. Consequently, $(1 - l)\|\cdot\| \leq |\cdot|_{\widetilde{Y}}$, which contradicts (8.2) because $1 - l > \lambda$. Therefore, $\delta > 0$.

Let $\eta > 0$ be such that $\eta < \min\{l, \delta\}$. Let ω_1 denote the first uncountable ordinal. We will find an uncountable system in $\{x_\alpha, \phi_\alpha\}_{1 \leq \alpha < \omega_1} \subset B_X \times B_{X^*}$ such that $|\phi_\alpha(x_\alpha)| \geq 1 - \eta + \eta^2$ for all $1 \leq \alpha < \omega_1$, while $|\phi_\alpha(x_\beta)| \leq 1 - \eta$ for all $\alpha \neq \beta$. Indeed, fix $x_1 \in B_X$ and $\phi_1 \in B_{X^*}$ such that $\phi_1(x_1) = 1$. Suppose for an ordinal $1 < \mu < \omega_1$ that x_α, ϕ_α have been chosen as prescribed for all $\alpha < \mu$. We denote

$$F_\mu := \{x \in S_X; |\phi_\alpha(x)| \leq 1 - l \text{ for all } \alpha < \mu\}$$
$$\text{and} \quad X_\mu := \overline{\text{span}}(\{x_\alpha; \alpha < \beta\}). \tag{8.4}$$

Because $\eta < \delta$ as defined in (8.3), we can choose $x_\mu \in F_\mu$ such that $d(x_\mu, X_\mu) > \eta$. Now select $x^*_{\mu,1} \in S_{X^*}$ such that $x^*_{\mu,1}(x_\mu) = 1$, and choose $x^*_{\mu,2} \in S_{X^*}$ such that $x^*_{\mu,2}(x_\mu) > \eta$, while $x^*_{\mu,2}(X_\mu) = 0$. Let $\phi_\mu = (1 - \eta)x^*_{\mu,1} + \eta x^*_{\mu,2}$. Then $\phi_\mu(x_\mu) > 1 - \eta + \eta^2$; $|\phi_\mu(x_\alpha)| \leq 1 - \eta$ for all $\alpha < \mu$ because $x_\alpha \in X_\mu$; and $|\phi_\alpha(x_\mu)| \leq 1 - l < 1 - \eta$ for $\alpha < \mu$ because $x_\mu \in F_\mu$. By transfinite induction, we construct a sequence as we claimed. Scaling the ϕ_α's so that $\phi_\alpha(x_\alpha) = 1$ produces a system as in (g), where $a = (1 - \eta)/(1 - \eta + \eta^2)$.

(g) $\Rightarrow$ (a) This is an immediate consequence of the separation theorem. □

As a further consequence, we provide a characterization of separable Asplund spaces in terms of w^*-separability of w^*-compact convex sets in the second dual.

Corollary 8.20. *For a separable Banach space X, the following are equivalent:*

(a) *X^* is separable.*
(b) *Every dual ball in X^{**} is w^*-separable.*
(c) *Every w^*-closed convex subset in X^{**} is w^*-separable.*

Proof. (a)$\Rightarrow$(c) follows from the separability of X^*, and (c)$\Rightarrow$(b) is trivial. To prove (b)$\Rightarrow$(a), we suppose X^* is not separable. Then X^* contains an uncountable biorthogonal system (Corollary 4.34), and according to Theorem 8.19, there is an equivalent dual ball in X^{**} that is not w^*-separable. □

In particular, let us point out that, for $X = \ell_1$, every double dual ball on X^{**} is w^*-separable but there are other w^*-closed balls in X^{**} that are not w^*-separable. This can be done without using the full power of the corollary.

Indeed, since $\ell_1(\aleph_1) \subset \ell_\infty$, we have that ℓ_∞ has an uncountable biorthogonal system. Modifying the technique of Lemma 8.17, one can easily build an equivalent (nondual) norm on ℓ_∞ whose ball is not a countable intersection of half-spaces. Then, by Proposition 8.14, the dual ball is not w^*-separable.

The following theorem adds to the list of conditions equivalent to those given in Theorem 8.19. For this we will declare that a bounded family $\{x_\alpha; 1 \leq \alpha < \omega_1\}$ is a *convex right-separated ω_1-family* if $x_\alpha \notin \overline{\text{conv}}(\{x_\beta; \alpha < \beta < \omega_1\})$.

Theorem 8.21 (Granero et al. [GJMMP03]). *For a Banach space X, the following are equivalent.*

(a) *Any of the equivalent conditions in Theorem 8.19.*
(b) *X has a convex right-separated ω_1-family.*
(c) *There is a convex subset of X^* that is not w^*-separable.*

We refer the reader to [GJMMP03, §7] for the proof of this theorem; however, let us note that (a) readily implies (b) and (c), and it is not difficult to show the equivalence of (b) and (c). However, (b) or (c) implies (a) is rather delicate. The next theorem shows that conditions from Theorems 8.19 and 8.21 are implied by ω_1-independence and in fact a formally stronger property is obtained.

Theorem 8.22 (Granero et al. [GJMMP03]). *Suppose X is a Banach space that has an uncountable ω-independent family $\{x_\alpha\}_{1 \leq \alpha < \omega_1}$. Then, for each $0 < \eta < 1$, there exists an uncountable sequence $\{\alpha_i\}_{i<\omega_1}$ and a bounded uncountable sequence $\{f_i\}_{i<\omega_1} \subset X^*$ such that*

$$f_i(x_{\alpha_i}) = 1 \ (i < \omega_1) \qquad \text{and} \qquad |f_i(x_{\alpha_j})| < \eta \ (i \neq j, i, j < \omega_1),$$

and, moreover, $f_i(x_{\alpha_j}) = 0$ for $j < i < \omega_1$. In particular, X has an ω_1-polyhedron.

The proof will use the following lemma.

Lemma 8.23 ([GJMMP03]). *Let X be a Banach space, $\{x_i\}_{1 \leq i < \omega_1} \subset X$ an uncountable bounded ω-independent family, $H \subset X$ a closed separable subspace, and $N \in \mathbb{N}$. Then there exist ordinal numbers $\rho < \gamma < \omega_1$ such that $x_\rho \notin \overline{\text{conv}}(H \cup \{\pm Nx_i\}_{\gamma < i < \omega_1})$.*

Proof. Without loss of generality, suppose that $\|x_i\| \leq 1$ for all $i < \omega_1$. Assume that for every pair of ordinals ρ, γ such that $\rho < \gamma < \omega_1$ we have $x_\rho \in \overline{\text{conv}}(H \cup \{\pm Nx_i\}_{\gamma < i < \omega_1})$. For $n \in \mathbb{N}$ and $\rho < \gamma < \omega_1$, define $D_\gamma = \text{conv}(\{\pm Nx_i\}_{\gamma \leq i < \omega_1})$ and

$$H(\rho, \gamma, n) = \left\{(u, \lambda) \in H \times (0, 1]; \exists v \in D_\gamma \text{ with } \|\lambda u + (1-\lambda)v - x_\rho\| < \frac{1}{2n}\right\}.$$

If $\rho < \gamma < \gamma' < \omega_1$ and $n \geq 1$, then by hypothesis and definition we have $H(\rho, \gamma, n) \neq \emptyset$ and $H(\rho, \gamma, n+1) \subset H(\rho, \gamma, n) \supset H(\rho, \gamma', n)$. For $\beta < \omega_1$ and $n \geq 1$, let

$$H(\beta, n) = \mathrm{cl}\left(\bigcup\{H(\rho, \gamma, n) : \beta \le \rho < \gamma < \omega_1\}\right),$$

where "cl" means the closure in $H \times (0, 1]$. Then, for $\beta < \beta'$ and $n \ge 1$, one has

$$\emptyset \ne H(\beta', n) \subset H(\beta, n) \supset H(\beta, n + 1).$$

Because $H \times (0, 1]$ is hereditarily Lindelöf, for each $n \ge 1$, there exists $\beta_n < \omega_1$ such that, for every $\beta_n \le \beta < \omega_1$, one has $(u, \lambda) \in H(\beta, n)$; this implies that there exists $\beta \le \rho < \gamma < \omega_1$ and $v \in D_\gamma$ such that

$$\|x_\rho - (\lambda u + (1 - \lambda)v)\| < 1/n.$$

Let $\beta_0 = \sup_{n \ge 1} \beta_n$ and fix $\beta_0 \le \rho < \gamma < \omega_1$ and $n \ge 1$. Choose $(u, \mu) \in H(\rho, \gamma, n)$ and $w \in D_\gamma$ such that $\|x_\rho - (\mu u + (1 - \mu)w)\| < 1/(2n)$. Because $(u, \mu) \in H(\beta_0, n) = H(\gamma, n)$, there exist $\gamma \le \sigma < \theta < \omega_1$ and $v \in D_\theta$ such that $\|x_\sigma - (\mu u + (1 - \mu)v)\| < 1/n$.

Define $T = x_\sigma - (\mu u + (1 - \mu)v)$. Then $\mu u = x_\sigma - T - (1 - \mu)v$ and

$$\left\|x_\rho - (x_\sigma - T - (1 - \mu)v + (1 - \mu)w)\right\| < \frac{1}{2n}.$$

Because $\|T\| < 1/n$, one obtains

$$
\begin{aligned}
\|x_\rho &- (x_\sigma - (1 - \mu)v + (1 - \mu)w)\| \\
&= \|x_\rho - (x_\sigma - T - (1 - \mu)v + (1 - \mu)w) - T\| \\
&\le \|x_\rho - (x_\sigma - T - (1 - \mu)v + (1 - \mu)w)\| + \|T\| \\
&< \frac{1}{2n} + \frac{1}{n} = \frac{3}{2n}.
\end{aligned}
$$

Because $x_\sigma, v, w \in E_\gamma$, where $E_\gamma := \overline{\mathrm{span}}\{x_i\}_{\gamma \le i < \omega_1}$, we deduce that $x_\rho \in E_\gamma$ by letting $n \to \infty$ with ρ and γ fixed; in particular, this implies that $E_{\beta_0} = E_\beta$ for all $\beta_0 \le \beta < \omega_1$. Let $S = x_\rho - (x_\sigma - (1 - \mu)v + (1 - \mu)w)$. Then

$$x_\rho = S + \mu v + (1 - \mu)w + x_\sigma - v.$$

That $\mu v - (1 - \mu)w, -v \in D_\gamma$, $x_\sigma \in (1/N)D_\gamma$ and $\|S\| < 3/(2n)$ imply that $x_\rho \in \mathrm{cl}((1 + 1/N)D_\gamma + D_\gamma) = \mathrm{cl}((2 + 1/N)D_\gamma)$. Letting $F_\gamma = (2 + 1/N)D_\gamma$, we conclude that x_ρ is an accumulation point of F_γ. Consequently, every x_i with $\beta_0 \le i < \omega_1$ is an accumulation point of every F_γ for $\gamma < \omega_1$.

Let $(a_n)_{n \ge 1}$ be a sequence of positive numbers such that $\sum a_n = \infty$ and $\lim a_n = 0$. Using the proof of Theorem 1.58, one can inductively construct a sequence $\{\epsilon_n\}$ of signs, a sequence $\{\lambda_r^n\}_{n \ge 1, 1 \le r \le k(n)}$ of real numbers, and a sequence $\{\gamma_r^n\}_{n \ge 1, q \le r \le k(n)}$ of ordinals such that:

(i) $\sum_{r=1}^{k(n)} |\lambda_r^n| \le 2N + 1$ for every $n \ge 1$,

(ii) $\tau < \gamma_1^n < \ldots < \gamma_{k(n)}^n < \gamma_1^{n+1} < \ldots < \omega_1$ for every $n \ge 1$, and

(iii) $x_\tau + \sum_{n \ge 1} a_n \epsilon_n y_n = 0$, where $y_n = \sum_{r=1}^{k(n)} \lambda_r^n x_{\gamma_r^n}$.

One can consult [GJMMP03, Proof of Lemma 3.1] for further details on this construction.

Finally, the series $x_\tau + \sum_{n \geq 1} a_n \epsilon_n (\sum_{r=1}^{k(n)} \lambda_r^n x_{\gamma_r^n})$ converges to 0. This provides the contradiction that $\{x_i\}_{i<\omega_1}$ is not ω-independent. Consequently, $\rho < \gamma < \omega_1$ can be chosen so that $x_\rho \notin \overline{\text{conv}}(H \cup \{\pm N x_i\}_{\gamma \leq i < \omega_1})$. □

Proof of Theorem 8.22. Let $\{x_i\}_{1 \leq i < \omega_1}$ be an ω-independent family in X, and we may suppose without loss of generality that $\|x_i\| \leq 1$ for all $i < \omega_1$. Let $N \in \mathbb{N}$, satisfy $1/N \leq \eta$.

Next we will construct by induction two subsequences $\{i_\alpha, j_\alpha\}_{\alpha < \omega_1}$ of ordinal numbers with $i_\alpha < j_\alpha \leq i_\beta < j_\beta < \omega_1$ for $\alpha < \beta < \omega_1$ such that

$$x_{i_\alpha} \notin \overline{\text{conv}}(\text{span}\{x_{i_\beta} : \beta < \alpha\} \cup \{\pm N x_j\}_{j_\alpha \leq j < \omega_1}). \tag{8.5}$$

For this, let $\alpha < \omega_1$ and assume that $\{i_\beta, j_\beta\}_{\beta < \alpha}$ satisfying (8.5) have been chosen. Let $H := \overline{\text{span}}\{x_{i_\beta}\}_{\beta < \alpha}$ and $\nu = \sup_{\beta < \alpha}\{j_\beta\}$ (if $\alpha = 1$, set $H = \{0\}$ and $\nu = 1$). According to Lemma 8.23, there exist $\nu \leq \rho < \gamma < \omega_1$ such that $x_\rho \notin \overline{\text{conv}}(H \cup \{\pm N x_i\}_{\gamma \leq i < \omega_1})$. Thus we let $i_\alpha = \rho$ and $j_\alpha = \gamma$ to complete the induction.

According to (8.5), $x_{i_\alpha} \notin \overline{\text{conv}}(\text{span}\{x_{i_\beta} : \beta < \alpha\} \cup \{\pm N x_{i_j}\}_{\alpha < j < \omega_1})$. The Hahn-Banach theorem now ensures the existence of $f_\alpha \in X^*$ so that

$$1 = f_\alpha(x_{i_\alpha}) > \sup\{f_\alpha(x) : x \notin \overline{\text{conv}}(\text{span}\{x_{i_\beta} : \beta < \alpha\} \cup \{\pm N x_{i_j}\}_{\alpha < j < \omega_1})\}.$$

Consequently, $f_\alpha(x_{i_\beta}) = 0$ if $\beta < \alpha$, and $|f_\alpha(x_{i_\beta})| < 1/N$ if $\alpha < \beta < \omega_1$. The proof is completed by observing that there is an uncountable subsequence $A \subset \omega_1$ such that $\{\|f_\alpha\| : \alpha \in A\}$ is bounded. □

Let us note that Sersouri [Sers89] was the first to prove that *a Banach space with an uncountable ω-independent family must have an ω_1-convex right-separated family.* As a consequence of this, the space $C(K)$, where K is the compact space in Theorem 4.41, does not have an uncountable ω-independent family.

Finally, the following is a summary of some relations among conditions from this and the previous section.

Theorem 8.24. *Consider the following conditions:*

(a) X *admits an uncountable biorthogonal system.*
(b) X *has an uncountable ω-independent system.*
(c) X *admits an equivalent norm so that B_{X^*} is not w^*-separable (see Theorems 8.19 and 8.21 for other conditions equivalent to this).*
(d) X *has a support set.*
(e) X *is not separable.*

In ZFC: (a) *implies each of the other conditions, each of the conditions implies* (e), *and also* (a)$\Rightarrow$(b)$\Rightarrow$(c).

Under MM*:* (a) *through* (e) *are equivalent* (see Theorems 4.48 and 8.12 and [Todo06]).

Under CH *or* ♣*:* (d) *does not imply* (c) (Kunen's $C(K)$ space [Negr84] or $C(K)$ from Theorem 4.41).

It is consistent in ZFC *that* (e) *does not imply* (d) (see Todorčević [Todo06] and Koszmider [Kosz04a] for precise details).

8.3 Norm-Attaining Operators

The classical Bishop-Phelps theorem says that the set of all linear functionals in X^* that attain their norm on B_X is dense among all linear functionals in X^*. In order to study the analogous property for operators, we declare that a bounded linear operator $T : X \to Y$ *attains its norm* if there is an $x \in B_X$ such that $\|Tx\| = \|T\|$. This section surveys some results connected to the property that the set of all operators between given Banach spaces X and Y that attain their norm is dense in $\mathcal{L}(X, Y)$. In particular, it is shown that if X has a fundamental biorthogonal system, then X can be renormed in this fashion, as was shown by Godun and Troyanski.

We now introduce a condition on $(X, \|\cdot\|)$ that is useful in showing the denseness of norm-attaining operators in $\mathcal{L}(X, Y)$, where Y is any Banach space. A Banach space $(X, \|\cdot\|)$ is said to have *property* (α, λ) if there is a system $\{x_i; f_i\}_{i \in I}$ in $X \times X^*$ and $0 \le \lambda < 1$ such that $\|x_i\| = \|f_i\| = 1$ for all $i \in I$, $f_i(x_i) = 1$ for all $i \in I$ and $|f_i(x_j)| \le \lambda$ whenever $i \ne j$ and the absolutely convex closed hull of $\{x_i\}$ is equal to B_X. In addition, if

$$\inf_j \max\{|f_j(x_i)|, |f_i(x_j)|\} = 0 \qquad \text{for all } i \in I,$$

then $(X, \|\cdot\|)$ is said to have *strict property* (α, λ). If $(X, \|\cdot\|)$ has (strict) property (α, λ) for some $\lambda \in (0, 1)$ and we are not concerned with the particular value of λ, we will say that X has *(strict) property* α. Also, we will sometimes refer to the system $\{x_i; f_i\}_{i \in I}$ as an *α-system*.

The following property is useful on the range space for determining when norm-attaining operators are dense. A Banach space $(X, \|\cdot\|)$ is said to have *property* (β, λ) if there is a system $\{x_i; f_i\}_{i \in I} \subset S_X \times S_{X^*}$ and $\lambda < 1$ such that $\|x\| = \sup_i f_i(x)$ for every $x \in X$, $f_i(x_i) = 1$ for every $i \in I$, and $|f_i(x_j)| \le \lambda$ for every $i \ne j$. If, additionally, B_{X^*} is the closed convex hull of $\{\pm f_i\}$, we will say that $(X, \|\cdot\|)$ has *strong property* (β, λ). Again, when we are not concerned with the value λ, we will say $(X, \|\cdot\|)$ has *(strong) property* β, and we may refer to the system $\{x_i, f_i\}_{i \in I}$ as a *β-system*.

Before examining which Banach spaces can be renormed to have these properties, we point out a few applications—beginning with operators attaining their norm—that have stimulated interest in these properties. As has become standard in this subject, we will say a Banach space $(X, \|\cdot\|)$ has *property A* if for any Banach space Y the norm-attaining operators are dense in $\mathcal{L}(X, Y)$, while $(Y, \|\cdot\|)$ is said to have *property B* if for any Banach space X the norm-attaining operators are dense in $\mathcal{L}(X, Y)$.

Theorem 8.25 (Lindenstrauss [Lind63]). *Let X be a Banach space.*

(a) *If $(X, \|\cdot\|)$ has property α, then all operators from X into any Y that attain their norm are dense in all operators; that is, $(X, \|\cdot\|)$ has property A.*

(b) *If $(Y, \|\cdot\|)$ has property β, then $(Y, \|\cdot\|)$ has property B.*

Proof. (a) Suppose $T : X \to Y$ is a continuous linear operator, $T \neq 0$, and $\epsilon > 0$. Let $\{x_i; f_i\}_{i \in I}$ be an α-system, and fix $i \in I$ such that $\|T(x_i)\| > \|T\|(1 + \epsilon\lambda)/(1 + \epsilon)$. Define the operator $\widetilde{T}$ by $\widetilde{T}(x) = T(x) + \epsilon f_i(x)T(x_i)$. Then $\|\widetilde{T}(x_i)\| > \|T\|(1+\epsilon\lambda)$, while for $j \neq i$, $\widetilde{T}(x_j)\| \leq \|T\|(1+\lambda\epsilon)$. Therefore, $\widetilde{T}$ attains its supremum at x_i, and also $\|T - \widetilde{T}\| \leq \epsilon\|T\|$.

(b) Let $\{y_i; f_i\}_{i \in I} \subset Y \times Y^*$ be as in the definition of property β. Given $\epsilon > 0$, and $T : X \to Y$ with $T \neq 0$, fix $x_0 \in S_X$ and i such that $\|f_i(Tx_0)\| > \|T\|(1+\epsilon\lambda)/(1+\epsilon)$ and define $\widetilde{T}$ by $\widetilde{T}(x) = T(x) + \epsilon f_i(Tx)y_i$; then check that the remaining details follow as in (a). $\quad\square$

The *(Dixmier) characteristic* of a subspace Z of a dual Banach space X^* was introduced in Chapter 2 (see formula (2.1)) as the supremum $r(Z)$ of all numbers r such that $B_{X^*} \cap Z$ is w^*-dense in rB_{X^*}.

Definition 8.26. *For a nonreflexive Banach space X, we denote $\mathcal{R}(X) := \sup\{r(Z); Z$ is a closed proper subspace of $X^*\}$.*

Clearly $\mathcal{R}(X) \leq 1$, and it is not hard to show that $\mathcal{R}(X) \geq 1/2$; see Exercise 8.7. We will see that strong property β has applications in this direction.

Theorem 8.27 ([FiSc89], [GoTr93], [More97]). *Let $(X, \|\cdot\|)$ be a Banach space.*

(a) *If $(X, \|\cdot\|)$ has property (α, λ) under the system $\{x_i; f_i\}_{i \in I}$, then the points $\{\pm x_i\}$ are uniformly strongly exposed by $\{\pm f_i\}$. That is, if $x \in B_X$ and $f_i(x) \geq 1 - (1 - \lambda)\epsilon$, then $\|x - x_i\| \leq 2\epsilon$ (and hence if $-f_i(x) \geq 1 - (1 - \lambda)\epsilon$, then $\|x - (-x_i)\| \leq 2\epsilon$).*

(b) *If $(X, \|\cdot\|)$ has property α, then the denting points of B_X are $\{\pm x_i\}_{i \in I}$, where $\{x_i; f_i\}_{i \in I}$ is an α-system.*

(c) *Suppose $(X, \|\cdot\|)$ has strong property (β, λ). Then $\mathcal{R}(X) \leq (1 + \lambda)/2$.*

(d) *Property α implies that $\|\cdot\|$ has no points of local uniform rotundity.*

Proof. (a) Let $x \in \text{conv}(\{\pm x_i\}_{i \in I})$ be such that $f_i(x) \geq 1 - (1 - \lambda)\epsilon$. Then we write $x = (1 - a)x_i + \sum_{k=1}^{n} \lambda_k \epsilon_k x_{i_k}$, where $\epsilon_k = \pm 1$, $\epsilon_k x_{i_k} \neq x_i$, $\sum_{k=1}^{n} \lambda_k = a$, and $a \geq 0$. Then $\|x - x_i\| \leq 2a$; lastly $(1 - a) + \lambda a \geq f_i(x) \geq 1 - (1 - \lambda)\epsilon$ and so $\epsilon \geq a$ as desired.

(b) Let $\{x_i; f_i\}_{i \in I}$ be an α-system with $\lambda < 1$ in the definition of property (α, λ). Note then that $\|x_i - x_j\| \geq 1 - \lambda$ for $i \neq j$. Suppose x_0 is a denting point of B_X. Let S_n be slices of B_X containing x_0 of diameter α_n, where $\alpha_n \to 0$ and so $\alpha_n < 1 - \lambda$ for large n. Consequently, there is exactly one $i_0 \in I$ such that $x_{i_0} \in S_n$ for all n. Thus $x_0 = x_{i_0}$.

(c) Let $\Phi \in S_{X^{**}}$, and let $Z \subset X^*$ be the kernel of Φ. Let $\epsilon > 0$, and choose $f_{i_0}^*$ such that $\langle \Phi, f_{i_0} \rangle > 1 - \epsilon$. Now, if $x^* \in B_{X^*}$ and $x^*(x_{i_0}) > 1 - (1 - \lambda)(1 - \epsilon)/2$, then by part (a), $\|x^* - x_{i_0}^*\| \leq 1 - \epsilon$. Therefore, $x^* \notin Z$. Consequently, if $\phi \in Z \cap B_{X^*}$, then $|\phi(x_{i_0})| \leq 1 - (1 - \lambda)(1 - \epsilon)/2 \leq (1 + \lambda + \epsilon)/2$. Consequently, $r(Z) \leq (1 + \lambda)/2$, where $r(Z)$ denotes the characteristic of the subspace Z. Since Z was arbitrary, we are done.

(d) See Exercise 8.10. □

Proposition 8.28 (Schachermayer [Scha83]). *The following properties of the space c_0 hold:*

(a) *For $K > 1$, there is a norm $\|\| \cdot \|\|$ on c_0 with $\| \cdot \| \geq \|\| \cdot \|\| \geq K^{-1} \| \cdot \|$ and such that $(c_0, \|\| \cdot \|\|)$ has property α.*
(b) *c_0 endowed with its usual norm does not have property A.*
(c) *c_0 endowed with a strictly convex norm does not have property B.*

In particular, neither property A nor property B are invariant under isomorphisms.

Proof. (a) Let $\{z_n\}_{n=1}^\infty$ be a dense sequence in the unit ball of c_0. Define x_n by

$$x_n(i) = z_n(i) \ \text{ if } i \neq n, \qquad x_n(i) = K \ \text{ if } i = n.$$

Let B be the absolutely closed convex hull of $\{x_n\}_{n=1}^\infty$ and let $\|\| \cdot \|\|$ be the Minkowski functional of B. Clearly, $\|\| \cdot \|\| \geq K^{-1} \| \cdot \|$. On the other hand, if $\|x\| \leq 1$, fix a subsequence $\{x_{n_j}\}$ with $n_1 < n_2 < n_3 < \ldots$ such that $x_{n_j} \to x$. Let $u_m = (z_{n_1} + \ldots + z_{n_m})/m$. Then $u_m \in B$ and $u_m \to x$. Therefore, $\|\| \cdot \|\| \leq \| \cdot \|$. Now let $f_n = K^{-1} e_n$, where e_n is the n-th coordinate vector in ℓ_1. Then $f_n(z_n) = 1$, while $f_n(z_m) \leq K^{-1}$ if $n \neq m$. This proves (a).

Both (b) and (c) follow from the fact that c_0 has an equivalent strictly convex norm, but its unit ball under its usual norm has no extreme points. Note that the remark that property B is not invariant under isomorphisms also relies on the next theorem. □

Theorem 8.29 (Partington [Part82]). *Let $(X, \| \cdot \|)$ be a Banach space. Then for any $K > 3$ there is a norm $\|\| \cdot \|\|$ such that $\| \cdot \| \leq \|\| \cdot \|\| \leq K \| \cdot \|$ and $(X, \|\| \cdot \|\|)$ has property β.*

Proof. Let γ be the density character of X and fix a set $\{u_\alpha; \alpha < \gamma\}$ with $\|u_\alpha\| = 1$ for each α that is norm dense in S_X. Now select $\{u_\alpha^*; \alpha < \gamma\} \subset X^*$ such that $\|u_\alpha^*\| = u_\alpha^*(u_\alpha) = 1$ for each α. Let s be a constant such that $1 > s > 2/(K - 1)$. By transfinite induction, we may find for each $\alpha < \gamma$ in turn $y_\alpha^* \in S_{X^*}$ and $y_\alpha \in S_X$ such that $y_\alpha^*(y_\beta) = y_\alpha^*(u_\beta) = 0$ for all $\beta < \alpha$, and $y_\alpha^*(y_\alpha) > s$.

Let M be a constant with $K - 1 > M > 2/s$. Since $(u_\alpha^* + My_\alpha^*)(u_\alpha) + (u_\alpha^* - My_\alpha^*)(u_\alpha) = 2$, we may choose a set of signs $\epsilon_\alpha = \pm 1$ such that $g_\alpha = u_\alpha^* + \epsilon_\alpha My_\alpha^*$ satisfies $|g_\alpha(u_\alpha)| \geq 1$ for all α. Now choose constants r and D such that

$$1 > r > (M+1)/K \qquad \text{and} \qquad D > (2 + M(1-s))/(1-r).$$

We may now select a subset A of γ and a set of modulus-1 scalars, $\{\delta_\alpha; \alpha \in A\}$, by transfinite induction, satisfying the condition that $0 \in A$ and, for $\alpha > 0$, $\alpha \in A$ if and only if $|g_\beta(u_\alpha)| \leq r$ for all $\beta \in A$ with $\beta < \alpha$. We choose δ_α such that ϵ_α and $\delta_\alpha g_\alpha(u_\alpha)$ have the same sign. Let $z_\alpha = y_\alpha + \delta_\alpha Du_\alpha$.

Now, if $\alpha, \beta \in A$ and $\beta < \alpha$, then

$$|g_\alpha(z_\alpha)| = |u_\alpha^*(y_\alpha) + \epsilon_\alpha My_\alpha^*(y_\alpha) + \delta_\alpha Dg_\alpha(u_\alpha)| \geq Ms + D - 1$$

since $y_\alpha^*(y_\alpha) > s$, $|g_\alpha(u_\alpha)| \geq 1$, and ϵ_α and $\delta_\alpha g_\alpha(u_\alpha)$ have the same sign;

$$|g_\beta(z_\alpha)| = |u_\beta^*(y_\alpha) + \epsilon_\beta My_\beta^*(y_\alpha) + \delta_\alpha Dg_\beta(u_\alpha)| \leq 1 + M + rD$$

since $|g_\beta(u_\alpha)| \leq r$; and

$$|g_\alpha(z_\beta)| = |u_\alpha^*(y_\beta + \delta_\beta Du_\beta)| \leq 1 + D$$

since $y_\alpha^*(y_\beta) = y_\alpha^*(u_\beta) = 0$.

Moreover, $Ms + D - 1 > 1 + M + rD$ by the choice of D, and $Ms + D - 1 > 1 + D$ by the choice of M. Now let $f_\alpha = g_\alpha/r$ for $\alpha \in A$, and define $\|| \cdot \||$ by $\||x\|| = \sup_{\alpha \in A} |f_\alpha(x)|$ for $x \in X$. Then $\||x\|| \leq (1+M)\|x\|/r$ and, given $\alpha < \gamma$, either $\alpha \in A$, so that $\||u_\alpha\|| \geq 1/r$, or else $\alpha \notin A$, so that

$$|f_\beta(u_\alpha)| > 1 \qquad \text{for some } \beta \in A, \beta < \alpha.$$

Because $\{u_\alpha; \alpha < \gamma\}$ is dense in S_X, it follows that $\|| \cdot \|| \geq \| \cdot \|$, and consequently $\| \cdot \| \leq \|| \cdot \|| \leq K\| \cdot \|$. Moreover, letting $x_\alpha = z_\alpha/f_\alpha(z_\alpha)$, we see that $(X, \|| \cdot \||)$ has property β with the system $\{x_\alpha; f_\alpha\}_{\alpha \in A}$, where

$$\lambda = \max\{1 + D, 1 + M + rD\}/(Ms + D - 1)$$

as desired. $\qquad\qquad\qquad\qquad\qquad\qquad\qquad\qquad\qquad\qquad\qquad\qquad\square$

The main existence theorem we will present on property α is as follows.

Theorem 8.30 (Godun and Troyanski [GoTr93]). *If X has a biorthogonal system with cardinality equal to* dens X, *then for each $\epsilon \in (0,1)$, X admits an equivalent norm $|\cdot|$ such that $(X, |\cdot|)$ has property (α, ϵ).*

Proof. Let $Z = \overline{\text{span}}\{x_i\}_{i \in I}$. Then Z has a fundamental biorthogonal system. Therefore, Z has a quotient space $E = Z/Y$ such that dens $E = $ dens Z, and E has a separable projectional decomposition (see Theorem 4.15), i.e., there exists a transfinite set of projections $P_\alpha : E \to E$, $\alpha < \alpha_0$, such that

(i) $\|P_\alpha\| = 1$,
(ii) $P_\alpha P_\beta = P_{\min(\alpha,\beta)}$,
(iii) the space $E_\alpha = Q_\alpha E$ is infinite-dimensional and separable for every $\alpha < \alpha_0$, where $Q_\alpha = P_{\alpha+1} - P_\alpha$, and
(iv) $\overline{\text{span}} \left(\bigcup_{\alpha < \alpha_0} E_\alpha \right) = E$.

As in the proof of Lemma 4.17, we get that, for any $\eta > 0$, there is a $(1 + \eta)$-bounded fundamental biorthogonal system $\{e_{\alpha,n}; e^*_{\alpha,n}\}_{n \in \mathbb{N}}$ such that the system $\{e_{\alpha,n}\}_{n \in \mathbb{N}}$ is not equivalent to the unit vector basis of ℓ_1. Now let $f_{\alpha,n} = Q^*_\alpha e^*_{\alpha,n}$. Then $\{e_{\alpha,n}; f_{\alpha,n}\}_{\alpha < \alpha_0, n \in \mathbb{N}}$ is a $(2 + 2\eta)$-bounded fundamental biorthogonal system in E.

Let $T : X \to E$ be the quotient map. According to Godun's lifting theorem ([Godu83b]; see Lemma 4.18), there exists a system $\{u_{\alpha,n}\}_{\alpha < \alpha_0, n \in \mathbb{N}}$ in Y such that $Tu_{\alpha,n} = e_{\alpha,n}$, $\{u_{\alpha,n}\}_{n \in \mathbb{N}}$ is not equivalent to the unit vector basis of ℓ_1 for each $\alpha < \alpha_0$ and $\{u_{\alpha,n}; T^* f_{\alpha,n}\}_{\alpha < \alpha_0, n \in \mathbb{N}}$ is a $(4 + 5\eta)$-bounded fundamental biorthogonal system on Y.

Using the Hahn-Banach theorem, we can find norm-preserving extensions of the dual functionals, and relabel this as the biorthogonal system $\{y_{i,n}; y^*_{i,n}\}_{(i,n) \in I \times \mathbb{N}}$, where $|I| = $ dens X such that

$$\|y_{i,n}\| = 1, \quad \|y^*_{i,n}\| \le c, \quad (i,n) \in I \times \mathbb{N}, \tag{8.6}$$

where $c = 4 + 5\eta$ and, for all $i \in I$,

$$\{y_{i,n}\}_{n \in \mathbb{N}} \text{ is not equivalent to the usual basis of } \ell_1. \tag{8.7}$$

Using (8.7), it follows that, furnishing $y_{i,n}$ with the proper sign, we can assume that, for any $i \in I$, there exists a sequence of numbers $\{c_{k,i,n}\}$ such that

$$c_{k,i,n} \ge 0, \quad \sum_n c_{k,i,n} = 1, \quad \lim_{k \to \infty} \left\| \sum_n c_{k,i,n} y_{i,n} \right\| = 0. \tag{8.8}$$

Let $\epsilon > 0$ and $\delta = \epsilon/c(1 + \epsilon)$, denote $x_{i,n} = \delta z_i + y_{i,n}$, $(i,n) \in I \times \mathbb{N}$, and let

$$V = \overline{\text{conv}}(\{\pm x_{i,n}\}_{(i,n) \in I \times \mathbb{N}}),$$

where $\{z_i\}_{i \in I}$ is a dense subset of B_X. Clearly $V \subset (1 + \delta)B_X$ and, from (8.8), $\delta U_X \subset V$, so that the Minkowski functional of V is an equivalent norm $|\cdot|$ on X. Note that

$$y^*_{i,n}(x_{i,n}) \ge 1 - c\delta > 0, \quad (i,n) \in I \times \mathbb{N}. \tag{8.9}$$

Therefore, the functionals $x_{i,n}^* = y_{i,n}^*/y_{i,n}^*(x_{i,n})$, $(i,n) \in I \times \mathbb{N}$ are well defined, and

$$x_{i,n}^*(x_{i,n}) = 1, \quad (i,n) \in I \times \mathbb{N}. \tag{8.10}$$

If $(i,n) \neq (j,m)$, then (8.9) implies

$$|x_{i,n}^*(x_{j,n})| = |y_{i,n}^*(\delta z_j + y_{j,n})|/y_{i,n}^*(x_{i,n}) \leq \epsilon. \tag{8.11}$$

Obviously, $|x_{i,n}| \leq 1$ for all $(i,n) \in I \times \mathbb{N}$, and using (8.10) we obtain that $|x_{i,n}^*| \geq 1$ for all $(i,n) \in I \times \mathbb{N}$. Because V is the closed convex hull of $\{\pm x_{i,n}\}_{(i,n)\in I\times\mathbb{N}}$, we have that $|f| = \sup_{i,n} |f(x_{i,n})|$ for every $f \in X^*$. According to (8.10) and (8.11), it follows that $|x_{i,n}^*| \leq 1$, which in turn implies $|x_{i,n}| = |x_{i,n}^*| = 1$, and we conclude that $(X, |\cdot|)$ has property (α, ϵ). $\qquad \square$

The technique of norm-attaining operators was applied by Lindenstrauss in [Lind63] to obtain pioneering results on the Fréchet differentiability of convex functions in infinite-dimensional spaces.

We close this section by stating a renorming result for strong property β. Godun and Troyanski [GoTr93] proved that *if X has a fundamental biorthogonal system and* dens $X =$ dens X^*, *then X admits an equivalent norm $|\cdot|$ such that $(X, |\cdot|)$ has strong property β.* Let us note that a precursor to this was given in [FiSc89] in the case where X is a separable Asplund space, and, in particular, these results showed that *there are nonreflexive spaces that can be renormed so that $\mathcal{R}(X) < 1$* (see Theorem 8.27(c)).

8.4 Mazur Intersection Properties

We begin this section with a characterization of the Mazur intersection property that was shown by Giles, Gregory, and Sims. Using that characterization, we prove the following results, all of which are due to Jiménez-Sevilla and Moreno. A Banach space X can be renormed by a norm with the Mazur intersection property whenever X^* admits a fundamental biorthogonal system such that the coefficient functionals belong to X; consequently, there are non-Asplund spaces that can be renormed to have the Mazur intersection property. Also, under the additional axiom ♣, there are Asplund spaces that cannot be renormed to have the Mazur intersection property. The section concludes with some results of Valdivia and Rychtář concerning DENS Asplund spaces.

Definition 8.31. *A Banach space is said to have the* Mazur intersection property *if every bounded closed convex set can be represented as an intersection of balls.*

This property depends on the particular norm and is characterized as follows.

Theorem 8.32 (Giles, Gregory, and Sims [GGS78]). *Let X be a Banach space. Then the following are equivalent.*

(a) X has the Mazur intersection property.
(b) For every $\epsilon > 0$, there is a norm dense set in S_{X^*} each of whose points is in a w^*-slice of B_{X^*} having diameter $\leq \epsilon$.
(c) The w^*-denting points of B_{X^*} are norm dense in S_{X^*}.

The following lemma will be used in the proof.

Lemma 8.33 (Mazur [Mazu33]). Let C be a closed bounded convex set in a Banach space X, and suppose $0 \notin C$. If ϕ is a w^*-denting point of B_{X^*} and $\inf_C \phi > 0$, then there is a ball B such that $C \subset B$ and $0 \notin B$.

Proof. Let $\alpha > 0$ be such that $\inf_C \phi \geq \alpha$ and $x \in S_X$, $\delta > 0$ such that $\phi \in S(B_{X^*}, x, \delta)$ and the diameter of $S(B_{X^*}, x, \delta) \leq \epsilon$, where $\epsilon = \frac{\alpha}{3}$. For each $n > 1$, let $D_n = B_{(n-1)\epsilon}(n\epsilon x)$. If $u \in D_n$, then $\|u\| \geq \epsilon > 0$ so $0 \notin D_n$. Suppose that $C \not\subset D_n$ for each $n > 1$. Thus we can choose $y_n \in C \setminus D_n$ and we fix $g_n \in S_{X^*}$ such that

$$g_n(n\epsilon x - y_n) = \|n\epsilon x - y_n\| \geq (n-1)\epsilon.$$

Thus

$$g_n(n\epsilon x) \geq (n-1)\epsilon + g_n(y_n).$$

Now $g_n(y_n) \geq -K$, where $C \subset B_K$ for some $K > 0$. Therefore

$$g_n(x) \geq \frac{n-1}{n} - \frac{K}{n\epsilon} \to 1 \quad \text{as} \quad n \to \infty.$$

Consequently, $g_n \in S(B_{X^*}, x, \delta)$ for large n. Moreover,

$$\begin{aligned}(\phi - g_n)(y_n) &= \phi(y_n) + g_n(n\epsilon x - y_n) - \epsilon n g_n(x) \\ &\geq \alpha + (n-1)\epsilon - \epsilon n g_n(x) \\ &= \alpha - \epsilon + n\epsilon(1 - g_n(x)) \\ &\geq \alpha - \epsilon = 2\epsilon.\end{aligned}$$

This contradicts the fact that the diameter of $S(B_{X^*}, x, \delta) \leq \epsilon$. Therefore $C \subset D_n = B_{(n-1)\epsilon}(n\epsilon x)$ for some $n > 1$ and $0 \notin B_{(n-1)\epsilon}(n\epsilon x)$. $\square$

We now prove Theorem 8.32.

Proof. (a)$\Rightarrow$(b) Let $f \in S_{X^*}$, and let $\epsilon \in (0, 1)$. It suffices to show that there is a w^*-slice $S(B_{X^*}, u, \alpha)$ with diameter $\leq \epsilon$ such that $g \in S(B_{X^*}, u, \alpha)$ implies $\|f - g\| < \epsilon$. Let $K = f^{-1}(0) \cap B_X$. Choose $x_0 \in B_X$ such that $f(x_0) > 1 - \frac{\epsilon}{6}$. Then $x_0 \notin \overline{K + B_{1-\frac{\epsilon}{6}}}$. Thus, using (a), we can find $z_0 \in X$ and $r > 0$ such that $x_0 \notin B_r(z_0)$, while $K + B_{1-\frac{\epsilon}{6}} \subset B_r(z_0)$. Now $\|x_0 - z_0\| = r + 2\delta$ for some $\delta > 0$. We let $u = (x_0 - z_0)/\|x_0 - z_0\|$, $\alpha = \delta/\|x_0 - z_0\|$, and suppose $g \in S(B_{X^*}, u, \alpha)$. Then $g(x_0 - z_0) \geq \|x - z\| - \delta > r$. Hence

$$K + B_{1-\frac{\epsilon}{6}} \subset B_r(z_0) \subset \{v; g(v) \leq r + g(z_0)\} \quad \text{and} \quad g(x_0) > r + g(z_0),$$

which means $1 - \frac{\epsilon}{6} \leq r + g(z_0) < g(x_0) \leq 1$. Consequently,

$$|g(x_0) - f(x_0)| < \frac{\epsilon}{6}. \tag{8.12}$$

Moreover, if $k \in K$, then $g(k) + 1 - \frac{\epsilon}{6} \leq r + g(z_0) < g(x_0) \leq 1$. Thus $g(k) \leq \frac{\epsilon}{6}$, and because K is symmetric, this means

$$|g(k)| \leq \frac{\epsilon}{6} \quad \text{for all} \quad k \in K. \tag{8.13}$$

Now, if $\|g\| = 1$, applying (8.12) and (8.13), the parallel hyperplane lemma asserts that $\|f - g\| < \frac{\epsilon}{3}$. Now any $g \in S(B_{X^*}, u, \alpha)$ has the property that $\|g\| \geq g(x_0) \geq 1 - \frac{\epsilon}{6}$, and then for any such g, $\|f - g\| < \frac{\epsilon}{2}$, and from that we also obtain that the diameter of $S(B_{X^*}, u, \alpha) \leq \epsilon$.

(b)$\Rightarrow$(c) The sets $O_n = \{g \in S_{X^*}; g \text{ is in a } w^*\text{-slice of } B_{X^*} \text{ with diameter} < 1/n\}$ are relatively open and norm dense in S_{X^*}. Their intersection is norm dense in S_{X^*}, and each point in the intersection is a w^*-denting point.

(c)$\Rightarrow$(a) By translation, it suffices to show that given a bounded closed convex set C with $0 \notin C$, there is a ball B for which $C \subset B$ but $0 \notin B$. Because the w^*-denting points are norm dense in S_{X^*}, there is a w^*-denting point ϕ with $\inf_C \phi > 0$. The result now follows from Lemma 8.33.   □

Thus, for example, $\mathbb{R}^n$ endowed with the p-norm for $1 < p < \infty$ possesses the Mazur intersection property, where if $p = 1$ or $p = \infty$ it does not have the Mazur intersection property. We now list several consequences of Theorem 8.32.

Corollary 8.34 (Mazur [Mazu33]). *Suppose the norm on X is Fréchet differentiable. Then X has the Mazur intersection property.*

Proof. Use Theorem 8.32, Šmulyan's theorem, and the Bishop-Phelps theorem.   □

Corollary 8.35. *Suppose X can be renormed to have the Mazur intersection property. Then $\operatorname{dens} X = \operatorname{dens} X^*$.*

Proof. Use Theorem 8.32 to map a norm-dense set in S_X of the appropriate cardinality onto a norm-dense set in S_{X^*}.   □

Corollary 8.36 ([JiMo97]). *Suppose that X^* is not separable and that X has the Mazur intersection property. Then X has an uncountable set $\{x_\alpha\}_{\alpha \in A}$ such that $x_\alpha \notin \overline{\operatorname{conv}}(\{x_\beta\}_{\beta \in A, \beta \neq \alpha})$ for all $\alpha \in A$.*

Proof. Choose an uncountable set $\{\phi_\alpha\}_{\alpha \in A} \subset S_{X^*}$ that are w^*-denting points of B_{X^*} and $\|\phi_\alpha - \phi_\beta\| > \delta > 0$ when $\alpha \neq \beta$. For each $\beta \in A$, choose $\epsilon_\beta \in \{1/n; n \in \mathbb{N}\}$ and $x_\beta \in S_X$ such that $\phi_\alpha \notin S(B_{X^*}, x_\beta, 1 - \epsilon_\beta)$ if $\alpha \neq \beta$. Let $A_n = \{\alpha \in A; \epsilon_\alpha = 1/n\}$. For some n_0, A_{n_0} is uncountable, and we have $\phi_\alpha \notin S(B_{X^*}, x_\beta, 1 - 1/n_0)$ whenever $\alpha, \beta \in A_{n_0}$ and $\alpha \neq \beta$. Then, for $\beta \in A_{n_0}$, $\phi_\beta(x_\beta) > 1 - 1/n_0$, while $\phi_\beta(x_\alpha) < 1 - 1/n_0$ for all $\alpha \in A_{n_0}$, $\alpha \neq \beta$.   □

The previous corollary immediately yields the following one.

Corollary 8.37 ([JiMo97]). (♣) *The space $C(K)$, with K the compact space defined in Theorem 4.41, is an Asplund space that cannot be renormed to have the Mazur intersection property.*

Although the Mazur intersection property does not characterize Asplund spaces, the following classical theorem shows the relation in separable spaces.

Theorem 8.38 (Phelps [Phel60]). *For a separable Banach space X, X can be renormed to have the Mazur intersection property if and only if X^* is separable.*

Proof. If X^* is separable, then X admits an equivalent Fréchet differentiable norm (see, e.g., [Fa~01, Thm. 11.23]), and thus, by Corollary 8.34, X has the Mazur intersection property. The converse implication follows from Corollary 8.35 or Corollary 8.36.   □

Note that there are WCG spaces that are not Asplund, while they can be renormed to have the Mazur intersection property; see Example 8.45 below.

Lemma 8.39 (Rychtář [Rych04]). *Let E be a Banach space such that dens $E^* = \Gamma$ and $Y \subset E$ be a closed subspace. Assume that there is a fundamental biorthogonal system $\{f_\gamma; x_\gamma\}_{\gamma \in \Gamma} \subset Y^* \times Y$. Then there is a fundamental biorthogonal system $\{q_\gamma; x_\gamma\}_{\gamma \in \Gamma} \subset E^* \times E$.*

Proof. First, by relabeling and rescaling, we may have a fundamental system $\{f_\gamma^n; x_\gamma^n\}_{\gamma \in \Gamma, n \in \mathbb{N}} \subset Y^* \times Y$ such that, for every $\gamma \in \Gamma, \lim_n \|f_\gamma^n\| = 0$. By the Hahn-Banach theorem, consider $f_\gamma^n \in E^*$. Let $\{g_\gamma\}_{\gamma \in \Gamma}$ be a dense set of $B_{E^*} \cap Y^\perp$.

Next, we claim that $A = \{g_\gamma + f_\gamma^n\}_{\gamma \in \Gamma, n \in \mathbb{N}}$ is linearly dense in E^*. Indeed, let $G \in E^{**}$ be such that $G(f) = 0$ for every $f \in A$. Then $G(g_\gamma) = \lim_n G(g_\gamma + f_\gamma^n) = 0$, and thus $G \in (Y^\perp)^\perp = Y^{**}$. Hence $G = 0$, as $\{f_\gamma^n\}_{\gamma \in \Gamma, n \in \mathbb{N}}$ is linearly dense in Y^*. It follows that $\{g_\gamma + f_\gamma^n; x_\gamma^n\}_{\gamma \in \Gamma, n \in \mathbb{N}} \subset E^* \times E$ is a fundamental biorthogonal system.   □

Corollary 8.40 ([JiMo97]). *Every Banach space X can be embedded into a Banach space Z with a biorthogonal system $\{z_i; z_i^*\}_{i \in I}$ such that $\mathrm{span}(\{z_i^*\}_{i \in I})$ is norm dense in Z^*.*

Proof. Let $\Gamma = \mathrm{dens}\, X^*$; apply Lemma 8.39 to $E = X \oplus \ell_2(\Gamma)$.   □

We shall not prove the following result.

Theorem 8.41 (Valdivia [Vald93b]). *Let X be a Banach space. Let Y be a WLD subspace such that $\mathrm{dens}\, Y \geq w^*\text{-}\mathrm{dens}\, X^*$. Then there exists a total biorthogonal system $\{y_\gamma; y_\gamma^*\}$ in $X \times X^*$ such that $\overline{\mathrm{span}}\{y_\gamma; \gamma \in \Gamma\} = Y$.*

The following result shows how to construct norms on spaces with the Mazur intersection property using appropriate biorthogonal systems.

Theorem 8.42 ([JiMo97]). *Let $(X^*, \| \cdot \|^*)$ be a dual Banach space with biorthogonal system $\{x_i; f_i\}_{i \in I} \subset X^* \times X$, and let $X_0 := \overline{\operatorname{span}}(\{x_i\}_{i \in I})$. Then X^* admits an equivalent dual norm $|\cdot|^*$ that is locally uniformly rotund at the points of X_0. In particular, if X_0 is dense in X^*, then X can be renormed to have the Mazur intersection property.*

Proof. We follow [JiMo97, Lemma 2.3]. By normalizing, we assume $\|f_i\| = 1$ for each $i \in I$. Let $\Delta = \{0\} \cup \mathbb{N} \cup I$. Define a map $T : X_0 \to \ell_\infty(\Delta)$ by

$$
T(x)(\delta) = \begin{cases} \|x\|^* & \text{if } \delta = 0, \\ 2^{-n} G_n(x) & \text{if } \delta = n \in \mathbb{N}, \\ f_i(x) & \text{if } i \in I, \end{cases}
$$

for every $x \in X^*$ and $\delta \in \Delta$, where

$$
F_A(x) = \sum_{i \in A} |f_i(x)|,
$$
$$
E_A(x) = \operatorname{dist}(x, \operatorname{span}(\{x_i\}_{i \in A})) \qquad A \subset I, |A| < \infty,
$$

and

$$
G_n(x) = \sup_{|A| \leq n} \{E_A(x) + n F_A(x)\}.
$$

Then $T(X^*) \subset \ell_\infty(\Delta)$ and $T(X_0) \subset c_0(\Delta)$. On the other hand, because $2^{-n}(1 + n^2) \leq 2$ for each $n \in \mathbb{N}$, we have $\|x\|^* \leq \|T(x)\|_\infty \leq 2\|x\|^*$. Notice also that the map T_δ is w^*-lower semicontinuous for every $\delta \in \Delta$.

Let p be the Day norm on $\ell_\infty(\Delta)$, and consider in X^* the map $n(x) := p(T(x))$, $x \in X^*$. It follows that $n(\cdot)$ is an equivalent norm on X^*, and it is given by

$$
n(x)^2 = \sup \left\{ \sum_{i=1}^n \frac{|T_{\delta_i}(x)|^2}{4^i}; (\delta_1, \delta_2, \ldots, \delta_n) \subset \Delta, \delta_i \neq \delta_j, n \in \mathbb{N} \right\}.
$$

Now, $n(\cdot)$ is a dual norm because it is w^*-lower semicontinuous, and we will denote it as $|\cdot|^*$. The norm p defined on $\ell_\infty(\Delta)$ is locally uniformly rotund at points of $c_0(\Delta)$ (see the proof of Theorem 3.48).

We now prove that $|\cdot|^*$ is locally uniformly rotund at the points of X_0. Indeed, let $x \in X_0$ and $\{x_m\}_{m=1}^\infty \subset X^*$ be such that $\lim_m |x_m|^* = |x|^* = 1$ and $\lim_m |x_m + x|^* = 2$. Then $\lim_m p(T(x_m)) = p(T(x)) = 1$ and $\lim_m P(T(x_m) + T(x)) = 2$. By the locally uniform rotundity of p on $c_0(\Delta)$, $\lim_m P(T(x_m) - T(x)) = 0$, and consequently

$$
\lim_m \|T(x_m) - T(x)\|_\infty = 0. \tag{8.14}
$$

Now let $A := \{i \in I; f_i(x) \neq 0\}$ and $M := \max\{\|x_i\|; i \in A\}$, and choose $N \geq \max\{M, |A|\}$. For every $B \subset I$ with $|B| < \infty$, we have

$$E_B(x) + NF_B(x) \le M \sum_{i \in A \setminus B} |f_i(x)| + N \sum_{i \in B \cap A} |f_i(x)|$$

$$\le N \sum_{i \in A} |f_i(x)| = NF_A(x).$$

This shows that $G_N(x) = NF_A(x)$. According to (8.14), there is $m_0 \in \mathbb{N}$ such that $|G_N(x_m) - G_N(x)| < \epsilon$ and $|NF_A(x_m) - NF_A(x)| < \epsilon$ for every $m \ge m_0$. Consequently,

$$G_N(x_m) - NF_A(x_m) \le 2\epsilon + G_N(x) - NF_A(x) = 2\epsilon$$

and thus

$$E_A(x_m) \le G_N(x_m) - NF_A(x_m) \le 2\epsilon.$$

This implies $\lim_m \text{dist}\,(x_m, \text{span}(\{x_i\}_{i \in A})) = 0$, and therefore we can choose a sequence $\{z_m\}_{m \in \mathbb{N}} \subset \text{span}(\{x_i\}_{i \in A})$ such that $\lim_m \|x_m - z_m\|^* = 0$. Using (8.14), we obtain

$$\lim_m \max_{i \in A} |f_i(z_m) - f_i(x)| = \lim_m \max_{i \in A} |f_i(x_m) - f_i(x)| = 0,$$

and consequently

$$\lim_m \|x_m - x\|^* = \lim_m \|z_m - x\|^* = 0,$$

which completes the proof of the local uniform rotundity properties. The "in particular" part of the theorem now follows from Theorem 8.32.   □

Corollary 8.43 ([JiMo97]). *Every Banach space can be isomorphically embedded in a Banach space with the Mazur intersection property.*

Proof. The corollary is a consequence of the previous two results.   □

Corollary 8.44. *The Banach space $C[0, \omega_1]$ has a renorming with the Mazur intersection property.*

Proof. This follows from Lemma 8.39 and Theorem 8.42. Note that this also follows because $C[0, \omega_1]$ admits a Fréchet differentiable norm.   □

From this, we immediately obtain the following example.

Example 8.45. The Banach space $\ell_1 \times \ell_2(c)$ can be renormed to have the Mazur intersection property. In particular, JL$_2^*$ can be renormed to have the Mazur intersection property, and WCG spaces with the Mazur intersection property need not be Asplund spaces.

The next results concern the existence of "large and nice" subspaces in Asplund spaces, which incidentally have implications for the Mazur intersection property.

Let us recall that a Banach space X is called DENS whenever $\text{dens}\, X = w^*\text{-}\text{dens}\, X^*$ (see Definition 5.39).

Theorem 8.46 (Rychtář [Rych04]). *Let E be a Banach space. Then the following are equivalent:*

(a) *There is a subspace $Y \subset E$ with a shrinking M-basis $\{x_\gamma; f_\gamma\}_{\gamma \in \Gamma}$.*
(b) *There is an Asplund space $X \subset E$ that is a DENS space and $\operatorname{dens} X = \operatorname{card} \Gamma$.*
(c) *There is a subspace $Z \subset E$ that is a DENS space, $\operatorname{dens} Z = \operatorname{card} \Gamma$, and admits a Fréchet smooth norm.*

Moreover, if one of the above occurs with $\operatorname{card} \Gamma = \operatorname{dens} E^*$, then

(d) *E can be renormed to have the Mazur intersection property.*

Proof. Let us prove first the implication (a) $\Rightarrow$ (d) (under the extra assumption that $Y \subset E$ is a subspace with a shrinking M-basis $\{y_\gamma; y_\gamma^*\}_{\gamma \in \Gamma}$, where $\operatorname{card} \Gamma = \operatorname{dens} E^*$). If this is the case, Lemma 8.39 implies that $E^* \times E$ has a fundamental biorthogonal system, and so, by the Remark following Proposition 8.42, E can be renormed to have the Mazur intersection property.

We now turn to the proof of the equivalence of (a), (b), and (c).

(a)$\Rightarrow$(c) If Y has a shrinking M-basis $\{y_\gamma; y_\gamma^*\}_{\gamma \in \Gamma}$, then Y admits a Fréchet differentiable norm and is weakly compactly generated; see, e.g., [Fa~01, Theorem 11.23]. Every weakly compactly generated Banach space is DENS (see Proposition 5.40), so $\operatorname{dens} Y = w^*\text{-}\operatorname{dens} Y^*$. Moreover, $\operatorname{dens} Y^* = \operatorname{card} \Gamma$, as $\{y_\gamma^*; y_\gamma\}_{\gamma \in \Gamma}$ is a fundamental system in $Y^* \times Y$ (see Exercise 4.1).

(c)$\Rightarrow$(b) Every space with a Fréchet smooth norm is an Asplund space (see, e.g., [DGZ93a, Theorem 5.3]).

(b)$\Rightarrow$(a) is a consequence of the following result, which says something a little bit more precise. □

Theorem 8.47 (Valdivia [Vald96]). *Let X be a DENS Asplund space, and let $\operatorname{dens} X = \operatorname{card} \Gamma$ for some infinite set Γ. Then X contains a subspace Y with a shrinking M-basis $\{y_\gamma; y_\gamma^*\}_{\gamma \in \Gamma}$ in $Y \times Y^*$.*

Proof (J. Rychtář). The proof goes in the spirit of [LiTz77, Theorem 1.a.5] and [Gode95]. We will use the concept of the Jayne-Rogers selector; see [DGZ93a, Chapter 1]. The Jayne-Rogers selection map $\mathcal{D}^X$ on an Asplund space X is a multivalued map that satisfies the following:

1. $\mathcal{D}^X(x) = \{D_n^X(x); n \in \mathbb{N}\} \cup D_\infty^X(x) \subset X^*$,
2. D_n^X, for $n \in \mathbb{N}$, are continuous functions from X to X^*,
3. $D_\infty^X(x) = \lim_{n \to \infty} D_n^X(x)$ for every $x \in X$,
4. $D_\infty^X(x)(x) = \|x\|^2 = \|D_\infty^X(x)\|^2$,
5. $X^* = \overline{\operatorname{span}} \mathcal{D}^X(X)$.

Such a selector exists by [DGZ93a, Theorem 1.5.2].

Let μ be the first ordinal of cardinal $\operatorname{card} \Gamma$. In order to construct the Y sought, we will define, by transfinite induction, vectors $y_{\alpha+1} \in X$, subspaces $Y_\alpha \subset X$, and subsets $F_\alpha \subset X^*$ for all $\alpha < \mu$. Put $Y_0 := \{y_0\}$, where $y_0 := 0$, and set $F_0 = \{0\}$. Pick an arbitrary nonzero y_1 in X. Then put $Y_1 := \operatorname{span}\{y_1\}$, and let $F_1 := \{\mathcal{D}^X(y); y \in Y_1\}$. Assume that for some ordinal $1 \leq \alpha < \mu$ we already defined Y_α with $\mathrm{dens}\, Y_\alpha \leq \mathrm{card}\,\alpha$. Then put $F_\alpha := D^X(Y_\alpha)$. From the $\|\cdot\|\text{-}\|\cdot\|$-continuity of D_n^X, for all $n \in \mathbb{N}$, we obtain

$$\mathrm{dens}\, F_\alpha \leq \aleph_0\, \mathrm{dens}\, Y_\alpha \leq \aleph_0\, \mathrm{card}\,\alpha = \mathrm{card}\,\alpha < \mathrm{card}\,\Gamma,$$

so F_α is not w^*-dense in X^*. We can then find $y_{\alpha+1} \neq 0$ in $(F_\alpha)_\perp \subset X$. Notice that $y_{\alpha+1} \notin Y_\alpha$. Put $Y_{\alpha+1} := \mathrm{span}\{Y_\alpha \cup \{y_{\alpha+1}\}\}$, a closed subspace of X. Set $F_{\alpha+1} := D^X(Y_{\alpha+1})$. If $\alpha \leq \mu$ is a limit ordinal, put $Y_\alpha := \overline{\bigcup_{\beta<\alpha} Y_\beta}$ and $F_\alpha := D^X(Y_\alpha)$.

We carry on this inductive construction for $0 \leq \alpha \leq \mu$. Notice that $Y_\alpha = \overline{\mathrm{span}}\{y_{\beta+1}\}_{\beta<\alpha} = \overline{\bigcup_{\beta<\alpha} Y_{\beta+1}}$ for $1 \leq \alpha \leq \mu$. Put $Y := Y_\mu = \overline{\mathrm{span}}\{y_{\beta+1}\}_{\beta<\mu}$. From now on, we shall work in $\langle Y, Y^* \rangle$, writing F_α for the set of restrictions to Y of elements in $F_\alpha \subset X^*$.

Claim. $Y_\alpha \oplus (F_\alpha)_\perp = Y$, and $P_\alpha : Y \to Y_\alpha$, the canonical projection associated to the decomposition, has norm 1.

In order to prove the claim, first use Lemma 3.33 to check that $Y_\alpha \oplus (F_\alpha)_\perp$ is a topological direct sum. Given $y^* \in Y_\alpha^\perp \cap \overline{F_\alpha}^{w^*}$ and $\beta < \mu$, if $\beta < \alpha$, then $y_{\beta+1} \in Y_\alpha$, so $\langle y_{\beta+1}, y^* \rangle = 0$; otherwise, $y_{\beta+1} \in (F_\alpha)_\perp$, so again $\langle y_{\beta+1}, y^* \rangle = 0$ and we conclude that $y^* = 0$. An application of Lemma 3.34 finishes the proof of the claim. Observe that, in particular, F_μ is w^*-dense in Y^*.

Now $(P_{\alpha+1}^* - P_\alpha^*)Y^* = \big((P_{\alpha+1} - P_\alpha)Y\big)^*$, so we can choose $y_{\alpha+1}^* \in (P_{\alpha+1}^* - P_\alpha^*)Y^*$ with $\|y_{\alpha+1}^*\| = 1$ for $\alpha < \mu$. Obviously, the system $\{y_{\alpha+1}; y_{\alpha+1}^*\}_{\alpha<\mu}$ is biorthogonal and fundamental in $Y \times Y^*$. We shall prove that it is a shrinking M-basis. To that end, it will be enough to prove the following claim.

Claim. $(P_\alpha^*)_{\alpha\leq\mu}$ is a shrinking family of projections (i.e., for any limit ordinal $\alpha \leq \mu$, $P_\alpha^* Y^* = \overline{\bigcup_{\beta<\alpha} P_\beta^* Y^*}$).

To prove the claim for $\alpha \leq \mu$ a limit ordinal, put $Z := P_\alpha Y$. Then $Z^* = P_\alpha^* Y^*$. Z is an Asplund space, so $Z^* = \overline{\mathrm{span}}\{D^Z(Z)\}$, where D^Z is the restriction of D^X to Z (so D^Z is the Jayne-Rogers selection map for Z). Take $z^* \in Z^*$. Fix $\varepsilon > 0$. We can then find n and m in $\mathbb{N}$ with $n \leq m$, $z_1, \ldots, z_m$ in Z, $k_1, \ldots, k_n$ in $\mathbb{N}$, and $\lambda_1, \ldots, \lambda_m$ in $\mathbb{R}$ such that

$$\left\| x^* - \left(\sum_{i=1}^n \lambda_i D_{k_i}^Z(z_i) + \sum_{i=n+1}^m \lambda_i D_\infty^Z(z_i) \right) \right\| < \varepsilon.$$

Because $D_p^Z(z_i) \xrightarrow{\|\cdot\|} D_\infty^Z(z_i)$, when $p \to \infty$ for every $i = n+1, \ldots, m$, we can find $k_{n+1}, \ldots, k_m$ in $\mathbb{N}$ such that

$$\left\| x^* - \sum_{i=1}^m \lambda_i D_{k_i}^Z(z_i) \right\| < \varepsilon.$$

We have $P_\alpha Y = \overline{\bigcup_{\beta<\alpha} P_\beta(Y)}$, so we can find $\beta < \alpha$ and z_i' in $P_\beta Y$, $i = 1, 2, \ldots, m$, such that

$$\left\| z^* - \sum_{i=1}^{m} \lambda_i D_{k_i}^Z (z_i') \right\| < \varepsilon.$$

Recalling that $P_\beta^* Y^* = \overline{F_\beta}^{w^*} \supset D^Z Y_\beta$ and that $\varepsilon > 0$ was taken as arbitrary, we get the conclusion. $\qquad \square$

Corollary 8.48 (Valdivia [Vald96]). *Let X be a DENS Asplund space and let* $\operatorname{dens} X := \operatorname{card} \Gamma$ *for some infinite set Γ. Let Y be a closed subspace of X such that w^*-$\operatorname{dens} Y^* = \operatorname{dens} X$. Then there exists a fundamental biorthogonal system $\{x_\gamma^*; z_\gamma\}_{\gamma \in \Gamma}$ in $X^* \times X$ such that $z_\gamma \in Y$ for all $\gamma \in \Gamma$.*

Proof. We have

$$\operatorname{dens} Y \leq \operatorname{dens} X = w^*\text{-}\operatorname{dens} Y^* \leq \operatorname{dens} Y,$$

so Y is a DENS Asplund space and $\operatorname{dens} Y = \operatorname{dens} X$. Apply Theorem 8.47 to get a closed subspace Z of Y with a shrinking M-basis $\{z_\gamma; z_\gamma^*\}_{\gamma \in \Gamma}$ in $Z \times Z^*$. Now use Lemma 8.39 for $Z \subset X$ in order to extend the shrinking M-basis to a fundamental biorthogonal system $\{x_\gamma^*; z_\gamma\}$ in $X^* \times X$. $\qquad \square$

8.5 Banach Spaces with only Trivial Isometries

In this section, we use total biorthogonal systems to show that every Banach space can be renormed so that the only isometries are $\pm$Identity (Jarosz [Jaro88]). This extends the result for separable spaces (Bellenot [Bell86]), which in turn extended the result for Hilbert spaces (Davis [Davi71]).

Theorem 8.49 (Jarosz [Jaro88]). *Every Banach space can be renormed to have only $\pm$Identity as isometries.*

Note that the proof given in [Jaro88], which we follow, works for the complex case, too. We shall denote the scalar field by $\mathbb{K}$ and shall use a few intermediate results as follows.

Proposition 8.50. *Let Γ be a set and X be a Banach space such that $c_0(\Gamma) \subset X \subset \ell_\infty(\Gamma)$. Then there is a norm $\|\| \cdot \|\|$ on X, equivalent to the original sup norm of X, such that a linear map $T : X \to X$ is both a $\| \cdot \|$- and a $\|\| \cdot \|\|$-isometry if and only if $T = \lambda I$, where $|\lambda| = 1$.*

Proof. Let $T : X \to X$ be a $\| \cdot \|$-isometry. Observe that $x, y \in X$ with $\|x\| = \|y\| = 1$ do not have disjoint supports if and only if there exist $u \in X$, with $\|u\| \leq 1$ and scalars α, β with $|\alpha| = |\beta| = 1$ such that

$$\|x + \alpha y + \beta u\| > 1, \text{ and } \|x + \lambda u\| \leq 1, \ \|y + \lambda u\| \leq 1, \text{ for all } |\lambda| = 1.$$

Because the property above depends only on linear and metric properties of X, it is preserved by T. Therefore, T maps elements of X with disjoint supports onto elements with disjoint supports. It then follows that T satisfies $T(e_\gamma) = \epsilon_\gamma e_{\pi(\gamma)}$ for $\gamma \in \Gamma$, where $\pi : \Gamma \to \Gamma$ is a permutation and $|\epsilon_\gamma| = 1$ for $\gamma \in \Gamma$.

Now fix a well-ordering $<$ on Γ, and for $x \in X$ define

$$\|\|x\|\| := \max\{\|x\|, \sup\{|2x(\gamma) + x(\beta)| : \gamma < \beta \in \Gamma\}\}.$$

Suppose now that T is a $\|\| \cdot \|\|$-isometry. We will show that π is the identity on Γ. For this it is enough to see that π preserves order, so assume by way of contradiction that $\gamma < \gamma'$ but $\pi(\gamma) > \pi(\gamma')$. Then $\|\|2e_\gamma + e_{\gamma'}\|\| = 5$. On the other hand,

$$\|\|T(2e_\gamma + e_{\gamma'})\|\| = \|\|\epsilon_\gamma e_{\pi(\gamma)} + \epsilon_{\gamma'} e_{\pi(\gamma')}\|\| = \max\{2, |2\epsilon_{\gamma'} + 2\epsilon_\gamma|\} \le 4,$$

which is a contradiction showing that π is the identity on Γ. To complete the proof, we show that $\epsilon_\gamma = \epsilon_{\gamma'}$. Indeed, if this were not true, then $\|\|e_\gamma + e_{\gamma'}\|\| = 3$, but

$$\|\|T(e_\gamma + e_{\gamma'})\|\| = \|\|\epsilon_\gamma e_\gamma + \epsilon_{\gamma'} e_{\gamma'}\|\| = \max\{2, |2\epsilon_\gamma + \epsilon_{\gamma'}|\} < 3.$$

Therefore, $\epsilon_\gamma = \epsilon_{\gamma'}$ for all $\gamma, \gamma' \in \Gamma$.   □

Proposition 8.51. *Let $(X, \| \cdot \|)$ be a Banach space, x_0 a nonzero element of X, p a continuous norm on $(X, \| \cdot \|)$, G_1 the group of all isometries of $(X, \| \cdot \|)$, and G_2 the group of all isometries T of (X, p) such that Tx_0 and x_0 are linearly independent. Then there is a norm $\| \cdot \|_w$ on $Y = X \oplus K$ such that $\| \cdot \|_w$ and $\| \cdot \|$ coincide on X and the group of all isometries of $(Y, \| \cdot \|_w)$ is isomorphic to $G_1 \cap G_2$.*

Proof. Let $p'(\cdot) = p(\cdot) + \| \cdot \|$. Observe that the norm p' is equivalent to $\| \cdot \|$ and that a linear map $T : X \to X$ preserves both $\| \cdot \|$ and p' if and only if it preserves $\| \cdot \|$ and p. Hence we may assume that the norms $\| \cdot \|$ and p are equivalent. By multiplying p by an appropriate constant, we assume

$$1000\|x\| \le p(x) \quad \text{for } x \in X$$

and that

$$\|x_0\| \le 0.1.$$

We let

$$A := \{(x, \alpha) \in X \oplus K = Y : \max\{\|x\|, |\alpha|\} \le 1\},$$

$$C := \{(x + x_0, 2) \in X \oplus K : p(x) \le 1\},$$

and let $\| \cdot \|_w$ be the norm whose unit ball W is the closed balanced convex set generated by $A \cup C$.

Observe that $\|(x, \alpha)\|_w = \|x\|$ for all $(x, \alpha) \in Y$, where $|\alpha| \le \|x\|$. Hence the norm $\| \cdot \|_w$ coincides with the original norm on X. Also, if $T : X \to X$ preserves both norms $\| \cdot \|$ and p and $Tx_0 = \lambda x_0$, where $|\lambda| = 1$, then $T \oplus I_K$ is an isometry of Y.

Assume now that $T : Y \to Y$ is a $\| \cdot \|_w$-isometry. The proposition will be proved by showing that there is a λ, $|\lambda| = 1$ so that

(i) T maps X onto X,

(ii) $T|_X$ preserves both $\|\cdot\|$ and p, and

(iii) $T(x_0, 0) = (\lambda x_0, 0)$ and $T(0, 1) = (0, \lambda)$, where $|\lambda| = 1$.

Notice that C, as well as all of its rotations λC, $|\lambda| = 1$, are faces of W. We distinguish two types of points in the boundary of W:

(1°) points interior to a segment I contained in the boundary of W whose length with respect to the W norm is at least 0.1, and the limits of such points;

(2°) all other points.

Because these types of points are metrically defined, they are preserved by T. On the other hand, it is easy to see that the points of type (1°) cover all of the boundary of W except for the relative interiors of the faces λC. Thus, $T(x_0, 2) \in \lambda C$ with $|\lambda| = 1$. Replacing T by $\bar\lambda T$, we can assume that $T(x_0, 2) \in C$, and since T maps the face C onto a face of W, we have $TC = C$. To prove that T maps X onto X, we let $x \in X$ with $p(x) \le 1$. Then

$$T(x, 0) = T((x + x_0, 2) - (x_0, 2)) = T(x + x_0, 2) - T(x_0, 2) \in C - C \subset X,$$

and because $\{x : p(x) \le 1\}$ contains a ball in X, this is true for all $x \in X$; that is, $TX \subset X$, and by symmetry $TX = X$. Because $\|\cdot\|_w$ agrees with $\|\cdot\|$ on X, it follows that $T|_X$ is a $\|\cdot\|$-isometry.

Because $TC = C$, the function $T|_X$ maps $B := \{x \in X : p(x) \le 1\}$ onto itself. Hence, for any $x \in X$ with $p(x) \le 1$, one has the following implications:

$$x_0 \pm x \in B \Rightarrow Tx_0 \pm Tx \in B \Rightarrow p((Tx_0 - x_0) \pm Tx) \le 1 \Rightarrow$$

$$p(Tx) \le \frac{1}{2}(p(Tx + T(x_0 - x_0)) + p(Tx - (Tx_0 - x_0))) \le 1.$$

By symmetry, we get $p(x) = p(Tx)$, evidently $Tx_0 = x_0$, and consequently $T(0, 1) = (0, 1)$. □

Proof of Theorem 8.49. Let $(X, \|\cdot\|)$ be a Banach space, and let $Y = X \oplus K$. We will construct a norm on Y that coincides with the original norm on $X \equiv X \oplus \{0\} \subset X \oplus K = Y$ such that Y has only trivial isometries. Because X has a total bounded biorthogonal system (Theorem 4.12), there is an injective map $J : X \to \ell_\infty(\Gamma)$ such that $c_0(\Gamma) \subset \overline{J(X)}$. Now let $\|\|\cdot\|\|$ be a norm on $E := \overline{J(X)}$ as given by Proposition 8.50. Fix $\gamma \in \Gamma$. Then

$$E \equiv \{e \in E : e(\gamma) = 0\} \oplus_\infty K;$$

thus, according to Proposition 8.51 and Proposition 8.50, there is a continuous norm $\tilde p$ on E such that $(E, \tilde p)$ has only trivial isometries. Define a continuous norm on p on X by

$$p(x) := \tilde p(J(X)), x \in X.$$

Now, $(J(X), \tilde p)$, and hence (X, p), have only trivial isometries. Applying Proposition 8.51 again, there is a norm on $Y = X \oplus K$ with only trivial isometries that coincides with $\|\cdot\|$ on X. □

We conclude this chapter with some comments and notes related to its contents. For additional information on support sets, we refer the reader to [Role78], [Laza81], [Mont85], [Kutz86], [BoVa96], [GJM98], and especially [Kosz04a], [Todo], and [Todo06], which show among other things that the existence of support sets is undecidable in ZFC.

Although we highlighted many of its main results, the reader is referred to [GJMMP03] for further interesting results involving Kunen-Shelah properties.

We touched upon only a very narrow part of the subject of norm-attaining operators and did not discuss related topics such as the numerical radii of operators and Bishop-Phelps theorems for multilinear forms. In addition to the seminal paper of Lindenstrauss [Lind63], let us mention that Bourgain [Bour77] showed that a certain Bishop-Phelps property is equivalent to the RNP. We recommend the survey paper [Acos06] and the references therein for an account of the various properties related to norm attaining operators, and the survey paper [KMP06] for an account focused on the numerical index of Banach spaces.

In addition to the papers referenced in the theorems on Mazur intersection properties, let us mention that there has been focus on differentiability properties on Banach spaces with the Mazur intersection properties ([Geor88], [KeGi91], [More98], [GGJM00]); related properties such as the ball-generated property [GoKa89]; Mazur intersection properties for compact convex or weakly compact convex sets ([Sers88], [Sers89], [WhZi87a], [WhZi87b], [Zizl86], [Vand98]); and a unified approach to various such properties [ChLi98]. In a different direction, [GMP04] investigates various questions concerning the stability of collections of sets that are intersections of closed balls. For further information on the Mazur intersection property, we recommend the survey paper [GJM04].

To our knowledge, the following is an open question related to the Mazur intersection property. If X has a Fréchet differentiable bump function, does X isomorphically have the Mazur intersection property? Kunen's $C(K)$ space presents an interesting dichotomy here: either it is an Asplund space with no Fréchet differentiable bump function or it is a Banach space with a Fréchet differentiable bump function that cannot be renormed to have the Mazur intersection property. Let us note further that if X has a Fréchet differentiable bump function and additionally has the RNP, then X can be renormed to have the Mazur intersection property; see [DGZ93b].

8.6 Exercises

8.1. Show that if a compact topological space is not hereditarily Lindelöf, then it has a closed non-G_δ subset.

Hint. The complement of a G_δ-subset in a compact space is σ-compact.

8.2. Show that the space $C(K)$, where K is the compact space in Theorem 4.41, does not have an LUR norm.

Hint. Look at the weak hereditarily Lindelöf property of $C(K)$.

8.3. Show that support sets and uncountable biorthogonal systems are pulled back by quotients.

8.4. (a) Show that every weakly compact convex subset of a Banach space X can be written as a countable intersection of half-spaces if and only if X^* is weak*-separable.
(b) Find a Banach space X with a long sequence of closed half-spaces $\{H_\alpha\}_{\alpha<\omega_1}$ such that (i) $\{0\} = \bigcap_{\alpha<\omega_1} H_\alpha$, (ii) $\{0\}$ is not the intersection of any countable subcollection of the H_α, and (iii) there is a countable collection of closed half-spaces L_n such that $\{0\} = \bigcap_{n=1}^\infty L_n$.

Hint. (a) For one direction, if X^* is not weak*-separable, then $\{0\}$ is not a countable intersection of half-spaces because there is no countable total subset in X^*.
(b) One example can be found using coordinate functionals on $\ell_1(\aleph_1)$ and noting that $\ell_1(\aleph_1) \hookrightarrow \ell_\infty$ and thus is weak*-separable.

8.5. Suppose $\{x_i; f_i\}_{i\in I}$ is an uncountable biorthogonal system. Show that the ball $B := 2B_X \cap \{x; |f_i(x)| \leq 1\}$ cannot be represented as a countable intersection of half-spaces.

Hint. Easy using Lemma 8.17 on the uncountable sequence $\{2x_i\}$.

8.6. Does there exist a nonseparable compact space K such that the dual ball of $C(K)^*$ is weak*-separable?

Hint. Yes; see [Tala80b].

8.7. Let X be a nonreflexive Banach space. Show that $\mathcal{R}(X) \geq 1/2$, where $\mathcal{R}(X)$ is defined in Definition 8.26.

Hint. See [DuSi76, Proposition 1.1].

8.8. Prove parts (b) and (c) of Proposition 8.28. In particular, suppose that $T : X \to Y$ is one-to-one and that Y is strictly convex. If T attains its norm at $x_0 \in S_X$, show that x_0 is an extreme point of B_X.

Hint. For parts (b) and (c) of Proposition 8.28, use the fact that the unit ball of c_0 endowed with its usual norm has no extreme points.

8.9. Suppose $(X, \|\cdot\|)$ has property α with $\lambda = 0$. Prove that $(X, \|\cdot\|)$ is isometric to $\ell_1(\Gamma)$.

Hint. See [More96, Proposition 2.3].

8.10. Let C be a closed bounded convex set, and let $x \in C$. The point x is said to be a *strong vertex point* of C if there exists a closed bounded convex subset $D \subset C$ with $x \notin D$ such that $C = \text{conv}(\{x\} \cup D)$.

(a) Show that every strong vertex point is strongly exposed. Conclude that if $\{x_i, x_i^*\}$ is an α-system, then $\{x_i\}_{i \in I}$ is the set of strong vertex points in B_X.

(b) Show that a strong vertex point is not a point of local uniform rotundity, and hence if X has property α, its norm is nowhere locally uniformly rotund.

Hint. See [More97].

8.11. Assume X^* is weak*-separable. Is X necessarily isomorphic to a subspace of ℓ_∞?

Hint. No; consider the space JL_2 in [JoLi74].

References

[Acos06]   M.D. Acosta, *Denseness of norm-attaining mappings*, RACSAM Rev. R. Acad. Cienc. Exactas Fís. Nat. Ser. A Mat. **100** (2006), no. 1-2, 9–30.

[AAP96]   M.D. Acosta, F. Aguirre, and R. Payá, *A new sufficient condition for the denseness of norm-attaining operators*. Rocky Mountain J. Math. **26** (1996), no. 2, 407–418.

[AKP99]   G. Alexandrov, D. Kutzarova, and A.N. Plichko, *A separable space with no Schauder decomposition*, Proc. Amer. Math. Soc. **127** (1999), no. 9, 2805–2806.

[AlPl]   G. Alexandrov and A.N. Plichko, *Connection between strong and norming Markushevich bases in nonseparable Banach spaces*, preprint.

[AlAr92]   D.E. Alspach and S.A. Argyros, *Complexity of weakly null sequences*, Dissertationes Math. (Rozprawy Mat.) **321** (1992), 44 pp.

[AJO05]   D.E. Alspach, R. Judd, and E. Odell, *The Szlenk index and local ℓ_1 indices of a Banach space*, Positivity **9** (2005), 1–44.

[AmLi68]   D. Amir and J. Lindenstrauss, *The structure of weakly compact sets in Banach spaces*, Ann. Math. **88** (1968), 35–44.

[AnCaa]   C. Angosto and B. Cascales, *The quantitative difference between countable compactness and compactness*, to appear.

[AnCab]   C. Angosto and B. Cascales, *Distances to spaces of Baire one functions*, preprint.

[Ar82]   S.A. Argyros, *On nonseparable Banach spaces*, Trans. Amer. Math. Soc. **270** (1982), 193–216.

[Ar96]   S.A. Argyros, *Weakly Lindelöf determined Banach spaces not containing ℓ_1*, unpublished typescript, University of Athens, 1996.

[Ar01]   S.A. Argyros, *A universal property of reflexive hereditarily indecomposable Banach spaces*, Proc. Amer. Math. Soc. **129** (2001), 3231–3239.

[ArBe87]   S.A. Argyros and Y. Benyamini, *Universal WCG Banach spaces and universal Eberlein compacts*, Israel J. Math. **58** (1987), 305–320.

[ArBZ84]   S.A. Argyros, J. Bourgain, and T. Zachariades, *A result on the isomorphic embeddability of $\ell_1(\Gamma)$*, Studia Math. **77** (1984), 77–91.

[ACGJM02] S.A. Argyros, J.F. Castillo, A.S. Granero, M. Jiménez-Sevilla, and J.P. Moreno, *Complementation and embeddings of $c_0(I)$ in Banach spaces*, Proc. London Math. Soc. **85** (2002), 742–768.

[ArDo]   S.A. Argyros and P. Dodos, *Genericity and amalgamation of classes of Banach spaces*, preprint.

[ArFa85]   S.A. Argyros and V. Farmaki, *On the structure of weakly compact subsets of Hilbert spaces and applications to the geometry of Banach spaces*, Trans. Amer. Math. Soc. **289** (1985), 409–427.

[ArMe93]   S.A. Argyros and S. Mercourakis, *On weakly Lindelöf Banach spaces*, Rocky Mountain J. Math. **23** (1993), 395–446.

[ArMe05a]   S.A. Argyros and S. Mercourakis,*Examples concerning heredity problems of WCG Banach spaces*, Proc. Amer. Math. Soc. **133** Day's result(2005), 773–785.

[ArMe05b]   S.A. Argyros and S. Mercourakis, *A note on the structure of WUR Banach spaces*, Comment. Math. Univ. Carolin. **46** (2005), no. 3, 399–408.

[AMN89]   S.A. Argyros, S. Mercourakis, and S. Negrepontis, *Functional-analytic properties of Corson-compact spaces*, Studia Math. **89** (1988), 197–229.

[ArTod05]   S.A. Argyros and S. Todorčević, *Ramsey Methods in Analysis*. Advanced Courses in Mathematics, CRM Barcelona. Birkhäuser Verlag, Basel, 2005.

[ArTo04]   S.A. Argyros and A. Tolias, *Methods in the theory of hereditarily indecomposable Banach spaces*, Mem. Amer. Math. Soc. **170** (2004), no. 806.

[ArTs82]   S.A. Argyros and A. Tsarpalias, *Isomorphic embeddings of $\ell_1(\Gamma)$ into subspaces of $C(\Omega)^*$*, Math. Proc. Cambridge Philos. Soc. **92** (1982), no. 2, 251–262.

[Arth67]   C.W. McArthur, *On a theorem of Orlicz and Pettis*, Pacific J. Math. **22** (1967), 297–302.

[AzFe02]   D. Azagra and J. Ferrera, *Every closed convex set is the set of minimizers of some C^∞-smooth convex function*, Proc. Amer. Math. Soc. **130** (2002), 3687–3892.

[BaRo71]   G.F. Bachelis and H.P. Rosenthal, *On unconditionally converging series and biorthogonal systems in a Banach space*, Pacific J. Math. **37** (1971), 1–5.

[Bana32]   S. Banach, *Théorie des Opérations Linéaires*, Chelsea Publishing Co., New York, 1955.

[Bell00]   M. Bell, *Universal uniform Eberlein compact spaces*, Proc. Amer. Math. Soc. **128** (2000), 2191–2197.

[BGT82]   M. Bell and J. Ginsburg, and S. Todorčević, *Countable spread of expY and λY*, Topology Appl. **14** (1982), no. 1, 1–12.

[BeMa]   M. Bell and W. Marciszewski, *On scattered Eberlein compacts*, to appear.

[Bell86]   S.F. Bellenot, *Banach spaces with trivial isometries*, Israel J. Math. **56** (1986), 89–96.

[BHO89]   S.F. Bellenot, R. Haydon, and E. Odell, *Quasi-reflexive and tree spaces constructed in the spirit of R. C. James*. Contemp. Math. **85** (1989), 19–43.

[BeLi00]   Y. Benyamini and J. Lindenstrauss, *Geometric Nonlinear Functional Analysis*, Volume 1, American Mathematical Society Colloquium Publications, **48**. American Mathematical Society, Providence, RI, 2000.

[BRW77]   Y. Benyamini, M.E. Rudin, and M. Wage, *Continuous images of weakly compact subsets of Banach spaces*, Pacific J. Math. **70** (1977), 309–324.

[BeSt76]   Y. Benyamini and T. Starbird, *Embedding weakly compact sets into Hilbert spaces*, Israel J. Math. **23** (1976), 137–141.

[Bess58]   C. Bessaga, *A note on universal Banach spaces of a finite dimension*, Bull. Pol. Acad. Sci. **6** (1958), 97–101.

[Bess72]   C. Bessaga, *Topological equivalence of nonseparable reflexive Banach spaces, ordinal resolutions of identity and monotone bases*, Ann. Math. Studies, **69** (1972), 3–14.

[BesPe60]   C. Bessaga and A. Pełczyński, *Spaces of continuous functions (IV)*, Studia Math. **19** (1960), 53–62.

[BesPe79]   C. Bessaga and A. Pełczyński, *Some aspect of the present theory of Banach spaces*, in S. Banach, Travaux sur L'Analyse Fonctionnelle, PWN, Warszawa, 1979.

[Bohn41]   F. Bohnenblust, *Subspaces of $\ell_{p,n}$ spaces*, Amer. J. Math. **63** (1941), 64–72.

[BoFi93]   J.M. Borwein and S. Fitzpatrick, *A weak Hadamard smooth renorming of $L_1(\Omega, \mu)$*, Canad. Math. Bull. **36** (1993), no. 4, 407–413.

[BMV06]   J.M. Borwein, V. Montesinos, and J. Vanderwerff, *Boundedness, differentiability and extensions of convex functions*, J. Convex Anal. **13**, (2006), to appear.

[BoVa96]   J.M. Borwein and J. Vanderwerff, *Banach spaces which admit support sets*, Proc. Amer. Math. Soc. **124** (1996), 751–756.

[BoVa04]   J.M. Borwein and J. Vanderwerff, *Constructible convex sets*, Set-Valued Analysis **12** (2004), 61–77.

[Boss93]   B. Bossard, *Codages des espaces de Banach séparables. Familles analytiques ou coanalytiques d'espaces de Banach*, C. R. Acad. Sci. Paris **316** (1993), 1005–1010.

[Boss02]   B. Bossard, *A coding of separable Banach spaces. Analytic and coanalytic families of Banach spaces*, Fund. Math. **172** (2002), no. 2, 117–152.

[Bour77]   J. Bourgain, *On dentability and the Bishop-Phelps property*, Israel J. Math. **28** (1977), 265–271.

[Bour79]   J. Bourgain, *The Szlenk index and operators on C(K) spaces*, Bull. Soc. Math. Belg. **31** (1979), 87–117.

[Bour80a]   J. Bourgain, *On separable Banach spaces, universal for all separable reflexive spaces*, Proc. Amer. Math. Soc. **79** (1980), 241–246.

[Bour80b]   J. Bourgain, *On convergent sequences of continuous functions*, Bull. Soc. Math. Belg. **32** (1980), 235–249.

[BFT78]   J. Bourgain, D.H. Fremlin, and M. Talagrand, *Pointwise compact sets of Baire measurable functions*, Amer. J. Math. **100** (1978), 845–886.

[BRS81]   J. Bourgain, H.P. Rosenthal, and G. Schechtman, *An ordinal L^p-index for Banach spaces, with application to complemented subspaces of L^p*, Ann. Math. **114** (1981), 193–228.

[CMR]   B. Cascales, W. Marciszewski, and M. Raja, *Distance to spaces of continuous functions*, Topology Appl. **153** (2006), no. 13, 2303–2319.

[CGPY01]   J.M.F. Castillo, M. González, A.N. Plichko and D. Yost, *Twisted properties of Banach spaces*, Math. Scand. **89** (2001), 217–244.

[ChLi98]   D. Chen and B.L. Lin, *Ball separation properties in Banach spaces*, Rocky Mount. J. Math. **28** (1998), 835–873.

[CoNe82]   W.W. Comfort and S. Negrepontis, *Chain Conditions in Topology*, Cambridge Tracts in Mathematics, **79**. Cambridge University Press, Cambridge-New York, 1982.

[CoLi66]   H.H. Corson and J. Lindenstrauss, *On weakly compact subsets of Banach spaces*. Proc. Amer. Math. Soc. **17** (1966), 476–481.

[CoDa72]   W. Courage, and W.J. Davis, *A characterization of M-bases*, Math. Ann. **197** (1972), 1–4.

[DaLi73]   F.K. Dashiell and J. Lindenstrauss, *Some examples concerning strictly convex norms on C(K) spaces*, Israel J. Math. **16** (1973), 329–342.

[Davi71]   W.J. Davis, *Separable Banach spaces with only trivial isometries*, Rev. Roumaine Math. Pures Appl. **16** (1971), 1051–1054.

[DFJP74]   W.J. Davis, T. Figiel, W.B. Johnson and A. Pełczyński, *Factoring weakly compact operators*, J. Funct. Anal. **17** (1974), 311–327.

[DaJo73a]   W.J. Davis and W.B. Johnson, *On the existence of fundamental and total bounded biorthogonal systems in Banach spaces*, Studia Math. **45** (1973), 173–179.

[DaJo73b]   W.J. Davis and W.B. Johnson, *Basic sequences and norming subspaces in non-quasireflexive Banach spaces*, Israel J. Math. **14** (1973), 353–367.

[DaSi73]   W.J. Davis and I. Singer, *Boundedly complete M-bases and complemented subspaces in Banach spaces*, Trans. Amer. Math. Soc. **175** (1973), 187–194.

[Day62]   M.M. Day, *On the basis problem in normed spaces*, Proc. Amer. Math. Soc. **13** (1962), 655–658.

[Day73]   M.M. Day, *Normed Linear Spaces*, Third edition. Ergebnisse der Mathematik und ihrer Grenzgebiete, Band **21**. Springer-Verlag, New York-Heidelberg, 1973.

[Della77]   C. Dellacherie, *Les derivations en theorie descriptive des ensembles et le theorie de la borne*, Séminaire de Probabilités, XI (Univ. Strasbourg, Strasbourg, 1975/1976), pp. 34–46. Lecture Notes in Math., Vol. 581, Springer, Berlin, 1977.

[Dev87]   R. Deville, *Un théorem de transfert pour la propriété des boules*, Canad. Math. Bull. **30** (1987), 295–300.

[DeGo93]   R. Deville and G. Godefroy, *Some applications of projectional resolutions of identity*, Proc. London Math. Soc. **67** (1993), 183–199.

[DGZ93a]   R. Deville, G. Godefroy, and V. Zizler, *Smoothness and Renormings in Banach Spaces*, Pitman Monographs and Surveys in Pure and Applied Mathematics **64**, Longman Scientific and Technical, New York, 1993.

[DGZ93b]   R. Deville, G. Godefroy, and V. Zizler, *Smooth bump functions and the geometry of Banach spaces*, Mathematika **40** (1993), 305–321.

[Dies75]   J. Diestel, *Geometry of Banach spaces—selected topics*, Lecture Notes in Mathematics, Vol. 485. Springer-Verlag, Berlin-New York, 1975.

[Dies84]   J. Diestel, *Sequences and Series in Banach Spaces*, Graduate Texts in Mathematics **92**, Springer-Verlag, New York, 1984.

[DGJ00]   S.J. Dilworth, M. Girardi, and W.B. Johnson, *Geometry of Banach spaces and biorthogonal systems*, Studia Math. **140** (2000), no. 3, 243–271.

[DGK95]   S.J. Dilworth, M. Girardi, and D. Kutzarova, *Banach spaces which admit a norm with the uniform Kadets-Klee property*, Studia Math. **112** (1995), 267–277.

[DoFe]   P. Dodos and V. Ferenczi, *Some strongly bounded classes of Banach spaces*, preprint.

[DLT98]   P.N. Dowling, C.J. Lennard, and B. Turett, *Asymptotically isometric copies of c_0 in Banach spaces*, J. Math. Anal. Appl. **219** (1998), 377–391.

[Dugu66]   J. Dugundji, *Topology*, Allyn and Bacon Inc., Boston, 1966.

[DuSi76]   D. van Dulst and I. Singer, *On Kadets-Klee norms on Banach spaces* Studia Math. **54** (1976), 205–211.

[DuSch]   N. Dunford and J.T. Schwartz, *Linear Operators, Part I*, Interscience Publishers, Inc., New York, 1967.

[Dut01]   Y. Dutrieux, *Lipschitz quotients and the Kunen-Martin theorem* Comment. Math. Univ. Carolin. **42** (2001), 641–648.

[Eb47]   W.F. Eberlein, *Weak compactness in Banach spaces, I.* Proc. Nat. Acad. Sci. USA **33** (1947), 51–53.

[EdWh84]   G.A. Edgar and R.F. Wheeler, *Topological properties of Banach spaces*, Pacific J. of Math. **115** (1984), 317–350.

[Emm86]   G. Emmanuele, *A dual characterization of Banach spaces not containing ℓ^1*. Bull. Polish Acad. Sci. Math. **34** (1986), no. 3-4, 155–160.

[Enfl73]   P. Enflo, *A counterexample to the approximation property in Banach spaces*, Acta Math. **130** (1973), 309–317.

[EnRo73]   P. Enflo and H.P. Rosenthal, *Some results concerning $L^p(\mu)$-spaces*, J. Functional Analysis **14** (1973), 325–348.

[Eng77]   R. Engelking, *General Topology*, Monografie Matematyczne, Tom 60. [Mathematical Monographs, Vol. 60] PWN—Polish Scientific Publishers, Warsaw, 1977.

[EHMR84]   P. Erdős, A. Hajnal, A. Mate, and R. Rado, *Combinatorial Set Theory: Partitions Relations for Cardinals*, North-Holland, Amsterdam, 1984.

[Fab87]   M. Fabian, *Each weakly countably determined Asplund space admits a Fréchet differentiable norm*, Bull. Austral Math. Soc. **36** (1987), 367–374.

[Fab97]   M. Fabian, *Gâteaux Differentiability of Convex Functions and Topology—Weak Asplund Spaces*, John Wiley & Sons, Interscience, New York, 1997.

[FaGo88]   M. Fabian and G. Godefroy, *The dual of every Asplund space admits a projectional resolution of identity*, Studia Math. **91** (1988), 141–151.

[FGHZ03]   M. Fabian, G. Godefroy, P. Hájek, and V. Zizler, *Hilbert-generated spaces,* J. Functional Analysis **200** (2003), 301–323.

[FGMZ04]   M. Fabian, G. Godefroy, V. Montesinos, and V. Zizler, *Inner characterization of weakly compactly generated Banach spaces and their relatives*, J. Math. Anal. Appl. **297** (2004), 419–455.

[FGZ01]   M. Fabian. G. Godefroy, and V. Zizler, *The structure of uniformly Gâteaux smooth Banach spaces*, Israel J. Math. **124** (2001), 243–252.

[Fa~01]   M. Fabian, P. Habala, P. Hájek, V. Montesinos, J. Pelant, and V. Zizler, *Functional Analysis and Infinite-Dimensional Geometry*, CMS Books in Mathematics **8**, Springer-Verlag, New York, 2000.

[FHMZ05]   M. Fabian, P. Hájek, V. Montesinos, and V. Zizler, *A quantitative version of Krein's theorem*, Rev. Mat. Iberoamericana, **21** (2005), 237–248.

[FHMZ]   M. Fabian, P. Hájek, V. Montesinos, and V. Zizler, *Weakly compact generating and shrinking Markushevich bases*, Serdica Math. J. **32**, 4 (2006), 277-288.

[FHZ97]   M. Fabian, P. Hájek, and V. Zizler, *Uniform Eberlein compacta and uniform Gâteaux smooth norms*, Serdica Math. J. **23** (1997), 351–362.

[FMZ02a]   M. Fabian, V. Montesinos, and V. Zizler. *Pointwise semicontinuous smooth norms.* Arch. Math. **78** (2002), 459–464.

[FMZ02b]   M. Fabian, V. Montesinos, and V. Zizler, *Weakly compact sets and smooth norms in Banach spaces*, Bull. Austral. Math. Soc. **65** (2002), 223–230.

[FMZ04a]   M. Fabian, V. Montesinos, and V. Zizler, *A characterization of subspaces of weakly compactly generated Banach spaces*, J. London Math. Soc. **69** (2004), 457–464.

[FMZ04b]   M. Fabian, V. Montesinos, and V. Zizler, *The Day norm and Gruenhage compacta*, Bull. Austral. Math. Soc. **69** (2004), 451–456.

[FMZ05]   M. Fabian, V. Montesinos, and V. Zizler, *Biorthogonal systems in weakly Lindelöf spaces*, Canad. Math. Bull. **48** (2005), 69–79.

[Farm87]   V. Farmaki, *The structure of Eberlein, uniformly Eberlein and Talagrand compact spaces in $\Sigma(\mathbb{R}^\Gamma)$*, Fund. Math. **128** (1987), no. 1, 15–28.

[Fin89]   C. Finet, *Renorming Banach spaces with many projections and smoothness properties*, Math. Ann. **284** (1989), 675–679.

[FiGo89]   C. Finet and G. Godefroy, *Biorthogonal systems and big quotient spaces*, Banach space theory (Iowa City, IA, 1987), 87–110, Contemp. Math., **85**, Amer. Math. Soc., Providence, RI, 1989.

[FMP03]   C. Finet, M. Martín, and R. Payá, *Numerical index and renorming*, Proc. Amer. Math. Soc. **131** (2003), 871–877.

[FiSc89]   C. Finet and W. Schachermayer, *Equivalent norms on separable Asplund spaces*, Studia Math. **92** (1989), 275–283.

[FoSi65]   C. Foias and I. Singer, *On bases in $C[0,1]$ and $L[0,1]$*, Rev. Roumaine Math. Pures Appl. **10** (1965), 931–960.

[FMS88]   M. Foreman, M. Magidor, and S. Shelah, *Martin's Maximum, saturated ideals, and non-regular ultrafilters. Part I*, Ann. Math. **127** (1988), 1–47.

[FGK]   R. Frankiewicz, M. Grzech, and R. Komorowski, to appear.

[Frem84]   D.H. Fremlin, *Consequences of Martin's axioms*, Cambridge University Press, Cambridge, 1984.

[FrSe88]   D.H. Fremlin and A. Sersouri, *On ω-independence in separable Banach spaces*, Quarterly J. Math. **39** (1988), 323–331.

[Geor88]   P.G. Georgiev, *Mazur's intersection property and a Krein-Milman type theorem for almost all closed, convex and bounded subsets of a Banach space*, Proc. Amer. Math. Soc. **104** (1988), 157–168.

[GGJM00]   P.G. Georgiev, A.S. Granero, M. Jiménez-Sevilla, and J.P. Moreno, *Mazur intersection properties and differentiability of convex functions in Banach spaces*, J. London Math. Soc. **61** (2000), 531–542.

[GGS78]   J.R. Giles, D.A. Gregory, and B. Sims, *Characterization of normed linear spaces with Mazur's intersection property*, Bull. Austral. Math. Soc. **18** (1978), 471–476.

[GiSci96]   J.R. Giles and S. Sciffer, *On weak Hadamard differentiability of convex functions on Banach spaces*, Bull. Austral. Math. Soc. **54** (1996), 155–166.

[God80]   G. Godefroy, *Compacts de Rosenthal*, Pacific J. Math. **91** (1980), 293–306.

[Gode95]   G. Godefroy, *Decomposable Banach spaces*, Rocky Mountain J. Math. **25** (1995), 1013–1024.

[Gode01]   G. Godefroy, *The Szlenk index and its applications*, General topology in Banach spaces, 71–79, Nova Sci. Publ., Huntington, NY, 2001.

[Gode02]   G. Godefroy, *Banach spaces of continuous functions on compact spaces*, Recent progress in general topology, II, 177–199, North-Holland, Amsterdam, 2002.

[Gode06]   G. Godefroy, *Universal spaces for strictly convex Banach spaces*, RACSAM Rev. R. Acad. Cienc. Exactas Fís. Nat. Ser. A Mat. **100** (2006), 136–146.

[GoKa89]   G. Godefroy and N.J. Kalton. *The ball topology and its applications.* Contemporary Math. **85** (1989), 195–237.

[GKL00]   G. Godefroy, N.J. Kalton, and G. Lancien, *Subspaces of $c_0(\mathbb{N})$ and Lipschitz isomorphisms*, Geom. Funct. Anal. **10** (2000), no. 4, 798–820.

[GKL01]   G. Godefroy, N.J. Kalton, and G. Lancien, *Szlenk index and uniform homeomorphisms*, Trans. Amer. Math. Soc. **353** (2001), 3895–3918.

[GoLo89]   G. Godefroy and A. Louveau, *Axioms of determinacy and biorthogonal systems*, Israel J. Math. **67** (1989), 109–116.

[GoTa82]   G. Godefroy and M. Talagrand, *Espaces de Banach representables*, Israel J. Math. **41** (1982), 321–330.

[Godu77]   B.V. Godun, *Weak* derivatives of transfinite order for sets of linear functionals*, (Russian) Sibirsk. Mat. Ž. **18** (1977), no. 6, 1289–1295, 1436.

[Godu78]   B.V. Godun, *On weak* derivations of sets of linear functionals*, Math. Zametki **23** (1978), 607–616.

[Godu81]   B.V. Godun, *On norming subspaces in some conjugate Banach spaces*, Math. Zametki **29** (1981), 549–555.

[Godu82a]   B.V. Godun, *On Markushevich bases*, Dokl. Akad. Ukr. SSR, Ser. A **266** (1982), 11–14.

[Godu82b]   B.V. Godun, *Bounded and unbounded complete biorthogonal systems in a Banach space*, Sibirsk. Mat. J. **23** (1982), 190–193.

[Godu83a]   B.V. Godun, *Biorthogonal systems in spaces of bounded functions*, Dokl. Akad. Ukr. SSR Ser. A **3** (1983), 7–9.

[Godu83b]   B.V. Godun, *On complete biorthogonal systems in Banach spaces*, Funct. Anal. Appl. **17** (1983), 1–5.

[Godu83c]   B.V. Godun, *On fundamental biorthogonal systems in some conjugate Banach spaces*, Comment. Math. Univ. Carolin. **24** (1983), 431–436.

[Godu84]   B.V. Godun, *Quasicomplements and minimal systems in ℓ_∞*, Math. Zametki **36** (1984), 117–121.

[Godu85]   B.V. Godun, *A special class of Banach spaces*, Math. Notes **37** (1985), 220–223.

[Godu90]   B.V. Godun, *Bases of Auerbach in Banach spaces isomorphic to $\ell_1[0,1]$*, C. R. Bulg. Acad. Sci, **43** (1990), 19–21. (Russian).

[GLT93]   B.V. Godun, B.L. Lin, and S. Troyanski, *On Auerbach bases*, Contemp. Math. **144** (1993), 115–118.

[GoKa80]   B.V. Godun and M.I. Kadets, *Banach spaces without complete minimal systems*, Funct. Anal. Appl. **14** (1980), 301–302.

[GoKa82]   B.V. Godun and M.I. Kadets, *On norming subspaces, biorthogonal systems and predual Banach spaces*, Sibirsk. Math. J. **23** (1982), 44–48.

[GoTr93]   B.V. Godun and S. Troyanski, *Renorming Banach spaces with fundamental biorthogonal systems*, Contemp. Math. **144** (1993), 119–126.

[GoMo]   A. González, V. Montesinos, *A note on WLD Banach spaces*, to appear.

[Gr98]   A.S. Granero, *On the complemented subspaces of $c_0(I)$*, Atti Semin. Mat. Fis. Univ. Modena **96** (1998), 35–36.

[Gr06]   A. S. Granero, *An extension of the Krein-Šmulyan Theorem*, Rev. Mat. Iberoamericana **22** (2006), 93–110.

[GHM04]   A.S. Granero, P. Hájek, and V. Montesinos, *Convexity and w^*-compactness in Banach spaces*, Math. Ann., **328** (2004), 625–631.

[GJMMP03]   A.S. Granero, M. Jiménez-Sevilla, A. Montesinos, J.P. Moreno, and A.N. Plichko, *On the Kunen-Shelah properties in Banach spaces*, Studia Math. **157** (2003), 97–120.

[GJM98]   A.S. Granero, M. Jiménez-Sevilla, and J.P. Moreno, *Convex sets in Banach spaces and a problem of Rolewicz*, Studia Math. **129** (1998), 19–29.

[GJM99]   A.S. Granero, M. Jiménez-Sevilla, and J.P. Moreno, *Geometry of Banach spaces with property β*, Israel J. Math. **111** (1999), 263–273.

[GJM02]   A.S. Granero, M. Jimémez-Sevilla, and J.P. Moreno, *On ω-independence and the Kunen-Shelah poperty*, Proc. Edinburg Math. Soc. **45** (2002), 391–395.

[GJM04]   A.S. Granero, M. Jiménez-Sevilla, and J.P. Moreno, *Intersections of closed balls and geometry of Banach spaces*, Extracta Math. **19** (2004), 55–92.

[GMP04]   A.S. Granero, J.P. Moreno, and R.R. Phelps, *Convex sets which are intersections of balls*, Adv. Math. **183** (2004), 183–208.

[Gras81]   R. Grzaślewicz, *A universal convex set in Euclidean space*, Colloq. Math. **45** (1981), no. 1, (1982), 41–44.

[Grot52]   A. Grothendieck, *Critères de compacité dans les espaces fonctionnels généraux*, Amer. J. Math. **74** (1952), 168–186.

[Grot73]   A. Grothendieck, *Topological Vector Spaces*, Gordon and Breach, Sc. Pub. Ltd. London, 1973.

[Grot53]   A. Grothendieck, *Sur les applications lineaires faiblement compactes d'espaces du type CK)*, Canad. J. Math., **5** (1953), 129–173.

[Grue84]   G. Gruenhage, *Covering properties of $X^2 \backslash \Delta$, W-sets, and compact subsets of Σ-products*, Topology Appl. **17** (1984), no. 3, 287–304.

[Grue87]   G. Gruenhage, *A note on Gul'ko compact spaces*, Proc. Amer. Math. Soc. **100** (1987), 371–376.

[Grun58]   B. Grünbaum, *On a problem of S. Mazur*, Bull. Res. Counc. of Israel **7F** (1958), 133–135.

[Gul77]   S.P. Gul'ko, *On properties of subsets of Σ-products*, Sov. Mat. Dokl. **18** (1977), 14–38.

[Gul90]   S.P. Gul'ko, *On complemented subspaces of Banach spaces of the weight continuum*, Ekstremalnye Zadachi Teor. Funkts. **8** (1990), 34–41.

[GO75]   S.P. Gul'ko and A.V. Oskin, *Isomorphic classification of spaces of continuous functions on totally ordered sets* Funkc. Anal. Pril. **9** (1975), 61–62.

[Gura66]   V.I. Gurarii, *Bases in spaces of continuous functions on compacta and some geometrical questions* (in Russian), Izv. Akad. Nauk SSSR Ser. Mat. **30** (1966), 289–306.

[GuKa62]   V.I. Gurarii and M.I. Kadets, *Minimal systems and quasicomplements in Banach spaces*, Sov. Math. Dokl. **3** (1962), 966–968.

[HHZ96]   P. Habala, P. Hájek, and V. Zizler, *Introduction to Banach spaces I, II*, Matfyzpress, Prague (1996).

[Hag73]   J. Hagler, *Some more Banach spaces which contain ℓ_1*, Studia Math. **46** (1973), 35–42.

[Hag77]   J. Hagler, *Nonseparable "James tree" analogues of the continuous functions on the Cantor set*, Studia Math. **61** (1977), 41–53.

[Hag77b]   J. Hagler, *A counterexample to several questions about Banach spaces*, Studia Math. **60** (1977), 289–308.

[HaJo77]   J. Hagler and W.B. Johnson, *On Banach spaces whose dual balls are not w^*-sequentially compact*, Israel J. Math. **28** (1977), 325–330.

[HagO78]   J. Hagler and E. Odell, *A Banach space not containing ℓ_1 whose dual ball is not weak* sequentially compact*, Illinois J. Math.**22** (1978), 290–294.

[HagSt73] J. Hagler and Ch. Stegall, *Banach spaces whose duals contain complemented subspaces isomorphic to $C[0,1]$*, J. Functional Analysis **13** (1973), 233–251.

[Haj94]   P. Hájek, *Polynomials and injections of Banach spaces into superreflexive spaces*, Arch. Math. **63** (1994), 39–44.

[Haj96]   P. Hájek, *Dual renormings of Banach spaces*, Comment. Math. Univ. Carolin. **37** (1996), 241–253.

[Haj98]   P. Hájek, *Smooth functions on c_0*, Israel J. Math. **104** (1998), 89–96.

[HaJo04]   P. Hájek and M. Johanis, *Characterization of reflexivity by equivalent renorming*, J. Funct. Anal. **211**, 1 (2004), 163–172.

[HaLa]   P. Hájek and G. Lancien, *Various slicing indices on Banach spaces*, Mediterranean J. Math. To appear.

[HaLaMo]   P. Hájek, G. Lancien, and V. Montesinos, *Universality of Asplund spaces*, to appear in Proc. Amer. Math. Soc.

[HaLaP]   P. Hájek, G. Lancien, and A. Procházka, preprint.

[HaRy05]   P. Hájek and J. Rychtář, *Renorming James tree space*, Trans. Amer. Math. Soc. **357** (2005), 3775–3788.

[HaZi95]   P. Hájek and V. Zizler, *Remarks on symmetric smooth norms*, Bull. Austral. Math. Soc. **52** (1995), 225–229.

[Ha77]   R. Haydon, *On Banach spaces which contain $\ell_1(\tau)$ and types of measures on compact spaces*, Israel J. Math. **28** (1977), 313–324.

[Ha78]   R. Haydon, *On dual L^1-spaces and injective bidual Banach spaces*, Israel J. Math. **31** (1978), 142–152.

[Ha80]   R. Haydon, *Non-separable Banach spaces*, in Functional Analysis: Surveys and Recent Results II, North Holland, Amsterdam, 1981, 19–30.

[Ha81]   R. Haydon, *A non-reflexive Grothendieck space that does not contain ℓ_∞*, Israel J. Math. **40** (1981), 65–73.

[HSZ98]   P. Holický, M. Šmídek, and L. Zajíček, *Convex functions with non-Borel set of Gâteaux differentiability points*, Comment. Math. Univ. Carolin. **39** (1998), 469–482.

[How73]   J. Howard, *Mackey compactness in Banach spaces*, Proc. Amer. Math. Soc. **37** (1973), 108–110.

[Huff80]   R. Huff, *Banach spaces which are nearly uniformly convex*, Rocky Mount. J. Math. **10** (1980), 743–749.

[Jam50]   R.C. James, *Bases and reflexivity of Banach spaces*, Ann. Math. **52** (1950), 518–527.

[Jam64]   R.C. James, *Weak compactness and reflexivity*, Israel J. of Math. **2** (1964), 101–119.

[Jam72a]   R.C. James, *Superreflexive spaces with bases*, Pacific J. Math. **41** (1972), 409–419.

[Jam72b]   R.C. James, *Some self-dual properties of normed linear spaces*, Symposium on Infinite-Dimensional Topology (Louisiana State Univ., Baton Rouge, La., 1967), pp. 159–175. Ann. of Math. Studies, **69**, Princeton Univ. Press, Princeton, N.J., 1972.

[Jam72c]   R.C. James, *Quasicomplements.* Collection of articles dedicated to J. L. Walsh on his 75th birthday, VI (Proc. Internat. Conf. Approximation Theory, Related Topics and their Applications, Univ. Maryland, College Park, Md., 1970). J. Approximation Theory **6** (1972), 147–160.

[Jame74]   G.J.O. Jameson, *Topology and Normed Spaces*, Chapman and Hall, London, 1974.

[Jaro88]   K. Jarosz, *Any Banach space has an equivalent norm with trivial isometries*, Israel J. Math. **64** (1988), 49–56.

[Je78]   T. Jech, *Set Theory.* Academic Press, New York, 1978.

[JiMo97]   M. Jiménez-Sevilla and J.P. Moreno, *Renorming Banach spaces with the Mazur Intersection Property*, J. Funct. Anal. **144** (1997), 486–504.

[JoRy]   M. Johanis and J. Rychtář, *On uniformly Gâteaux smooth norms and normal structure*, to appear.

[JoZi74a]   K. John and V. Zizler, *Smoothness and its equivalents in weakly compactly generated Banach spaces*, J. Funct. Anal. **15** (1974), 161–166.

[JoZi74b]   K. John and V. Zizler, *Some remarks on nonseparable Banach spaces with Markushevic bases*, Comment. Math. Univ. Carolin. **15** (1974), 679–691.

[JoZi77]   K. John and V. Zizler, *Some notes on Markushevich bases in weakly compactly generated Banach spaces*, Compositio Math. **35** (1977), 113–123.

[John70a]   W.B. Johnson, *No infinite-dimensional P-space admits a Markushevich basis*, Proc. Amer. Math. Soc. **26** (1970), 467–468.

[John70b]   W.B. Johnson, *A complementary universal conjugate Banach space and its relation to the approximate problem*, Israel J. Math. **13** (1972), 301–310.

[John70c]   W.B. Johnson, *Markushevich basis and duality theory*, Trans. Amer. Math. Soc. **149** (1970), 171–177.

[John71a]   W.B. Johnson, *Factoring compact operators*, Israel J. Math. **9** (1971), 337–345.

[John71b]   W.B. Johnson, *On the existence of strongly series summable Markushevich basis in Banach spaces*, Trans. Amer. Math. Soc. **157** (1971), 481–486.

[John72]   W.B. Johnson, *No infinite-dimensional P-space admits a Markushevich basis*, Proc. Amer. Math. Soc. **26** (1970), 467–468.

[John73]   W.B. Johnson, *On quasicomplements*, Pacific J. Math. **48** (1973), 113–118.

[John77]   W.B. Johnson, *On quotients of L_p which are quotients of ℓ_p*, Compositio Math. **34** (1977), 69–89.

[JoLi74]   W.B. Johnson and J. Lindenstrauss, *Some remarks on weakly compactly generated Banach spaces*, Israel J. Math. **17** (1974), 219–230.

[JoLi01h]   W.B. Johnson and J. Lindenstrauss, editors, *Handbook of the Geometry of Banach spaces*, Elsevier, Amsterdam, 2001.

[JoLi01]   W.B. Johnson and J. Lindenstrauss, *Basic concepts in the geometry of Banach spaces*, Handbook of the Geometry of Banach spaces, Vol.1, Eds. W. B. Johnson and J. Lindenstrauss, Elsevier, Amsterdam, 2001, p. 1–84.

[JoRo72]   W.B. Johnson and H.P. Rosenthal, *On w^*-basic sequences and their applications to the study of Banach spaces*, Studia Math. **43** (1972), 74–92.

[JRZ71]   W.B. Johnson, H.P. Rosenthal, and M. Zippin, *On bases, finite-dimensional decompositions and weaker structures in Banach spaces*, Israel J. Math. **9** (1971), 488–506.

[JoSz76]   W.B. Johnson and A. Szankowski, *Complementably universal Banach spaces*, Studia Math. **58** (1976), 91–97.

[JoZi89]   W.B. Johnson and M. Zippin, *Extension of operators from subspaces of $c_0(\Gamma)$ into $C(K)$ spaces,* Proc. Amer. Math. Soc. **107** (1989), 751–754.

[Jos75]   B. Josefson, *Weak sequential convergence in the dual of a Banach space does not imply norm convergence*, Ark. Math. **13** (1975), 79–89.

[Jos78]   B. Josefson, *Bounding subsets of $\ell_\infty(A)$*, J. Math. Pures et Appl. **57** (1978), 397–421.

[Jos02]   B. Josefson, *Subspaces of $\ell_\infty(\Gamma)$ without quasicomplements*, Israel J. Math. **130** (2002), 281–283.

[Juh83]   I. Juhász, *Cardinal functions in topology—ten years later*, Second edition. Mathematical Centre Tracts, **123**. Mathematisch Centrum, Amsterdam, 1980.

[Kad71]   M.I. Kadets, *On complementably universal Banach spaces*, Studia Math. **40** (1971), 85–89.

[KMP06]   V. Kadets, M. Martin, and R. Payá, *Recent progress and open questions on numerical index of Banach spaces*, RACSAM Rev. R. Acad. Cienc. Exactas Fís. Nat. Ser. A Mat. **100** (2006), no. 1-2, 155–182.

[Kal00a]   O. Kalenda, *Valdivia compact spaces in topology and Banach space theory*, Extracta Math. **15**, 1 (2000), 1–85.

[Kal00b]   O. Kalenda, *Valdivia compacta and equivalent norms*, Studia Math. **138**(2) (2000), 179–181.

[Kal02]   O. Kalenda, *M-bases in spaces of continuous functions on ordinals*, Colloq. Math. **92** (2002), 179–187.

[Kalt74]   N.J. Kalton, *Mackey duals and almost shrinking bases*, Proc. Cambridge Philos. Soc. **74** (1973), 73–81.

[Kalt77]   N.J. Kalton, *Universal spaces and universal bases in metric linear spaces*, Studia Math. **61** (1977), 161–191.

[Kalt89]   N.J. Kalton, *Independence in separable Banach spaces*, Contemp. Math., **85** (1989), 319–324.

[Kech95]   A. Kechris, *Classical Descriptive Set Theory*, Springer-Verlag, New York, 1995

[KeLou90]   A. Kechris and Louveau, *A classification of Baire class one functions*, Trans. Amer. Math. Soc. **318** (1990), 209–236.

[KeGi91]   P.S. Kenderov and J.R. Giles, *On the structure of Banach spaces with Mazur's intersection property*, Math. Ann. **291** (1991), 463–478.

[Khu75]   S.S. Khurana, *Extension of total bounded functionals in normed spaces.* Math. Ann. **217** (1975), 153–154.

[Kirk73]   R. Kirk, *A note on the Mackey topology for* $(C^b(X)^*, C^b(X))$, Pacific J. Math. **45** (1973), 543–554.

[Kis75]   S.V. Kislyakov, *Classification of spaces of continuous functions on ordinals*, Sibirsk. Mat. Ž. **16** (1975), 293–300.

[Klee58]   V. Klee, *On the borelian and projective types of linear subspaces*, Math. Scand. **6** (1958), 189–199.

[KOS99]   H. Knaust, E. Odell, and T. Schlumprecht, *On asymptotic structure, the Szlenk index and UKK properties in Banach spaces*, Positivity **3** (1999), 173–199.

[Kosz04a]   P. Koszmider, *A problem of Rolewicz about Banach spaces that admit support sets*, preprint.

[Kosz04b]   P. Koszmider, *Banach spaces of continuous functions with few operators*, Math. Ann. **330** (2004), 151–183.

[Kosz05]   P. Koszmider, *A space* $C(K)$ *where all nontrivial complemented subspaces have big densities*, Studia Math. **168** (2005), 109–127.

[Ko66]   G. Köthe, *Hebbare Lokalkonvexe Räume*, Math. Ann. **165** (1966), 181–195.

[Ko69]   G. Köthe, *Topological Vector Spaces I*, Springer Verlag, 1969.

[KKM48]   M.G. Krein, M.A. Krasnosel'skiĭ, and D.P. Milman, *On the defect number of linear operators in a Banach space and on certain geometrical questions*, Trud. Inst. Matem. Akad. Nauk Ukrain, SSR **11** (1948), 97–112.

[Kub]   W. Kubiś, *Linearly ordered compacta and Banach spaces with a projectional resolution of the identity.* To appear in Topol. Appl.

[Kun81]   K. Kunen, *A compact L-space under CH*, Topology Appl. **12** (1981), 283–287.

[Ku74]   J. Kupka, *A short proof and generalization of a measure theoretic disjointization lemma*, Proc. Amer. Math. Soc. **45** (1974), 7–72.

[Kutz86]   D. Kutzarova, *Convex sets containing only support points in Banach spaces with an uncountable minimal system*, C. R. Acad. Bulg. Sci. **39**, 12 (1986), 13–14.

[KuTr82]   D. Kutzarova and S.L. Troyanski, *Reflexive Banach spaces without equivalent norms which are uniformly convex or uniformly differentiable in every direction*, Studia Math. **72** (1982), 91–95.

[La72]   E. Lacey, *Separable quotients of Banach spaces*, An. Acad. Brasil Ciènc. **44** (1972), 185–189.

[Lanc93]   G. Lancien, *Dentability indexes and locally uniformly convex renorming*, Rocky Mount. J. Math. **23** (1993), 633–647.

[Lanc95]   G. Lancien, *On uniformly convex and uniformly Kadets-Klee renormings*, Serdica Math. J. **21** (1995), 1–18.

[Lanc96]   G. Lancien, *On the Szlenk index and the weak* dentability index* Quart. J. Math. Oxford **47** (1996), 59–71.

[Lanc06]   G. Lancien, *A survey on the Szlenk index and some of its applications*, RACSAM Rev. R. Acad. Cienc. Exactas Fís. Nat. Ser. A Mat. **100**, 1,2 (2006) 209–235.

[Laza81]   A.J. Lazar, *Points of support for closed convex sets*, Illinois J. Math. **25** (1981), 302–305.

[LeSo84]   A.G. Leiderman, G.A. Sokolov, *Adequate families of sets and Corson compacta,* Comment. Math. Univ. Carolin. **25** (1984), 233–246.

[LePe63]   A. Levin and Y. Petunin, *Some questions connected with the concept of orthogonality in Banach spaces*, Usp. Mat. Nauk **18**, 3 (1963), 167–170.

[Lind63]   J. Lindenstrauss, *On operators which attain their norms*, Israel J. Math. **1** (1963), 139–148.

[Lind68]   J. Lindenstrauss, *On subspaces of Banach spaces without quasicomplements*, Israel J. Math. **6** (1968), 36–38.

[Lind71a] J. Lindenstrauss, *On James's paper 'Separable Conjugate Spaces'*, Israel J. Math. **9** (1971), 279–284.

[Lind71b] J. Lindenstrauss, *Decomposition of Banach spaces*, Indiana Univ. Math J. **20** (1971), 917–919.

[LiTz77]   J. Lindenstrauss and L. Tzafriri, *Classical Banach Spaces I, Sequence Spaces*, Springer–Verlag, Berlin, 1977.

[Mack46]   G. Mackey, *Note on a theorem of Murray*, Bull. Amer. Math. Soc. **52** (1946), 322–325.

[Mar03]   W. Marciszewski, *On Banach spaces $C(K)$ isomorphic to $c_0(\Gamma)$*, Studia Math. **156**, 3 (2003), 295–302.

[Mark43]   A.I. Markushevich, *On a basis in the wide sense for linear spaces*, Dokl. Akad. Nauk. **41** (1943), 241–244.

[Mazu33]   S. Mazur, *Über schwache Konvergentz in en Räumen L^p*, Studia Math. **4** (1933), 128–133.

[Merc87]   S. Mercourakis, *On weakly countably determined Banach spaces*. Trans. Amer. Math. Soc. **300** (1987), 307–327

[MeNe92]   S. Mercourakis and S. Negrepontis, *Banach spaces and Topology II*, Recent progress in general topology (Prague, 1991), 493–536, North-Holland, Amsterdam, 1992.

[MeSt02]   S. Mercourakis and E. Stamati, *Compactness in the first Baire class and Baire-1 operators*, Serdica Math. J. **28**, 1 (2002), 1–36.

[MeSt]   S. Mercourakis and E. Stamati, *A new class of weakly $\mathcal{K}$ analytic Banach spaces*, Comment. Math. Univ. Carolin. **47** (2006), no. 2, 291–312.

[MiRu77]   E. Michael and M.E. Rudin, *A note on Eberlein compacts*, Pacific, J. Math. **72** (1977), 487–495.

[Milm70a] V.D. Milman, *Geometric theory of Banach spaces, Part I*, Russ. Math. Surveys **25** (1970), 111–170.

[Milm70b] V.D. Milman, *Geometric theory of Banach spaces, Part II*, Russ. Math. Surveys **26** (1970), 79–163.

[MOTV]   A. Moltó, J. Orihuela, S. Troyanski, and M. Valdivia: *A nonlinear transfer technique*, preprint.

[Mont85]   V. Montesinos, *Solution to a problem of S. Rolewicz*, Studia Math. **81** (1985), 65–69.

[More96]   J.P. Moreno, *On the geometry of Banach spaces with property α*, J. Math. Anal. Appl. **201** (1996), 600–608.

[More97]   J.P. Moreno, *Geometry of Banach spaces with (α, ϵ)-property or (β, ϵ)-property*, Rocky Mount. J. Math. **27** (1997), 241–256.

[More98]   J.P. Moreno, *On the weak* Mazur intersection property and Fréchet differentiable norms on dense open sets*, Bull. Sci. Math. **122** (1998), 93–105.

[Muj97]   J. Mujica, *Separable quotients of Banach spaces*, Rev. Mat. Univ. Complut. Madrid **10** (1997), 2, 299–330.

[Murr45]   F.J. Murray, *Quasicomplements and closed projections in reflexive Banach spaces*, Trans. Amer. Math. Soc. **58** (1945), 77–95.

316   References

[Nam85]   I. Namioka, *Eberlein and Radon-Nikodým compact spaces*, Lecture notes, University College London, Autumn 1985, unpublished typescript.

[Negr84]   S. Negrepontis, *Banach Spaces and Topology*, Handbook of set-theoretic topology, 1045–1142, North-Holland, Amsterdam, 1984.

[NeTs81]   S. Negrepontis and A. Tsarpalias, *A non-linear version of the Amir-Lindenstrauss method*, Israel J. Math. **38** (1981) 82–94.

[Niss75]   A. Nissenzweig, w^* sequential convergence, Israel J. Math. **22** (1975), 266–272.

[Od04]   E. Odell, *Ordinal indices in Banach spaces*, Extracta Math. **19** (2004), 93–125.

[OdSc98]   E. Odell and Th. Schlumprecht, *On asymptotic properties of Banach spaces under renormings*, J. Amer. Math. Soc. **11** (1998), 175–188.

[OdSc02]   E. Odell and Th. Schlumprecht, *Trees and branches in Banach spaces*, Trans. Amer. Math. Soc. **354** (2002), 4085–4108.

[OdSc06]   E. Odell and Th. Schlumprecht, *A universal reflexive space for the class of uniformly convex Banach spaces*, RACSAM Rev. R. Acad. Cienc. Exactas Fís. Nat. Ser. A Mat. **100**, 1,2 (2006) 295–323.

[Ori92]   J. Orihuela, *On weakly Lindelöf Banach spaces*, in Progress in Functional Analysis, edited by K.D. Bierstedt, J. Bonet, J. Horváth, and M. Maestre, Elsevier Science Publishers. B.V., Amsterdam, 1992.

[OrVa89]   J. Orihuela and M. Valdivia, *Projective generators and resolutions of identity in Banach spaces*. Congress on Functional Analysis (Madrid, 1988). Rev. Mat. Univ. Complut. Madrid **2** (1989), suppl., 179–199.

[Orn91]   P. Ørno, *On J. Borwein's concept of sequentially reflexive Banach spaces*, Banach Space Bulletin Board, 1991.

[Ost76]   A.J. Ostaszewski, *On countably compact, perfectly normal spaces*, J. London Math. Soc. **14** (1976), 505–516.

[Ost87]   M.I. Ostrovskij, w^*-derivations of transfinite order in the dual Banach space, Dokl. Akad. Nauk USSR **10** (1987), 9–12.

[OvPe75]   R.I. Ovsepian and A. Pełczyński, *The existence in separable Banach space of fundamental total and bounded biorthogonal sequence and related constructions of uniformly bounded orthonormal systems in L_2*, Studia Math. **54** (1975), 149–159.

[Part80]   J.R. Partington, *Equivalent norms on spaces of bounded functions*, Israel J. Math. **51** (1980), 205–209.

[Part82]   J.R. Partington, *Norm attaining operators*, Israel J. Math. **43** (1982), 273–276.

[Part83]   J.R. Partington, *On nearly uniformly convex Banach spaces*, Math. Proc. Cambridge Philos. Soc.**93** (1983), no. 1, 127–129.

[Pelc65]   A. Pełczyński, *On strictly singular and stricty cosingular operators I*, Bull. Polish Acad. Sci. Math. **13** (1965), 31–36

[Pelc68]   A. Pełczyński, *On Banach spaces containing $L^1(\mu)$*, Studia Math. **30** (1968), 231–246.

[Pelc68b]   A. Pełczyński, *On C(S)-subspaces of separable Banach spaces*, Studia Math. **31** (1968), 231–246.

[Pelc69]   A. Pełczyński, *Universal bases*, Studia Math. **32** (1969), 247–268.

[Pelc71]   A. Pełczyński, *Any separable Banach space with the bounded approximations property is a complemented subspace of a Banach space with a basis*, Studia Math. **40** (1971), 239–242.

[Pelc76]   A. Pełczyński, *All separable Banach spaces admit for every $\epsilon > 0$ fundamental and total biorthogonal sequences bounded by $1 + \epsilon$*, Studia Math. **55** (1976), 295–304.

[PeSz65]   A. Pełczyński and W. Szlenk, *An example of a nonshrinking basis*, Rev. Roumaine Math. Pures Appl. **10** (1965), 961–966.

[PeWo71]   A. Pełczyński and P. Wojtaszczyk, *Banach spaces with finite-dimensional expansions of identity and universal bases of finite-dimensional subspaces*, Studia Math. **40** (1971), 91–108.

[Phel60]   R.R. Phelps, *A representation theorem for bounded convex sets*, Proc. Amer. Math. Soc. **11** (1960), 976–983.

[Phel93]   R.R. Phelps, *Convex Functions, Monotone Operators and Differentiability* (2nd ed.), Lecture Notes in Mathematics **1364**, Springer-Verlag, Berlin, 1993.

[PlRe83]   A. Plans and A. Reyes, *On the geometry of sequences in Banach spaces*, Arch. Math. **40** (1983), 452–458.

[Plic75]   A.N. Plichko, *Extension of quasicomplementarity in Banach spaces*, Funkt. Anal. Pril. **9** (1975), 91–92.

[Plic77]   A.N. Plichko, *M-bases in separable and reflexive Banach spaces*, Ukrain. Mat. Ž. **29** (1977), 681–685.

[Plic79]   A.N. Plichko, *The existence of bounded Markushevich bases in WCG spaces*, Theory Funct. Funct. Anal. Appl. **32** (1979), 61–69.

[Plic80a]   A.N. Plichko, *A Banach space without a fundamental biorthogonal system*, Dokl. Akad. Nauk USSR **254** (1980), 450–453.

[Plic80b]   A.N. Plichko, *Construction of bounded fundamental and total biorthogonal systems from unbounded systems*, Dokl. Akad. Nauk USSR **254** (1980), 19–23.

[Plic80c]   A.N. Plichko, *Existence of a bounded total biorthogonal system in a Banach space*, Teor. Funks. Funkt. Anal. Pril. **33** (1980), 111–118.

[Plic80d]   A.N. Plichko, *A Banach space without a fundamental biorthogonal system*, Dokl. Akad. Nauk USSR **254** (1980), 450–453.

[Plic81a]   A.N. Plichko, *Some properties of the Johnson-Lindenstraus space*, Funct. Anal. Appl. **15** (1981), 88–89.

[Plic81b]   A.N. Plichko, *A selection of subspaces with special properties in a Banach space and some properties of quasicomplements*, Funct. Anal. Appl. **15** (1981), 82–83.

[Plic82]   A.N. Plichko, *On projective resolutions of the identity operator and Markuševič bases*, Sov. Math. Dokl. **25** (1982), 386–389.

[Plic83]   A.N. Plichko, *Projection decompositions, Markushevich bases and equivalent norms*, Mat. Zametki **34**, 5 (1983), 719–726.

[Plic84a]   A.N. Plichko, *Bases and complements in nonseparable Banach spaces*, Sibirsk. Mat. Ž. **25** (1984), 155–162.

[Plic84b]   A.N. Plichko, *On bases and complemented subspaces in nonseparable Banach spaces*, Sibirsk. Mat. Ž. **25**, (1984), 155–162.

[Plic86a]   A.N. Plichko, *On bounded biorthogonal systems in some function spaces*, Studia Math. **84** (1986), 25–37.

[Plic86b]   A.N. Plichko, *On bases and complemented subspaces in nonseparable Banach spaces II*, Sibirsk. Mat. Ž. **27** (1986), 149–153.

[Plic95]   A.N. Plichko, *On the volume method in the study of Auerbach bases of finite-dimensional normed spaces*, Colloq. Math. **69**, 2 (1995), 267–270.

[PlYo00]   A.N. Plichko and D. Yost, *Complemented and uncomplemented subspaces of Banach spaces*, Extracta Math. **15** (2000), 335–371.

[PlYo01]   A.N. Plichko and D. Yost, *The Radon-Nykodým property does not imply the separable complementation property*, J. Functional Analysis **180**, (2001), 481–487.

[Pol77]   R. Pol, *Concerning function spaces on separable compact spaces*, Bull. Acad. Polon. Sci. Sér. Sci. Math. Astronom. Phys. **25**, 10 (1977), 993–997.

[Pol79]   R. Pol, *A function space $C(X)$ which is weakly Lindelöf but not weakly compactly generated*, Studia Math. **64** (1979) 279–285.

[Pol80]   R. Pol, *On a question of H.H. Corson and some related problems*, Fund. Math. **109** (1980), 143–154.

[Pol84]   R. Pol, *On pointwise and weak topology in function spaces*, Warszaw University preprint 4/84, 1984.

[Prus83]   S. Prus, *Finite-dimensional decompositions with p-estimates and universal Banach spaces*, Bull. Pol. Acad. Sci. **31** (1983), 281–288.

[Prus87]   S. Prus, *Finite-dimensional decompositions of Banach spaces with (p,q)-estimates*, Dissertationes Math. (Rozprawy Mat.) **263** (1987), 1–41.

[Prus89]   S. Prus, *Nearly uniformly smooth Banach spaces,* Boll. Un. Mat. Ital. **7** (1989), 507–521.

[Pt63]   V. Pták, *A combinatorial lemma on the existence of convex means and its applications to weak compactness*, Proc. Symp. Pure Math. **7** (1963), 437–450.

[Raja03]   M. Raja, *Weak* locally uniformly rotund norms and descriptive compact spaces*, J. Functional Analysis **197** (2003), 1–13.

[Raja]   M. Raja, *Dentability indices with respect to measures of noncompactness*, preprint.

[Reif74]   J. Reif, *A note on Markushevich basis on weakly compactly generated spaces*, Comment. Math. Univ. Carolin. **15** (1974), 335–340.

[Riba87]   N.K. Ribarska, *Internal characterization of fragmentable spaces*, Mathematica **34** (1987), 243–257.

[Rod94]   B. Rodríguez-Salinas, *On the complemented subspaces of $c_0(I)$ and $\ell_p(I)$ for $1 < p < \infty$*, Atti Sem. Mat. Fis. Univ. Modena **92** (1994), 399–402.

[Role78]   S. Rolewicz, *On convex sets containing only points of support*, Comment. Math., Tomus specialis in honorem Ladislai Orlicz, I, 1978, 279–281.

[Rose68a] H.P. Rosenthal, *On complemented and quasicomplemented subspaces of quotients of $C(S)$ for Stonian S*, Proc. Nat. Acad. Sci. USA **60** (1968), 1165–1169.

[Rose68b] H.P. Rosenthal, *On quasicomplemented subspaces of Banach spaces*, Proc. Nat. Acad. Sci. USA **59** (1968), 361–364.

[Rose69a] H.P. Rosenthal, *On quasicomplemented subspaces of Banach spaces, with an appendix on compactness of operators form $L^p_{(\mu)}$ to $L^r_{(\nu)}$*, J. Functional Analysis **4** (1969), 176–214.

[Rose69b] H.P. Rosenthal, *On totally incomparable Banach spaces*, J. Functional Analysis **4** (1969), 167–175.

[Rose70a] H.P. Rosenthal, *On injective Banach spaces and the spaces $L^\infty(\mu)$ for finite measure μ*, Acta Math. **124** (1970), 205–247.

[Rose70b]   H.P. Rosenthal, *On relatively disjoint families of measures, with some applications to Banach space theory*, Studia Math. **37** (1970), 13–30.

[Rose70c]   H.P. Rosenthal, *On the subspaces of L^p (p > 2) spanned by sequences of independent random variables*, Israel J. Math. **8** (1970), 273–303.

[Rose72]   H.P. Rosenthal, *On factors of $C[0,1]$ with nonseparable dual*, Israel, J. Math. **13** (1972), 361–378.

[Rose73]   H.P. Rosenthal, *On subspaces of L^p*, Ann. Math. **97** (1973), 344–373.

[Rose74]   H.P. Rosenthal, *The heredity property for weakly compactly generated Banach spaces*, Compositio Math. **2** (1974), 83–111.

[Rose77]   H.P. Rosenthal, *Pointwise compact subsets of the first Baire class*, Amer. J. Math. **99** (1977), 362–378.

[Ruck70]   W.H. Ruckle, *Representation and series summability of complete biorthogonal sequences*, Pacific J. Math. **34** (1970), 511–518.

[Rych00]   J. Rychtář, *Uniformly Gâteaux differentiable norms in spaces with unconditional bases*, Serdica Math. J. **26** (2000), 353–358.

[Rych04]   J. Rychtář, *On biorthogonal systems and Mazur's intersection property*, Bull. Austral. Math. Soc. **69** (2004), 107–111.

[Sam83]   C. Samuel, *Indice de Szlenk des $C(K)$*, Seminar on the geometry of Banach spaces, Vol. I, II (Paris, 1983), 81–91, Publ. Math. Univ. Paris VII, **18**, Univ. Paris VII, Paris, 1984.

[Scha83]   W. Schachermayer, *Norm attaining operators and renormings of Banach spaces*, Israel J. Math. **44** (1983), 201–212.

[Sche75]   G. Schechtman, *On Pełczyński paper Universal bases*, Israel J. Math. **22** (1975), 181–184.

[ScWh88]   G. Schlüchtermann and R.F. Wheeler, *On strongly WCG Banach spaces*, Math. Z. **199** (1988), 387–398.

[ScWh91]   G. Schlüchtermann and R.F. Wheeler, *The Mackey dual of a Banach space*, Note Mat. **11** (1991), 273–287.

[Sema60]   Z. Semadeni, *Banach spaces non-isomorphic to their cartesian squares. II*, Bull. Acad. Pol. Sci. **8** (1960), 81–84.

[Sema82]   Z. Semadeni, *Schauder bases in Banach spaces of continuous functions*, Lecture Notes in Mathematics, **918**, Springer-Verlag, Berlin, 1982.

[Sers87]   A. Sersouri, *ω-independence in non separable Banach spaces*, Contemp. Math. **85** (1987), 509–512.

[Sers88]   A. Sersouri, *The Mazur property for compact sets*, Pacific J. Math. **133** (1988), 185–195.

[Sers89]   A. Sersouri, *Mazur's intersection property for finite-dimensional sets*, Math. Ann. **283** (1989), 165–170.

[Shel85]   S. Shelah, *Uncountable constructions for B.A., e.c. groups and Banach spaces*, Israel J. Math., **51** (1985), 273–297.

[Sing70a]   I. Singer, *Best Approximation in Normed Linear Spaces by Elements of Linear Subspaces*, Springer-Verlag, Berlin, 1970.

[Sing70b]   I. Singer, *Bases in Banach Spaces I*, Springer-Verlag, Berlin, 1970.

[Sing71]   I. Singer, *On biorthogonal systems and total sequences of functionals*, Math. Ann. **193** (1971), 183–188.

[Sing73]   I. Singer, *On biorthogonal systems and total sequences of functionals* II. Math. Ann. **201** (1973), 1–8.

[Sing74]   I. Singer, *On the extension of basic sequences to bases*, Bull. Amer. Math. Soc. **80** (1974), 771–772.

[Sing81]   I. Singer, *Bases in Banach Spaces II*, Springer-Verlag, Berlin, 1981.

[Sob04]   D. Sobecki, *A characterization of strongly weakly compactly generated Banach spaces.* Rocky M. J. Math. **34**, 1503–1505.

[Soko84]   G.A. Sokolov, *On some classes of compact spaces lying in Σ-products*, Comment. Math. Univ. Carolin. **25** (1984), 219–231.

[Steg73]   C. Stegall, *Banach spaces whose duals contain $\ell_1(\Gamma)$ with applications to the study of dual $L_1(\mu)$ spaces*, Trans. Amer. Math. Soc. **176** (1973), 463–477.

[Steg75]   C. Stegall, *The Radon-Nikodým property in conjugate Banach spaces*, Trans. Amer. Math. Soc. **206** (1975), 213–223.

[Szl68]   W. Szlenk, *The nonexistence of a separable reflexive Banach space universal for all separable reflexive Banach spaces*, Studia Math. **30** (1968), 53–61.

[Tala79]   M. Talagrand, *Espaces de Banach faiblement K-analytiques*, Ann. of Math. **119** (1979), 407–438.

[Tala80a]   M. Talagrand, *Un nouveau $C(K)$ qui possede la propriete de Grothendieck*, Israel J. Math. **37** (1980), 181–191.

[Tala80b]   M. Talagrand, *Separabilité vague dans l'espace des mesures sur un compact*, Israel. J. Math. **37** (1980), 171–180.

[Tala81]   M. Talagrand, *Sur les espaces de Banach contenant $\ell_1(\tau)$*, Israel J. Math. **40** (1981), 324–330.

[Tala86]   M. Talagrand, *Renormages de quelques $C(K)$*, Israel J. Math. **54** (1986), 327–334.

[Tang99]   W.K. Tang, *On Asplund functions*, Comment. Univ. Math. Carolin. **40** (1999), 121–132.

[Tere79]   P. Terenzi, *On bounded and total biorthogonal systems spanning given subspaces*, Rend. Accad. Naz. Lincei **67** (1979), 1–11.

[Tere83]   P. Terenzi, *Extension of uniformly minimal M-basic sequences in Banach spaces*, J. London Math. Soc. **27** (1983), 500–506.

[Tere90]   P. Terenzi, *Every norming M-basis of a separable Banach space has a block perturbation which is a norming strong M-basis.* Actas del II Congreso de Análisis Funcional, Jarandilla de la Vera, Cáceres, 20-27 Junio, 1990, Extracta Math. (1990), 161–169.

[Tere94]   P. Terenzi, *Every separable Banach space has a bounded strong norming biorthogonal sequence which is also a Steinitz basis*, Studia Math. **111** (1994), 207–222.

[Tere98]   P. Terenzi, *A positive answer to the basis problem*, Israel J. Math. **104** (1998), 51–124.

[Todo95]   S. Todorčević, *The functor $\sigma^2 X$*, Studia Math. **116** (1995), 49–57.

[Todo06]   S. Todorčević, *Biorthogonal systems and quotient spaces via Baire category theory*, Math. Ann. **335** (2006), 687–715.

[Todo]   S. Todorčević, *A generic function space $C(K)$ with no support set*, in preparation.

[Troy71]   S. Troyanski, *On locally uniformly convex and differentiable norms in certain nonseparable Banach spaces*, Studia Math. **37** (1971), 173–180.

[Troy72]   S. Troyanski, *On equivalent norms and minimal systems in nonseparable Banach spaces*, Studia Math. **43** (1972), 125–138.

[Troy75]   S. Troyanski, *On nonseparable Banach spaces with a symmetric basis*, Studia Math. **53** (1975), 253–263.

[Troy77]   S. Troyanski, *On uniform convexity and smoothness in every direction in nonseparable Banach spaces with an unconditional basis* (Russian), C. R. Acad. Sci. Bulg. **30** (1977), 1243–1246.

[Troy85]   S. Troyanski, *On a property of the norm which is close to local uniform convexity*, Math. Ann. **271** (1985), 305–313.

[Vald77]   M. Valdivia, *On a class of Banach spaces*, Studia Math. **60** (1977), 11–13.

[Vald88]   M. Valdivia, *Resolutions of identity in certain Banach spaces*, Collect. Math. **39** (1988), 127–140.

[Vald89]   M. Valdivia, *Some properties of weakly countably determined Banach spaces*, Studia Math. **93** (1989), 137–144.

[Vald90a]   M. Valdivia, *Projective resolutions of identity in $C(K)$ spaces*, Arch. Math. **54** (1990), 493–498.

[Vald90b]   M. Valdivia, *Resoluciones proyectivas del operador identidad y bases de Markushevich en ciertos espacios de Banach*, RACSAM Rev. R. Acad. Cienc. Exactas Fís. Nat. Ser. A Mat. **84** (1990), 23–34.

[Vald90c]   M. Valdivia, *Topological direct sum decompositions of Banach spaces*, Israel J. Math. **71** (1990), 289–296.

[Vald91]   M. Valdivia, *Simultaneous resolutions of the identity operator in normed spaces*, Collect. Math. **42** (1991), 265–284.

[Vald93a]   M. Valdivia, *Fréchet spaces with no subspaces isomorphic to ℓ_1*. Math. Japonica **38** (1993), 397–411.

[Vald93b]   M. Valdivia, *On certain total biorthogonal systems in Banach spaces*. Generalized functions and their applications (Varanasi, 1991), 271–280, Plenum, New York, 1993.

[Vald94]   M. Valdivia, *On certain classes of Markushevich bases*, Arch. Math. **62** (1994), 493–498.

[Vald96]   M. Valdivia, *Biorthogonal systems in certain Banach spaces*, Meeting on Mathematical Analysis (Spanish) (Avila, 1995). Rev. Mat. Univ. Complut. Madrid **9** (1996), Special Issue, suppl., 191–220.

[Vald97]   M. Valdivia, *On certain compact topological spaces*, Rev. Mat. Univ. Complut. Madrid **10** (1997), no. 1, 81–84.

[Vand95]   J. Vanderwerff, *Extensions of Markuševič bases*, Math. Z. **219** (1995), 21–30.

[Vand98]   J. Vanderwerff, *Mazur intersection properties of compact and weakly compact convex sets*, Canad. Math. Bull. **41** (1998), 225–230.

[VWZ94]   J. Vanderwerff, J. Whitfield, and V. Zizler, *Markuševič bases and Corson compacta in duality*, Canad. J. Math. **46** (1994), 200–211.

[Vas81]   L. Vašák, *On a generalization of weakly compactly generated Banach spaces*, Studia Math. **70** (1981), 11–19.

[Ve81]   *Weak topology of spaces of continuous functions*, Math. Notes **30** (1981), 849–854.

[Vers00]   R. Vershynin, *On constructions of strong and uniformly minimal M-bases in Banach spaces*, Arch. Math. **74** (2000), 50–60.

[Wal74]   R.C. Walker, *The Stone-Čech Compactification*, Springer-Verlag, Berlin, 1974.

[WhZi87a]   J.H.M. Whitfield and V. Zizler, *Mazur's intersection property of balls for compact convex sets*, Bull. Austral. Math. Soc. **35** (1987), 267–274.

[WhZi87b]   J.H.M. Whitfield and V. Zizler, *Uniform Mazur's intersection property of balls*, Canad. Math. Bull. **30** (1987), 455–460.

[Wilk]   D. Wilkins, *The strong WCD property for Banach spaces*, Internat. J. Math. Math. Sci. **18**, 1 (1995), 67–70.

[Will77]   N.H. Williams, *Combinatorial Set Theory*, Studies in Logic **91**, North-Holland, Amsterdam, 1977.

[Woj70]   P. Wojtaszczyk, *On a separable Banach space containing all separable reflexive Banach spaces*, Studia Math. **37** (1970), 197–202.

[Woj91]   P. Wojtaszczyk, *Banach Spaces for Analysts*, Cambridge Studies in Advanced Mathematics, **25**, Cambridge University Press, 1991.

[Y-V49]   A.C. Yesenin-Volpin, *On the existence of a universal bicompact of arbitrary weight*, Dokl. Akad. Nauk USSR **68** (1949), 649–652.

[Yost97]   D. Yost, *The Johnson-Lindenstrauss space*, Extracta Math. **12** (1997), 185–192.

[Zen80]   P. Zenor, *Hereditary m-separability and the hereditary m-Lindelöf property in product spaces and function spaces*, Fund. Math. **106** (1980), 175–180.

[Zip70]   M. Zippin, *Existence of universal members in certain families of bases of Banach spaces*, Proc. Amer. Math. Soc. **26** (1970), 294–300.

[Zip88]   M. Zippin, *Banach spaces with separable duals*, Trans. Amer. Math. Soc. **310** (1988), 371–379.

[Zizl84]   V. Zizler, *Locally uniformly rotund renorming and decomposition of Banach spaces*, Bull. Austral. Math. Soc. **29** (1984), 259–265.

[Zizl86]   V. Zizler, *Renorming concerning Mazur's intersection property of balls for weakly compact convex sets*, Math. Ann. **276** (1986), 61–66.

[Zizl03]   V. Zizler, *Nonseparable Banach spaces*, in Handbook of Banach Spaces, Volume II, p. 1743–1816, Ed. W.B. Johnson and J. Lindenstrauss, Elsevier Science Publishers B.V., Amsterdam, 2003.

Symbol Index

Subject Index

Entries in bold typeface correspond to the pages where the corresponding concepts are defined.

Author Index

Printed in the United States of America